ANIMAL SCIENCE, ISSUES AND PROFESSIONS

BATS: BIOLOGY, BEHAVIOR AND CONSERVATION

Animal Science, Issues and Professions

Additional books in this series can be found on Nova's website
under the Series tab.

Additional E-books in this series can be found on Nova's website
under the E-books tab.

ANIMAL SCIENCE, ISSUES AND PROFESSIONS

BATS: BIOLOGY, BEHAVIOR AND CONSERVATION

JAKOB L. ZUPAN
AND
SARA L. MLAKAR
EDITORS

Nova Science Publishers, Inc.
New York

Library of Congress Cataloging-in-Publication Data

Bats : biology, behavior, and conservation / editors, Jakob L. Zupan and
Sara L. Mlakar.
p. cm.
Includes bibliographical references and index.
ISBN 978-1-61209-749-7 (hardcover : alk. paper)
1. Bats. 2. Bats--Behavior. 3. Bats--Conservation. I. Zupan, Jakob L.
II. Mlakar, Sara L.
QL737.C5B3726 2011
599.4--dc22

2011004003

Published by Nova Science Publishers, Inc. † New York

CONTENTS

PREFACE

Bats are flying mammals in the order Chiroptera. The forelimbs of bats are webbed and developed as wings, making them the only mammals naturally capable of true and sustained flight. In this book, the authors present current research in the study of the biology, behavior and conservation of bats. Topics discussed in this compilation include: chiroptophobia (the fear of bats); the reproductive biology of male bats; bats and rabies in Brazil; postnatal development, wing morphology and flight performance of the short-nosed fruit bat; hematophagous bats; prolonged sperm storage in bats and metabolic adaptations in frugivorous bats.

Chapter 1 – Chiroptophobia is as widespread as bats are in nature. The modern fear of bats is almost exclusively the fear of rabies. This narrow view hinders our conservation efforts as well as discouraging research. Understanding bat-associated zoonoses is essential to study these ubiquitous creatures' longevity, immunology, epidemiology, and physiology. We need to learn how they interact with civilization, how we can safely interact in their environment, and, when necessary, how man and bat can safely share the same space together. A realistic assessment of the risks bats pose to the public and chiropterists will help us preserve bat populations and their habitats.

Bats are carriers of multiple viral pathogens. Neurological symptoms are associated with lyssaviruses (including rabies), Paramyxoviridae (Henipavirus genus: Nipah virus and Hendra virus), Flaviviridae (Japanese encephalitis virus, Kyasanur Forest disease virus, St. Louis encephalitis virus, West Nile virus), Toscana virus (Bunyaviridae), and Alphaviruses (Venezuelan equine encephalomyelitis, Eastern and Western Equine encephalitis viruses, and Sindbis virus (rash and arthritis in humans, encephalitis in mice)). Pulmonary disease is a prominent feature of infections with hantavirus infection, severe acute respiratory syndrome (SARS) coronaviruses, influenza A virus, Menangle virus, Tioman virus, parainfluenza virus type 2 (Paramyxoviridae causing flu-like illness), Kaeng Khoi virus (Bunyaviridae), Melaka virus (reovirus), and a previously unknown reovirus of bat origin related to Melaka and Pulau viruses. Hemorrhagic fevers result from Filoviridae (Marburg and Ebola viruses), Flaviviridae (Dengue and sometimes Kyasanur Forest virus), and Bunyaviridae (Hantavirus, Rift Valley fever virus). Chikungunya virus (Alphavirus-Togaviridae) infection is clinically similar to Dengue fever. The Tacaribe virus group (Arenavirus) includes hemorrhagic fever viruses. GBV-D virus (Flavivirus), a virus closely related to GBV-C, the cause of hepatitis G has been isolated from frugivorous bats (Pteropus giganteus).

Bat ectoparasites harbor *Bartonella henselae* (cat-scratch disease, endocarditis, bacillary angiomatosis/peliosis), a *Rickettsia* sp., *Borrelia* sp. (Lyme disease), *Neorickettsia* sp. (formerly *Ehrlichia* sp.), and *Trypanosoma* spp. (brain abscesses, Chagas' disease). Parasitic diseases include *Angiostongylus cantonensis* (eosinophilic meningitis), whose neurologic symptoms in bats may be confused with Australian bat lyssavirus infection.

Bat-origin *Histoplasma capsulatum* (pulmonary and disseminated disease) has caused disease in man, and microsporidia, a known group of gastrointestinal pathogens, has caused disease in bats. Other fungi carried by bats include *Blastomyces dermatitidis*, *Coccidioides immitis* and *Trichosporon* sp. Bats are asymptomatic carriers of *Leptospira* sp., and exposures to bats have resulted in human infection (sepsis, Weil's disease). In captivity, bats are susceptible to *Yersinia pseudotuberculosis* from rodents, also putting their keepers at risk for this infection (enterocolitis, pseudoappendicitis, Kawasaki-like syndrome, sepsis).

Cynad-associated amyotrophic lateral sclerosis-parkinsonism dementia in the Chamorro people of Guam is linked with the absorption of high concentrations of the cyanobacterial neurotoxins (β-methylamino-L-alanine) after consuming *Pteropus mariannus mariannus* (flying fox).

A picture of the risk of bat contact requires a dispassionate discussion of each pathogen's epidemiology, clinical presentation, prevention, and treatment. Understanding the relationship amongst man, chiroptera and these pathogens will assist us in developing prevention strategies (vaccine and vaccine applications), diagnosis and treatment, and at the same time allowing us to safely study and preserve bat populations.

Chapter 2 - Frugivorous bats are important members of tropical ecosystems and they play a key role in pollination and seed dispersal in tropical forests throughout the World. They evolved in a mutualistic exploitation system with plants in which they obtain food in the form of nectar, pollen or fruit while providing mobility of pollen grains or seeds. The interactions have been profoundly influenced the co-evolution between bats and plants. This Chapter deals about the foraging behaviour includes individual and group foraging, folivory and nectarivory behaviour, temporal variations in the foraging behaviour and influence of moonlight on the foraging behaviour of the short-nosed fruit bat, *Cynopterus sphinx* (Vahl, 1797). *Cynopterus sphinx* is a medium-sized fruit bat, common in many parts of India. It lives in small colonies, generally of three to four individuals. Its diurnal roosts include the underside of the dry fronds of palm trees *Borassus flabellifer*, the flower/fruit clusters of kitul palms *Caryota urens,* the stems of mast tree *Polyalthia longifolia* as well as under the roof of abandoned buildings.

Cynopterus sphinx feeds upon 23 species of fruits, 8 species of leaves, flowers / petals of *Coccinia indica*, and nectar from *Musa paradisiaca, Bassia latifolia* and *Ceiba pentandra. Cynopterus sphinx* begins to visit fruit-bearing trees about 30 min after sunset and often hovers nearer to food sources or lands on fruits or on nearby branches to remove whole or part of fruit by its mouth. These bats seldom remain in the fruit-bearing trees to feed, instead carry fruits to feeding roosts, repeating this behaviour several times throughout the night. Temporal distribution of the foraging behaviour of *C. sphinx* revealed two peaks of activity, one in the pre-midnight hours when bats fed mostly on "steady state" fruits and another during the post-midnight hours when bats fed on "big bang" fruits. Only solitary bats visited and fed on species with steady-state fruiting phenologies, whereas groups of bats regularly visited and fed on species with big bang fruiting phenologies.

Cynopterus sphinx fed the leaves mainly during the postmidnight hours. They snatched a leaf on flight without landing and hovering. The folivory in *C. sphinx* has been substantiated by collection of discarded bolus of the leaves of *Cassia fistula*, *Mimusops elengi* and *Coccinia indica* as fibrous pellets beneath the day roosts of *C. sphinx* and a mist-netted adult male bat had a fresh leaf of *C. fistula* in its mouth. *Cynopterus sphinx* exhibits folivory by ingesting only the protein-rich soluble contents of leaves and expelling the undigestible fibrous portions. Folivory in *C. sphinx* helps to meet their daily protein requirements.

Temporal variations have been observed on nectar feeding behaviour of *C. sphinx* when bats fed on flowers of a big bang tree *Bassia latifolia* and a steady state tree *Musa paradisiaca*. The bats foraged in-groups and solitarily upon the flowers of *B. latifolia* and *M. paradisiacal*, respectively. Group foraging facilitates an increased protection from predators, an increased sensory area and an increased knowledge on food patches. On the other hand, reduced feeding interference and reduced competition for food are the advantages with solitary foraging. The peak of bat-visits to the flowers of *M. paradisiaca* occurred at 2000 h and to flowers of *B. latifolia* it was at 2100 h. The peak visit coincided with the time of occurrence of maximum amount of nectar secretion and the relatively high concentration of sugar. The bats might acquire more energy by feeding upon the nectar, in addition to fruits. At the same time, the bats offer pollination service to the plants. These interactions have profoundly influenced the coevolution of bats and plants.

Moonlight plays an important role on the foraging behaviour of *C. sphinx*. Observations on bats were made when they foraged on the fruits of *Calophyllum inophyllum* and on the leaves of *Mimusops elengi* and *Cassia fistula*. The foraging activity gets modulated with different phases of the moon. During bright moonlit nights *C. sphinx* exhibited less activity as against that during new moon nights. The total number of feeding bouts was negatively correlated with percent moonlight. The number of feeding bouts decreased during the waxing phases of the moon. During the period of new moon the mean number of feeding bouts was significantly higher compared to the periods of other phases. During the first quarter and the last quarter phases, feeding occurred more during the dark part of the night, especially when the bats foraged upon *M. elengi* and *C. fistula*. While cloud cover enhanced the foraging activity, lightning and thunder suppressed it.

The radio-telemetry studies on the foraging behaviour of *C. sphinx* suggest that the mean commuting distance (0.22 ± 0.19 km) and mean foraging area (0.75 ± 0.27 km^2) of males comparatively lesser than the mean commuting distance (2.1 ± 1.0 km) and mean foraging area (0.83 ± 0.12 km^2) of females. Both males and females exhibit a high level of activity during early hours of the night soon after emergence and another activity peak during pre-dawn hours.

Chapter 3 - The chiropterans are the only mammals capable of true flight. They represent one of the largest and most diversified orders of mammals, inhabiting every continent, except Antarctica. This diversity is reflected in the anatomy and physiology of the reproductive organs that have been adapted for flight and geographic distribution, as seen in hibernating species inhabiting temperate zones and non-hibernating species generally inhabiting the tropics. Here we summarize the current knowledge of male bat reproductive strategies and genital organs morphophysiology. Similar to other mammals, the male genital system of Chiroptera is composed of the testis, efferent ductules, epididymis, vas deferens, urethra, penis and accessory sex glands, however a variety of peculiar features are observed in the morphophysiology of these organs. The bat testis presents a large variation in position, according to the presence or absence of scrotum, being scrotal, subcutaneous or abdominal. The testicular position may be permanent or transitory depending on the species considered. According to the acrossomic system method, ten to eleven stages of the cycle of seminiferous epithelium are recognized in bat seminiferous tubules. During regression, seminiferous tubules shelter only Sertoli cells and spermatogonia. Curiously, the Sertoli cells are located above the spermatogonia, forming a protective barrier for these cells that are responsible for re-colonization of the seminiferous tubules. The efferent ductules connect the rete testis to the epididymis. A description of *Artibeus lituratus* revealed 12 to 15 flexuous ductules arising separately from the *rete testis* and running parallel towards the epididymis, then anastomozing and forming multiple entries into the epididymal duct. This pattern of disposition is similar to that found for large eutherian mammals and men. The epididymis of bats may be anatomically divided into initial segment, caput, corpus and cauda. As shown in *A. lituratus*, the epididymal epithelium is composed of principal, basal, apical, narrow, clear and halo cells, with different distribution along the duct. In several hibernating species, the epididymis is responsible for storing viable sperm during long periods. Information about the male accessory sex glands of bats is scarce. The prostate and bulbourethral glands have been described in all species studied, whereas the ampullary gland and seminal vesicles are present in some species and absent in others. The bat penis is prominent and pendulous, presenting a well-developed glans. In some species, the penis presents a baculum (*os penis*), extending throughout the glans. In seasonal animals, environmental factors as photoperiod play an important role in gonadal activity. In these animals, the exposure to short days leads to decreased hypothalamic GnRH and pituitary gonadotropins (FSH and LH) releases, leading to a decrease in gonadal activity. Testosterone levels vary along the reproductive cycle of hibernating and non-hibernating bats. Besides androgens, it is now recognized that estrogens also play an important role in male reproduction. However, information about estrogen in male bats has been limited to one report about *A. lituratus*, a seasonal non-hibernating Neotropical species, in which both estrogen receptors ERα and ERβ were detected in the testis. During testicular regression, levels of estrogen receptors, especially ERβ, are increased, indicating that estrogens may be directly involved in the regulation of the bat male reproductive cycle.

Chapter 4 - Pteropodid bats are important aerial pollinators and seed dispersers in the Old World tropics and thus play a key role in forest dynamics and regeneration. In India, among the 13 species of pteropodid bats, the foraging ecology of four species *Cynopterus sphinx, C. brachyotis, Rousettus leschenaulti* and *Pteropus giganteus* were extensively studied. These species occur sympatrically and forage on resources like leaves, flowers and fruits, and inadvertently effect pollination and seed dispersal of the resource trees. These bats either consume fruits and flowers *in situ* or carry them to nearby roosts for feeding *ex situ* depend on the carrying capacity of bats and size and the nutritional quality of consuming resource. We suggest that small pteropodid bats aid short distance pollination and the dispersal of seeds of small sized fruits, whereas large pteropodid bats aid long distance pollination and the dispersal of seeds of large sized fruits with some exceptions that depended on the mass of the fruit. At least 300 plant species of nearly 200 genera rely mainly on these fruit bats for their propagation. Some of the plants are economically important and they produce approximately 500 economically valuable products including fruits, dyes, tannins, timber, medicines, fiber and fire wood. In this review, we discuss the foraging ecology of pterpodid fruit bats in the Old world tropics particularly with reference to chiropterophilly and zoochory by fruit bats.

Chapter 5 - The bats found in Brazil are included in nine families, 64 genera and 168 species (Reis et al., 2007). The most numerous family is *Phyllostomidae*, with 90 species registered, followed by *Molossidae* with 26, *Vespertilionidae* with 24, *Emballonuridae* with 15 species and the families *Noctilionidae*, *Mormoopidae*, *Natalidae*, *Furipteridae* and *Thyropteridae*, totaling another 13 species (Peracchi et al., 2006). With respect to feeding habit, the bats can show very varied feeding habits, and this diversity in diets is not found in any other group of mammals. Among the foods consumed are insects and other arthropods (such as scorpions), small fish, amphibians (frogs and tree frogs), birds, fruits, seeds, leaves, flowers, pollen, nectar, small vertebrates and blood (Gardner, 1977; Hill, Smith, 1988; Fenton, 1992; Peracchi et al., 2006).

Chapter 6 - The early life of bats has received very little attention, due to their secretive habits. Members of the Order Chiroptera are altricial and dependent upon the mothers for some weeks or even months following birth. Mother-infant interactions in mammalian species are generally characterized by a marked degree of synchrony in their behaviour. The type of habitat a flying animal chooses to live-in as well as its way of exploiting the habitat is closely related to its body size, wing shape, flight style, flight speed and flight energetics. This chapter deals about postnatal development and age estimation, wing morphology and flight development, and mother-young interactions in the short-nosed fruit bat, *Cynopterus sphinx*.

Adult individuals of *C. sphinx* breed twice in a year, i.e. during September – October and February – March. The young of *C. sphinx* were altricial at birth, their eyes had opened on fifth day from their birth, and the pinnae become unfolded between sixth and ninth day of age. At the age of three days, the mean forearm length and body mass were equivalent to 42.2% and 18.2%, respectively of the values of postpartum females. The length of forearm increased linearly until 36 days and attained 94.1% of mean forearm length of postpartum females at the age of 219 days. Body mass of pups increased linearly until 60 days and attained 72.7% of mean mass of postpartum females at the age of 219 days. Length of total epiphyseal gap of fourth metacarpal-phalangeal joint increased until 15 days of age and subsequently decreased linearly and closed at about 60 days of age. The age predicting equation based on the length of forearm is valid when its dimensions are between 29.4 mm and 52.4 mm (i.e. 3–36 days of age). Whereas the equation for predicting age based on the length of total epiphyseal gap is valid when its dimensions ranged from 47.0 µm to 6.0 µm (i.e. 15–60 days of age). Growth patterns of forearm length and body mass were best fitted with the logistic and Gompertz nonlinear growth models, respectively. There was no significant difference in the growth patterns of body mass and length of total epiphyseal gap with reference to lengths of forearm of captive and wild-grown pups.

Pups of *C. sphinx* were flightless until 25 days of age and they started fluttering at the mean age of 40 days. The wingspan and wing area increased linearly until 45 days of age by which the young bats exhibited clumsy flight with gentle turn. The mean wing area and wingspan of 45-day-old pups were equivalent 55.5% and 67.6%, respectively to the postpartum females. At birth, *C. sphinx* had less developed handwing compared to armwing, however the handwing developed faster than armwing during their postnatal period. Young bats achieved sustained flight at 55 days of age. Wing loading decreased linearly until 35 days of age and thereafter increased to a maximum of 12.82 Nm^{-2} at 125 days of age. The aspect ratio increased linearly until 55 days of age and thereafter stabilized. The logistic growth equation best fitted to the postnatal changes in wingspan and wing area. The predicted minimum power speed (V_{mp}) and maximum range speed (V_{mr}) decreased until onset of flight and thereafter the V_{mp} and V_{mr} increased linearly and approached 96.2% and 96.4%, respectively to the speed of postpartum females at 125 days of age. The requirement of minimum flight power (P_{mp}) and maximum range power (P_{mr}) increased until 85 days of age and thereafter stabilized. The minimum theoretical radius of banked turn (r_{min}) decreased until 35 days of age and thereafter increased linearly and attained 86.5% of r_{min} of postpartum females at the age of 125 days.

Under captivity the females of *C. sphinx* gave birth in the head-down roosting position. Soon after the birth, mothers grasped their infants, covered them with patagium and engaged in cleaning the infants by continuous licking. The newborn babies were found clinging to their mothers tightly using their thumbs and hind feet, and much of the time they firmly attached to a nipple. Mothers carried their newborns with them while foraging. Neonates produced chirping sounds almost immediately after birth. Mothers never approached an alien baby throughout the mother-young reunion experiment. During early postnatal period, the success of mother-young reunion was mainly by the mothers' effort. Mothers always smelt their babies before accepting them and soon after retrieving they licked them for a short period. The time taken for mother-young reunion was shorter during early lactation than mid and late lactations. During the early stage of development, the lactating females showed more care on their young, but the care-giving behaviour declined as the young grew. Such direct contact declined to 6% between 58 and 60 days of lactation. Pups started roosting separately adjacent to their mothers at the age of 30 days and at this stage they scampered frequently in the roost. Pups flew clumsily when they were about 45 days old. After the attainment of clumsy flight, the pups made foraging attempts by biting and licking fruit pieces. Between 45 and 55 days of age, the young bats engaged in both suckling and feeding upon fruits. The young bats exhibited independent foraging from 55 days of their age and weaned completely from their mothers after 65 days of age. After achieving independent foraging, juveniles typically departed from their roost and began to feed upon fruits earlier than adults. The young bats made more number of foraging bouts than their mothers in a night.

Chapter 7 - There is great interest in mammals due to their diversity, beauty, utility and also possible problems and diseases that they can cause. One of the orders most characteristic of mammals, because they have specialized structures for true flight, is the order *Chiroptera*.

The order *Chiroptera* is divided into two suborders; *Megachiroptera* and *Microchiroptera*. Among the microchiropters, there are 17 families, but only 9 families of bats occur in Brazil. Hematophagous bats are inserted in the family *Phyllostomidae*, subfamily *Desmodontinae* (Reis, 2007). The subfamily *Desmodontinae* comprises three species that possess a hematophagous feeding habit (Gardner, 1977), and they are *Desmodus rotundus* (E. Geofroy, 1810), *Diphylla ecaudata* (Spix, 1823) and *Diaemus youngi* (Jentink, 1893).

Chapter 8 - Prolonged sperm storage is a reproductive particularity in *vespertilionid* and *rhinolophid* bats. Several hypotheses have been proposed to explain this important reproductive phenomenon. The initial proposal that prolonged storage of spermatozoa is a natural consequence of hibernating hypothermia has now been refuted because many species of bats remain unusually active during hibernation period. In *Corynorhinus mexicanus* during winter only a daily torpor is present, but no a true hibernation period and sperm storage, this occurs both in the male (epididymides) and in the female (uterus). It is known that the presence of reactive oxygen species (ROS), affects the spermatozoa in complex ways, but it is also beneficial, even necessary, for normal sperm physiology. This becomes more interesting if we remember that some of the most important physiological properties of mammalian sperm cells develop gradually as the cells progress through the epididymis. In the *Corynorhinus mexicanus* the three enzymes related to ROS modulation are present in the cephalic and caudal region of the epididymides during the progress of the epididymis sperms and their maturation, but interesting to say the activity of the superoxide dismutase (SOD) is not present or perhaps it is inhibited in both epididymal segments in the post testicular phase of the sperm storage function. The SOD is present during the period of maturation and during the transport of the sperms, but it is totally or almost totally inhibited or not present in both epididymal segments during the post testicular phase of storage, and an absence in the sperm cells SOD activity (We try to avoid the possibility of producing peroxidative damage in the sperm cells storage during this last phase). It is interesting to observe that the glutathione peroxidase (GPX) presents a complete pattern contrary to that of the SOD; its activity is low during the period of production and maturation of the epididymal sperms which depend perhaps on the testicular activity, but present an important activity in both segments of the epididymis during the storage phase. The activity of catalase (CAT) is preserved relatively high during the whole reproductive cycle indicating its importance in the protection of the spermatic cells against the effect of H_2O_2. However it is important to mention that its activity is significantly higher in the tail of the epididymis during the maturation and progress process (January-July) and particularly during the storage process (August-September, and perhaps the first half of October). In sperm cells from the cephalic region, the CAT activity reaches its highest levels at the beginning of September and the lowest at the end of the same month, date in which the CAT activity in sperm from the caudal region reaches its highest level. In the epididiymal fluid the CAT activity shows a similar pattern in the cephalic region from that found for SOD and GPX in both regions. We observed high levels in September, and they remained low around the middle of September, rising gradually until they arrived at their highest level at the end of October. The redox balance associated with the microenvironment, through which the sperms pass, must be specific and differentially controlled to assure its adequate function.

Chapter 9 - Frugivorous bats like Artibeus lituratus and Artibeus jamaicensis are known to digest carbohydrate-rich meals rapidly and efficiently. These bats feed on variety of fruits with preference for figs, which presents an availability of 14.4 g of carbohydrate per 100 g wet mass. Ingesting the equivalent of their body weight every night, frugivorous bats have to deal with an overload of glucose daily and must have several physiological adaptations that include specialized gastrointestinal tract structures and hormonal response capable of avoiding post-prandial hyperglycemia.

It has been shown that frugivorous bats present high paracellular absorption of an analog of D-glucose (3-O-methyl-D-glucose) in their intestine. This pathway seems to contribute at least for 55% of total glucose uptake in Rousettus aegyptiacus and 70% in Artibeus lituratus. This mechanism must have an important role considering that these bats have shorter intestinal tracts than similarly sized non-flying mammals.

Besides the high intestinal absorptive capacity, some authors also pointed to a specialization of the endocrine pancreas of R. aegyptiacus as an additive mechanism to deal with high glucose influx. It has been demonstrated that the endocrine pancreas of R. aegyptiacus represents 9.1% of the total pancreas volume, far more than in some other mammals. In a similar way, A. lituratus presents an apparent large β-cell distribution with different islets sizes and shapes, similar to R. aegyptiacus. However, the pattern of β-cell distribution in A. lituratus differs from that found in other mammals. Their β cells seem to be located, mainly, at the periphery of the islets, a pattern not observed in R. aegyptiacus, rodents, humans, non-human primates or pigs, where β cells are located throughout the islets, but especially in the core. This pattern could reflect physiological and environmental adjustments to accommodate changing demands for insulin. This species also presents elevated glucose tolerance, high insulin sensitivity and efficient insulin signalization in the pectoral muscle and hepatic tissue. Another important adaptation in A. lituratus is the ability to maintain normoglycemia even after 6 days of starvation, possibly because of its high hepatic glycogen stores and increased hepatic gluconeogenesis.

In conclusion, and from an evolutionary point of view, the rapid and efficient digestion, high sensitivity to glucose and insulin and the presence of an apparent large β-cell distribution represent adequate adaptations that guarantee the proper control of glucose homeostasis in fruit-eating bats, constantly challenged by a high influx of glucose from its carbohydrate-rich diet.

Chapter 10 - Modern research in conservation biology of bats permits to establish the status of these mammals on the basis of their demography and population dynamics. Most concern goes to those species that are endangered or are more vulnerable due to evidence of the decline in their populations or the degradation of their natural habitats. Taking Europe as an example, nearly a third of its bat species have been decreasing in numbers during the last decade according to the IUCN Red List of Endangered Species. Researchers control population numbers year after year and try to identify the reason for the changes they observe. However, short term fluctuations in population numbers are common in all species, and the importance of these may be difficult to interpret. Tracking population numbers during longer periods of time might be the best means to establish real trends and to identify the possible causes of these changes. The data obtained from the bat remains found in fossil and sub-fossil localities provide the means for this long-term tracking of populations, some good examples available in published papers. In this chapter we present an overview of how bat fossils can be useful in conservation biology with some reference examples, and the limits of the use of these mammals for this purpose will be discussed.

Chapter 11 - Prolonged sperm storage is an adaptation to the reproductive cycle that is most highly developed in microchiroptera, where sperm may be stored up to ~198 days and still retain fertilizing capacity. The aim of this review is to describe the mechanism and molecules necessary for sperm to become efficiently stored in the female genital tract. Perpendicular orientation of sperm with their head towards the reproductive tract epithelium are characterized in all sperm-storing bat species so far investigated, suggesting that such relationships are an integral part of the mechanism of prolonged storage of sperm. Recent study suggests that the ultimate controls of the mechanism underlying sperm storage are hormonal. It has been demonstrated that sperm-storing female bats contain a high circulating level of androgen during the period of sperm storage. It is suggested that androgen creates a unique microenvironment that facilitates prolonged sperm storage. The identification of the specific molecules responsible for prolonged sperm storage may suggest a mechanism to keep sperm viable for a prolonged period for use in assisted reproductive techniques.

In: Bats: Biology, Behavior and Conservation
Editors: J. L. Zupan and S. L. Mlakar, pp. 1-90

ISBN 978-1-61209-749-7

Chapter 1

CHIROPTOPHOBIA: RABIES, SARS, AND OTHER BAT–RELATED HUMAN PATHOGENS

***Dennis J. Cleri**[*1]**, Anthony J. Ricketti**[≠1]* ***and John R. Vernaleo**[£2]*

[1]Internal Medicine Residency Program, St. Francis Medical Center, Trenton, New Jersey, and Seton Hall University School of Graduate Medical Education, South Orange, New Jersey

[2]Division of Infectious Diseases, Wyckoff Heights Medical Center, Brooklyn, New York

ABSTRACT

Chiroptophobia is as widespread as bats are in nature. The modern fear of bats is almost exclusively the fear of rabies. This narrow view hinders our conservation efforts as well as discouraging research. Understanding bat-associated zoonoses is essential to study these ubiquitous creatures' longevity, immunology, epidemiology, and physiology. We need to learn how they interact with civilization, how we can safely interact in their environment, and, when necessary, how man and bat can safely share the same space together. A realistic assessment of the risks bats pose to the public and chiropterists will help us preserve bat populations and their habitats.

Bats are carriers of multiple viral pathogens. Neurological symptoms are associated with lyssaviruses (including rabies), Paramyxoviridae (Henipavirus genus: Nipah virus and Hendra virus), Flaviviridae (Japanese encephalitis virus, Kyasanur Forest disease virus, St. Louis encephalitis virus, West Nile virus), Toscana virus (Bunyaviridae), and Alphaviruses (Venezuelan equine encephalomyelitis, Eastern and Western Equine encephalitis viruses, and Sindbis virus (rash and arthritis in humans, encephalitis in mice)). Pulmonary disease is a prominent feature of infections with hantavirus infection, severe acute respiratory syndrome (SARS) coronaviruses, influenza A virus, Menangle

* Address all correspondence, proofs, and reprint requests to: Dennis J. Cleri, MD: tSt. Francis Medical Center, Department of Medicine, Room B-158, 601 Hamilton Avenue, Trenton, New Jersey 08629-1986, [Tel: 609-599-6291 or 609-599-5061 FAX: 609-599-6232 E-mail: dcleri@stfrancismedical.org].

≠ Tel: 609-581-1400 or 609-599 5061 FAX: 609-585-5234 E-mail: allandpulm@aol.com.

£ Tel: 718-963-7764 FAX: 718-205-7307 E-mail: johnvernaleo@bqhcny.org.

virus, Tioman virus, parainfluenza virus type 2 (Paramyxoviridae causing flu-like illness), Kaeng Khoi virus (Bunyaviridae), Melaka virus (reovirus), and a previously unknown reovirus of bat origin related to Melaka and Pulau viruses. Hemorrhagic fevers result from Filoviridae (Marburg and Ebola viruses), Flaviviridae (Dengue and sometimes Kyasanur Forest virus), and Bunyaviridae (Hantavirus, Rift Valley fever virus). Chikungunya virus (Alphavirus-Togaviridae) infection is clinically similar to Dengue fever. The Tacaribe virus group (Arenavirus) includes hemorrhagic fever viruses. GBV-D virus (Flavivirus), a virus closely related to GBV-C, the cause of hepatitis G has been isolated from frugivorous bats (Pteropus giganteus).

Bat ectoparasites harbor *Bartonella henselae* (cat-scratch disease, endocarditis, bacillary angiomatosis/peliosis), a *Rickettsia* sp., *Borrelia* sp. (Lyme disease), *Neorickettsia* sp. (formerly *Ehrlichia* sp.), and *Trypanosoma* spp. (brain abscesses, Chagas' disease). Parasitic diseases include *Angiostongylus cantonensis* (eosinophilic meningitis), whose neurologic symptoms in bats may be confused with Australian bat lyssavirus infection.

Bat-origin *Histoplasma capsulatum* (pulmonary and disseminated disease) has caused disease in man, and microsporidia, a known group of gastrointestinal pathogens, has caused disease in bats. Other fungi carried by bats include *Blastomyces dermatitidis*, *Coccidioides immitis* and *Trichosporon* sp. Bats are asymptomatic carriers of *Leptospira* sp., and exposures to bats have resulted in human infection (sepsis, Weil's disease). In captivity, bats are susceptible to *Yersinia pseudotuberculosis* from rodents, also putting their keepers at risk for this infection (enterocolitis, pseudoappendicitis, Kawasaki-like syndrome, sepsis).

Cynad-associated amyotrophic lateral sclerosis-parkinsonism dementia in the Chamorro people of Guam is linked with the absorption of high concentrations of the cyanobacterial neurotoxins (β-methylamino-L-alanine) after consuming *Pteropus mariannus mariannus* (flying fox).

A picture of the risk of bat contact requires a dispassionate discussion of each pathogen's epidemiology, clinical presentation, prevention, and treatment. Understanding the relationship amongst man, chiroptera and these pathogens will assist us in developing prevention strategies (vaccine and vaccine applications), diagnosis and treatment, and at the same time allowing us to safely study and preserve bat populations.

Keywords: Aravan virus, Australian bat lyssavirus, Coronavirus, Cynad-associated ALS-parkinsonism dementia, *Desmodus rotundus,* Duvenhage virus, *Eptesicus fuscus,* European bat lyssavirus, Flying fox, Hendravirus Histoplasmosis, Irkut virus, Khujand virus, Lagos bat virus, *Lasionycteris noctivagans Lyssavirus* genus , Mokola virus, *Myotis dasycneme, Myotis daubentoni*, *Myotis myotis,* Nipah virus, *Pipistrellus subflavus*, *Pteropus alecto*, *Pteropus conspicillatus, Pteropus mariannus mariannus, Pteropus poliocephalus Pteropus scapulatus,* Rabies, *Rhinolophus blythi Rhinolophus ferrumequinum, Rousettus leschenaultia,* SARS, *Tadarida brasiliensis*, West Caucasian bat virus, Vampire bat

"Eye of newt, and toe of frog,
Wool of bat, and tongue of dog,
Adder's fork, and blind-worm's sting,
Lizard's leg, and howlet's wing,--
For a charm of powerful trouble,
Like a hell-broth boil and bubble."
[William Shakespeare: Macbeth]

INTRODUCTION

Except for Antarctica and a few oceanic islands, bats are found worldwide, including north of the Arctic Circle in the summer (but only in areas with trees). [1-4] Considering their high metabolic rates, they are long-lived mammals (some living to 41 years). [5-7] Of mammals, Order Chiroptera with nearly 1,200 species (20% of all mammalian species), is second only to Order Rodentia (over 2000 species) in evolutionary success. [4, 8]

Natural translocation of bats occurs in cold weather with migration. Unintentional translocation by ships, in shipping containers, and by aircraft, transport bats around the world. [9] Species of *Pteropus* (fruit bats/flying foxes – suborder Megachiroptera) have been tracked by satellite and found to have flown between Australia and New Guinea. One individual migrated over 3000 km. [10] Further complicating the epidemiology of bat-associated zoonoses are roosting patterns. Spatial segregation of single bat populations into modules (asymmetric use of trees for nesting) affects infection dynamics. [11]

Sixty-one percent of the 1,415 species of human pathogens are zoonotic. Forty-nine percent of emerging viral infections cause severe neurologic disease, and 80% of these are zoonotic.[12] Bats are a reservoir for 60 viruses, 59 of which are RNA viruses.[4] The most important of these, the Rhabdoviridae, are the most feared encephalopaths. Unique immunoglobulin heavy chain diversity (found in Pteropid bats) may have permitted bats to co-evolve successfully with these RNA viruses, allowing them to be persistently infected without disease (except for rabies virus and Australian bat lyssavirus). [8, 13, 14]

So feared are bats, a bat hovering over the world is the insignia of the GRU (Glavnoye Razvedyvatel'noye Upravleniye – the foreign military intelligence directorate of the General Staff of the Red Army and its successor in the Russian Federation, founded by Leon Trotsky, October 21, 1918 as the Registration Directorate). In modern times, worldwide chiroptophobia stems directly from the fear of contracting rabies. Bats are associated with other lyssaviruses (except Mokola virus), severe acute respiratory syndrome-like coronaviruses (CoV), order Mononegavirales – Paramyxoviridae family (Nipah virus, Hendra virus, Menangle virus, Tioman virus), Filoviridae (Marburg and Ebola viruses), Flaviviridae (Japanese encephalitis virus, Kyasanur Forest disease, virus, St. Louis encephalitis virus, West Nile virus), Bunyaviridae (Hantavirus, Rift Valley fever virus, Toscana virus), and, a respiratory disease caused by a reovirus of bat origin.[8, 14-17]

Bat ectoparasites harbor *Bartonella henselae*. [18, 19] Bats have been associated with *Borrelia* spp., *Yersinia pseudotuberculosis*, *Y. enterocolitica*, *Leptospirosis* sp. a *Rickettsia* sp., *Trypanosoma* spp., *Leishmania* spp, histoplasmosis, blastomycosis, coccidiomycosis, *Trichosporon* sp., and others. [19-30] Cycad-associated amyotrophic lateral sclerosis (ALS)-Parkinsonism dementia in the Chamorro people of Guam may be related to consuming *Pteropus mariannus mariannus* (flying fox). [31, 32]

We will discuss the major bat-associated human pathogens, their microbiology, epidemiology, and clinical presentations.

Once more unto the breach, dear friends, once more;
Or close the wall up with our English dead.
In peace there's nothing so becomes a man
As modest stillness and humility;
But when the blast of war blows in our ears,

Then imitate the action of the tiger:
Stiffen the sinews, summon up the blood.
[William Shakespeare, Henry V]

RABIES AND LYSSAVIRUSES

"...autem omnis morsus habet fere quoddam virus..."
[Aurelius Cornelius Celsus (25 BC-45 AD): *De re medicina*]

History

The word rabies is derived from the Sanskrit 'rabhas,' the god of death and his dog (India, 13th century BC – Vedic period). [33] The disease was first mentioned in the Eshnunna law code (~2300 BC). Rabies is referred to in sixth century Chinese literature, and *The Iliad* (700 BC). Democritus (500 BC) described canine rabies, and Celsus labeled the disease 'hydrophobia.' Cordamus, a contemporary of Celsus was first to suggest the disease was transmittable through saliva. In 1804, Georg Gottfried Zinke transmitted the disease from one dog to another by instilling saliva from an infected dog into incisions in the foreleg of another animal. On July 6, 1885, fourteen-year-old Joseph Meister (1871-1940) was the first person to receive 13 injections of Louis Pasteur's rabies vaccine that consisted of dried spinal cord of rabid rabbit. [33]

Microbiology/Virology

Rabies virus is a negative-stranded enveloped RNA virus with a 12 kb genome size that contains only five genes. [34, 35] It multiplies in vertebrates and insects, and is believed to have evolved from an insect rhabdovirus that infected insectivorous bats 7,000 to 12,000 years ago. [33] The present canine rabies virus originated from a common ancestor within the past 1,500 years. Present rabies viruses in non-flying mammals are from six geographically and phylogenetically distinct clades. Their ancestry is believed to be from domestic dogs in the southern Indian subcontinent. Older canine rabies viruses that circulated in the Middle East more than 2,000 years ago have not survived, were caused by bat rabies virus, or were a different lyssavirus genotype. [34]

Rabies (lyssavirus type 1) and the other lyssaviruses are members of the order Mononegavirales whose families include Rhabdoviridae, Paramyxoviridae (human parainfluenza viruses, mumps, measles, respiratory syncytial viruses, Nipah virus – Henipaviruses and others), Bornaviridae (Borna disease virus – fatal encephalomyelitis of horses and sheep with symptoms similar to rabies, staggering disease in cats, and avian bornavirus causing proventricular dilatation disease in psittacine birds), and Filoviradae (Marburg and Ebola viruses).[33, 36-39] In 1985 Borna disease virus antibodies were found in 1.6% of psychiatric patients tested. Repeated studies in other settings have led to inconsistent results. [36-42]

The Rhabdoviridae family that infect animals includes the *Lyssavirus*, the *Vesiculovirus* (vesicular stomatitis virus, Chandipura virus encephalitis, mandarin fish *Siniperca chuatsi* rhabdovirus), and the *Ephemerovirus* (bovine ephemeral fever) genera. Other Rhabdoviridae include *Cytorhabdovirus* (i.e., lettuce necrotic yellow virus), *Nucleorhadovirus* (i.e., potato yellow dwarf virus), *Dichorhabdovirus* (orchid fleck virus) and *Novirhabdovirus* (viral hemorrhagic septicemia virus, infectious hematopoietic necrosis virus) genera. [43-51]

The major *Lyssavirus* genus members are: rabies virus (found on all continents except Australia and Antarctica, and absent from a few remote island nations); Lagos bat virus; Mokola virus; Duvenhage virus; European bat lyssaviruses 1 and 2, Australian bat lyssavirus, and putative species Aravan virus, Khujand virus, Irkut virus, and West Caucasian bat virus.[43, 52] It should be noted that two cases of imported rabies were reported in Australia, one in 1987 and a long-incubation case in 1990. [53, 54]

Rochambeau virus, although listed as a putative *Lyssavirus* has been shown to have no relatedness to the remainder of the group, and is now considered part of the genus *Ephemerovirus*. [52, 55] Obodhiang virus (isolated from the mosquito *Mansonia uniformis* in the Sudan along the Nile) and Kotonkan virus (isolated from *Culicoides* sp. midges in Nigeria), are rhabdoviruses of the genus *Ephemerovirus*. Kolongo and Sandjimba viruses are found in birds and, based upon phylogenetics, may be considered a separate genus grouped with Tupaia virus. Mount Elgon bat virus, Kern canyon virus, and Oita virus are grouped into a monophyletic cluster (putative genus). [55, 56] The most recently described addition to Rhabdoviridae is the Moussa virus, isolated from the *Culex decens* mosquito in Cote d'Ivoire. [57] These Rhabdoviridae may become important to the chiropterist, as insects are the staple diet for many bats. The phylogenetic relationships among Rhabdoviridae are reviewed by Bourhy et al. [58]

Bats are exceedingly long-lived and are important reservoirs of rabies and other lyssaviruses. Not all experimentally infected bats (*Eptesicus fuscus*) succumb to infection and long-term repeated exposure to the virus (as may occur in bat colonies) confers immunologic memory and reduced susceptibility to rabies infection.[59]

Epidemiology

Rabies: United States

Principle terrestrial rabies wildlife reservoirs in the U.S. in 2008 were: raccoon (raccoon rabies virus variant) over the entire eastern U.S.; skunk rabies in the central U.S. and California; fox rabies in Texas, Arizona and Alaska; and, mongoose rabies in Puerto Rico. [60] The first U.S. bat-transmitted human rabies was reported in 1951. [61] With the control of canine rabies in the U.S., even in the face of widespread rabies in raccoons, skunks and other terrestrial mammals, insectivorous bats are the predominant source of indigenous human rabies infections. [60, 61] The overall incidence of bat rabies viral infections in the US is 0.5-1% of tested bats. The risk of rabies infection in a bat that appears ill or injured is 7%-50%. [62] Between 1995 and 2009 there were 44 cases of human rabies and an additional case of abortive human rabies after a bat exposure. [63, 64] Thirty-three were bat-associated, 8 were dog-associated (all dog exposures occurring outside of the U.S.), and one exposure each to a mongoose, raccoon and fox. [63]

Between 2007 (7,060 cases of animal rabies and one case of human rabies) and 2008 (6,841 cases of animal rabies and two cases of human rabies), there was a 3.1% decrease in the infection rate. Only 7% of the cases were in domestic animals. The majority of the cases were in raccoons (2,389 cases – 34.9%), but bats represented the second most frequently infected animals (1,589 cases – 26.4%), followed by skunks and foxes. [65] There was a 6.7% decrease in the number of bat rabies cases reported in 2008 as compared to 2007. In 2007, 6.4% of bats tested were infected as compared to 6% in 2008. In 2008, bat rabies was reported in 47 of the lower 48 states. No bat rabies was reported in Alaska, Hawaii, New Mexico, and Puerto Rico. Texas, California, New York, and Illinois reported nearly half of the bat rabies cases. Idaho, Illinois, Indiana, Mississippi, Nevada, Oregon, Utah, Washington, and Wisconsin reported bat rabies but no rabies in terrestrial mammals. [65]

In 2009, there were 5,343 cases of animal rabies and 4 cases of human rabies. No cases were reported in Massachusetts, Wisconsin (not reportable), Delaware, District of Columbia, South Carolina, Alabama, Louisiana, Arizona (not reportable), Colorado, Hawaii, Washington, American Samoa (not reportable), Guam, and the Commonwealth of the Northern Mariana Islands. The largest number of animal rabies cases was reported from the South Atlantic states (2,103 cases). The four human rabies cases came from Virginia, Texas, Indiana, and Michigan. [66]

The bats identified as most frequently infected in 2008 were: the big brown bat (*Eptesicus fuscus* – 63.8%); Brazilian (Mexican) free-tailed bat (*Tadarida brasiliensis* – 10.1%); hoary bat (*Lasiurus cinereus* – 4.4%); red bat (*L. borrealis* – 4.2%); western pipistrelle (*Pipistrellus hesperus* – 4.0%); little brown bat (*Myotis lucifugus* – 3.4%); and, silver-haired bat (*Lasionycteris noctivagans* – 1,9%).[65] Since 1990, bat-associated human rabies cases in the U.S. involved the silver-haired or eastern pipistrelle bats (*L. noctivagans* or *Pipistrellus subflavus* – 17 cases), *Myotis* species (1 case), Brazilian (Mexican) free-tailed bat (*Tadarida brasiliensis* – 10 cases), the big brown bat (*E. fuscus* – 1 case), and unknown bat species (4 cases). [63] The one patient who succumbed to rabies being bitten in Mexico by a dog and fox was infected with Brazilian (Mexican) free-tailed bat-variant rabies virus rather than a canine variant. [67]

Big brown bats (*E. fuscus*) are found throughout North and Central America. Rabies occurs in this species throughout the continental U.S. The majority of bat-rabies diagnosis (passive surveillance by state health departments) identified big brown bats as the principle species. [68] In one study of 35 wild big brown bats captured around Fort Collins, Colorado, all animals showed evidence of wing bites. Yet, between 1958 and 2006, there was only one indigenous case of human rabies attributed to big brown bat rabies virus variant. [68]

On the other hand, *L. noctivagans* and *P. subflavus* rabies variants account for 75% of human rabies cases and 70% of cryptic rabies. Both bats are uncommon (*L. noctivagans* and *P. subflavus* were <5% of bats submitted for testing in southeastern US, and *L. noctivagans* was <12% of bats submitted for testing in the northwest U.S.). The prevalence of rabies in these species is low or similar to rabies prevalence in more common species. Human encounters with these two species are unusual. Infection with *L. noctivagans* or *P. subflavus* rabies virus variants in other bat species (spillover infection) is unusual, and these two virus variants are rare. In spite of that, rabies deaths of terrestrial mammals are more frequently associated with these rabies virus variants in the southeastern and northwestern U.S., rather than more common rabies virus variants. Outside of these areas, more common rabies virus variants account for terrestrial mammalian rabies. [69]

In the northwest, *L. noctivagans* rabies virus variant is responsible for 57% of bat-associated rabies in terrestrial mammals and 80% of bat-associated human rabies. In the southeast, *P. subflavus* rabies virus variants account for 63% of bat-associated terrestrial mammal rabies and 89% of bat-associated human rabies. *P. subflavus* represents only 2% of all rabies-positive bats. [64, 65, 69]

Increased infectivity may account for these disparities. In neuroblastoma cells, *L. noctivagans* rabies virus variant and rabies virus variants from domestic dogs and coyote replicate equally well. In fibroblast and epithelial cells, especially at 34°C, *L. noctivagans* rabies virus variant replicates to higher titers than the other viruses. [69]

In America, all rabies virus isolates belong to genotype 1 (serotype) lyssavirus (common rabies virus). [70] In the United States and Canada, silver-haired bat rabies viruses are the leading cause of death from rabies. [71] In Europe, rabies virus (genotype 1) European bat lyssavirus 1 (EBL1) (genotype 5), and European bat lyssavirus 2 (EBL2) (genotype 6) are the circulating lyssaviruses that cause human rabies. In Europe, the bat lyssaviruses are limited to bats, principally *Eptesicus serotinus* and *Myotis* species. The EBL1 and EBL2 are immunologically very similar to rabies virus and standard prophylaxis should be administered upon exposure. [72]

Rabies: Canada

In 2008, Canada reported 235 cases of animal rabies: 80% in wild animals; 6.8% in livestock; and, 6.4% in domestic companion animals. This was a decrease of 13.9% from the 273 cases reported in 2007. [65] Rabies in raccoons, bats and cattle decreased while increases were seen in skunks, dogs, and foxes. Since 1924, there were 24 documented cases of human rabies reported in Canada, including one case in 2007. Since 1970, bat-variant rabies virus caused 6 of the 7 human rabies cases. Passive surveillance of bats from western Canada (1985-1989) revealed 4.8% were positive for rabies virus. [73] No human rabies was reported in Canada in 2008. [65]

Rabies: Mexico

Ninety percent of the 3,600 cases of human rabies reported in Mexico between 1939 and 2003 were transmitted from dogs. [74] A national rabies control program beginning in 1990 resulted in the administration of 150 million doses of rabies vaccine between 1990 and 2005. [75] Control of dog rabies reduced the number of human cases from an average of 70 per year (1939-1989) to a total of 320 cases (1990-2003). [74] Dog-mediated human rabies decreased from 60 cases per year in 1990 to 0 cases in 2000. Canine rabies decreased from 3,049 cases in 1990 to 70 cases in 2007. [75]

Between 1998 and 2005, antigenic characterization of human rabies viruses revealed that 30% of the cases were of bat-origin viruses. [74] The major species of bats associated with rabies are the hematophagous bat, *Desmodus rotundus*, and the insectivorous bat *T. brasiliensis*. [61] D. *rotundus* is a non-migratory colonial bat that roosts in natural or human shelters. They populate areas where food is available, principally cattle. [76]

Most recently, there was a decrease of 19.4% in the number of animal rabies cases between 2008 (232 cases) and 2007 (288 cases). Thirteen percent of animal rabies occurred in dogs and 78.9% occurred in cattle. Only two cases of wildlife rabies were reported. Exposure to vampire bats accounted for all 3 cases of human rabies reported in Mexico in 2008. [65]

Two cases of human rabies associated with bats (probably vampire bats) in Jalisco were reported in 2009. [77]

Rabies: Latin America

Since 1975, more than 500 cases of human rabies and 100,000 cases/year of fatal bovine rabies have been directly attributed to the common vampire bat, *D. rotundus*. [78] Campaigns throughout Latin America have resulted in rabies vaccination for 42-44 million dogs per year. Of 39 cases of human rabies in 2003, 27 cases were canine-transmitted human rabies. [79-81] This represents over a 90% reduction in canine rabies. [79, 81]

In 2004, 79 cases of rabies were reported including eight cases from the US. [81] Of the 71 cases from Latin America, 46 cases were transmitted by vampire bats. [82] In 2005, 55 vampire bat-transmitted rabies cases were reported in Latin America as the result of several outbreaks, with 41 of these cases in the Amazon region of Brazil. Analysis of one outbreak identified the following risk factors: (1) age < 17 years; (2) loss of electric power; and (3) not receiving rabies prophylaxis. The village was characterized as having few farm animals for the vampire bats to feed upon. [83]

Of the three species of vampire bats found in Latin America (*Diphylla ecaudata* – hairy legged vampire bat, *Diaemus youngi* – white-winged vampire bat, and *D. rotundus* – the common vampire bat), only *D. rotundus* is known to feed on mammals. *Diphylla ecaudata* and *Diaemus youngi* feed exclusively on the blood of wild birds. [84] Antigenic characterization of rabies virus isolates from several outbreaks revealed rabies virus variant 3, which is associated with *D. rotundus*. A study conducted in southeastern Peru revealed a 10.3% antibody-positive serology for rabies in both vampire and non-vampire bats, although all brain tissues tested for evidence of active disease by direct fluorescence antibody (DFA) were negative. Vampire bats (*D. rotundus*) and non-vampire bat species (*Carollia* spp., *Artibeus* spp., *Anoura caudifera*, *Choeroniscus minor*, *Dermanura andersoni*) had statistically similar seropositivity rates. [85]

Brazil: In Brazil, cattle rabies viruses fall into two phylogenetic groups, dog- and vampire bat-rabies virus variants. Isolates from affected cattle are 99.2% vampire bat-rabies virus variants. [76] Bovine rabies appears to increase in the wet season, when vampire bats are giving birth and lactating. When no longer dependent on their mothers, juvenile males are forced out of the principal colony by dominant males. The juvenile males join satellite male colonies where fighting occurs. This influx of susceptible males contributes to spreading of rabies amongst the bats and results in the seasonal increase in rabies in cattle. [86]

Although vampire bat populations in colonies remain stable, they often visit neighboring roosts within 10-20 km. This interaction allows for the relatively slow moving spread of rabies in vampire bat populations. Geographically, rabies outbreaks amongst vampire bats move approximately 40 km per year. [76]

In another Brazilian study examining 7,393 bats collected between 1997 and 2002, fluorescent antibody testing for rabies virus antigen, and mouse inoculation tests, found 1.3% positive for infection. One-third of the positive bats were *Artibeus lituratus*. [87] Rabies in *A. lituratus* was reported for the first time in 2009 in Montes Claros, State of Minas Gerais, Brazil. [88] *Eptesicus* sp., and *Myotis* sp. were the most frequent positive species of the Vespertilionidae. *Molossus rufus* and *M. molossus* together represented ~14% of positive bats. In another study, Vespertilionidae (24 species) represented 29.3% of infected bats, but as a group, 50% were infected. Similarly, Molossidae (26 species) represented 26.8% of infected bats, but as a group, 42.3% were infected. [61] There was no statistical difference in the infection rates between male (51.5%) and female bats (48.5%). [87] Another Brazilian study looking at 10,579 samples collected between 1993 and 2007 described an urban rabies cycle predominated by canine rabies (1993 to 1997), and a rural cycle predominated by bats and herbivores (beginning 1998). [89]

Of the 1,120 species of bats, 64 genera and 167 species reside in Brazil. Of these, 41 species in 25 genera and 3 families were infected with rabies. Between 2001 and 2007, 49.1% of rabies-infected bats were non-hematophagous, 29% were hematophagous, and the species were undetermined in the remainder. [61]

Chile: In Chile, the last canine rabies occurred in 1972. Of the seven genera of bats found in Chile, four genera (*Myotis*, *Lasiurus*, *Tadarida*, and *Desmodus*) have species that carry rabies in Americas. de Mattos et al characterized 105 rabies isolates (all from urban centers) from cats, dogs, 2 cows, 1 pig, and 90 insectivorous bats (mouse-eared bat - *Myotis chiloensis*, red bat - *Lasiurus borealis* and Brazilian free-tailed bat - *T. brasiliensis*) collected in Chile between 1977 and 1998. All bat specimens were obtained from animals captured during the daytime, inside and outside of residences and public buildings, and displaying "anomalous behavior." [90] The first case of human rabies since 1972 was reported in March 1996. As is frequent in the US, the patient had no known exposure to a rabid animal. Characterization of the rabies virus identified it as a variant from *T. brasiliensis*. [91]

No isolates appeared to be related to terrestrial rabies reservoirs or related to rabies viruses isolated from the common vampire bat (*D. rotundus*). The Brazilian free-tailed bat and red bats (*Lasiurus borealis*) were identified as the reservoirs for the most frequently isolated rabies genetic variant in Chile. [90, 91]

Rabies: Europe

In 2009, The World Health Organization reported 2,545 domestic animals, 4,144 terrestrial animals, 38 bats, and 11 humans infected with rabies in Europe. Rabies in Europe is principally found in wildlife, with 80% of cases found in the red fox (*Vulpes vulpes*). Control of sylvatic rabies, especially in this species has been accomplished through oral vaccination programs. Oral vaccines (either recombinant or attenuated live rabies viruses) are placed in sachets encased in bait (fishmeal, fat and paraffin) to immunize these canines and other animals. [92] Canine rabies is not present in the countries of the European Union as of 2005. [93] As of 2005, Ireland, the United Kingdom, Sweden, Norway, Finland, Denmark, the Netherlands, Luxembourg, Belgium, France, Switzerland, the Czech Republic, Italy, Spain and Portugal are free of rabies in terrestrial animals. [93]

European bat lyssavirus I and II (EBLV-1 and EBLV-2) are present in a limited number of countries and have been found in one third of native European bat species. EBLV-1 is believed to have arisen in bats 100 to 500-750 years ago. EBLV variants (EVLB-1a and EVLB-1b) have different places of origin. [94, 95] EBLV-1 is most commonly found in serotine bats (*Eptesicus serotinus*). [96] Evidence of infection was found in *Tadarida teniotis*, *Myotis dasycneme*, *M. daubentonii*, *M. myotis* (EBLV-1b), *M. nattereri* (EBLV-1b), *Pipistrellus pipistrellus*, *P. nathusii*, *Vespertilio murinus* (EBLV-1a), *Nyctalus noctula*, *Miniopterus schreibersii* (EBLV-1b), and *Rhinolophus ferrumequinum* (EBLV-1b). [97, 98] EBLV-2 has been isolated less frequently. The virus has been found in bats from the United Kingdom, The Netherlands, Germany, Finland, and Switzerland in *Myotis* species (*M. daubentoni* and *M. dasycneme*). [96, 98-100] EBLV-2 was isolated from a Daubenton's bat

(*M. daubentoni*) that displayed abnormal behavior (including unprovoked aggression). [99, 101] Seroprevalence of EBLV antibodies in bat populations (*Eptesicus serotinus*, *Tadarida teniotis*, *M. myotis*, *M. nattereri*, *Miniopterus schreibersii*, and *Rhinolophus ferrumequinum*) is 60%. [96]

EBLV infections have been identified in man, bats (where it is not necessarily fatal), sheep, and stone martens (*Martes foina*). [97] There were at least three documented cases of human EBLV infections reported: EBLV-1 (Belgorod, Russia, 1985); and, EBLV-2 (Helsinki, Finland, 1985; Dundee, Scotland, 2002). Two suspected cases (Voroshilovgrad – currently Lugansk, Ukraine, 1977; Molodgvardeysk, Ukraine, 2002) were also reported. Positive antibody titers have been detected in two domestic cats of 152 cats tested (Denmark), and one sheep of 2,179 tested (Denmark). [96]

Current antirabies vaccines provide protection against EBLV-1, EBLV-2, and Australian bat lyssavirus infections, but not emerging Eurasian lyssavirus strains (Aravan, Khujand, Irkut, and West Caucasian bat viruses), or lyssaviruses belonging to phylogroup II. [102, 103] EBLV-2 appears less virulent than EBLV-1, and both are less virulent than rabies virus. [103]

Rabies: Asia

The World Health Organization estimates that of the 55,000 annual deaths from rabies each year, 31,000 occur in Asia, mostly in children. [104] In Africa and Asia, the principle zoonotic reservoir putting humans at risk is domestic dogs. [105-107]

Of the worldwide annual death toll from rabies, India accounts for approximate 20,000 deaths. [108] Of patients interviewed exiting anti-rabies centers, 92% suffered dog bits, 3.2% monkey bites, 1.8% cat bites, and 0.4% fox bites. More than half of the bites were unprovoked (64.3%) or by stray animals (64.7%). [109] In another survey, only 1% of stray dogs and 16% of pet dogs had protective antirabies antibodies. [110] Public health estimates indicate that in order to control the epidemic of human rabies in India: (1) 70% of the dog population (estimated at 25 million) needs to be vaccinated in a short period of time; (2) immunity must be maintained; and, (3) the population must be protected from spillover infection by control of the movement of dogs. [108]

China: Between 1950 and 2004, there were 108,412 rabies deaths. Periodic re-emergence of human rabies appears to be related to high susceptible dog population density.[111] The seroprevalence or rabies in Chinese bats appears to be low. Of 685 bats of eight species tested from 10 sites, only 15 bats (2.2%) of three species (*Rousettus leschenaultii*, *Rhinolophus blythi*, and *Rhinolophus ferrumequinum*) were positive.[112]

Aravan virus: Of the five lyssaviruses identified after the examination of over 3,000 bats in the former Soviet Union, EBLV-1 was isolated from bats in the Ukraine. Lyssaviruses isolated in Asiatic Soviet Union are Aravan virus (southern Kyrgyzstan, 1991), Khujand virus (northern Tajikistan, 2001), Irkut virus (Irkutsk region, 2002), and West Caucasian bat virus (Kyrgyzstan, 1991). [113, 114] Aravan virus was isolated from the lesser mouse-eared bat (*M. blythi*) from the Osh region of Kyrghyzstan, and the *Pteropus giganteus* (giant Indian flying fox) from Bangladesh. *M. blythii* ranges from northern Africa, the Mediterranean, southern Europe, Crimea, Caucasus, Israel, Palestine, southwest Asia, and parts of central and eastern Asia. [115, 116] The virus also circulates among insectivorus bats in Europe and Africa. [115] Aravan virus is related to Khujand virus and was similar to genotypes 4 (Duvenhage virus), 5 (European bat lyssavirus-1) and to a lesser extent, genotype 6 (European bat lyssavirus-2). Khujand virus is more closely related to genotype 6. [114]

Khujand virus was isolated from *M. mystacinus* and *Pteropus giganteus*; Irkut **virus** was isolated from *Murina leucogaster*; and, West Caucasian bat virus was isolated from *Miniopterus schreibersii*. [116, 117] Standard pre, and post-exposure rabies prophylaxis protocols in Syrian hamster and ferret experimental infections showed reduced protection. This was most pronounced for West Caucasian bat virus. [118]

In Thailand, canine rabies is endemic. As of 2005, although no animal or human deaths have been reported from these new lyssaviruses, neutralizing antibodies were detected for Aravan virus, Khujand virus, Irkut virus, and Australian bat lyssavirus in 4% (16) of 394 serum samples. Although serum was collected from *Pteropus lylei*, *Eonycteris spelaea*, *Hipposideros armiger*, and *Rousettus leschennaulti*, only samples from *P. lylei* and *E spelaea* were reactive. Of 16 dead bats (2 - *P. lylei* and 14 - *P. hypomelanus*), none were positive by fluorescein isothiocyanate-conjugated anti-rabies monoclonal antibody fluorescent microscopic examination (DFA) or by mouse-intracerebral inoculation testing. [119]

West Caucasian bat virus: Experimental intramuscular (neck) injections with West Caucasian bat virus in big brown bats (*E. fuscus*) resulted in clinical rabies in 3 bats with death within 10 to 18 days, recoverable virus from brains, and viral RNA detectable in one oral swab and other tissue. The rabid bats did not develop antibodies. Four surviving bats that were intramuscularly injected into the masseter muscles developed viral antibodies. At 6 months, brains, blood pellets, salivary glands, and mouth swabs were negative for virus. No viral carrier state could be demonstrated. [120]

Orally infected bats neither became ill or developed antibodies to West Caucasian bat virus. [120] In another study, viral RNA was detected in oral swabs < 5 days prior to the onset of clinical rabies. [121] Neutralizing antibody for West Caucasian bat virus in *Miniopterus* sp. bats from Kenya was found in 17-26% of bats sampled from 4 of 5 locations and varying from roost to roost. Seroprevalence in females (26%) was higher than in males (19%). [122]

Irkut virus: Bats previously infected with Irkut virus maintained their anti-Irkut viral antibodies for 12 months whether or not they were infected with West Caucasian bat virus. [120]

Rabies: Australia

Australia, along with New Zealand, the United Kingdom, Japan, and a number of small islands, are considered free of rabies. [123] In 1867, several dogs on the Australian island state of Tasmania developed rabies, but the epizootic was quickly stopped. [124] The first laboratory-confirmed human or animal rabies case was reported in 1987. The patient had traveled to India, Pakistan, Nepal, Singapore, and Thailand between February and October 1986. Sixteen months earlier, he was bitten on the finger by a wild monkey he was feeding at a marketplace in northern India. He became ill in June 1987 with encephalitis, and atypical Guillain-Barre syndrome. He died 23 days later. [125] The second case, a 10-year-old girl of Vietnamese origin who had lived in Hong Kong, developed clinical rabies in 1990 after living in Australia for 5 years. Further investigation revealed the virus was more closely related to those found in China, and not the Australian bat lyssavirus. The incubation period was originally believed to be <6.5 years, but newer calculations placed the incubation period between 4.5-6 years. [126, 127]

Australian bat lyssavirus was first reported in fruit bats in 1996. The virus infects all four species of Australian frugivorous Megachiroptera (gray-headed flying fox (*Pteropus poliocephalus*), the black flying fox (*P. alecto*), the little red flying fox (*P. scapulatus*), and the spectacled flying fox (*P. conspicillatus*)) and three species of Microchiroptera (insectivorous bats). Flying foxes range widely including moving between northern Australia to Papua New Guinea. [128]

The first bat (1996) was a 5-month-old female black flying fox (*P. alecto*) that was found under a tree unable to fly. The second bat (1995 – retrospective tissue analysis), a juvenile female *P. alecto* was noted to be more aggressive than usual. Although these bats were found in Ballina in northern New South Wales, Australian bat lyssavirus has been found in all four species of flying foxes throughout their ranges, and in *Saccolaimus flaviventris* (yellow-bellied, sheath-tail bat) in Queensland. [129]

The virus is similar to classic rabies virus and is considered a member of lyssavirus serogroup 1, but genetically different enough to be classified as genotype 7. Two strains of the virus circulate in Australia – one in flying fox bats and the other in insectivorous bats. [129]

Two cases of human infection occurred. The first case (1996) was an animal handler who was scratched or bitten by a yellow-bellied sheath-tailed bat who subsequently died of encephalitis 5 weeks later. [129] The second case (1998) involved a 37-year-old woman who, while removing a flying fox from the back of a child, was bitten on her finger. Twenty-seven months later, she became symptomatic, and she died on day 19 of her illness. [130]

In Australia, one study found 6% sick, injured, or orphaned bats were antibody-positive for Australian bat lyssavirus. Normal bats have been found to be antibody-positive, but virus isolation has been limited to bats exhibiting neurologic signs. Most commonly, symptomatic bats are lethargic, but a minority is overly aggressive. The central nervous system pathology is similar to that of rabies – nonsuppurative meningoencephalitis and ganglioneuritic involvement. [131] Salivary glands in bats with positive central nervous system pathology were not always positive for evidence of infection or virus. [129]

A description of naturally occurring Australian bat lyssavirus infection in a juvenile black flying fox begins with the discovery of two bats, 2-3 weeks old found on an unusually low tree roost without their mother. For 5 weeks, the bats behaved normally. The first sign of neurologic disease in one bat began during week 6 with sudden and progressive aggression toward its companion bats. On that first day, the bat frothed at the mouth, had repeated lordotic spasms, and vocalized loudly. On day 2, the bat was calmer, but still vocal, was eating little but biting objects. The bat was treated with antibiotics for pharyngitis, and dexamethasone. It appears that the treatment with steroids, temporarily improved its neurologic state and the bat began to eat solid food. The next day, the bat was dysphagic. The bat progressively worsened, could not roost, lay supine, had diarrhea, dysphagia, lost weight, and died on day 9. The bats that were in contact with this animal remained healthy, were antibody negative, and brain impression smears taken 11 weeks later were negative for the virus. [132]

Standard rabies prophylaxis will illicit a protective antibody response against Australian bat lyssavirus. [129] All bat bites should be considered at high risk and these individuals should receive standard post-exposure prophylaxis. Any individual who may potentially handle bats should receive pre-exposure prophylaxis and have titers checked at regular intervals. [133] After the realization of the importance of rabies prophylaxis for bat exposures, Australian bat exposures accounted for 37.5% of all post-exposure prophylaxis. [129, 134]

Rabies: Africa

In Africa, there is approximately 24,000 deaths/year from principally canine rabies. [35] In South Africa, there were 4,767 reported veterinary rabies cases (1993-2005) with a 79% increase in the number of domestic animal cases and a 59% increase in rabies in domestic dogs. Overall, there was a 19% increase in the number of cases as compared to the period 1980-1994. There appeared to be a spread to provinces neighboring KwaZulu-Natal, where most rabies cases were concentrated. [135]

African rabies viruses are classified as follows: genotype/serotype 2, Lagos bat virus; genotype/serotype 3, Mokola virus; and genotype/serotype 4, Duvenhage virus. [43, 70] Natural infections with Lagos bat virus, Mokola virus and Duvenhage virus are only found in Africa. [136]Lagos bat virus has been isolated from the straw-colored fruit bat, *Eidolon helvum* and the Gambian epaulet bat, *Epomophorus gambianus*, and *Epomops buettikoferi* (Ghana).[136, 137] Kolente virus was isolated from *Hipposideros* sp. (roundleaf) bats in Guinea and is considered an original type of Lagos bat virus. [138, 139] Duvenhage virus has been isolated from *Miniopterus* sp. [70, 115]

The primary reservoirs for rabies virus are dogs and mongoose, but jackals are also important for viral maintenance. Rabies represents a significant risk to the survival of two endangered species – African wild dogs and Ethiopian wolves. Duvenhage and Lagos bat viruses are well adapted to bats while Mokola virus is found in rodent species, although a bat reservoir is strongly suspected. [140-142] Five non-typical rabies virus isolates were reported in 2001 from Zimbabwe (honey badgers (*Mellivora capensis*), African civets (*Civettictis civetta*), and mongoose (*Herpestidae* sp.)). It is believed that these are rare wildlife strains probably maintained in the slender mongoose (*Galerella sanguinea*) population. [143]

Four putative species from bats, first found in Eurasia have been added to the *Lyssavirus* genus: Aravan virus, Khujand virus, Irkut virus and West Caucasian bat virus. [116, 144, 145] Neutralizing antibodies for West Caucasian bat virus were found in serum samples from *Miniopterus* insectivorous bats from Kenya. [122] Fruit bats, *Rousettus aegyptiacus*, that shared caves with *Miniopterus* species tested, had positive serology for Lagos bat virus, but not West Caucasian bat virus. [122] Aravan virus was isolated from the lesser mouse-eared bat (*M. blythi*). [145] From Bangladesh, the giant Indian flying fox (*Pteropus giganteus*) sera were found to neutralize activity against Khujand, and Aravan viruses. [116] Of great concern for all chiropterists is the finding that commercially available human or veterinary vaccine, or an experimental vaccinia-rabies glycoprotein recombinant vaccine, or the combined effect of rabies immunoglobulin and vaccine after exposure afforded significantly decreased protection (based upon Syrian hamster and ferret experiments). Pre- and post-exposure prophylaxis administered to animals infected with West Caucasian bat virus, considered "the most divergent", provided no significant protection. [118, 122]

A new species of lyssavirus, Shimoni bat virus was isolated in 2009 from the brain of a dead Commerson's leaf-nosed bat (*Hipposideros commersoni*) found in a cave in coastal Kenya. Phylogenetic analysis places the virus in phylogroup II closest to Lagos bat virus. Whether present pre- and post exposure rabies prophylaxis will protect against this virus remains to be seen. [146].

Transmission of Rabies

Rabies is commonly transmitted by the bite of an infected animal usually through virus excreted in saliva. Estimates of the risk of acquiring the infection by animal bite is 5-80%, while the risk from a scratch is 0.1-1%. [33] Overall, the prevalence of rabies in bats was estimated at 15%. The risk of acquiring rabies in the U.S. was estimated as 5% after a superficial bite to the hand; 0.1% after contact with rabid saliva on a recent wound; and, 0% after contact with rabid saliva on a wound older than 24 hours. [147]

Rabies has been transmitted by inhalation of infected materials in bat caves, laboratory accidents that aerosolized infected material and handling and skinning infected carcasses. [33, 148, 149] In 2002, a Scottish bat conservationist died of clinical rabies caused by European bat lyssavirus 2 infection. Johnson, Phillpotts, and Fooks demonstrated in a mouse model that rabies virus and European bat lyssavirus 2 could be infectious through inhalation in a dose-dependent manner. [149]

Rabies virus has been isolated from saliva, respiratory secretions, sputum, nasal swabs, pharyngeal swabs, eye swabs, tears, cerebrospinal fluid, urine, blood, serum, peripheral nerves, skeletal muscle, skin, heart, liver, adrenal glands, kidneys, lung, spleen, pancreas and brain. [33] Infected cornea has transmitted the disease to human transplant recipients. [33]

Contracting rabies after cornea transplantation has occurred 8 times in the past. [150] In 2005, 4 organ recipients (liver, 2 kidneys, and an iliac artery graft) developed rabies encephalitis after receiving organs from a patient who died of rabies after a bat exposure. [150, 151] Six transplant recipients received organs from a single donor who was later found to have been infected with rabies and died of encephalitis. The two cornea transplant recipients and the liver transplant recipient did not become ill. The liver transplant recipient had received rabies prophylaxis 20 years ago. One kidney and one lung transplant patient died within days, while the other kidney transplant recipient died 7 weeks later. All recipients who initially survived received post-exposure prophylaxis and antiviral therapy. [152]

Reports of human-to-human transmission from family contact, lactation, kissing, a human bite, intercourse, and providing health care to a patient with rabies are documented in the literature. [33, 153] Oral transmission of rabies is well documented in animals. It is believed to be the cause of rabies in foxes (*Vulpes vulpes*) and striped skunks after consumption of dead or dying vampire bats. [33, 154] Rabies has occurred in mature and immature red foxes (*V. vulpes*) and striped skunks (*Mephitis mephitis*) after feeding them rabies infected white mice. [33]

Clinical Presentations of Rabies

Clinical Presentations of Rabies in Bats

Epidemiological factors increasing the risk of a bat being infected with rabies include: (1) location of the bat; (2) circumstances of capture; (3) sick bats discovered by citizens; (4) inability to fly; (5) flying in the daylight; (6) lethargy; and, (7) paralysis. [61, 155] In a study from Brazil, 52.59% of rabies-infected bats were found in an atypical location. [61] Of bats found infected with rabies: 11.34% of bats found inside houses; 8.24% of bats hanging in buildings; 1.03% of bats roosting on roofs (although the first bat from the colony was captured on the ground); bats captured while feeding on a human; bats flying near a child; 3-15% of obviously ill bats; and, 1% asymptomatic bats. [61]

Bats infected with EBLV-1 have been found to excrete that virus while asymptomatic. EBLV-2 infected bats have displayed the same symptoms of rabies virus infected bats (uncoordinated, muscle spasms, unprovoked aggression, flying during daylight, flying into a tree, weakness, and refusing food and water). [102] Long term studies of EBLV-1 infection on mouse-eared bats (*M. myotis*) found no increase in mortality in naturally infected bat colonies. [156] During these same long-term (12 year) studies, none of the bats captured displayed any behavior that could be attributed to rabies infection. [157]

Big brown bats (*E. fuscus*) experimentally infected with big brown bat variant rabies virus via intramuscular injection, 80% (16 of 20) developed clinical rabies. Four infected bats survived. Incubation periods were 13-52 days (mean: 24 days). Clinical disease developed in 12 of the 16 bats between 13 and 17 days of infection. Four bats had incubation periods of 52 days. Common symptoms were acute weight loss, reclusive behavior, ataxia, and paresis persisting for more than one day. [68]

Naturally infected big brown bats in captivity have been discovered laying in their food bowel, becoming sensitive to noise, progressively vocal and aggressive, biting aggressively when handled, attacking cage mates, and refusing water. [158] Bats from the same colony of the first rabid bat, captured that same year had a low seroprevalence for rabies antibodies (2% in adult females and 3% in volant juveniles). The following year, 23% of adult females and 14% of volant juvenile bats were seropositive. Three of 22 bats tested both years seroconverted. Of the 51 bats sampled during the follow-up year, 14 were found to be seropositive (27%). [158]

D. rotundus experimentally infected with a vampire bat-sourced rabies variant virus had 25-60% mortality depending on the dose of the inoculum. Incubation periods ranged from 5-41 days (mean: 16.8 days), and the bats remained ill for 18-48 hours. Rabies virus was detected in the salivary glands of 60% of the infected bats. Ten bats died. Eight of the 10 were symptomatic with paralytic rabies. The first signs were difficulty standing on their feet and thumbs and lack of coordination of posterior limbs. This was followed by weakness, muscular tremors, and spasms, irritability to light, wind and sounds, paralysis, and prostration. One bat was incontinent of urine and two displayed purulent conjunctivitis. Although there were no signs of furious rabies, other studies have observed abnormal aggressiveness, biting or viciously attacking other animals in the cage, dashing against the wire mesh, and darting violently from one side of the cage to the other. Two days before death, bats experienced a significant loss of weight (11.5-22.6%) and some bats had no brown fat on post mortem examination. More than half of the bats that survived the infection gained weight. Two asymptomatic bats died of rabies. Post mortem examination revealed a 1% loss of body weight in each animal. [159]

The first sign of rabies in experimentally infected vampire bats (*D. rotundus*) by intramuscular challenge with high doses of vampire bat rabies virus variant, (bat mortality 79%) was decrease in blood consumption 4 days prior to death, resulting in dehydration. Three days prior to death, some, but not all infected bats remained quiet in the corner of their cage. Forty-eight hours prior to death, some bats developed paralysis of the wings and could not remain upright. Paralysis of the hind-legs was evident in bats hanging from one leg. Tremors were evident in some bats 48 hours prior to death. Salivary secretion of virus was detected only intermittently in survivors. [78]

Intramuscularly injected insectivorous bats (*E. fuscus*) with European bat lyssavirus type-1 developed rabies after an extended incubation period. Intracranially injected bats developed disease with a mean time to death of 11 days. No mortality was seen with intranasal or oral virus administration. Virus shedding in saliva was seen in sick bats, particularly prior to the onset of illness. The highest concentration of virus and viral RNA in extraneural tissue was seen in the thyroid. The authors speculated that this caused subacute thyroiditis, abnormal adrenocortical activity affecting the clinical presentation of rabies in these subjects. [160]

Rabies in other Animals

There is evidence that dogs may recover from natural rabies infection and excrete virus before symptoms occur. Salivary excretion of virus may be intermittent for months. In experimentally infected dogs, 18% died without showing signs of rabies, and, 20% of symptomatic animals recovered without any intervention. [33, 161-164] Less than 50% to 78% of dogs develop 'furious' rabies and 18% develop 'dumb' rabies. [33, 165]

The description of rabies in other animals is similar. [33, 166-169] Whereas cross-species transmission of RNA viruses has been attributed to their high mutation rates and high contact rates, the establishment of rabies virus in bats that is infective to man may be more due to similarity in host defenses. [170, 171] Two thousand rabid bats are collected annually in the U.S. after human or domesticated animal contact. [171]

North American bats have been evolving from 3-60 million years. Rabies viral lineages are directly related to specific hosts. Streicker et al point out that, "This species association of viral lineages enables identification of the species origins of relatively rare CST [cross-species transmission] events from bats to humans or domesticated animals or within the bat community." [171] New rabies virus lineages have been discovered in *Lasiurus intermedius floridanus*, *L. seminolus*, and *Myotis yumanensis*. [171]

The authors postulated that, "Nearly all viruses from cross-species infections were...no more genetically divergent than donor-lineage viruses, suggesting that they were more likely to be dead-end infections than infections occurring within stuttering chains of transmission in the recipient species." Analysis of their results indicates that cross-species transmission occurred between the most closely related North American bat species and lack of sustained transmission in more distantly related species. They conclude that rapid evolution cannot overcome "phylogenetic barriers" for new viral clades during (1) initial infection and, (2) sustained transmission. In more distantly related bat species, the decrease in cross-species infection may be due to lower contact rates or reduced risk of infection after exposure to the virus.[171]

Clinical Presentations of Rabies in Humans

Human rabies remains an almost universally fatal disease. As with any infection, recovery is dependent on clearing of the virus by the immune system. Rabies virus avoids these defenses by virus-mediated immune suppression with loss of cellular-mediated immunity. Immune effectors do not penetrate the blood-brain barrier. [172]

There are five stages of clinical human rabies: stage I – incubation; stage II – prodromal period; stage III – symptomatic rabies; stage IV – coma; and, stage V - recovery or death. [33]

Stage I – Incubation period (4 days -19 years)

The incubation period may be as short as 4 days to as long as 19 years. [33, 173, 174] The average incubation period is 2 months with 75% of patients developing symptoms 20-90 days after exposure. The shortest incubation periods (4 days) are associated with inoculation into highly innervated areas such as the face or brachial plexus. Children and vaccination failures also have shorter incubation periods. [33] Incubation periods of greater than one year occur in <5% of patients. [33, 126, 130] Long incubation periods (27 months, 4 years and 6 years) have been reported in patients infected with Australian bat lyssavirus. [33, 130] For Australian bat lyssavirus, incubation periods may be difficult to gauge as between 47.0% (1979-1992) and 57.2% (1993-2006) of fatal cases of encephalitis in Australia go without an etiologic diagnosis. [175] A number of patients received less than adequate post-exposure prophylaxis. [176]

Stage II – Prodromal period (1-10 days)

Patients initially complain of anxiety and/or depression. Some patients have predominantly gastrointestinal complaints (nausea, vomiting, diarrhea, and abdominal pain). Fever and chills occur in up to 50% of patients, and occasional patients complain of headache (23%). Itching, pain or paresthesia occurs at the site of the bite or in the proximal neural radiation. Myoedema (mounding of a part of the muscle when hit with a reflex hammer that subsequently disappears in seconds) is first seen during the prodrome and persists throughout the course of the illness. [33, 176]

Stage III – Symptomatic rabies (2-14 days; average survival 5-7 days)

Furious rabies: In one study, patients presented for care after being symptomatic for 5.4$\pm$3.1 days (range 1-15 days). In this same study, 72% of patients presented with paralytic symptoms and 28% had altered mental status or symptoms interpreted as psychiatric in nature. [176] Paralytic symptoms included Guillain-Barre syndrome, encephalomyeloradiculopathy, (acute disseminated encephalomyelitis), lumbosacral polyradiculopathy, and myelomeningitis with rapidly progressive weakness of the limbs and early bulbar involvement. Weakness was the initial symptom in patients with acute disseminated encephalomyelitis (ADEM). [176]

Patients are agitated, hyperactive, violent, have waxing and waning alertness, exhibit bizarre behavior, hallucinate, and display aggression with intermittent lucid periods. [33, 177] All patients eventually develop change in sensorium.[176] Autonomic dysfunction (piloerection, excessive salivation, sweating, priapism, repeated ejaculations, neurogenic pulmonary edema) may be prominent. [33]

Hydrophobia begins with difficulty in swallowing liquids followed by pharyngeal and laryngeal spasms, and aspiration. These spasms may last 1-5 minutes. Eventually, the sight of water produces these spasms. Similarly, aerophobia, produced by gently fanning the patient produces severe spasms. [33]

Paralysis of the palate and vocal cords produces a barking cough. Meningismus is common. Fasciculatuions are reported in 23% of patients. Unusual presentations include ocular myoclonus with hemichorea. [33, 176]

Psychiatric symptoms may mimic schizophrenia or delirium tremens. [33] Conversely, Cotard's syndrome (a major depressive disorder in which patients deny their own existence and that of the world around them) may present with hydrophobia. [33, 178]

The stress of exposure to rabies and the need for post-exposure prophylaxis has exacerbated long-standing psychosis. "Rabies hysteria" has presented as refusal to drink water. [33]

Seizures are not frequent, but are usually a pre-terminal event. [33]

Dumb rabies: Dumb rabies was associated with vampire bat-rabies after an epidemic on Trinidad prior to World War II. There was another outbreak of rabies attributed to vampire bats that presented with fever, headaches, hydrophobia, and death in 29 patients in the Peruvian Amazon. An analysis of the glycoprotein, nucleoprotein, and phosphoprotein gene sequences of virus isolated from two cases of paralytic human rabies and two cases of encephalitic human rabies found no single nucleotide- or amino acid-specific pattern for rabies virus to explain the difference in clinical presentation. In addition, there were no differences found within receptors responsible for pathogenicity or the immunodominant glycoprotein domain. [179]

Patients with paralytic rabies have relatively longer survival. Peripheral nerve dysfunction secondary to demyelination is believed to contribute to patients' weakness. In furious rabies, there is progressive focal denervation without weakness. It has been postulated that different sites of neural involvement and possibly neuronal gene expression may be involved in the differences in clinical presentation. Additionally, aberrant immune responses may contribute this difference. In paralytic rabies there is a lack of cellular immunity to rabies virus antigen and the absence of cerebrospinal fluid rabies neutralizing antibody in most patients. [180, 181]

Clinical presentation may entail weakness or paralysis of one limb or quadriplegia. Weakness is not uncommonly more severe in the limb where the virus was inoculated. Patients complain of pain in the affected muscles and fasciculation may be seen. Sensation usually remains intact although some mild sensory abnormalities have been recorded. Although patients develop meningeal signs, unlike furious rabies, patients do not become confused. As the disease progresses, cranial nerve involvement is evidenced by loss of facial expression. Combined motor and sensory deficits develop including urinary incontinence. Guillain-Barre syndrome occurs in 20% of patients. Patients may survive as long as one month without the need for a respirator. Unassisted, death occurs as the result of paralysis of the swallowing muscles and respiratory failure. [33]

Nonneurologic complications: Hypotension from hypovolemia or central nervous system involvement (usually brainstem) is common. Nausea, vomiting, diarrhea, ileus, and gastrointestinal bleeding occur. Patients in the intensive care unit receiving respiratory support succumb to myocarditis with arrhythmias or congestive heart failure. Intensive care unit patients have complications of pneumothorax, cerebral edema, diabetes insipidus, and diffuse axonal neuropathy. [33]

Stage IV – Coma (occurs immediately after symptoms appear to 14 days later)

Coma may develop immediately after symptoms appear, but often within the first 14 days of illness. [33]

Stage V – Recovery or Death (occurs usually within 18 days)

Death occurs within 18 days of the beginning of symptoms. Intensive care and respiratory support increases survival from 25 days to months. [33]

Of six cases of patients who recovered from rabies, five had received vaccine but not anti-rabies immunoglobulin. At least three patients had severe neurologic sequelae. Diagnosis was made based on exposure history, compatible symptoms, and detection of rabies virus-neutralizing antibodies. [33, 182-184]

In 2009, a 17-year-old girl developed encephalitis 2 months after exposure to a bat. She initially presented with frontal headache, photophobia, emesis, neck pain, dizziness, and facial paresthesia. She became febrile, disoriented, and developed nuchal rigidity. The patient was treated with antibiotics and steroids, recovered and was discharged home in 3 days. Approximately one week later, she returned with headache, photophobia, emesis, myalgia, particularly of the back and neck, limitation of vision, weakness, and loss of sensation in the upper extremity. Cerebrospinal fluid (CSF) revealed pleocytosis and her magnetic resonance imaging (MRI) showed enlarged ventricles. She was discharged 16 days later only to return in one week with a headache. Her headache improved after another lumbar puncture and she was discharged and lost to followup. Although no rabies virus antigens or RNA was detected, CSF tested positive for rabies antibodies (by immunofluorescent antibody test). Serum IgG (1:8192) and IgM (1:32) were reactive. The patient received one dose each of rabies immune globulin and rabies vaccine. This is believed to be the first reported case of abortive human rabies. [184]

Treatment of Rabies

In 2005, Willoughby et al reported a 15-year-old girl who contracted rabies after a 5-mm laceration to her left index finger from a bat. Although the wound was washed with peroxide, she received no post-exposure prophylaxis. One month later, she experienced fatigue and paresthesia of the left hand followed by diplopia, ataxia, and nausea and vomiting without fever. Her MRI and angiography were negative. She became febrile and her neurologic disease progressed. She was treated with an "antiexcitatory" and antiviral drug regimen (ketamine, midazolam, ribavirin, and amatadine), but did not receive any post-exposure prophylaxis. She developed serum and cerebrospinal fluid neutralizing antibodies to rabies virus, but no viral nucleic acid was detected. She was discharged 76 days later.

At day 131, she was able to smile, laugh, interact with her examiner, but had dysarthric speech, constant buccolingual choreoathetosis with generalized choreoathetosis, intermittent dystonia, ballismus, lurching gait, and fine-motor difficulties. She was able to attend high-school part time. Five months after treatment, she had dysarthria, weakness in the left hand and foot, bilateral extensor-plantar response, generalized choreoathetosis, intermittent dystonia and a lurching gait. One year after treatment, her generalized choreoathetosis completely resolved and 18 months later, her dysarthria and gait improved but she could not return to previous level of sports activity. On her last examination, she had normal affect, cognition, and cranial nerve examination. [182, 183] This therapy has been termed the "Milwaukee Protocol." [185] The main components of the Milwaukee Protocol have been used in 20 additional cases, all resulting in fatal outcomes. [186]

Limitation of immune access to the rabies-infected central nervous system (symptomatic patients) prevents the therapeutic effectiveness of provision of antiviral immunity. Mouse models using silver-haired bat rabies virus found that increasing the blood brain-barrier permeability and central nervous system inflammation resulted in increased viral clearance and improved survival. Reducing inflammation and decreasing blood-brain barrier permeability with steroids increased mortality. [71, 187]

Wild-type silver-haired bat rabies virus intracerebrally injected into mice produced little inflammatory response. Similarly injected laboratory-adapted virus B2C (attenuated rabies virus) resulted in an extensive inflammatory response with innate immune and antiviral responses (including alpha and beta interferon signaling pathways and inflammatory chemokines. This effect may be a significant mechanism by which vaccine protects the CNS. [188] We do not know whether this contributes to the pathogenesis of post-vaccination encephalomyelitis (acute disseminated encephalomyelitis (ADEM)). [189, 190]

Diagnosis of Rabies

The standard for the diagnosis of rabies is DFA direct staining or immunohistochemical localization of rabies viral antigen by indirect immunoperoxidase technique using polyclonal antibody directed against rabies viral nucleocapsid of biopsy specimens. [43, 176] Rabies virus localizes to hair follicles. In humans, a skin biopsy of the nape of the neck will be positive in 50% of patients during the first week of symptoms. Reverse transcriptase-polymerase chain reaction (Rt-PCR) will detect virus in the CSF, saliva, tissue, or decomposed brain. Immunofluorescent antibody staining of corneal epithelial cells (corneal impression test) is no longer commonly used. [43, 191]

Initial diagnosis of symptomatic rabies is complicated by the fact that the virus provokes a poor immunologic response. Very few patients have neutralizing antibody present upon clinical presentation. Antibody is only rarely present in the cerebrospinal fluid when it does appear in the serum. [192]

Rapid fluorescent focus inhibition test is used for detecting serum neutralizing antirabies antibody. [43] Nucleic acid-amplification tests may be used for the diagnosis of rabies and identifying different viral strains. Real-time PCR is preferred to conventional reverse-transcription PCR, and second-round PCR (nested or heminested) is useful in ante-mortem diagnosis to detect low levels of viral RNA. [193] Quantitative real-time reverse transcription PCR is more sensitive than viral isolation. [194] Nucleoprotein monoclonal antibodies are able to recognize viral antigen by immunofluroescence assays, western blotting, and immunohistochemistry. [195]

CSF findings are often but not invariably normal except in patients with meningismus. When abnormal, there is mild mononuclear pleocytosis. HIV negative patients with Guillain-Barre syndrome, bladder dysfunction, fever or hyponatremia have a cerebrospinal fluid lymphocytosis (<30 cells/dL) and an elevated CSF protein (<100 mg/dL). These patients may display signs of early rabies. [33]

Radiology: Chest x-rays often reveal bronchopneumonia secondary to aspiration and/or respiratory failure. CT scans of the head show nonenhancing symmetrical hypodensities of the basal ganglia. MRI in dumb and furious rabies is similar. There are nonenhancing, ill-defined, mild hyperintensity changes in the brain stem, hippocampus, hypothalamus, deep and subcortical white matter, and cortical gray matter (T-2 weighted) in the conscious patient. Gadolinium enhancement of the hypothalamus, brain stem nuclei, spinal cord, gray matter, and intradural cervical nerve roots is seen in comatose patients. In other studies, the pons was enhanced. Fluid attenuation inversion recovery MRI may differentiate rabies encephalitis from post-vaccinial ADEM. In a patient bitten on the arm, there was MRI enhancement of the brachial plexus. [33]

Prevention of Rabies

There must be no delay in the administration of post-exposure prophylaxis for true exposure. The most common cause of post-exposure prophylaxis failure is not administering rabies immune globulin. Rabies vaccines do not produce protective antibodies until 10-14 days after the first dose. HIV-infected patients will respond to vaccine if CD_4 counts are above 300 cells/μL. Failures have also be reported when rabies virus is inoculated directly into nerve endings. Pregnancy is not a contraindication to pre- or post-exposure prophylaxis. [33, 195-198]

In a recent study of 870 backpackers to Southeast Asia, only 18.1% had completed pre-exposure rabies prophylaxis. A significant number were potentially exposed to rabid animals (3.56% were licked and 0.69% were bitten). More than half (54%) of the exposures were within the first 10 days of their visits. [199]

Pre-exposure rabies prophylaxis should be administered to veterinarians, veterinarian students, and their assistants, zoo and animal handlers, rabies vaccine production workers, spelunkers, and researchers dealing with bats, rabies viruses, and their related variants. Pre-exposure rabies prophylaxis should be offered to missionaries, volunteers, and other individuals traveling in countries where rabies or its variants are endemic and access to immediate and appropriate medical care is problematic. [33, 200]

New vaccines under development include those from reconstituted and traditionally attenuated viruses, newer adjuvants, protein subunit vaccines, plasmid-vector/DNA vaccines, and recombinant viral vector vaccines (poxvirus vectors, herpesvirus vectors, adenovirus "Ads" vectors). [201]

Post-vaccination encephalomyelitis is a form of ADEM, a monophasic inflammatory demyelinating disease of the central nervous system. Worldwide, it is the most common demyelinating disease. Other forms of ADEM include multiple sclerosis, acute hemorrhagic leukoencephalitis, cerebellitis, optic neuropathy, acute transverse myelitis, brain stem encephalitis, multiphasic ADEM, and neuromyelitis optica (Devic's disease). Most cases (~75%) of ADEM are post-infectious (referred to as post-infectious encephalomyelitis – PIEM) and post-immunization where the febrile event closely coincides with neurologic symptoms. ADEM has occurred rarely after a bee sting. PIEM has occurred most frequently after measles, but also reported after infections with mumps, influenza A and B, hepatitis A and B, herpes simplex, human herpes virus E, varicella, rubella, Epstein-Barr, cytomegalovirus, HIV, *Mycoplasma pneumoniae*, *Chlamydia*, *Legionella*, *Campylobacter*, and *Streptococcus* infections. [189, 190, 202-204]

Vaccines associated with ADEM are rabies, diphtheria-tetanus-polio, smallpox, measles, mumps, rubella, Japanese B encephalitis, pertussis, influenza, hepatitis B, and the Hog vaccine. [189, 190, 202-204] The incidence following neural-based rabies vaccine is 1:300 to

1:7,000 as compared to 1:50,000 for non-neural-based rabies vaccines. Older rabies vaccines made from rabbit brain had a 1:400 incidence of neurologic complications. Semple rabies vaccine (made from infected sheep brains) recipients developed neuroparalytic complications in 1:600 to 1:1575 vaccinations. Duck embryo vaccine has an incidence rate of 1:25,000. [190, 204]

Human diploid cell vaccine (HDCV) is the preferred vaccine for rabies prophylaxis. Local reactions are reported in 60-89.5% of recipients. Systemic hypersensitivity occurs in 6% of patients receiving a booster. Other studies have shown no serious side effects. Rare neurologic adverse events have been recorded, but it remains doubtful that they are linked to the HDCV. These include four cases of Guillain–Barre syndrome, one case of seizure, and other central and peripheral nervous system disorders. [205]

Recommendations for pre- and post-exposure prophylaxis have been published by the Centers for Disease Control and Prevention, Preventive Medicine Residency, Office of Workforce and Career Development, CDC, USA. [205]

SEVERE ACUTE RESPIRATORY SYNDROME (SARS)

To-morrow, and to-morrow, and to-morrow,
Creeps in this petty pace from day to day,
To the last syllable of recorded time;
And all our yesterdays have lighted fools
The way to dusty death. Out, out, brief candle!
Live's but a walking shadow
[William Shakespeare: Macbeth]

SARS - Microbiology/Virology

The etiologic agent for SARS is the positive-sense single-stranded RNA human CoV group 2b (SARS-CoV) of animal origin. SARS-like CoV (SL-CoV) animal CoV virus has a 99% homology with SARS-CoV. [206] There are at least 26 species of CoVs infecting 36 animal species including dogs, cats, pigs, rabbits, cattle, horses (equine torovirus-EToV or Berne virus) mice (murine hepatitis virus), rats, chickens, pheasant, turkeys, whales and humans. [206-208] Animals and humans are infected by group 1 and 2 CoVs. Birds are infected with group 3 CoVs. [206]

The family Coronaviridae consists of *Coronavirus* genus and *Torovirus* genus. Human coronaviruses (HCoVs) during epidemics may be responsible for up to 35% of upper respiratory infections. CoVs cause respiratory, gastrointestinal, central nervous system, liver, and reproductive disease in rats, mice, chickens, turkeys, cattle, wild ruminants, dogs, cats, rabbits, and pigs. Human toroviruses do not grow on tissue culture. The Berne virus is a torovirus isolated from horses with diarrhea. [209]

Group 1 coronavirus was identified by RNA detection in 3 of 31 Canadian *M. lucifugus* from Manitoba. The coronavirus is related but distinct from the Rocky Mountain bat coronavirus from little brown bats (*M. lucifugus*). It is believed to be a variant of the *M. occultus* corona virus. [210]

Group 1 viruses include human coronavirus 229E (HCoV-229E), porcine epidemic diarrhea virus, and feline infectious peritonitis virus. Group 3 contains avian infectious bronchitis virus. Group 2 CoVs include human CoVs (HCoV-OC43 and HCV-HkU1), murine hepatitis virus, rat CoV, bovine CoV, porcine hemaglutinating encephalomyelitis virus, equine CoV, and canine respiratory CoV. [206, 208] Human CoVs include SARS-CoV (group 2b CoV), HCoV-229E and HCoV-NL63 (group 1 CoV), and HCoV-OC43 and HCoV-HKU1 (group 2 CoV). SL-CoV does not appear to cause disease in humans. [206, 208]

HCoV-NL63 and HCoV-HkU1 have worldwide distribution, principally infecting children. HCoV-NL63 may be responsible for up to 10% or childhood respiratory disease. It commonly causes mild upper respiratory illness (cough, fever, and rhinorrhea). More severe lower respiratory disease is seen in younger children (bronchiolitis and croup). [206, 211, 212] In France, HCoV-229E, HCoV-OC43, HCoV-NL63 and HCoV-HKU1 were isolated from 9.8% of respiratory specimens from hospitalized children with upper (50.5%) and lower (29.4%) respiratory tract infections and immunocompromised adults. [206, 213] Necrotizing enteritis in newborns has been associated with CoV-like agents. [206]

Bats are infected with the largest and most diverse number of CoVs and are older than all CoVs of other animal species. All CoVs found in other species were derived from bat CoVs. The most closely related bat CoV to the SARS-CoV diverged in 1986, 17 years prior to the SARS outbreak, suggesting transmission through an intermediate host. [214]

Interspecies transmission of CoVs is well documented with animals and birds acting as natural reservoirs for CoV-related diseases in domestic animals and humans. Even with this being the case, the Centers for Disease Control and Prevention (CDC) found no evidence of transmission of SARS-CoV from bats to humans among bat biologists who were "always" or "most of the time" (66-68% of test subjects) exposed to bat blood, saliva, tissue, bites, or scratches. [215]

Based upon virus sequence data, the masked (Himalayan) palm civet (*Paguma larvata*) acted as an "amplification host." Closely related viruses were isolated from raccoon dogs (magnut or tanuki - *Nyctereutes procyonoides*). Unlike rabies virus, the Coronaviridae family (Nidovirales order) have the largest known positive-strand RNA genome (27-33 kb), increasing the likelihood of genetic variation. In animals, these viruses cause mild to severe enteritis, respiratory infections, neurologic disease and systemic infections. [206, 211] SARS-CoV was identified as a new virus of animal origin after reviewing: (1) genetic sequencing; (2) retrospective human serologic studies finding no evidence of SARS-CoV or related viral infections; (3) during the 2002 to 2003 SARS epidemic, serologic surveys among Chinese market traders found a higher seroprevalence for antibodies against SARS-CoV or related viruses amongst animal traders than controls; (4) the earliest SARS cases lived near produce markets but not near farms, and almost half were food handlers with likely animal contact; and (5) SARS-CoVs isolated from animals in markets were almost identical to human isolates. [206]

Genome sequence analysis indicated that SARS-CoV-like virus was not circulating among masked palm civets for a long period. Bat CoVs with an 88-92% sequence homology with human and civet isolates were found in horseshoe bats (*Rhinolophus* sp.). The most important difference between these viruses was in the region encoding for the spike (S) protein that determines host range and tissue tropism. Bats were the natural reservoir for the ancestor to SARS-CoV. When the CoV crossed species from bats to palm civets and humans, it utilized a new receptor. Other animals may also be involved. SARS-CoV may infect raccoon dogs, red foxes, Chinese ferrets, mink, pig, wild boar, and rice field rats. [206]

SL-CoV has been isolated from Chinese horseshoe bats (*Rhinolophus pearsoni*, *R. macrotis*, *R. pussilus*, and *R. ferrumequinum*), and cave-dwelling fruit bats (*Rousettus leschenaultii* – dog faced fruit bats or flying foxes - Megachiroptera). [206] Serologic and PCR testing has found evidence of infection by closely related SARS-CoV viruses thousands of kilometers apart in China and in Hong Kong, in multiple species of bats. Other CoVs have been isolated from bats in China: *R. sinicus*, *R. pearsoni*, *R. ferrumequinum*, *R. macrotis*, *M. ricketti*, *Miniopterus magnater*, *M. pusillus*, *M. schreibersii*, *Scolophlus kuhlii*, *Tylonycteris pachypus*, *Pipistrellus abramus*, and *P. pipistrellus*. [206] In the United States in wild and zoo-kept animals, CoVs have been isolated from bats (*M. occultus*, *E. fuscus*), sambar deer (*Cevus unicolor*), white-tailed deer (*Odocoileus virginianus*), waterbuck (*Kobus ellipsipyrmnus*), elk (*Cervus elephus*), caribou (*Rangifer tarandus*), sitatunga (marshbuck - *Tragelaphus spekeii*), giraffe (*Giraffa camelopardalis*), and musk oxen (*Ovibos moschatus*). [206]

Bats are sold along side of other animals including civets and raccoon dogs in live Chinese markets. It is postulated that civets became infected while being kept in proximity to bats. Civets have not been observed becoming ill with naturally occurring infection. When injected with SARS-CoV, they become febrile, lethargic, lose aggressiveness, and have decreased appetite. Again, there is no evidence of infection in wild or farmed civets. Bats appear to tolerate the infection, as do experimentally infected hamsters, guinea pigs, young mice, rats, cats, and pigs. [206] Infected *Rhinolophus* species develop a self-limited infection. The bats who become carriers of the SARS-related bat CoV have lower body weights as compared to bats that are carriers of *Rhinolophus* Chinese horseshoe bat CoV HKU2. [216]

The three structural membrane proteins of SARS-CoV are the spike (S) protein, the membrane protein, and the envelope protein. SARS-CoV and SL-CoV of bats and civets membrane envelope proteins share 96% and 100% similarity. [206] Bat SL-CoVs share 88% to 92% sequence homology with human or civet isolates but with principle differences in the S protein. [207] Variation in the S protein is responsible for host range, interspecies transmission, and adaptation. [206]

Almost 10% of *Rhinolophus* Chinese horseshoe bats carry SARS-related CoV. They appear healthy and clear the virus in 2 weeks to 4 months. Genome sequencing revealed frequent recombination between strains and the possible generation of a civet SARS-related CoV. Molecular clock analysis revealed SARS-related CoVs emerged with the time of their most recent common ancestor in 1972, and diverged between civets and bats in 1995. Civet SARS-related CoV is a recombinant virus arising from the *Rhinolophus* Chinese horseshoe bat SARS-related CoV. Lau et al believe that the horseshoe bats serve as a reservoir for recombination between strains from different locations within the bats' foraging range. [216]

Bat CoVs have been detected in Germany, Africa, and North and South America. In a survey of bats from the Philippines, 55.8% had evidence of infections with group 1 (genus *Alphacoronavirus*) and group 2 (genus *Betacoronavirus*). All bats were healthy. The CoVs detected had a 95% sequence identity to Bat-CoV/China/A515/2005 and 83% sequence identity to Bat-CoV/HKU9-1China/2007. [217] In the Rocky Mountains, group 1 CoV has been detected in 17% of *E. fuscus* and 50% of *M. occultus*. [218] In Ghana, 9.76% of insectivorous bats (insect-eating leaf-nosed bats genus *Hipposideros*) were positive for CoV related to SARS-CoV and human CoV 229E. CoV was not detected in *Eidolon helvum* fruit bats. [219] In Kenya, 19% of bat fecal swabs tested for the presences of CoV were positive. Species involved included *Cardioderma cor*, *Chaerephon* sp., *Chaerephon pumilus*, *Eidolon helvum*, *Hipposideros commersoni*, *Miniopterus africanus*, *Miniopterus inflatus*, *Miniopterus minor*, *Miniopterus natalensis*, *Otomops martinsseni*, and *Rousettus aegyptiacus*. [220] In Slovenia, 38.8% of horseshoe bat *(Rhinolophus hipposideros)* fecal samples tested were positive by RT-PCR for SARS-like CoV. [221]

A 2006 review of the relationship of bats and CoVs found that: (1) none of the bat CoVs belonged to group 3; (2) with few exceptions, most bat CoVs are species-specific – different bat species from the same location carry different viruses, whereas, the same species from different locations carry viruses of the same genetic lineage; and, (3) there was difficulty in isolating live virus in spite of finding high levels of viral genetic material by quantitative PCR. [222]

SARS - Pathology

Post-mortem examinations reveal the highest concentration of virus in the lungs and less in the bowel. This is directly related to the SARS-CoV receptors. Alveolar epithelium followed by alveolar macrophages have the highest concentration of virus. There is little involvement of the bronchiolar epithelium and no involvement of the bronchial epithelium or regional lymph nodes. There is diffuse alveolar damage, pulmonary edema, and hyaline membranes. Vascular fibrin thrombi are common and pulmonary infarcts are often seen. Spleen and lymph nodes reveal lymphocyte depletion and white pulp atrophy in the spleen. Biopsy of the gastrointestinal tract reveals minimal mucosal lymphoid depletion. In patients with renal failure, there is acute tubular necrosis without glomerular disease. The CNS reveals edema and degeneration of neurons with evidence of viral infection. [206, 207]

Angiotensin-converting enzyme 2 (ACE2) is the principle receptor for SARS-CoV. [206, 207] L-SIGN (CD209L or DC-SIGNR) is expressed in liver, lymph nodes and placenta, acts with the liver and lymph node sinusoidal endothelial cell C-type lectin to enhance SARS-CoV infection. L-SIGN is the alternative receptor for SARS-CoV cell entry. [206, 207]

ACE2 from the horseshoe bat was unable to act as a receptor for SARS-CoV. ACE2s from *M. daubentoni* and *Rhinolophus sinicus* allow less efficient viral entry facilitated by SARS-CoV S protein than human ACE2 (hACE2). This suggests that these 2 bat species may be susceptible to SARS-CoV and may act as natural hosts for SARS-CoV. [223] In an *in vitro* model using modified cell lines, civet virus has a restricted host range and cannot utilize hACE2, whereas SARS Urbani (one of the human epidemic strains of SARS-CoV) S protein can utilize both the civet and human ACE2s. [224] In another in vitro model utilizing cell lines expressing human, civet and horseshoe bat ACE2 molecules, SL-CoV S protein was unable to utilize any of the 3 ACE2s as a receptor. SARS-CoV S protein failed to enter cells expressing bat ACE2. The SL-CoV S protein is compatible with SARS-CoV S protein in structure and function. [225] The SL-CoV and S-CoV gene products are highly conserved except for the N-terminal region of the S proteins, which have a 63-64% sequence identity. [226]

The major species barriers between humans and civets for SARS-CoV infections are the interactions between the virus and host receptor, specifically the cell surface peptidase, ACE2. The virus enters the cell using the S protein's receptor binding region (S1) and membrane fusion region (S2). The S1 region binds with the ACE2. Studies have shown that in order for SARS-CoV to infect humans, two major mutations must occur on the receptor-binding domain (residues 479 and 487). Residue 479 determines whether the virus can infect humans. Residue 487 determines whether there is a sustained infection. [227]

SARS-CoV infections in raccoons, domestic cats, ferrets, and monkeys have been observed. Human, cattle, orangutan, and monkey ACE2s will allow attachment of certain human strains of SARS-CoV but not civet strains or a select human strain. Cat and ferret ACE2s allow for attachment of certain human strains. Specific civet and raccoon ACE2s will allow for attachment of all four human and civet viral strains. Because of their wide range of compatibility with ACE2s, civets and raccoons may be the most important hosts for spread of disease, especially man. [227]

SARS - Epidemiology

From November 2002 to July 2003 SARS spread from Foshan (Shunde district), Guangdong Province in the People's Republic of China around the world to 33 countries and 5 continents. By July 2003, there were at least 8,447 cases with 21% occurring in health care workers and 813 deaths (9.6% overall mortality). In most countries, though health care workers only accounted for 2-3% of adult cases. [206, 207, 211, 228] Most cases were reported from China (5,327) and the Far East (Hong Kong: 1755; Taiwan: 678; and, Singapore: 206). Mortality for patients younger than 60 years of age was 13.2% and 50% for patient older than 60 years. Half of the patients that developed acute respiratory distress syndrome (ARDS) died. In Hong Kong and Hanoi, 46% and 63% of cases were in health care workers. Two hundred-fifty-two cases were reported in Canada between February 23 and June 12, 2003. Twenty-nine cases were reported in the United States during approximately the same period. [206]

Distinguishing features of SARS epidemiology include spread by close person-to-person contact by droplet or fomites; laboratory-acquired disease with subsequent spread to family members; and a few patients appeared to be "super-spreaders" of infection. [206, 207]

Two major outbreaks were described. The first outbreak beginning in late 2002 and early 2003 centered in Guangdong province, China with isolated clusters in Taiwan, Singapore, and mainland China after accidental release of the virus. The second outbreak beginning in late 2003 and early 2004 involved a different SARS-CoV. The second outbreak began again in Guangdong province in patients with animal contact. [206] The SARS-CoV from the second outbreak appeared more closely related to animal isolates than the human isolates from 2002-2003 suggesting a second independent species-crossing event. [208]

The risk of being infected with SARS-CoV has a calculated base-case reproduction number (R_0) of 2.2-3.7. Attack rates from 10.3% to 60% translated in 2.4 to 31.3 cases/1000 exposure-hours. [229] Superspreading events resulting from patients excreting high titers of virus, aerosolization, contamination of the environment by fomites, and close contact in health care settings has resulting in up to 300 infections from one patient. [229] Patients on dialysis have a higher risk of infection with SARS-CoV. [230]

SARS-CoV has been isolated from sputum, nasal secretions, serum, feces, and bronchial washings. Contact or droplets, fecal-oral spread or fecal-droplet spread may transmit SARS-CoV. Infected aerosols from indoor plumbing were detected in a high-rise apartment complex

(Amoy Gardens, Hong Kong) where more than 300 residents became infected.[206, 229, 231, 232] Virus may survive in sewage for 2 weeks at 4°C and for 2 days at 20°C.[213] Infected aerosols entered apartments through bathroom drains, and were blown by prevailing winds to other buildings. [206, 233, 234] Weather (ambient winds, low mixing heights preventing dispersion of aerosols, and a decrease in temperature enabling virus to survive for longer periods are believed to play a major role in the spread of SARS.[206, 235] Viral nasopharyngeal swab concentrations were directly proportionate to the distance from the index case. This suggested airborne spread, although the role of rodents and fomites could not be determined. [233, 236] Viral load and viral shedding peaks at 10-14 days. [227]

Further study indicates that there appears to be two peaks in SARS-CoV transmission. One peak occurs 2 days, and the second 10 days after the onset of symptoms. Index cases over the age of 60 years or with higher lactate dehydrogenase levels on admission to the hospital (indicating higher viral loads) were more likely to transmit the infection. [237] Transmission is increased by endotracheal intubation or noninvasive positive pressure ventilation such as continuous positive airway pressure (CPAP) or bi-level positive airway pressure (BiPAP). [227]

Risk factors for infection or complications from infection with SARS-CoV include HLA-B 4601 haplotype (Taiwanese patients), HLA-B 0703 and HLA-DRB1 0301 alleles (Hong Kong Chinese patients), low or deficient mannose binding lectin serum levels, and increased expression of the IP-10 gene (increased IP-10 concentrations). Patients who are L-SIGN homozygous for CLEC4M tandem repeats have a lower risk of infection. [206, 238, 239]

SARS - Clinical Presentations

Clinical features of SARS that appear to set it apart from more common respiratory infections are: (1) prolonged incubation period (4-6 days instead of 1-2 days with influenza or other respiratory viral infections); (2) slower course of illness; (3) nosocomial spread; (4) viral load and viral shedding crescendos at 10-12 days after the onset of symptoms followed by a slow decline; and, (5) patients become more contagious after admission to the hospital. [206, 216] In Singapore, 78% of cases was the result of nosocomial spread. [216]

Asymptomatic viral colonization was recorded in 11.5% of health care workers who had a positive second-nested reverse-transcriptase PCR (RT-PCR). These individuals did not seroconvert, develope disease, or have positive second and third tests. [206, 207]

Clinically, SARS presents with 2 phases. Phase I involves viral replication, which is characterized by increasing viral load, fever, myalgia and other systemic symptoms. Patients temporarily improve over a few days. Phase II is heralded by the recurrence of fever, hypoxemia, and progressive pneumonia. This is most likely due to immunologic injury as the viral load during this phase decreases. [207]

In symptomatic adult patients, fever was present in 99-100% of patients. Other symptoms included chills or rigors (55-90%), productive or nonproductive cough (43-100%), shortness of breath (10-80%), myalgia (20-60.9%), malaise or lethargy (35-70%), headache (11-70%), sputum production (10-29%), sore throat (23.2-30%), coryza (22.5%), nausea and vomiting (10-19.6%), and diarrhea (11-15% to 38.4%).[206] In the Amoy Gardens outbreak, 75% of patients had watery diarrhea. Some patients present with fever and bloody diarrhea without respiratory symptoms. [206]

SARS – Diagnosis

Radiology: Abnormal chest radiographs are found in 78.3-100% of adult patients. Pediatric patients may present with normal chest radiographs as often as 36.5% of the time, although another report indicated 97% had abnormal studies. More than half of the patients (56.4%) have unilateral disease and 90% have progressive disease. [206] Typical high-resolution computerized chest tomography (CT) reveals early solitary ground-glass opacification. [207]

Laboratory: Anemia is present in 49% of patients. Other findings include: evidence of hemolysis – 76%; lymphopenia early in the disease with low CD_4 and CD_8 counts seen in patients with poor prognosis; leukopenia 22-34.1% (with 64% of patients developing leukopenia in one study during the first week, and 2.5% developing transient neutropenia with WBC $<0.5x10^6$/dl); leukocytosis (in the second or third week – 61%); and, thrombocytopenia -33-44%. Serum chemistry abnormalities prominently included: hyponatremia (20.3-60%); hypokalemia (25.3-47%); hypocalcemia (60%); increased ALT (23.4%-56%); increased LDH (47-87%); increased CPK (19-56%); prolonged activated partial thromboplastin time (18-42.8%); and, increased D-dimer (45%). [206]

Sensitivity of RT-PCR is inadequate in the first 3 days of the illness. A six-item clinical score for emergency room triage during a SARS outbreak for otherwise healthy individuals is 92.5% sensitive and 71.2% specific in a noninfluenza season. [206]

Detection of specific SARS antibodies is problematic in early infections as IgG seroconversion usually occurs 20 days after the onset of symptoms. SARS-CoV nucleocapsid protein is found in high concentrations early in the infection. A recently developed antibody to SARS-CoV nucleocapsid protein that does not cross react with human CoV 229E or CoV OC43 shows promise for early diagnosis. [240]

SARS – Treatment

Treatment modalities have included ribavirin, protease inhibitors, interferons, and human monoclonal antibodies (animal models – ferrets), intravenous gammaglobulin and pentaglobulin, and convalescent plasma. Ribavirin did not appear to be useful and was associated with anemia, increased liver transaminases and bradycardia. Lopinavir/ritonavir combinations was associated with decreased mortality (2.3% VS 15.6%). Interferons and human monoclonal antibodies have shown promise in animal studies. Studies show that early steroid therapy may prolong viremia, and steroids should be reserved for rescue purposes later in the course of disease. Cleri et al, Hue et al, and Stockman et al have reviewed therapeutic options. [206, 207, 241, 242] A 2006 review of the reported treatment attempts during the 2002-2003 SARS outbreaks failed to identify any regimen that was unequivocally beneficial. [242]

SARS - Prevention

The cornerstone of prevention is appropriate patient infection control measures (respiratory isolation). Important measures include isolation of symptomatic health care workers, and personal protective measures. Post-mortem examinations should only be performed in biosafety laboratory level 3 (BSL-3) laboratories. In one study, none of the health care workers on a unit caring for SARS patients employing "meticulous" infectious control precautions developed symptomatic infection. [227] Rates of transmission of SARS-CoV varies from patient to patient. Severely ill intubated patients represent a significant risk for health care workers. Risk factors included presence in the room during fiberoptic intubation or EKG recording, unprotected eye contact with secretions, patient APACHE II score ≥ 20, and a patient PaO_2 of ≤ 59. [243]

PARAMYXOVIRIDAE

Henipavirus Genus: Hendra Virus and Nipah Virus

Paramyxoviridae are a large family of single-stranded negative sense RNA viruses that belong to the order Mononegavirales. Paramyxoviridae include the agents for measles virus, Rinderpest virus, canine distemper, Newcastle disease, mumps virus, human respiratory syncytial virus, metapneumovirus, human parainfluenza viruses 2, 4A and 4B, Avulavirus, and others. [244-246] *Pteropus* species have been identified as the natural hosts for both Hendra virus and Nipah virus. Habitat loss and hunting are believed to have affected these bat populations and resulted in the emergence of these two Henipavirus infections in livestock and man. [247]

Hendra virus (formerly Equine morbillivirus): Hendra virus first came to clinical attention in August 1994 when a Mackay, Queensland (Australia) sugar cane farmer who assisted in the necropsy of 2 horses died of encephalitis. In September that same year in Hendra, near Brisbane, 1000 km away, during an outbreak of acute respiratory disease in 18 horses, 14 horses and a horse trainer died after he suffered a percutaneous exposure from the first horse. The horse trainer died of pneumonia with evidence of myocarditis at autopsy. [244, 245, 247-249]

Epidemiology: Between 1994 and 2004, 23 horses were infected resulting in the death of 17 horses and 2 humans. Subsequently (2004) two female veterinarians became ill 1 week after performing an autopsy on a horse infected with Hendra virus. One developed fever, dry cough, sore throat, cervical lymphadenopathy, myalgias and malaise. Both were serologically diagnosed and fully recovered. A 2008 an outbreak in five horses resulted in the infection of 2 veterinary workers and 1 death. [244, 245, 247-249]

Serologic surveys in the area revealed no horses or pigs similarly infected. [244, 245, 247-249] Viruses obtained from black (*Pteropus alecto*), grey-headed (*P. poliocephalus*), little red (*P. scapulatus*) and spectacled (*P. conspicillatus*) flying foxes were identical to those isolated from the infected horses and humans. Serologic evidence of infection was reported in Papua New Guinea (bare-backed fruit bat - *Dobsonia moluccense* and Bismarck flying fox - *Pteropus neohibernicus*), and Port Moresby and New Britain (*D. andersoni*, *P. capistratus*, *P. hypomelanus*, and *P. admiralitatum*). Twelve to 53% of samples examined were serologically positive. [244] Serologic surveys of over 5000 samples from 46 species of wild and domestic animals found no other natural reservoir. [245, 249]

Serologic surveys of 296 contacts with stables, sick horses, patients from the Mackay and Brisbane outbreaks and wildlife rehabilitators who cared for sick, injured or orphaned bats revealed no evidence of infection. [244, 245] *P. poliocephalus* intentionally infected with Hendra virus did not become ill. Antibody developed in 6 of 8 bats, while 2 bats developed vascular lesions containing viral antigen. The disease did not appear highly contagious as uninoculated bats and horses housed with the infected bats did not acquire the infection. Experimentally infected horses (75%) develop evidence of infection in urine, kidneys, saliva. [244] Virus was also isolated from the kidneys, and urine of experimentally infected cats. [129, 244] A possible vector linking bats, horses and humans is the *Ixodes holcyclus* tick, which is parasitic of flying-foxes. [250]

In horses, natural infection has an incubation period of 8-11 days, but is shorter in experimentally infected animals. Horses develop depression, loss of appetite, fever, ataxia, tachycardia, tachypnea, and copious frothy nasal discharge. The animals pace aimlessly and develop hemorrhagic nasal discharge and swelling of the head. Respiratory distress is the usual terminal event. Surviving horses suffer with localized myoclonic twitches. Asymptomatic infection in horses has been reported. [248]

Clinical disease in humans: Hendra virus infection in humans may be as mild as a self-limited influenza-like illness to a fatal pneumonia. Incubation in humans is 5-8 days. Short-lived illness has been reported with dry cough, sore throat, cervical lymphadenopathy, myalgias, fatigue and fever lasting only 8 days. Illness may be prolonged (6 weeks with fever, myalgias, headache, lethargy, and vertigo, followed by recovery). Rapid demise was seen in a patient who presented with an influenza-like picture. He rapidly developed dehydration, acidosis, arterial thrombosis of the leg, and cardiac irritability before death. In another patient, death from encephalitis occurred 14 months after complete recovery from an initial bout of encephalitis. [251]

Antibodies are detected by immunofluorescence, enzyme-linked immunoabsorbant assay (ELISA) or rapid immune plaque assay. Viral genome is detected by RT-PCR. Viral cultures are performed on Vero cell cultures and virus is detected by electron-microscopy. [247, 249]

Treatment consists of supportive therapy. Ribavirin has *in vitro* activity against Hendra virus and does cross the blood-brain barrier. [251]

Experimental models: In experimental models, guinea pigs develop vasculitis (lung, brain, kidney, spleen, and placenta), non-suppurative encephalitis, and syncytia in transitional epithelium of the bladder. Infected cats succumb to severe interstitial pneumonia, vasculitis (lung, gastrointestinal tract and lymph tissue), and syncytia in the endothelium. Horses develop severe pulmonary disease (pulmonary edema, interstitial pneumonia), vasculitis (lung and brain including meninges), and mild lymphocytic meningitis. Fruitbats develop mild vascular lesion in the placenta. [249]

Nipah virus: An initial outbreak of encephalitis spread through pig farms in Malaysia between September 1998 and April 1999. This was accompanied by an outbreak of encephalitis among pig farmers. [246, 248, 249, 251] In 1998, there were 109 deaths from Nipah virus encephalitis in Malaysia. [248, 249, 251] Between September 1998 and June 1999, there were more than 250 cases of encephalitis with 40% mortality. [245] Importation of sick animals caused the disease to spread to Singapore. Nipah virus is the cause of respiratory disease and encephalitis in pigs and encephalitis humans. Pigs initially develop fever and respiratory symptoms with a 1-5% to 5-15% mortality. [246, 248,249, 251] Young pigs develop dyspnea, loud barking cough and occasionally neurologic signs (uncoordinated gait, trembling, and hind leg paresis. Older animals developed mucopurulent or bloody nasal discharges and more generalized neurologic signs. Early stage abortion, 40% mortality in piglets because of inability of sows to nurse, and sudden death without prior signs of infection have all been reported. [249]

Epidemiology: The virus is transmitted to pigs by *Pteropus* species frugiverous bats and insectivorous bats, and then is spread horizontally amongst pig herds. Island flying foxes (small or variable flying fox - *P. hypomelanus*) and Malayan flying foxes (greater flying fox - *P. vampyrus*) are reservoirs in Malaysia. Other Malaysian bat species with serologic evidence of Nipah virus infection include Megachiroptera (fruit bats) *P. lylei* (Lyle's flying foxes), *P. hypomelanus*, *P. vampyrus*, *Cynopterus brachyotis* (lesser dog-faced or short-nosed fruit bat), *Eonycteris spelaea* (cave nectar bat), and Microchiroptera (insectivorous) *Scotophilus kuhlii* (lesser Asiatic yellow bat or house bat). [252-256] *P. giganteus* is the reservoir in Bangladesh and India. [249]. The overall infection rate in pigs approaches 100%. [251] Horses, when exposed, develop antibodies, and there is one report of a horse developing dilated meningeal vessels. [129, 248]

Clinical disease in humans: In humans, patients develop fever, headache, myalgia, and rapidly progressive neurologic compromise with coma and death. The initial mortality rate in Malaysia was 38%, and some patients who recovered, relapsed and died. During a subsequent outbreak in Bangladesh, there was a 73% human mortality. The mean time from onset of symptoms to death was 7 ± 4.6 days. Sequelae from the infection include persistent fatigue up to five months post infection in most patients, and 32% of post-encephalitis patients had significant persistent neurologic deficits. Half of the recovered patients less than 16 years of age had persistent behavioral disturbances. Control of the outbreak required forbidding transportation of pigs, and the culling of 1.1 million animals. [248, 251]

From 2001 to 2008, Nipah virus encephalitis has occurred annually (except for 2002 and 2006) in Bangladesh. Differences between the outbreaks in Bangladesh and India, and Malaysia were: (1) absence of an intermediate host; (2) person-to-person spread; and (3) higher case-fatality rates. In 2001 in India, nosocomial spread played a significant role. This was not the case in Bangladesh. Risk factors include exposure to sick cows, consumption of contaminated raw date palm sap, contact with another encephalitis patient, and climbing trees. One study could not identify contact with an intermediate host as a risk factor. [251, 257]

Seasonal Nipah virus infection in Bangladesh coincides with date palm sap collection season. Raw date palm sap is a Bengali delicacy. Gachhis (date palm sap collectors) report that fruit bats often urinate and defecate into the sap, which reduces its commercial value. The Gachhis seldom use methods to reduce the access of bats to the sap. [238]

Experimental models: Guinea pigs are an unreliable model. Intranasal inoculation resulted in no clinical disease while intraperitoneal injection resulted in both weight loss and transient fever followed by recovery or progressive weight loss and death. Chicken embryos displayed degeneration and necrosis of neurons, degeneration of vascular endothelia, CNS hemorrhage, arteritis, and neuronal syncytia. Golden hamsters developed generalized vasculitis (brain, lung, liver, kidney, heart) and meningitis. Infected cats developed pulmonary edema and hemorrhage, generalized vasculitis, syncytia in endothelium and respiratory epithelium, necrosis of lymphatic tissue, degeneration of neurons, and meningitis. Pigs displayed lymphoctyic meningo-encephalitis with vasculitis and rare endothelial syncytia, mild lesions in respiratory epithelium, interstitial pneumonia, lymphoid necrosis and/or lymphocyte depletion, and mild vasculits of the lung and spleen. Fruit bats appear to be the least affected with only mild vasculits in the intestine without antigen detection by immunohistochemistry. [249]

Rubulavirus Genus: Menangle Virus and Tioman Virus

Menangle virus: In 1997, Menangle virus was detected in Australian bats after rates of farrowing (decreased from 82-60%) and piglet deformities and survivorship changed in the New South Wales commercial swine industry. Antibodies to Menangle virus was detected in the majority of swine and a few farm workers. [244, 246, 259] Sera from a breeding colony of *P. poliocephalus* and *P. scapulatus* were positive while serosurveys of other animals in the vicinity of the affected farm were negative. [244, 246]

Clustered nonrandom sampling of bats in Australia found neutralizing antibodies in 46% of black flying foxes (*P. alecto*), 41% of grey-headed flying foxes (*P. poliocephalus*), 25% of spectacled flying foxes (*P. conspicillatus*), and 1% of little red flying foxes (*P. scapulatus*). Attempts to isolate the virus from feces and tissues of the *Pteropus* species were unsuccessful. [260]

In pigs, occasional abortions occurred. Stillborn piglets had severe degeneration of the brain and spinal cord, arthrogryposis (arthrogyrposis multiplex congenita), brachygnathia, and rarely, fibrinous body cavity effusion and pulmonary hypoplasia. [246, 261]

In 1997, human disease was described in 2 piggery workers exposed to infected pigs. One patient with frequent contact with birthing pigs, developed sudden onset of malaise, chills, sweats and fever. He had severe headaches and myalgia without respiratory or gastrointestinal symptoms and required bed rest for 10 days. A day after taking amoxicillin, he developed a spotty red rash. Examination on day 4 revealed lymphadenopathy and rubellaform rash. He returned to work 14 days later but reported a 10-kg weight loss and easy fatigability. Reevaluation 2 months later was normal except for mild tenderness in the right hypochondrium with normal routine blood tests. Imaging revealed the liver at the upper limits of normal and an enlarged spleen.

The second patient also develop sudden onset of fever, chills, rigors, drenching sweats, malaise, back pain, severe frontal headache and photophobia. He too, had no respiratory or gastrointestinal symptoms. The headache resolved in 4-5 days but he developed a spotty nonpruritic rash that persisted for 7 days. He recovered in 10 days, noting a 3-kg weight. Reevaluation 2 months later revealed elevated liver function studies. Ultrasonography revealed mild hepatomegaly and the spleen at the upper limit of normal. The patient was hepatitis C-antibody positive and IgA-deficient. [262]

Acute and convalescent serum antibodies are useful in the diagnosis. Newly developed monoclonal antibodies against the nucleocapsid protein are expected to be a useful tool in immunohistological diagnosis. [263]

Tioman virus: Tioman virus was isolated from pooled urine samples from the Tioman Island, Malaysia, flying fox (*P. hypomelanus*). [264] ELISA seropositive *P. rufus* and *Rousettus madagascariensis* for Tioman, Nipah and Hendra viruses were identified on Madagascar. In the same study, *Eidolon dupreanum* was found to be positive for Nipah and Hendra viruses, but not for Tioman virus. [265] The virus is genetically related to Menangle virus (62%). It replicates in mouse neurons and human neuroblastoma cells. Human infection probably occurs as 1.8% of the inhabitants of Tioman Island have neutralizing antibodies against Tioman virus. Bat-to-human transmission has not been demonstrated. [266] It is suspected that the virus is ingested with contaminated fruits as 2 of the 3 confirmed Tioman virus neutralizing antibody-positive subjects had eaten fruit partially consumed by bats. [266, 267]

As with Menangle virus, Yaiw et al have suggested that the pig is an amplifying host. Experimentally infected pigs either were asymptomatic or developed fever. Virus was isolated from oral swabs, but not from the urine. Viral antigen was detected in lymphocytes in the thymus, tonsils, spleen, lymph nodes and ilial Peyer's patches, in addition to the tonsillar epithelium and thymic epithelioreticular cells. [267]

Mouse brain inoculations caused plaque-like necrotic areas, with preferential replication in the neocortex and limbic system, and involvement of inflammatory cells. Intraperitoneal injection did not result in spread of infection to the central nervous system. [268] Although both Menangle and Tioman viruses replicate in human neuronal cells, they differ in cytopathology. Intracellular ultrastructure in infected neuronal cells allows them to be differentiated by electron microscopy. [269]

Other Viruses

Henipavirus infections in bats have been reported from Australia, Malaysia, Thailand, Cambodia, Indonesia, Bangladesh, India, and Madagascar most often in *Pteropus* genus fruit eating bats. These bats are not found in Africa. In 2009, evidence of three novel putative Henipaviruses' RNA was identified in the feces of African straw-colored fruit bats (*Eidolon helvum*) from Ghana. [270]

Mapuera virus, a rubulavirus was isolated from an asymptomatic fruit bat, *Stumira lilium* from the Brazilian rainforest in 1979. [271] As of yet, it is not pathogenic in man, but mouse intracerebral injections are fatal. Mapuera virus differs from other rubulaviruses in its V protein. The V protein blocks interferon signaling from several species of mammalian cells. This would not be a barrier to cross-species infection. [272] Mapuera virus appears to be closely related to porcine rubulavirus. [271, 273]

Tuhoko viruses 1, 2, and 3 (paramyxoviruses) have been isolated from Leschenault's rousettes fruit bats (*Rousettus leschenaultii*). [274]

REOVIRIDAE

Reoviruses: Three new viruses linked to bats have been isolated in Malaysia: the paramyxovirus, Tioman virus (1999); and the reoviruses, Pulau virus (1999) and Melaka virus (2006). [275] Important human reovirus pathogens (family Reoviridae) include: Coltiviruses (Colorado tick fever virus, Salmon River virus, Eyach virus); Seadornaviruses (Banna virus); Orthoreovirus (SARS-CoV, and Melaka virus, Kampar virus, other respiratory and gastrointestinal pathogens); Orbivirus (Kampar virus Antigenic Complex, Oklahoma tick fever, Orungo virus, Lebombo virus, Changuinola Virus, and African Horse Sickness Virus); and Rotavirus. [276-278] Reoviruses (**r**espiratory **e**nteric **o**rphan **v**iruses) are nonenveloped double-stranded RNA viruses. Orthoreoviruses are found in mammals, birds and reptiles. [17]

Pulau virus was isolated while looking for the natural host reservoir for Nipah virus during the 1998-1999 outbreak. Like Nipah and Tioman viruses, Pulau virus is a syncytia-forming virus related to the Nelson Bay orthoreovirus, which was isolated from the heart blood of a fruit bat (*P. poliocephalus* – New South Wales) in 1968. Nelson Bay virus was the first reovirus isolated from a bat. [229] PCR analysis indicates that an African grey parrot reovirus clusters with the Nelson Bay virus from bats. [280]

Melaka virus is the third virus of the Nelson Bay virus species group was isolated from a patient in Melaka, Malaysia. [17] Melaka virus has not yet been found in the Philippines. [281]

The index case presented with 1 day of high fever, cough, coryza, sore throat, headache, myalgia, malaise, loss of appetite, body weakness and prostration. There were no complaints of rash, gastrointestinal, visual or arthritic symptoms. Over 1 week, the cough worsened and became productive of yellow sputum. The patient experienced severe swallowing difficulties of both solids and liquids, and the sensation of tightness in the pharynx and neck. There was no shortness of breath or hemoptysis. He defervesced in 4 days but weakness persisted for 2 more weeks. His only physical findings were an injected and hyperemic throat, enlarged tonsils without exudate, and single mobile tender enlarged (1x1.5 cm) lymph nodes bilaterally in the anterior triangles. Virus was isolated from a throat swab on Madin-Darby canine kidney (MDCK) cells. [17]

Interestingly, 1 week prior to the onset of his illness, a bat flew into his living room, frantically flying without attaching to the ceiling or making any squeaking noises. Six and 7 days after the onset of his illness, two of his children became ill. This case history implicates both the role of bats and person-to-person spread of this virus. [17]

Kampar virus, the fourth member of the Nelson Bay virus species group, is a Melaka virus-like reovirus originally isolated from a throat swab from a man in Kampar, Perak, Malaysia with symptomatic infection (fever, respiratory disease, vomiting and watery diarrhea). The index case had accompanying cough, headache, sore throat, and loss of appetite. His physical examination revealed conjunctivitis, enlarged and injected tonsils and a blanching body erythema most prominent on the face and upper trunk. His illness resolved in 4 days although he remained weak and lethargic. Syncytial cytopathic effect was note on MDCK and Vero cells. Subsequently, the patient's wife and the medical officer attending the patient became ill. Investigation revealed fruit bats feeding in close proximity to the living room window and feeding in fruiting trees (particularly Ketapang trees) surrounding the house. [282]

Antibodies to Kampar virus were found in few *P. vampyrus* (1/55) and *P. hypomelaunus* (2/77) bats. The more numerous smaller fruit bats, *Eonycteris spelaea* and *Cynopterus brachyotis* needed to be tested for evidence of infection. [282]

Orbivirus Genus of the Family Reoviridae

Orbiviruses are part of the Reoviridae family. These viruses include African horse sickness virus, Bluetongue virus, Changuinola virus, Chenuda virus, Chobar Gorge virus, Corriparta virus, Epizootic hemorrhagic disease virus, Equine encephalosis virus, Eubenangee virus, Great Island virus, Ieri virus, Lebombo virus, Orungo virus, Palyam virus, Peruvian horse sickness virus, St. Croix River virus, Umatilla virus, Wad Medani virus, Wallal virus, Warrego virus, Wongorr virus, Yunnan orbivirus, Andasibe virus, Codajas virus, Ife virus, Itupiranga virus, Japanaut virus, Kammavanpettai virus, Lake Clarendon virus, Matucare virus, Tembe virus and Tracambe virus. [248, 276] Orbivirus has been isolated from *Eidolon helvum* bats from Nigeria, Cameroon and the Central African Republic. [283]

Bluetongue virus infects cattle, sheep, goats and wild ungulates in all tropical and subtropical regions. Bluetongue virus has been isolated from bats in the Republic of Guinea. [138] The principal vector is *Culicoides* sp. biting midges. Other potential vectors include the argasid tick, *Ornithodoros coriaceus*, the sheep ked (brown hairy wingless fly), *Melophagus ovinus*, and *Aedes lineatopennis* and *Anopheles vagus*. The virus is found in semen making venereal transmission possible but unlikely.

In sheep, fever persists for 5-6 days. There is nasal discharge, mucopurulent blood stained salivation, swelling of the lips, gums, dental pad and tongue, and excoriation of the buccal mucosa. Diarrhea and dysentery may occur, as well as foot lesions. Partial or complete loss of fleece is common. Death usually occurs within 6 days. Animals that recover take months to convalesce. In cattle, fewer animals develop clinical disease, although the apparent illness is similar to that in sheep. [248, 276]

African horse sickness virus infects horses, zebras, mules, donkeys, and dogs in Africa, and the Arabian and Iberian peninsulas. Vectors include hematophagous insects, most importantly midges (*Culicoides imicola*, *C. bolitinos*, *C. variipennis*, *C. pulicaris*, and *C. obsoletus*.). Other vectors include ticks (camel tick, *Hyalomma dromedarii* and the brown dog tick, *Rhipicephalus sanguineus*). Incubation period is 5-7 days. [248, 276]

The disease presents in an acute or pulmonary form, a subacute or cardiac form, a mixed form and a mild form. The acute pulmonary form (dunkop) presents with fever, labored breathing, paroxysms of coughing, and profuse nasal discharge. There is 95% mortality. The subacute cardiac form (dikkop) has a 3-week incubation period, edema of the head especially in the temporal fossa, eyelids, lips and later the chest, and, blue oral mucosa with petechiae of the tongue. The horses develop pericardial effusion, endocarditis and pulmonary edema. There is a 50% fatality rate with most deaths within 2 weeks. Recovery is prolonged. The mixed form has signs of both the pulmonary and cardiac forms of disease. The mild form, horse sickness fever, occurs in animals with some immunity or with low virulence viruses. The animals are febrile for 3 days, have poor appetites, slight conjunctivitis, and labored breathing. [248]

An accidentally aerosolized freeze-dried virus infected 4 vaccine workers. Three developed frontotemporal encephalitis, and all four developed uveochorioretinitis. [276]

Epizootic hemorrhagic disease **virus** naturally infects deer and cattle. Sheep can be experimentally infected, while goats and pigs are resistant to disease. Transmission is by *Culicoides* sp. midges, gnats and mosquitoes. In North American deer, morbidity is 90% and mortality is 60%. Pathology is similar to fatal cases of bluetongue in sheep with fever, hemorrhage and death. [248]

Orbivirus human disease: Human disease with orbiviruses is rare. Human disease is limited to 4 serogroups: Kemerovo antigenic complex (Kemerovo, Lipovnik, and Tribec viruses from Russia and Eastern Europe), Orungo virus (sub-Saharan Africa), Lebombo virus (South Africa and Nigeria) and Changuinola virus (Central America).

Kemerovo virus antigenic complex is transmitted by *Ixodes* sp. ticks. Meningitis and meningoencephalitis has been reported. A serologic survey in the Czech Republic found 18%

seropositivity. Lipovnik virus has been associated with meningoencephalitis and polyradiculitis. Oklahoma tick fever (myalgia, vomiting and severe abdominal pain) has been serologically linked to Kemerovo-group related viruses (Sixgun City and Lipovnik virus). [276]

Orungo virus is transmitted by *Aedes*, *Culex,* and *Anopheles* spp. mosquitoes. Acute febrile illness (fever, headache and myalgia) and encephalitis (with seizures and flaccid paralysis) in a child has been reported. Seroprevalence in endemic areas is 24-34%. [276]

Lebombo virus is transmitted by *Aedes* and *Mansonia* spp. mosquitoes. It has been linked to a febrile illness in a child. [276]

Changuinola virus is probably transmitted by *Phlebotomus* sp. flies. Prevalence rates are high and it has been linked to a febrile illness. [276]

Fomede virus (Chobar Gorge virus) has been isolated from *Nycteris gambiensis* (Gambian slit-faced bats) and *Nycteris nana* (dwarf slit-faced bat) from Guinea. Chobar Gorge virus was originally isolated from *Ornithodoros* sp. ticks from Chobar Gorge, Nepal in 1970. [138, 139, 284]

OTHER VIRUSES

Astrovirus

Astroviruses, large positive sense single-stranded RNA viruses infect man, bats, cats, cheetah, mink, sheep, pigs, chickens, and turkeys. Avian astroviruses are in the *Avastrovirus* genera, and mammalian astroviruses are in the *Mamastrovirus* genera. They commonly cause diarrhea, malaise, headache and nausea, and less frequently vomiting. Most disease is in infants and children and gastroenteritis in animals. [285, 286] Astroviruses have been isolated from Rhinolophidae (great leaf-nosed bat – *Hipposideros armiger*, intermediate leaf-nosed bat – *H. larvatus*, pomona leaf-nosed bat – *H. pomona*, greater horseshoe bat – *Rhinolophus ferrumequinum*, Pearson's horseshoe bat – *R. pearsonii*), Vespertilionidae (great evening bat – *Ia io*, Schreiber's long-fingered bat – *Miniopterus schreibersii*, Rickett's big-footed bat – *M. ricketti*, *Myotis* spp. Japanese pipistrelle – *Pipistrellus abramus*), Emballonuridae (black-bearded tomb bat – *Taphozous melanopogon*), and Megadermatidae (greater false vampire bat-*Megaderma lyra*) families of bats in China. [285]

Of the bats captured in the wild and in a mine cave in Hong Kong, 45% were infected with astrovirus. RT-PCR testing of rectal and throat specimens were positive for *Miniopterus magnater*, *Miniopterus pusillus*, *Miniopterus schreibersii*, *Myotis chinensis*, *M. ricketti*, *Pipistrellus abramus*, and *Rhinolophus rouxi* (Rufous horseshoe bat). Eleven *Cynopterus sphinx* fruit bats and 10 *Hipposideros armiger* insectivorous bats were negative. In the Hong Kong study, during single sampling occasions, a wide phylogenetic variety of viruses were identified with no significant clustering. [287]

Papillomavirus

The Papillomaviridae family is nonenveloped double stranded DNA icosahedral capsid viruses. Patients often have subclinical and latent infections. Papillomavirus causes common warts (71% of cutaneous warts, present in 4-20% of children), plantar warts (34% of cutaneous wars, most frequent in adolescents and young adults), and juvenile or flat warts (4% of cutaneous warts, most common in children). Butchers, meat packers, and fish handlers are also at increased risks for cutaneous warts. [288]

Epidermodysplasia verruciformis (Levandowsky-Lutz syndrome) is an autosomal recessive disorder (chromosome 17). It is characterized by the appearance at 5-7 years of age of benign macular, verruca plana-like or pityriasis versicolor-like lesions that frequently undergo malignant transformation. [289, 290] A large number but very specific human papillomaviruses are associated with this condition. [288] It has also been reported in patients with immune dysregulation including a patient with cutaneous lymphoma. [290].

Papillomavirus causes anogenital warts, recurrent respiratory papillomatosis, oral squamous cell papillomas, oral condyloma acuminatum, oral verrucae vulgaris, and Heck's disease (focal epithelial hyperplasia of the oral cavity). Human papillomaviruses have been found in epidermoid cysts, seborrheic keratoses, squamous and basal cell skin carcinomas, and aerodigestive carcinomas. [288]

The oncogenic potential of animal papillomaviruses as well as human papillomavirus as the cause of human cervical cancer is well established. [288]

A 5 year-old female Egyptian fruit bat (*Rousettus aegyptiacus*) had araised pigmented mass removed from the lateral canthus. Six other raised smooth to cauliflower-like skin masses were observed and 4 were removed from the left wing membranes. The masses were basosquamous carcinomas and papillomas. Masses removed 6 and 12 months later revealed bony invasion and squamous differentiation. Intranuclear staining was positive for bovine papillomavirus antibody. PCR revealed a 450 base-pair segment analogous to the L1 region of human papillomavirus types 96 and 5. [291] Human papillomavirus type 5 is considered a high-risk epidermodysplasia verruciformis human papillomavirus for skin carcinomas. [292] Another study has found this virus commonly present in the human population. [293]

Polyomavirus

Polyoma viruses are not usually associated with acute disease in the normal host. Some aggressive viruses cause tumors (Merkel polyomavirus causing Merkel cell skin cancer in humans; mouse polyomavirus and pneumotropic virus causes disease in newborn mice).[210] The human polyomavirus JC causes progressive multifocal leukoencephalopathy (PML) in immunocompromised patients, especially those with advanced HIV infection and hematologic malignancies. An inflammatory form of PML is seen in patients with immune reconstitution inflammatory syndrome (patients with HIV treated with highly active antiretroviral therapy (HAART)). Demyelination of white matter in the cerebellum occurs when JC virus infects granule cell neurons. JC virus also causes encephalitis with infection of the cortical pyramidal neurons and initial lesions restricted to the grey matter on the MRI. JC virus also causes meningitis accompanied by diplopia. [294]

Asymptomatic childhood infection with BK polyomavirus is common. Immunosuppressive therapy results in 1-10% of renal transplant patients developing BK virus-associated nephropathy. Patients with BK virus seropositivity, old age, and high anti-BK IgG levels before transplantation are at increased risk. Other complications include silent ureteral stenosis and hemorrhagic and nonhemorrhagic cystitis. Both allogeneic and autologous bone marrow transplant patients suffer with hemorrhagic cystitis within 2 months of transplantation (10-25%). Interstitial nephritis is also a complication of BK virus infections in geriatric, diabetic and HIV-infected patients. [294]

An unidentified polyomavirus was identified from 4 female little brown bats (of 31 bats *Myotis lucifugus*) and 2 *Myotis californicus*. This polyomavirus did not cause disease in the adult *Myotis lucifugus*. The virus was detected in healthy bats in a mating swarm. [210]

Viral Hemorrhagic Fevers

Human viral hemorrhagic fevers are primarily caused by 4 families of RNA viruses: Arenaviridae (Argentine, Bolivian, Brazilian, and Venezuelan hemorrhagic fevers, and Lassa fever); Bunyaviridae (Hantavirus genus, Congo-Crimean hemorrhagic fever (CCHF) from the Nairovirus genus, and Rift Valley fever from the Phlebovirus genus); Filoviridae (Ebola and Marburg viruses); and Flaviviridae (dengue and yellow fever). [295]

Arenaviruses commonly cause asymptomatic infections in rodents. The principal hosts for Lassa fever are *Mastomys natalensis*, and *Mus musculus*. The host for Guanarito virus, the causative agent for Venezuelan hemorrhagic fever is the cane mouse, *Zygodontomys brevicaudia*. The Tacaribe group of Arenaviruses infects fruit bats on Trinidad and Tobago. [296]

The Bunyaviridae was first isolated from *Aedes* mosquitoes. The *Bunyavirus* genus infects asymptomatic vertebrate amplifying hosts: La Cross virus – adult chipmunks, squirrels, foxes, and woodchucks; Jamestown virus; snowshoe hare virus – snowshoe hares, and white-tailed deer.

The *Phlebovirus* genus (Rift Valley fever) infects cattle and sheep, and spreads by *Aedes* sp. mosquito. Rift Valley fever is transmitted transovarially by the mosquitoes to their drought-resistant eggs. The virus is spread by inoculation and aerosol. Virus is present in milk, feces and aborted fetuses. [248]

Nairovirus genus is principally spread by ticks, although some disease is spread by *Culicoides* flies and mosquitoes: CCHF is spread by ticks (some species of *Hyalomma, Dermacentor*, and *Rhipicephalus* species) and flies; and, Nairobi sheep disease by ticks, flies and mosquitoes. Vertebrates including ground feeding birds become infected and a source of virus for ticks. Sheep, goats, cattle, ostriches, large wild herbivores, hares, and hedgehogs amplify CCHF, and sheep and goats are amplifying hosts for other Nairoviruses. [248, 295]

Hantavirus genus of the Bunyaviridae causes hemorrhagic fever with renal syndrome (HFRS) and Hantavirus pulmonary syndrome (HPS). [295] Mice and rats are the common reservoirs. *Apodemus agrarius* (the field mouse) and *Rattus rattus* and *Rattus norvegicus* are hosts for the HFRS virus, and the deer mouse (*Peromyscus maniculatus*) in North America, and *Akodon azarae* rodent (seroprevalence of 9.3%) are reservoirs of HPS. Most Hantavirus infections are asymptomatic. Infection usually results from inhalation of contaminated excreta. [295]

Rift valley fever was first reported in Kenya, but epizootics have occurred throughout sub-Saharan Africa, Egypt, Madagascar, Mauritania and the Arabian Peninsula. Rift valley fever virus causes deaths in young lambs and calves, as well as abortions in cattle and sheep herds. [248] Camels, domestic buffalo, monkeys, mice, rats, ferrets and hamsters are susceptible to infection. Goats are moderately susceptible, while pigs, rabbits, guinea pigs and poultry are resistant to infection. Mortality among ruminants is 10-30% and 1% in humans. Outbreaks occur during the rainy seasons when the mosquito population increases. Clinically, most patients develop a nonspecific febrile illness after a 2-6 day incubation period. There is a sudden onset of fever, headache, joint and muscle pains, conjunctivitis, and photophobia. This illness is brief, but returns and most patients have a prolonged recovery. Retinal disease (macular exudates, retinal hemorrhages and vasculitis) occurs in 5-10% of patients. Half of these patients have permanent visual impairment. Central nervous system complications occur in 1-5% of patients, and 1% develops viral hemorrhagic fever. Complicated patients have been treated with ribavirin and, where available convalescent human immune serum. [248, 295]

Rift valley fever virus has been isolated (in Guinea) from *Micropteropus pusillus*, *Miniopterus schreibersii*, and *Hipposideros caffer* bats. [139]

Filoviridae: Filoviridae are a family of viruses that belong to the order Mononegavirales (which includes Bornaviridae, Rhadoviridae, and Paramyxoviridae families). There are 2 genera of Filoviridae: Ebolavirus and Marburgvirus. Ebolavirus (EBOV) genus has 5 species (Zaire ebolavirus, Sudan ebolavirus, Reston ebolavirus, Cote d'Ivoire ebolavirus, Bundibugyo ebolavirus), and Marburgvirus consists of one species, Lake Victoria marburgvirus (MARV). [297] Outbreaks of Marburg and Ebola virus hemorrhagic fevers carry 80-90% mortality in humans and great apes. [298] Between 1967 and 2008, there were 9 outbreaks of Marburgvirus hemorrhagic fever, and 22 outbreaks of Ebolavirus hemorrhagic fever. Mortality rates for outbreaks ranged between 0 and 100%. Large outbreaks had between 53% to 80-90% mortality. [297]

Transmission of the disease is by person-to-person spread. Aerosol spread to nonhuman primates has been demonstrated. [295]

In October 1998, there was an outbreak of Marburg hemorrhagic Fever in an Democratic Republic of Congo (DRC) gold-mining village. More than half of the cases (52%) were in young male miners. Of these, only 27% had contact with other sick individuals. Sixty-seven percent of nonminers had contact with sick individuals. Of the sick miners, 94% worked underground. Housewives represented 21% and health care workers 5 % of patients. The outbreak continued into 2000 as the result of "short chains of human-to-human transmission". [16, 299]

In July and September 2007, miners working in Kitaka Cave lead mine, Uganda developed Marburg hemorrhagic fever. Bats from a colony of over 112,000, 411 *R. aegyptiacus* and 407 *Hipposideros* spp. bats were PCR and serologically tested in August 2007, and 200 *R. aegyptiacus* and 200 *Hipposideros* spp. were similarly tested in April-May 2008. The first and second rounds of testing revealed 5.6% and 4.5% of *R. aegyptiacus* infected with MARV respectively. No evidence of MARV infection in *Hipposideros* spp. tested during April-May 2008 was found. Serologic testing revealed 2.4% (13 adult bats of 546) of *R. aegyptiacus* for Marburg virus-specific IgG antibodies. No *Hipposideros* spp. was seropositive. All bats were healthy and able to forage for food. Testing of pregnant females, juveniles, and placentas revealed no evidence of vertical transmission in *R. aegyptiacus*. [300]

Examination of *R. aegyptiacus* fruit bats found infection rates (presence of viral RNA) of 1.4% in Gabon and 3.1% in the DRC. In the DRC, MARV infects 3.0% of *Miniopterus inflatus* and 3.6% of *Rhinolophus eloquens* (eloquent horseshoe bat). In Uganda, only 1 of 609 (*Hipposideros* sp.) insectivorous bats was infected. Examination of a healthy pregnant *R. aegyptiacus* from Kitum Cave, Kenya, found evidence of infection without histologic evidence of liver involvement. Attempts to isolate the virus were unsuccessful. [301].

Clinically, patients present with an abrupt onset of chills, fever, myalgia and malaise. This is followed by lethargy, nausea, vomiting, abdominal pain, anorexia, diarrhea, cough, headache and hypotension. Neurologic manifestations (headache, confusion, seizure, and coma) occur, although not all patients have hemorrhagic signs. Survivors have a prolonged convalescence. Early symptoms are similar between survivors and those with fatal illness. In the late stage of disease, patients with tachypnea are more likely to have a fatal outcome. Survivors complain of myalgia, arthralgia, weakness, hepatitis, ocular disease, myelitis, hearing loss and in some patients, psychosis. [297]

Severe disease caused by these viruses is attributed to: (1) rapid viral replication; (2) viral suppression of the host immune response; and, (3) vascular dysfunction. Much of the pathophysiology appears to be mediated through infected macrophages. [297] Patients dying of these infections have necrosis of parenchymal cells of the liver, spleen, kidneys, ovaries and testes. Often, in fatal cases there is no antibody response. [295]

The only mammals in which filoviruses have been detected are bats and nonhuman primates. [301] During the Ebola outbreaks between 2001 and 2003 in Gabon and the DRC, over 1000 small vertebrates were tested. Only three species of frugivorous bats showed evidence of asymptomatic Ebola virus infection. [302] Two theories have been proposed. In the first, long-term persistence in an infrequently encountered reservoir is the source of infection. The second theory postulates recent viral introduction and spread through the susceptible populations. [303]

Bats live in large colonies, and it would be expected that seropositivity rates would be high. Lyssavirus seroprevalence (Lagos bat virus – Kenya) in some bat species is between 60-70%. However, seroprevalence of MARV neutralizing antibodies in colonies of *R. aegyptiacus* bats in which PCR positive bats were found was 2.4-12%. [301]

Long-term survival (at least 13 months) was documented for a migratory fruit bat (*Eidolon helvum*) that was seropositive for Zaire ebolavirus and Lagos bat virus. The bat was captured in Accra, Ghana, fitted with a radio transmitter and released after testing. Of concern is that this species lives in large urban colonies and is a source of bushmeat in some areas. [304] Direct exposure to fruit bats was documented in the Occidental Kasai province of the DRC in 2007, where the first victim bought freshly killed bats from hunters to eat. [305]

Marburg virus: The first cases of Marburg virus hemorrhagic fever were reported in Europe in 1967 in laboratory workers who handled tissue and blood samples from nonhuman primates from Africa. [301] In one study of 1142 bats of 10 species from Gabon and the DRC, of 1138 tested by RT or nested PCR, only 4 of 283 Egyptian fruit bats (*R. aegyptiacus*) were positive by PCR. Of 438 bats tested serologically, 29 of 242 *R. aegyptiacus* had positive serology. [298] Human cases reported in sub-Saharan Africa were associated with visits to caves (i.e., Kitum Cave at Mount Elgon, Kenya). [301]

Ebola and Marburg viruses may be found among bats in the same area. *Casinycteris argynnis, Epomops franqueti, Hypsignathus monstrosus, Myonycteris torquata, Micropteropus pusillus, Mops condylurus, Rousettus aegyptiacus*, were sampled from three regions of Gabon and the Ebola epidemic region of north Congo for antibodies for Zaire ebolavirus (ZEBOV) and marburgvirus (MARV). ZEBOV antibodies were found in 4% of six species of bats (*Epomops franqueti* (4%), *Hypsignathus monstrosus* (7%), *Myonycteris torquata* (3%), *Micropteropus pusillus* (2%), *Mops condylurus, Rousettus aegyptiacus* (8%), and Microchiroptera (including *Mops condylurus* and *Hipposideros gigas*-12%)). *Casinycteris argynnis* had no antibodies detected for either virus. MARV antibodies were found in 1% of *R. aegyptiacus* (7%) and *H. monstrosus* (1%), *Micropteropus pusillus* (0.6%), and *E. franqueti* (0.3%). [306]

The prevalence of MARV antibodies was highest in *R. aegyptiacus* captured inside caves. There was no difference in MARV antibody prevalence found according to age or gender. ZEBOV antibodies were found more frequently in pregnant females than nonpregnant females. Of seropositive *R. aegyptiacus*, 42% were positive for MARV, 48% were positive for ZEBOV, and 10% were positive for both viruses. Dual infection, antigenic stimulation, and/or cross reactions between Ebola and Marburg antigens are all possible explanations for this finding. Gabon was the only country where both viruses were simultaneously present in bats. [306]

In another study in the DRC, *R. aegyptiacus* (8%) and the Microchiroptera *Rhinolophus eloquens* and *Miniopterus inflatus* were seropositive for MARV. Bats of different species tend to frequent the same trees, promoting interspecies virus transmission or antigenic stimulation. This could occur by infected saliva on commonly partially consumed fruit. [306]

Ebola virus: In Central Africa, frugivorous bats (*Hypsignathus monstrosus, Epomops franqueti, Myonycteris torquata*) appear to be a reservoir for Ebola virus (serology and PCR positive). [307] These 3 bat species, caught near affected villages on the Congo-Gabon border, were asymptomatically infected with Zaire ebolavirus. Seven bats had viral sequences that closely matched the viruses responsible for human disease outbreaks. Interestingly, the Zaire ebolaviruses from the 3 bat species had a common ancestry. [308]

An explanation for this phenomenon is that the virus reached a "genetic bottleneck." The first possibility is that the total number of bats within the infected area became so small that only one of the viral lineages survived. A second explanation is the infected bats introduced the virus into the affected area. The third explanation is that the virus was introduced to these bats the same time it was to other wildlife and humans. High seroprevalence rates (16.7%) with low PCR viral detection (3.2%) may indicate that viral replication in bats is restricted and taking place just prior to the host immune response. Infections may be synchronized by some external trigger. [308]

E. franqueti, H. monstrosus, and *Myonycteris torquata* seropositivity rates of 5% for Zaire ebolavirus (ZEBOV) uniformly across epidemic and nonepidemic areas in Gabon and the DRC indicate that infected bats were present in nonepidemic regions of Central Africa. ZEBOV rates decreased to 1% after outbreaks. This suggests that the proportion of infected bats is directly related to the risk of an outbreak. The large number of ZEBOV seropositive adult bats and pregnant *H. monstrosus* females suggests viral transmission within bat populations through fighting and sexual contact. [309]

Flaviviruses: Flaviviridae are positive-sense single-stranded RNA viruses with about 70 members. Flaviviridae is made up of 3 genera: (1) Flavivirus (dengue virus, Ilheus virus,

Japanese encephalitis virus, Kyasanur Forest disease virus, Kunjin virus, louping ill virus, Murray Valley encephalitis virus, Omsk hemorrhagic fever virus, Powassan virus, Rocio virus, St. Louis encephalitis virus, tick-borne encephalitis virus, tick-borne encephalitis virus, Wesselsbron disease, West Nile fever, yellow fever, and Zika disease virus); (2) Hepacivirus (hepatitis C virus); and, (3) Pestivirus (bovine viral diarrhea viruses, border disease virus in sheep, classic swine fever virus (hog cholera), and various animal pestiviruses). [248, 310-312] The most important human pathogens are Japanese encephalitis, West Nile, Murray Valley encephalitis, tick-borne encephalitis, Kyasanur Forest disease, and dengue viruses. Other significant pathogens in this group are Louping ill, Omsk hemorrhagic fever, Langat, Powassan, Royal Farm, and Gadgets Gully viruses. The newest flaviviruses are New Mapoon virus (Australia), Sitiawan virus (Malaysia), and ThCaAr virus (Thailand). Nineteen viruses are mosquito-borne, six are tick-borne and 4 are isolates from rodents or bats. [310-312]

Kyasanur Forest Disease virus (Flaviviridae genus) (KFDV): KFDV is characterized by an incubation period of 3-8 days, chills, frontal headache, body aches and high fever for 5-12 days. There is $\geq$ 30% mortality. KFDV causes high death rates in the black-faced langur (*Semnopithecus entellus* a.k.a. *Prebytis entellus*) and the red-faced bonnet monkey (*Macaca radiata*). In India, over the past 50 years, there have been continuing deaths in monkeys and 400 to 500 human deaths per year.[311] Humans at most risk are those who visited forests to collect firewood, grass, and other forest products. The usual vectors are ticks (*Haemaphysalis spinigera* from central India and Sri Lanka, and 7 species of *Dermacentor* and *Ixodes* ticks). The disease is transmitted to birds, white-tailed rats, white-bellied rats, shrews and bats by these ticks. KFDV has been identified in India, Sri Lanka, Saudi Arabia and China. [311]

Evidence of KFVD infection in bats where the disease has been transmitted to humans and monkeys was presented by Bhat, Sreenivasan, Goverdhan, et al. *Rousettus leschenaultii*, *Rhinolophus rouxi*, and *Cynopterus sphinx* are bat species shown to be infected. [312-318]

Transmission of flavivirus disease is usually by mosquito. *Aedes* species (most commonly *A. aegypti*) is the usual vector for dengue. Tree-hole breeding mosquitoes (*Haemagogus janthinomys*, *Haemagogus* sp. *Sabethes chloropterus*, *Aedes* sp.) usually transmit yellow fever virus. Kyasanur Forest disease is usually tick-borne (primarily *Haemaphysalis spinigera*, and ixodid ticks). The host for Omsk hemorrhagic fever is the water vole - *Arvicola amphibius* (formerly *Arvicola terrestris*), and the vectors are *Dermacentor reticulates* and *Ixodes apronophorus* ticks. [295]

St. Louis encephalitis (*Flaviviridae* genus) is a mosquito-borne disease (*Culex* sp.: *Cx. pipiens* and *Cx. quinquefasciatus* (formerly *Cx. fatigans*) in the east, *Cx. nigripalpus* in Florida, and *Cx. tarsalis* and members of the *Cx. pipiens* complex in western states) responsible for multiple outbreaks in the United States. Most cases occur in temperate climate though disease has been reported from Canada to Argentina. Although most infections are asymptomatic, symptomatic patients (<1% of infections) develop fever, headache, nausea, vomiting and malaise. Older individuals (90% of symptomatic patients) are more likely to develop debilitation from encephalitis. Risk of a fatal outcome (overall 5-15% in diagnosed cases) increases with age. [310, 319]

St. Louis encephalitis virus infection presents with a febrile prodrome (malaise, fever, headache, myalgias, with or without upper respiratory or abdominal complaints. This evolves over days to 1 week into three distinct syndromes, the severity of the clinical presentation increasing with age. Most commonly, patients have (1) febrile headaches (constitutional symptoms and headache. Patients may present with or progress to (2) aseptic meningitis, or (3) fatal encephalitis.

Progression of disease is noted by lethargy, periodic confusion, tremors, clumsiness, ataxia, and commonly, vomiting and diarrhea. Some patients have dysuria, urgency and incontinence. Meningeal signs are not always present. Generalized motor weakness is more common than focal signs. Most patients do not develop a deep coma. Overall mortality is 8%, but >20% for patients over 60 years. [310]

In a survey of bats in Guatemala, St. Louis encephalitis neutralizing antibodies were found in fruit bats (*Artibeus intermedius*, *A. jamaicensis*, *A. lituratus*, *A. phaeotis*), Pallas' long tongued bat (*Glossophaga soricina*), and the little yellow-shouldered bat (*Sturnira lilium*).[320]

In Guinea, Saboya virus a subspecies of yellow fever virus was isolated from the Gambian slit-faced bat, Nycteris gambiensis. [139]

GB viruses (GBV) including genus *Hepacivirus* (GBV-B and Hepatitis C virus (HCV)): GBVs found in man and Old and New World primates are believed to have coevolved with their host species over long periods. GBV-A and GBV-C are unassigned flaviviruses. GBV-A infects new world primates, but not man. Specific GBV-A genotypes are associated with specific monkey species (*Saguinus*, *Callithrix* (Callitrichidae family), and *Aotus* (Aotidae family), but does not cause hepatitis in these animals. GBV-C (formerly called Hepatitis G virus) infects man in many regions of the world, and African wild chimpanzees (*Pan troglodytes*), but does not cause hepatitis. GBV-B causes hepatitis in primates (tamarins – *Saguinus* sp.) and is closely related to HCV. [321, 322,]

HCV infection is found worldwide with 1-2% prevalence in the general population. Modes of transmission include percutaneous, nosocomial, sexual, and maternal-infant. It causes acute hepatitis, fulminant hepatitis (more common in Japan), and chronic hepatitis developing in 50-85% of persons with acute HCV infection. Patients with chronic hepatitis may develop insulin resistance, type 2 diabetes mellitus, steatosis, and cirrhosis (with 10-20% decompensating in 5 years – esophageal varices, ascites, coagulopathy, encephalopathy, and hepatocellular carcinoma). In patients with cirrhosis, 32% develop hepatocellular carcinoma at a rate of 3.9% per year. Immune complications of HCV infection are essential mixed cryoglobulinemia, membranoproliferative glomerulonephritis, and porphyria cutanea tarda. [321]

While studying Nipah virus, 98 free-ranging frugivorous *Pteropus giganteus* bats from a colony of 1800 bats from the Faridpur district of Bangladesh were studied. RNA was extracted from blood. A new GBV was identified, designated GBV-D. GBV-D is believed to be "ancestral to GBV-A and –C clades." [322]

Arenaviruses

Arenaviridae contains 22 species and 9 other virus species that taxonomically remain undecided. They are further subdivided by antigenic properties: Lassa-Lymphocytic choriomeningitis serocomplex (Lassa virus, lymphocytic choriomeningitis virus, Ippy virus, Mobala virus, and Mopeia virus), and the Tacaribe serocomplex. In North America, the Tacaribe virus complex consists of Bear Canyon virus, Tamiami virus, and Whitewater Arroyo virus. On Trinidad, there is Tacaribe virus. Allpahuayo virus, Amapari virus, Cupixi virus, Flexal virus, Guanarito virus, Junin virus, Latino virus, Machupo virus, Oliveros virus, Parana virus, Pichinde virus, Pirital virus, and Sabia virus are found in South America. Catarina virus is a provisional member of the Tacaribe serocomplex group. Members of the rodent family Cricetidae are the principal carriers of the Tacaribe serocomplex. [323-325]

Arenaviruses commonly cause asymptomatic infections in rodents. The principal hosts for Lassa Fever are *Mastomys natalensis*, and *Mus musculus*. The host for Guanarito virus, the causative agent for Venezuelan hemorrhagic fever is the cane mouse, *Zygodontomys brevicaudia*. The Tacaribe group of Arenaviruses infects fruit bats on Trinidad and Tobago, the white-throated woodrat (*Neotoma albigula*) in New Mexico, and the white-toothed woodrats in Mexico. [296, 324-326]

Humans become infected with direct contact with infected rodents through bites, and inhalation of infected excreta and secreta.[323, 325, 327[Viral hemorrhagic fever viruses in this group are Lassa fever virus (western Africa), Junin virus (Argentina), Machupo virus (Bolivia), Guanarito virus (Venezuela), and Sabia virus (Brazil). Lymphocytic choriomeningitis virus causes meningitis and meningoencephalitis, congenital malformations and infections in transplant recipients. [323]

Yet, clade B viruses (Amapari, Cupixi and Tacaribe viruses) have not been found to cause human disease although Tacaribe virus may have caused a mild febrile and central nervous system symptoms in a laboratory worker. *Artibeus* sp. of fruit bats on Trinidad and Tobago has been suggested as nonrodent hosts for Tacaribe virus. [296]

ALPHAVIRUSES

Venezuelan Equine Encephalomyelitis, and Eastern and Western Equine Encephalitis Viruses (Togaviridae)

Alphaviruses are part of the family Togaviridae, positive-strand RNA viruses that include the 3 mosquito-borne agents: Venezuelan equine encephalomyelitis (VEE); and, Eastern and Western equine encephalitis viruses. Eastern equine encephalitis (EEE) causes disease in horses and passerine birds in addition to humans. Its animal reservoir is birds, and the human vectors are *Aedes* sp., *Coquillettidia* spp., and *Culiseta melanura* mosquitoes. Western equine encephalitis (WEE) causes fever in birds, especially emus (*Dromaius novaehollandiae*), and horses. Birds are the animal reservoirs, and *Culex tarsalis* is the principal human vector. Venezuela equine encephalitis (VEE) causes fever and encephalitis in horses. Horses and other animals are reservoirs. *Psorophora* sp., *Culex* sp., *Mansonia, Deinocerites*, and *Aedes* sp mosquitoes are the human vectors. [328]

VEE, EEE, and WEE manifests itself in horses with fever, muscle fasciculation, depression, head pressing, incoordination, recumbency, opisthotonos, padding and death. [248]

In humans infected with EEE virus, there is a 5-10 day prodrome of headache, high fever, chills, nausea, vomiting and/or diarrhea. Those who progress to encephalitis become confused, somnolent, and photophobic. Seizures (tonic-clonic or partial complex) occur more frequently in younger patients.

WEE has a shorter prodrome (1-4 days) with similar signs and symptoms as EEE. Prodrome may resolve without neurologic complications. Neurologic complications are seen in 30% of younger patients.

VEE has an incubation period of 1-6 days followed by fever and malaise commonly accompanied by chills, myalgia, and headache with or without photophobia, and infrequently a sore throat. Fever may disappear one day only to return the next. Overall case-fatality rate is 1% but 20% in patients with encephalitis. [328]

Neutralizing antibodies were identified in the following bats in Guatemala: *Artibeus intermedius* (EEE); *A. jamaicensis* (VEE, EEE); *A. lituratus* (EEE); *Carollia brevicauda* (VEE); *Carollia subrufa* (VEE); *Glossophaga soricina* (EEE); *Phyllostomus discolor* (VEE); *Rhynchonycteris naso* (EEE); *Sturnira lilium* (VEE, EEE); and, *Sturnira ludovici* (VEE). [320]

Chikungunya Virus (Togaviridae) and Chikungunya Fever

Chikungunya virus is a member of the Togaviridae family. It is part of the Semliki Forest virus group and is similar to O'nyong-nyong virus. It is a single stranded positive-sense RNA virus that is transmitted to humans by mosquito bite. Outbreaks of chikungunya fever have been recorded in Eastern and Central Africa. Most recently outbreaks (February 2005 to June 2006) in the Indian Ocean (Reunion Island, the Seychelles, Madagascar, Mauritius, and Mayotte, and simultaneously in India and several cases in the Emilia-Romagna Region of central Italy have occurred. [328, 329]

After an incubation period of 1-12 days, patients experience rapid onset of fever, and crippling polyarthralgia, myalgia and rash. Most patients recover completely in a few weeks, but 5-10% have chronic symptoms that last more than 1 year. On Reunion Island, some patients experienced severe neurologic symptoms with 255 fatalities. [328, 329]

Epidemiology: In Africa, chikungunya virus is maintained in a sylvatic cycle between monkeys and baboons (*Cercopithecus* sp.) and other animals. In Senegal, the virus was isolated from *Cercopithecus aethiops* monkeys, *Galago senegalensis* galago (primates-bushbabies or nagapies), and, *Xerus erythropus* palm squirrel. Vectors in Africa include *Aedes furcifer*, *Aedes taylori*, *Aedes leucocephalus*, *Aedes dalzieli*, *Aedes africanus*, and *Aedes neoafricanus*. In Asia, the vectors are *Aedes aegypti* and *Aedes albopictus* (Asian tiger mosquito) and are maintained in an urban cycle with man. [330, 331] In Gagnik, Gosas, and Rao, Senegal, virus was isolated from *Scotophilus* sp. bats. [331] Diallo M et al state, "Isolates from squirrel, chiroptera, and ticks (*Alectorobius sonrai*), as well as the presence of antibodies specific for chikungunya virus in rodents and birds, support the assumption that secondary wild cycles exist (Cornet M and others, unpublished data). [331]

Vesicular Stomatitis Virus

Vesicular stomatitis (VS) virus is a negative-sense RNA (Rhabdovirus). There are 2 serotypes (VS-New Jersey and VS-Indiana), and 3 subtypes (VS-Fort Lupton, VS-Alagos

(Brazil), and VS-Cocal (Trinidad). VS-New Jersey serotype is the most common and virulent. [248, 332]

The disease is limited to the Western Hemisphere and is endemic in Mexico, from Panama to Venezuela and Peru, and the Ossabaw Island off the Georgia (U.S.A.) (VS New Jersey serotype). On Ossabaw Island, antibodies have been detected only from feral swine, cattle, horses, donkeys, deer, and raccoons. Occasionally it is seen in the United States, Brazil, and Argentina. [248, 332]

Cattle, horses and donkeys are the most susceptible, but disease occurs in pigs, camelids and humans, and possibly sheep, and goats. Morbidity in dairy herds is 5-10% to 80%.

Sandflys and blackflies transmit the disease both transovarially and to susceptible animals, although transovarial infection is infrequent. Radostits et al have suggested that this is indicative of another natural wildlife reservoir.[248]

Wildlife in Central America is frequently infected. There is evidence of windborne infection in the United States and Mexico. Saliva of infected animals is highly infectious. Insect vectors include blackflies (*Simulium vittatum* and *S. notatum*), *Culicoides* spp., flies (*Musca domestica* and *M. autumnalis*), phlebotamine sandfly (*Lutzomyia shannoni* the vector on Ossabaw Island for the feral pig), *Lutzomyia trapidoi*, mosquitoes, and eye gnats (*Hippelates* spp.). [248, 332]

Contact with infected animals is the most important risk factor for human disease. In humans, most disease is mild or subclinical. Some patients develop an influenza type illness with fever, malaise, myalgias, vomiting, and pharyngitis. Oral vesicular lesions have been rarely reported. One patient's illness lasted 3 weeks with oral lesions, lymphadenopathy and a significant weight loss. In a laboratory accident, fever, chills, retro orbital pain, myalgias, nausea and vomiting, and diarrhea began after 30 hours and resolved in 3 days. Vesicular stomatitis virus causes central nervous system disease in baby mice. Meningoencephalitis has been reported in children. [248, 332]

Surveys of bats in Guatemala found neutralizing antibodies against, either or both VS-New Jersey or VS-Indiana in *Artibeus jamaicensis*, *A. lituratus*, *A. phaeotis*, *Desmodus rotundus*, *Sturnira lilium*, and *Vampyrodes caraccioli*. [320] The highest prevalence of antibody was against VS-New Jersey. [320]

ADENOVIRUSES

Adenoviruses are nonenveloped DNA viruses that commonly cause human respiratory tract disease (5% of upper respiratory tract and 10% of pneumonias in children) pharyngoconjunctival fever (follicular conjunctivitis, fever, pharyngitis and cervical adenopathy), gastroenteritis, acute hemorrhagic cystitis especially in children, meningitis and meningoencephalitis usually as a complication of pneumonia, and rarely acute myocarditis in children, myositis with rhabdomyolysis, arthritis, and pancreatitis. Disseminated life-threatening infections occur in hematopoietic stem cell and solid organ transplant patients. [333]

In China, 8% of bat fecal samples tested were positive for adenoviruses closely related to tree shrews and canines. Bat species tested included *Hipposideros armiger* (6.5% prevalence), *M. horsfieldii*, *M. ricketti* (40% prevalence), *Myotis* sp., and *Scotophilus kuhlii*. [334] In

Germany, of 55 bats tested, 2 common pipistrelles (*Pipistrellus pipistrellus*) were positive for genus *Mastadenovirus* of the family Adenoviridae. Bats were found moribund and no other pathogen was found to explain their condition. The authors conclude the virus was passed between bats living in close proximity. [335]

Two novel adenoviruses were isolated from the common noctule (*Nyctalus noctula*) and the greater horseshoe bat (*Rhinolophus ferrumequinum*). Janoska et al believe that the phylogenetic analysis suggests that canine adenoviruses originated in vespertilionid bats. [336].

HERPESVIRUSES

Herpesviridae family is divided into 3 subfamilies: Alphaherpesviruses (herpes simplex virus types 1 and 2, varicella-zoster virus, and herpes B virus); Betaherpesviruses (cytomegalovirus, human herpesvirus 6 and 7); and, Gammaherpesviruses (Epstein-Barr virus, Kaposi's sarcoma associated virus/human herpes 8 virus (rhadinovirus)). [337]

Alphaherpesviruses: These viruses are characterized by latent neuron sensory ganglion infections, and are rapidly lytic in cell cultures. They characteristically cause mucocutaneous infections in healthy patients and may disseminate especially in immune compromised patients. [337-340]. Herpes B virus (cercopithecine herpesvirus 1 previously known as herpesvirus simiae) causes disease in macaque monkeys similar to disease caused by herpes 1 disease in man, but human infection with this virus has a 70% mortality.

Betaherpesviruses: These viruses typically establish latent infections. Infection rates for cytomegalovirus are 60-70% in the U.S. and 100% in Africa. Infection is often asymptomatic, may be self-limited (cytomegalomononucleosis syndrome), chronic, or single or multisystem in the immunocompromised patient. [341] Both human herpes types 6 and 7 causes exanthema subitum (roseola infantum or sixth disease), fever and febrile seizures. Human herpes type 6 has caused a mononucleosis syndrome and chronic or fulminant hepatitis. [342]

Gammaherpesviruses: These viruses typically establish latency. Adults are 90-95% seropositive for Epstein-Barr virus. Primary infection in children is usually asymptomatic. Older individuals develop a self-limited mononucleosis syndrome. Other manifestations include hemophagocytic lymphohistiocytosis, oral hairy leukoplakia in AIDS patients, and malignancy (Hodgkin's lymphoma, Burkitt's lymphoma, nasopharyngeal carcinoma, nasal NK and T-cell lymphomas). [343, 344] Kaposi's sarcoma-associated herpes virus infection results in Kaposi's sarcoma, primary effusion lymphoma, multicentric Castleman's disease especially in immune compromised and AIDS patients. [344]

The first bat herpesvirus was reported in 1996 by Tandler. He identified enlarged viral particles in acinar cells in little brown bat (M. lucifugus) submandibular glands. He noted that unlike cytomegalovirus in other animal species, the virus particles within the cytoplasmic vacuoles lacked capsomeres, but these capsomeres were prominent in particles free in the cytosol. [345, 346]

In 2007, Wibbelt et al reported on the herpesvirus identification in European bats: bat gammaherpesvirus 1 (batGHV-1) from serotine bat (Eptesicus sertinus), Natterer's bat (M. nattereri), Nathusius' pipistrelle bat (*Pipistrellus nathusii*) and the common pipistrelle bat (Pipistrellus pipistrellus); bat gammaherpesvirus 2 (batGHV-2) from Natterer's bat (M.

nattereri), and the greater mouse eared bat (M. myotis); bat gammaherpesvirus 3 (batGHV-3) from the noctule bat (Nyctalus noctula), Natterer's bat (M. nattereri), and the greater mouse eared bat (M. myotis); bat gammaherpesvirus 4 (batGHV-4) from the noctule bat (Nyctalus noctula), and Natterer's bat (M. nattereri); bat gammaherpesvirus 5 (batGHV-5) and (batGHV-6) 6 from the common pipistrelle bat (P. pipistrellus); bat gammaherpesvirus 7 (batGHV-7) from the common long eared bat (Plecotus auritus); bat betaherpesvirus 1 (batBHV-1) from the common pipistrelle bat (P. pipistrellus). BatGHV-1, BatGHV-2, BatGHV-5, and BatGHV-7 had a significant percent identity with Herpesvirus saimiri (Saimiriine herpesvirus-2, a gamma 2-herpesvirus (rhadinovirus) of non-human primates (squirrel monkeys)) by partial DNA polymerase gene sequence (DPOL).[346, 347] BatBHV-1 shows some relatedness to Tupaia herpesvirus 1 (45% by DPOL), a virus previously isolated from malignant lymphomas and from degenerating lung or spleen cell cultures of tree shrews (Tupaia spp.). [346, 348] BatGHV4 is related to chimpanzee herpesvirus RHV2 (67% by DPOL). Phylogenetic analysis of the bat gammaviruses revealed a similar distance from equine herpesvirus 2 and gammaherpesviruses from badger (Percavirus that includes *Equid herpesvirus 2, Equid herpesvirus 5, Mustelid herpesvirus 1)* and Rhadinovirus. [346]

A jaundiced anorexic serotine bat (Eptesicus serotinus) rescued in Hungary was found to be infected with a genus Rhadinovirus of the subfamily Gammaherpesvirinae. The bat died within 24 hours of being discovered. Autopsy revealed vacuolar degeneration of hepatocytes and white cells in the sinusoidal lumina. Electron microscopy revealed hydropic degeneration and apoptotic cells with pycnotic nuclei in the liver. [349] Watanabe et al described a gammaherpesvirus from an insectivorous bat (Hipposideros diadema) from Panay Island, the Phillippines.[350]

In Wakayama Prefecture, Japan, examination of insectivorous vespertilionid bats (Miniopterus fuliginosus), a novel bat betaherpesvirus 2 was identified. [351] Agua preta (gray short-tailed bat – Carollia subrufa) and Parixa (Thomas' nectar bat – Lonchophylla thomasi) herpesviruses were isolated from Brazilian frugivorous bats.[16, 352] An analysis of herpesviruses from African, South American and Malagasy and Cambodian frugivorous bats (families Pteropodidae and Phyllostomidae) were found to be genus Simplexvirus of the subfamily Alphaherpesvirinae. Isolates (throat swabs, organs, salavary gland, and blood) were from *Eidolon dupreanum*, *E. helvum*, and *Pteropus lylei* (Praxia virus) (family Pteropodidae), and *Lonchophylla thomasi* (family Phyllostomidae). [352]

Analysis of bat guano from Texas and California found viruses of eukaryotic origin including members of viral families that infect insects (Dicistroviridae, Iflaviridae, Tetraviridae and Nodaviridae and the subfamily Densovirinae), plants and fungi and mammals (Parvoviridae, Circoviridae, Picornaviridae, Adenoviridae, Poxviridae, Astroviridae, and Coronaviridae). No Herpesviridae or close relatives of human pathogens were identified. [353]

BACTERIA

Enteric Pathogens

The gastrointestinal contents of bats captured in Trinidad and Tobago were cultured for common enteric bacterial pathogens. Of 377 bats which included 12 species, 4 bats (1.1% were positive for *Salmonella* spp., 49 (13%) were positive for *Escherichia coli*. No bats were colonized with *E. coli* O157 or *Campylobacter* spp. *Salmonella* serogroups Rubislaw and Molade were isolated from the fish-eating greater bulldog bat, *Noctilio leporinus*. Insectivorous bats were colonized with *Salmonella* Caracas (*Molossus major*) and *Salmonella* Group 1 (*Molossus ater* – black mastiff bat). Eighty-two percent of *E. coli* was resistant to multiple antibiotics. Resistance to erythromycin (61%) and streptomycin (27%) was common, resistance to sulphamethoxazole/trimethoprim (2%) was rare, and no isolates were resistant to gentamicin. [354]

Kluyvera spp. are Enterobacteriaceae that have infrequently caused human infections (urinary tract infections, pyelonephritis, sepsis, diarrhea, soft tissue infections, cholecystitis, peritonitis, intra-abdominal abscess, mediastinitis following open-heart surgery, urethrorectal fistula, sepsis in an infant with umbilical vein and artery catheters, and meningitis in an infant with a ventriculoperitoneal shunt). [355-360] The organism is part of the normal flora of the gastrointestinal tract. [355] *K. cryocrescens* has been isolated from a septic preterm infant, and a multidrug resistant *K. ascorbata* was isolated from a compromised adult patient.[356-358] *K. cryocrescens* was isolated from young children with a central venous pressure catheters. [361, 362] The organism represents a significant risk for immunosuppressed and transplant patients. [363, 364]

Clinical isolates have included, blood, stool, urine, nasopharynx, bile, wounds, and intravascular devices. [363] Environmental sources include sewage, soil, kitchen food, water, food processing facilities, and hospital sinks. [361, 365] The organism has been isolated from asymptomatic Madagascan lemurs, rainbow trout (*Oncorhynchus mykiss* – *K. intermedia*), large-billed crows (*Corvus* spp.), snails, slugs, and *Rousettus aegyptiacus*, the Egyptian fruit-bat (flying fox) (*K. ascorbata*). [366-369]

Leptospirosis and Weil's Disease

Leptospirosis is a common zoonosis found worldwide except in Antarctica. [370] Incidence is higher in warmer rather than temperate climates. In the U.S., the highest incidence is in Hawaii. [371, 372]

Reports of human cases of leptospirosis after bat exposure are now more frequent.[373] Certain occupations represent significant risk factor. Those in direct contact with animals (farmers, veterinarians, abattoir workers, meat inspectors, and rodent exterminators) are at increased risk of infection. Indirect contact risk is seen in sewer workers, miners, soldiers, septic tank cleaners, fish farmers, gamekeepers, canal workers, rice field workers, taro farmers, banana farmers, and sugar cane cutters. [371]

Leptospirosis is caused by members of the genus *Leptospira*, which consists of 14 species and over 200 serovars. [372, 373] *Leptospira* colonize kidneys of infected rats, swine, dogs, cattle, bats and sea lions, and the organisms are excreted into the environment with the infected urine. [24, 373, 374] In the Amazon basin, evidence of infection has been noted in 29% of mammals tested (rodents – 20%, marsupials – 39%, bats – 35%).[375] On Trinidad and Grenada, leptospiral antibodies were found in bats, mongooses, opossums, peridomestic and forest rodents, lizards, and toads. [376] In Denmark, *Leptospira* sp. were identified in 3 insectivora (*Sorex araneus* – common shrew, *Neomys fodiens* – water shrew, *Erinaceus europaeus* - hedgehogs), 4 bats (*Myotis daubentonii* – red-grey bats, *Pipistrellus pipistrellus, Nyctalus noctula,* and *Eptesicus serotinus*), rodents and carnivores (*Vulpes vulpes* – red fox and *Mustela erminea* - short-tailed weasel). *M. daubentonii*, *P. pipistrellus* and *Nyctalus noctula* had high carrier rates (15-20%). [377]

Disease is normally acquired by contact with infected urine from carrier animals, contaminated environment (soil or water), or infected animal tissue.[24, 372] 16S rRNA gene sequences found in environmental water sources from rural areas in the Peruvian Amazon have been identified in pigs, cattle and bats.[378] Infection has occurred as the result of drinking water from contaminated fountains or swimming in contaminated water. [371]

In humans, most cases are believed to be asymptomatic. Self-limited systemic illness is seen in 90% of infections. [372] The incubation period of clinical disease is 10 days (5-14 days). Disease often begins with an acute septicemic phase lasting 5-7 days (remittent fever, chills. headache, rigors, myalgias, conjunctival suffusion, abdominal pain, anorexia, nausea, vomiting, diarrhea, cough, pharyngitis, and rarely a pretibial maculopapular rash). Less commonly, patients develop hepatosplenomegaly, and lympadenopathy.

During the acute phase, blood and cerebrospinal fluid (CSF) cultures may grow the organisms. After 5-7 days of illness, organisms may be grown from the urine. [372] Primary cultures for *Leptospira* should be held for 13 weeks. Pure subcultures in liquid media will grow in 10-14 days. [371]

The immune phase begins after a short period of improvement. It lasts 4-30 days, with bacteria present in the urine and detectable in most organs but are now absent from the blood and CSF. In the immune phase, patients may have symptoms similar to the acute phase. In addition, depending on the severity, patients develop jaundice, renal failure, cardiac arrhythmias, pulmonary symptoms, aseptic meningitis (80% of cases), conjunctival suffusion with or without hemorrhage, photophobia, eye pain, myalgia, adenopathy, and hepatosplenomegaly. Abdominal pain is uncommon and may indicate pancreatitis. [371, 372, 379, 380]

The most severe form of infection, icteric leptospirosis with renal failure, was first reported in 1886 by Adolf Weil.[371] Weil's disease may begin without any brief improvement after the acute disease. Patients develop high fever, fulminant liver and renal failure, hemorrhagic pneumonitis, cardiac arrhythmias (50% of monitored patients) and shock. Severe pulmonary hemorrhagic syndrome (SPHS) may occur in the absence of hepatic and renal dysfunction. Pathology of SPHS is similar to acute respiratory distress syndrome. Mortality of Weil's disease is 5-40%. [372]

Complications: Anicteric leptospirosis is usually associated with no mortality although an outbreak in China reported in 1965 noted 2.4% mortality from pulmonary hemorrhage. [371] Overall, pulmonary complications are seen in 20-70% of patients.[381] Respiratory complications may be seen in up to 67% of icteric disease. Cardiac involvement is seen in 10-40% if icteric disease. Conjunctival suffusion is seen in most patients as well as icterus in Weil's disease. Uveitis may be seen weeks or months after acute illness and visual problems may persist for more than 20 years after the acute illness. Late-onset uveitis may be an autoimmune reaction after a second exposure to Leptospira (similar to equine recurrent uveitis). Late-onset meningitis has been reported once. [371, 372]

Bats: Bats are known to carry Leptospira in regions of the Amazon and Australia. In the Peruvian Amazon, 3 of 598 bats (*Phyllostomus hastatus*, *Mimon crenulatum*, *Promops nasutus*) were culture positive and 20 of 589 (*Artibeus obscurus*, *A. planirostris*, *Carollia perspicilliata*, *Desmodus rotundus*, *Glossophaga soricina*, *Lonchophylla thomasi*, *Mimon crenulatum*, *Myotis riparius*, *Phyllostomus hastatus*, *Promops nasutus*, *Rhinophylla pumilio*, *Sturnira lilium*, *S. tildae*, *Uroderma bilobatum*) were were PCR positive. Of the species of *Leptospira* identified, *L. interrogans* serovar *icterohemorrhagiae* (normally found in peridomestic rats) suggested a rodent-bat infection cycle. [374] In an urban area, only 6 of 343 bats (182 insectivorous, 161 frugivorous or nectarivorous) bats were PCR positive, and none were seropositive. [382]

In Australia, 11% of 193 kidney samples and 39% or 46 urine samples from flying foxies were PCR positive for *Leptospira*. These included the spectacled flying fox (*Pteropus conspicillatus* 4%+), black flying fox (*Pt. alecto* – 16%+) grey-headed flying fox (*Pt. poliocephalus* – 5%+) and the little red flying fox (*Pt. scapulatus* – 29%+). Microscopic Agglutination Tests and PCR were compared for *Pt. alecto* from the Northern Territory. There was no correlation between PCR and serologic studies with only 47% of seropositive samples being PCR postive. [383] Serovar *L. interrogans* serovar *cynopteri* is believed to be found only in Australia. In another study of Australian flying foxes, *L. kirschneri* servar *australis* was the most frequently isolated. [384]

A study of California sea lions (*Zalophus californianus*) and sea lion pups from the Gulf of California suggest enzootic host-adapted *Leptospira* servars. Seroprevalence rates were 54%. Seronegative pups were often PCR positive. There were no reports of increased mortality or evidence of illness in the sea lion pups. Crowding of rookeries and the presence of bats and rodents (the fishing bat, *Myotis vivesi*, and endemic and introduced rodents *Peromyscus* spp, *Mus musculus*, *Rattus rattus*, and *R. norvergicus*) on some of the islands has been put forth as an explanation of this phenomenon. [385]

Bartonella Species

Human disease caused by *Bartonella* species includes Oroya fever (Carrion's disease) and verruga peruana (*B. bacilliformis*), endocarditis (*B. quintana, B. henselae*), bacillary angiomatosis/peliosis (*B. quintana, B. henselae*), cat-scratch disease (B. *henselae, B. clarridgeiae*), and HIV-associated meningoencephalitis (*B. quintana, B. henselae*). Oroya fever (Carrion's disease) is the bacteremic disease which clinically may range from a mild to fatal illness. Severe disease is manifested by fever, chills, diaphoresis, anorexia, prostration, headache, confusion, anemia, myalgias, abdominal pain, vomiting, jaundice, lymphadenopathy, seizures, meningoencephalitis, angina, and high mortality rates. With treatment, there is still 10% mortality. Verruga peruana is the designation given to the crops of skin lesions that develop weeks to months after resolution of the untreated bacteremic disease. [386]

B. quintana is the cause of "Trench fever" and has been responsible for outbreaks among homeless individuals in North America. It is associated with poor personal hygiene and sanitary conditions. *Pediculus humanus* (human body louse) is its vector. [386]

Endocarditits has been reported with *B. quintana, B. henselae*, and rarely with *B. elizabethae*. Bacillary angiomatosis is neovascular proliferation that involves the skin and/or internal organs, particularly in patients with advanced HIV infection.

Cat-scratch disease is caused by *B. henselae* and rarely by *Afipia felis*. Patients develop a primary papule or pustule followed by regional lymphadenopathy. Musculoskeletal symptoms occur in 10% of patients. Other complications are unilateral granulomatous conjucntivitis and preauricular lymph node swelling (Parinaud's oculoglandular syndrome), and neuroretinitis. [386]

DNA from *Bartonella* has been detected from ectoparasites of bats, including ticks, mites, cimicid bugs, bat flies and bat fleas. Novel *Bartonella* species have been isolated from *Trichobius major* (Diptera: Streblidae) and *Cimex adjunctus* (Hemiptera: Cimicidae), *M. austroriparius* (Rhoads) ectoparasites from 2 southeastern U.S. bat caves. *B. vinsonii* subsp. *vinsonii* was detected in *Ctenophthalmus pseudagyrtes* (flea from insectivores and small rodents). *B. tribocorum*, and 2 novel *Bartonella* genotypes have been isolated from *Xenopsylla cheopis*, *Sternopsylla texanus* (bat flea), and *Orchopeas howardi* (squirrel flea) collected from gray squirrels (*Sciurus carolinensis*). The bat tick, *Carios kelleyi* harbors *B. henselae* and a novel *Rickettsia* sp. [18, 387]

Q Fever, Rickettsia, and Borrelia

Q fever (Coxiella burnetii)**:** *Coxiella burnetii* is a pleomorphic bacterium with a Gram-negative cell wall. It is not true a rickettsiae as it passively enters cells. [388] It is endemic in birds, and wild and domestic mammals, and arthropods. The organism is found in high concentrations in placenta, amniotic fluid and parturition products of sheep, cattle and goats. Humans are infected by aerosols of parturition products. Domestic ungulates are the principal reservoir, although dogs, cats, and pigeons have been implicated as well as the consumption of raw milk. In humans, the incubation period is ~ 20 days. Disease may be asymptomatic, influenza-like, cause hepatitis, or fatal respiratory disease.[389] Less frequent complications include endocarditis, osteomyelitis, fever of unknown origin, and pneumonia. In infants, infection presents with fever of unknown origin, meningeal irritation, malaise and febrile seizures. Rare complications are bone marrow necrosis, histiocytic hemophagocytosis, hemolytic anemia, and may rarely be mistaken for lymphoma. Other hematologic complications are transient hypoplastic anemia, reactive thrombocytosis, thrombocytopenia, splenic rupture, optic neuritis and erythema nodosum. [388]

With the increase of suburban Q fever in French Guiana, a search for a wildlife reservoir was inconclusive. As compared to control groups, seeing a bat near one's home was a significant risk factor (P=0.002 univariate analysis; odds ratio=3.0). *Molossus molossus* (57 bats), *Phyllostomus hastatus* (17 bats) and other bat species (12 bats) were serologically examined for evidence of *Coxiella burnettii* infection but none was found to be positive. [389]

Rickettsial infections in bats (in Brazil)**:** Molossidae, Vespertilionidae, and Phyllostomidae bats were tested for serologic activity against rickettsial diseases found in

Brazil. Sera was reactive for *Rickettsia rickettsii* (8.6% - 39/451), *R. parkeri* (9.5% - 34/358), *R. amblyommii* (7.8%-28/358), and, *R. rhipicephali* (1.1%-4/358). No bats were reactive for *R. bellii* (0/358).[390]

Relapsing Fever (Borrelia sp.): Relapsing fever is transmitted by argasid ticks (genus *Ornithodoros*). Human disease is caused by *Borrelia hermsii*, *B. turicatae* and *B. parkeri.* Infections are usually acquired in rural/rustic cabins. *B. turicatae* is acquired in caves inhabited by the ticks' vertebrate host. Fourteen of 31 *Carios kelleyi* bat ticks collected from a house in Jones County, Iowa (built in 1857) were PCR positive for a unique *Borrelia* sp. more closely related to *B. turicatae* and *B. parkeri*, but distinct from all 3 *Borrelia* species known to cause relapsing fever. [391] Spirochetes reactive with *Borrelia*-specific monoclonal antibody were seen microscopically in coxal fluid and salivary glands from 1 *C. kelleyi* tick. Schwan et al have proposed *B. johnsonii* as the designation of this newly identified spirochete in honor of Dr. Russell C. Johnson. [392]

Carios kelleyi bat tick: Carios kelleyi tick is known to feed on humans and bats, and is found in bat habitats. Tick nymphs and adults collected from residential and community buildings in Jackson County, Iowa were tested by PCR for *Rickettsia*, *Borrelia*, *Bartonella*, *Coxiella*, and *Anaplasma* species. Rickettsia DNA was found in 28 of 31 live ticks. The flagella gene of a *Borrelia* sp. closely related to *B. turicatae* and the intergenic spacer region of *Bartonella henselae* were detected in 1 each of 31 live ticks. No evidence of *Coxiella* sp. or *A. phagocytophilum* infection was found. [393]

A larval *Argas vespertilionis* tick fatally infected a *Pipistrellus* sp. bat in the United Kingdom causing a fatal hepatitis and septicemia. The 16S rRNA gene identified the organism as *Borrelia* species that was close to but not identical with *B. recurrentis*, *B. duttonii* or *B. crocidurae*. [394]

Rickettsia and Borrelia: Eptesicus fuscus bats from Georgia were serologically tested for antibodies against Borrelia hermsii, Orientia tsutsugamushi, Rickettsia coronii, and Rickettsia rickettsii. Three of 56 bats tested serologically positive for *Borrelia* sp. and 1 of 56 bats tested serologically positive for spotted fever group Rickettsia. Feeding rickettsia-infected Carios kelleyi bat ticks to pathogen-free guinea pigs resulted in seroconversion in 14 days. The authors concluded that E. fuscus was naturally exposed to both pathogens. If the pathogens were transmitted by bat ticks, then individuals living in proximity to the bats and ticks would be at risk for infection. [26]

Potomac horse fever (aka equine neorickettsiosis/equine monocytic ehrlichiosis/Equine Ehrlichial colitis) (Neorickettsia (formerly Ehrlichia) risticii): Potomac horse fever is acquired by horses after ingesting aquatic insects including caddisflies (Dicosmoecus gilvipes) parasitized with trematode metacercaria infected with Neorickettsia risticii. The disease causes fever, depression,, anorexia, diarrhea, colic, and laminitis. Some disease may not be clinically apparent. Thoroughbreds are 3 times more likely to be seropositive than standard breads. Early treatment has reduced fatality rates from 30% to 7%. [395]

The species of trematode involved and how it is maintained in nature led to the examination of Eptesicus fuscus big brown bats and Myotis lucifugus little brown bats from Pennsylvania. Gravid *Acanthatrium oregonense* were recovered from 12 of 15 *Eptesicus fuscus*, and 8 of 9 *Myotis lucifugus*. Trematodes from 6 *Eptesicus fuscus* contained DNA from *N. risticii. N. risticii* was detected by PCR and immunofluorescence labeling with anti-*N. risticii* antibody indicating vertical transmission of infection in *A. oregonense*. [396]

FUNGI

Histoplasmosis

Histoplasma capsulatum, a dimorphic fungus of the Ascomycetes family, is the etiologic agent of histoplasmosis (also known as Cave disease, Darling's disease, Ohio valley disease, Spelunker's disease). The fungus has a worldwide distribution except Antarctica. [397-399] Bats are believed to be the major natural fungal reservoir. [400] *H. capsulatum* var *capsulatum* has been isolated from urban bats (*Eumops bonariensis*, Buenos Aires). These isolates were genetically close to human isolates from the same city as the bat isolates. [401] A related ascomycete (Chrysosporium state of *Renispora flavissima* (Gymnoascaceae)) that fluoresces with *H. capsulatum* fluorescent antibodies, that does not appear to be pathogenic was also discovered growing in bat guano. [402]

H. capsulatum is endemic in the Ohio and Mississippi river valleys and grows well in temperate moist soil. Disease is acquired by inhalation of infectious elements. Disruption of soil, construction especially where bats reside, agriculture, and similar outdoor activities put individuals at risk. [397]

A study of occupationally acquired histoplasmosis found that 73% of landfill workers and 50% of bridge workers had evidence of infection ($\geq$ 4-fold titer rise in complement fixation or positive urinary *H. capsulatum* antigen). The relative risk was highest for those seeing or having contact with bats (relative risk (RR) = 7.0) as compared to jack-hammering (RR = 4.0). [403] Only brief exposures may be necessary to acquire infection. [404] In the Mexican State of Guerrero, a high proportion of positive histoplasmin responses were seen in cave-tourist guides.[405] The increased risk for acquiring bat-associated histoplasmosis occurs near the entrance to or inside bat-caves.[27, 406, 407] Acute pulmonary histoplasmosis has been associated with inhalation of a contaminated aerosol of water and bat guano from a large colony of fruit bats near a waterfall. [408]

In Japan, histoplasmosis is considered an imported disease. Examination of 187 bat guano samples from 67 bat caves in 17 prefectures found no evidence of *H. capsulatum* by culture or PCR. [409]

The initial route of infection is through the lungs and is often asymptomatic (90%). Clinical forms include symptomatic acute primary infection with large inoculums. There is a 7-21 day incubation period followed by fever, headache, nonproductive cough, chills, and chest pain. Arthralgias, erythema nodosum and erythema multiforme occur in 6% of patients and another 6% develop acute pericarditis. A Ghon complex similar to pulmonary tuberculosis is seen on chest x-rays. Specific antifungal treatment is often unnecessary unless symptoms persist. Acute disease due to reinfection occurs to those exposed to a large inoculum. Onset is within 3 days of exposure, and the disease is commonly milder. [397]

Complications include histoplasmoma (a slowly enlarging mass lesion), mediastinal granuloma and fibrosis due to enlargement of mediastinal lymph nodes, chronic pulmonary histoplasmosis that may cavitate, acute, subacute and chronic disseminated histoplasmosis, and ocular disease. Disseminated disease in HIV patients may have skin and mucosal disease as well as hepatosplenomegaly. [397, 410, 411]

African histoplasmosis (caused by H. capsulatum var. duboisii): This disease typically presents with skin lesions (ulcers, nodules or psoriatic-like plaques that may spontaneously resolve), subcutaneous tender nodules, and osteolytic bone lesions in 50% of patients typically in the skull and ribs. *H. capsulatum* var. *duboisii* may cause progressive disseminated disease including disseminated disease in HIV-infected patients. [397]

Experimental nasal instillation of *H. capsulatum* in *Artibeu lituratus* bats results in pulmonary disease with dissemination to the spleen, liver and intestine within 2 weeks. Complement fixing antibodies are present in 3 weeks, while precipitating antibodies are not detected until 5 weeks. Delayed hypersensitivity is noted in 2-4 weeks. [412, 413]

Bat guano is made up of *H. capsulatum* and mites. Guano was sampled from the La Boca cave of Santiago, Mexico (with a colony of 5 million *Tadarida brasiliensis mexicana* insectivorous bats). *Sancassania* ca. *sphaerogaster* is the most abundant mite species from the guano. It feeds on and successfully reproduces in milieu rich in *H. capsulatum*. It feeds and kills ingested fungus, acting as a fungal control, while at the same time spreading fungus that attaches to its cuticle over short ranges. [414]

Pneumocystis Species in Bats

Pneumocystis jirovecii is a unicellular fungus of low pathogenicity that is acquired at an early age. Although it is believed that in the normal host early infection is asymptomatic, *P. jirovecii* may be demonstrated by molecular techniques in 16-32% of children presenting with respiratory symptoms. In immunocompromised patients (especially AIDS patients), it causes interstitial pneumonia. Other manifestations are disseminated disease, and immune reconstitution inflammatory syndrome which appears as shortness of breath, cough, and pulmonary infiltrates, but needs to be treated with anti- inflammatory agents instead of antimicrobials. [415]

Pneumocystis species are believed to have co-evolved with their specific mammalian hosts for 100 million years. Host specificity is absolute. Each specific mammalian host is parasitized by its own *Pneumocystis* species, although some mammals may be parasitized by more than one host-specific *Pneumocystis* species. [416] Examining *Pneumocystis* sp. from primates, rodents, carnivores, bats, lagomorphs, marsupials, and ungulates, mitochondrial large subunit (mtLSU) and small subunit (mtSSU) rRNA gene could be attributed to each species (co speciation of *P. carinii*), and no species was found to harbor human *Pneumocystis jirovecii* (then referred to as *P. carinii* f. sp. *hominis*). [417]

Pneumocystis species environmental sources have not been identified. What is known is that airborne transmission occurs between both immunocompetent and immunosuppressed hosts. Host specificity is important and attempts to induce infection across species have been unsuccessful. The nonspecific host rapidly eliminates the organisms. [416]

P. carinii airborne transmission between specific hosts, bat colonial cohabitation, and the ability to identify *P. carinii* by mtSSU and mtSLU rRNA gene sequencing, allowed *P. carinii* to be used as a phylogeographic marker. *Tadarida brasiliensis* (Argentina and Mexico), *Pipistrellus pipistrellus* (France), and *Glossophaga soricina* (French Guyana and Mexico) were collected. DNA was extracted from lung tissue and sequenced. Results demonstrated the usefulness of using this fungal parasite as a marker for chorology and migration patterns. [416]

Blastomycosis

Blastomyces dermatitidis is endemic in North America especially along the Mississippi and Ohio River basins, the midwestern states and Canadian provinces that border the Great Lakes, and on the New York and Canadian side of the St. Lawrence River. Less frequently, cases have been reported form Central and South America, India and the Middle East. Occasionally cases have been reported in nonendemic areas. The organism thrives in warm, moist soil, especially acidic soil, in forested areas where there is much debris and decaying vegetation. [418, 419]

Primary infection is by inhalation. The fungus causes acute pneumonia that often goes undiagnosed unless multiple cases are involved. Incubation is 30-45 day and symptoms are nonspecific. Cases resolve spontaneously, symptomatically in 4 weeks, but radiographs take longer to normalize. Chronic or recurrent pneumonia manifests as lobar or segmental alveolar infiltrates with or without cavitation. Occasionally mass lesions, multinodular lesions, perihilar masses or miliary disease are seen. [418]

Dissemination results in subcutaneous nodules, verrucous skin lesions, papulo-pusular rashes, ulcerative lesion, osteolytic bone lesions, and multiple ring-enhancing central nervous system disease. In men, 10-30% of cases involve the prostate and epididymis. Children generally develop pulmonary disease. Disease in pregnancy is rare, but dissemination is common. In the compromised host (AIDS and transplant patients) the infection is often disseminated and aggressive. [418]

Blastomycosis has been reported in wolves, ferrets, deer, polar bears, African lions, Atlantic bottlenose dolphins, and Steller's sea lions. Most cases are granulomatous infections in the lung.

Bats: A pregnant female Indian fruit bat was found dead. The bat was obtained from India at 1 month of age, first housed in South Carolina, then in Indiana with 14 other fruit bats. Autopsy revealed bilateral coalescing pulmonary nodules, microscopically identified as *B. dermatitidis*. [419] *B. dermatitidis* was isolated from 1 of 46 samples of *Rhinopoma hardwickei hardwickei* Gray (the lesser rat-tailed) bat captured on a single day from the basement of the Safdar-Jang Tomb monument in New Delhi. None of 581 other lesser rat-tailed bats, 3 insectivorous or 1 frugivorous bats from several sites in Delhi and New Delhi harbored the fungus. [420] One of 155 rat-tailed bats from an abandoned school had the fungus isolated from its lungs. [421] Orally infected *R. hardwickei* bats transiently shed *B. dermatitidis*. The fungus is found in reduced numbers throughout the gastrointestinal tract with the bat shedding fungus for 16-24 hours. [422]

Paracoccidioides

Paracoccidioides brasiliensis is found from Mexico to Argentina, although some Caribbean islands and Chile are not affected. The primary infection is pulmonary and is usually asymptomatic. When pulmonary disease progresses, patients present with interstitial infiltrates (64%), and mixed lesions with linear and nodular infiltrates and alveolar patterns. Pulmonary lesions may be confluent. Mucosal lesions are infiltrated, ulcerated and painful. One quarter of the patients have skin lesions, especially around orifices and lower limbs (most commonly the head and neck). Lymph node involvement may result in draining lesions. Adrenal glands are frequently involved (85% in autopsy series). Occasionally, nodular lesions of the spleen and liver, CNS masses, ulcers in the gastrointestinal tract, and granulomatous lesions of the male genitourinary tract, vascular system, bone and bone marrow may be present. [423]

Bats: Experimentally (intraperitoneal) infected *Artibeus lituratus* developed delayed hypersensitivity in 2 weeks, no precipitating antibodies up to 7 weeks later and eventually died. Intranasal instillation resulted in primary lung infection with spread to the spleen in 3 weeks and liver in 9 weeks. Antibodies were detected in 5 weeks. Fungal cells could not survive more than 8 hours in the digestive tract. Although there was one report of *P. brasiliensis* being recovered from *A. lituratus* (Columbia, South America), these experiments suggested that bats are not important in disease dissemination, although they may be a risk factor for bat handlers. [424-426]

Coccidioidomycosis

Coccidioides species are dimorphic fungi closely related to other ascomycetes, particularly *Blastomyces dermatitidis* and *Histoplasma capsulatum*. The natural habitat of the organism is limited to the Western Hemisphere to soils in the lower Sonoran life zone where there is little rainfall, hot summers and mild dry winters. Alkaline soils are preferred. Infection is caused by inhalation of airborne arthroconidia, especially during duststorms. The infectious arthroconidia may be accidentally transported in soil on fomites. [427]

More than half of infections are asymptomatic or very mild and never come to medical atention. Symptomatic patients usually become ill 7-21 days after exposure. Patients have cough, chest pain, shortness of breath, fever, malaise, weight loss and headache. A fine nonpruritic papular rash appears transiently. Erythema nodosum or erythema multiforme occasionally is seen, especially in women. Rarely the primary pneumonia after a high inoculum or in immunocompromised patients (especially HIV-infected individuals) progresses to respiratory failure or fulminant disease. [427]

Other complications include pulmonary nodules (4% of patients) or cavities. Chronic fibocavitary pneumonia is seen especially in diabetics or patients with pulmonary fibrosis. Dissemination (0.5% of patients) is not commonly associated with progressive pulmonary disease. Common involvement is seen in bones and joints, spine, skin lesions, and meningitis. [427]

C. immitis has been isolated from bat guano. Infection in humans and dogs occurred during a nine-banded armadillo (*Dasypus novemcinctus*) hunt. *C. immitis* was isolated in and around the armadillo burrow. This represents a potential risk for spelunkers and chiropterists. [28, 428, 429]

Human Dermatophytes in Bat Guano

Other human pathogenic dermatophytes found in bat guano include *Candida* sp. *Cladosporium* sp, *Cryptococcus neoformans*, *Microsporum gypseum*, *Trichophyton mentagrophytes*, *Trichophyton rubrum*, *Trichophyton terrestre*, and *Sporotrichum* sp. [428]

Basidiobolus ranarum

Basidiobolus ranarum is a *Zygomycetes* (order *Entomophthorales*) with a worldwide distribution, although infections are seen in India, Uganda, and other areas of Africa and Asia. [430, 431] It is cultured from soil, decaying vegetation, and the gastrointestinal tracts of amphibians, reptiles, fish and bats.[431] In Delhi, 7% of the intestinal contents of the insectivorous lesser rat-tailed bat (*Rhinopoma hardqickei hardwickei* Gray) contained *B. ranarum* Eidam. [432]

While other fungi in this class cause infection particularly in poorly controlled diabetics and immunosuppressed patients, this organism causes subcutaneous infections through accidental inoculation in normal individuals, especially in children. [431] Gastrointestinal disease and hepatic mass are rare. The disease has mimicked colon cancer. [430, 433] Basidiobolomycosis has presented as a subcutaneous swelling in the nose and face area. [434]

Bat White-Nose Syndrome (BWNS)

Newly introduced exotic pathogens have decimated wildlife populations (amphibians with chytridiomycosis – *Batrachochytrium dendrobatidis*; myxomatosis in rabbits (*Oryctolagus cuniculus*) in the United Kingdom; transmissible neuroendocrine cancer in Tasmanian devils; and West Nile virus in North American birds). Since the winter of 2006-2007, BWNS has caused a 75% decline in bat populations at survey hibernacula. [435, 436]

Affected bats have a white fungal growth on their muzzles, ears, and/or wing membranes. [437, 438] Hyphae infiltration into the wing membranes causes significant tissue destruction. In North America, infected bats suffer from premature arousal from hibernation, and have little or no fat reserves. [435, 439] Bats removed from hibernacula and prepared for shipping often arrive without the typical findings of BWNS. Skin changes are often inconsistent. There may be only patches of rough skin on the face, ears, forearms, wing membranes and feet with pinpoint white foci on the muzzle. In hibernating bats, little inflammation is present even with extensive fungal invasion. Bats collected outside the hibernacula in the spring with damaged wings display severe inflammation associated with the fungal infection. [440] This probably represents a state of immunosuppression during hibernation akin to human immune-suppression with the pathogen establishing infection during the immunosuppressed state. [441]

The disease is present in the northeastern and mid-Atlantic regions of the U.S. and Quebec and Ontario provinces of Canada. It is present in 115 bat hibernacula and has spread >1,300 km from the original site of discovery at Howe Cave near Albany, New York. It now occurs in 6 or 7 hibernating species of bats (big brown bat – *Eptesicus fuscus*, eastern small-footed bat – *Myotis leibii*, little brown bat – *M. lucifugus*, northern long-eared bat – *M. septentrionalis*, tricolored bat – *Perimyhotis subflavus*, and Indiana bat – *M. sodalis*). [435, 436]

BWNS has killed > 1 million bats and destroyed ~100% of some bat populations. [439] Computer modeling estimates that the affected bat populations will decrease by 99% (6.5 million to 65,000 bats) in less than 20 years. [435]

BWNS is caused by the psychrophilic fungus *Geomyces destructans*. [435, 436] There is evidence that the fungus was introduced from Europe. Reports of a similar appearing but non-fatal growth in European hibernating bats from Germany, Switzerland, France, Hungary, and the United Kingdom (21 bats of 5 species: *M. dasycneme*; *M. myotis*; *M. oxygnathus*; *M. brandtii*; and, *Rhinolophus ferrumequinum*) were colonized by *G. destructans*. [436, 439]

Geomyces pannorum: *G. destructans* is closely related to the human pathogen *Geomyces pannorum*, a ubiquitous psychrophilic keratinolytic fungus that has been isolated from soil, fur of wild mammals (France), floors of trains and ferryboats (Italy), boreal forests (Canada), air samples (Pavia), water sediments (India), and Artic environmental samples. Human infection with *G. pannorum* var. *pannorum* involves fingernails and superficial skin of those in close contact with the soil. [436, 442] It has caused recurrent skin infection in patients with ichthyosis. [443] Under experimental conditions, polyester polyurethane (PU) "coupons" buried for 5 months in acidic or neutral sandy loam soil predominantly grew *G. pannorum* along with *Phoma* sp. [444]

Puechmaille et al, and other investigators have put forth 3 theories: (1) *G. destructans* has been newly introduced into Europe and all bats are now at risk for fatal infection; (2) the fungus has been present in Europe for a long time and the European bats are immune to BWNS; and/or, (3) *G. destructans* is an opportunistic infection for bats already immunocompromised by another agent. [436, 439]

Trypanosoma Species

Trypanosoma cruzi is the etiologic agent for American trypanosomiasis. It is transmitted by triatomine insects (kissing bugs – *Rhodnius prolixus, Triatoma infestans* and *T. dimidiata*), transfusion and organ transplantation. [445, 446] Fossil evidence (discovery of *Triatoma dominicana* sp. n. (Hemiptera: Reduviidae: Triatominae)) in Dominican amber with *Trypanosoma antiquus* sp. n. in 2 fecal drops with mammalian hairs suggests that bats were the original vertebrate hosts for both triatomine and trypanosomatids – *T. cruzi*-like organisms. [447] Congenital infections occur in 5% of infants born to infected mothers. Acute disease is usually a disease of children, and most infections are mild. The first sign of infection appears about 1 week after the parasite has entered a break in the skin. The typical lesion of acute disease is the appearence of a chagoma (indurated area of erythema) with local lymph node involvement. Romana sign is painless edema of the palpebrae and periocular tissues when the conjunctiva is the portal of entry. Chronic disease manifests itself decades later most commonly with cardiomyopathy. Other complications include megaesophagus, hypersalivation and salivary gland hypertrophy and megacolon. [445, 446]

T. cruzi infections has been observed in mammalian hosts including Carnivora, Chiroptera, Didelphimorphia, Rodentia and Xernathra. [448, 449] Bats are infected with a variety of Trypanosomes including the subgenus *Schizotrypanum*. [450, 451] Bat Cimicidae (*Cimex* sp.) ectoparasites are vectors for *Trypanosma* spp. Brazilian bats were found to be hemoculture positive for *T. cruzi* (80% of *Phyllostomus hastatus* bats captured in one same buriti hollow palm tree). [448] Other bat families found to be infected (15%) were Molossidae, Noctilionidae, and Vespertilionidae. [446]

MICROSPORIDIOSIS

Microsporidia and eukaryotic obligate intracellular parasites cause opportunistic infections through zoonotic, waterborne, and foodborne transmission. Although previously classified as Protozoa, they are now classified as Fungi. [452, 453] Although there are 150 genera of Microsporidia, only *Enterocytozoon*, *Encephalitozoon*, *Pleistophora*, *Trachipleistophora*, *Vittaforma*, *Brachiola*, and *Nosema* cause disease in man. [453]

Encephalitozoon species are found in Psittaciformes birds (parrots, lovebirds, parakeets), Passeriformes, Columbiformes birds, urban pigeons (*Columba livia*), and common waterfowl and nonaquatic birds. [454-457] In man, it causes keratoconjunctivitis, sinusitis, pneumonia, nephritis, prostatitis, urethritis, cystitis, diarrhea, and disseminated infection. [452]

Disseminated microsporidiosis due to *Encephalitozoon hellem* resulted in the unexpected death of an adult female Egyptian fruit bat (*Rousettus aegyptiacus*) held in captivity. At autopsy, the kidneys were enlarged and the liver was mottled. Histopathology revealed intracytoplasmic spores with attendant inflammation. The free-ranging European brown hare (*Lepus europaeus*) when infected with *E. intestinalis* and *E. hellem* also develops kidney lesions (multifocal wedge-shaped chronic interstitial nephritis). [458, 459]

ANGIOSTRONGYLOSIS

Angiostrongylus cantonensis, the rat lung worm causes eosinophilic meningitis and encephalitis. Adult worms establish themselves in the rat lung. Eggs hatch in the rat lung, and larvae are swallowed, and defecated and infect its intermediate mollusk host (slugs, land snails and land planarian), freshwater prawns, land and coconut crabs, and frogs. [460]

An Australian study of 86 flying foxes (fruit bats) with neurologic disease found 16 with histologic evidence of CNS involvement with angiostrongylosis. In 10 of the cases, the worms were definitively identified as *A. cantonensis* 5th stage larvae. In 3 other bats a worm fragment and 3rd stage larvae were identified as *Angiostrongylus* sp. The flying foxes exhibited paresis, especially of the hind limbs, and depression. Brains with 5th stage larvae exhibited moderate to severe eosinophilic and granulomatous meningoencephalitis. No inflammation was seen in bats that died with 3rd stage larvae. There was no pulmonary involvement. The investigators noted that the neurologic disease cannot be differentiated from Australian bat lyssavirus infection and recommended that no attempt should be made to rehabilitate flying foxes with this symptom complex or other neurologic symptoms. [461]

BED BUGS AND BATS

Cimex lectularius (the common bed bug) infestations and bites have commonly been reported at homeless shelters, refugee camps, military barricks, ships, single family homes, hotels, and hostels. The insect family cimicidae contains 91 species of obligate hematophagous ectoparasites that feed on birds, bats (*C. pilosellus, C. adjunctus, C. hemipterus (C. rotundatus), C. pilosellus, C. pipistrella*) and terrestrial mammals.[462]· Cimicid subfamilies Primicimicinae and Latrocimicinae parasitize New World bats;

Cacodminae and Afrocimicinae are restricted to Old World bats; and, Haematosiphoninae are restricted to selected orders of New World birds. [463] *C. pipistrelli* has better survival than *C. lectularius* at higher temperatures, which may be indicative of long-term adaptation to bats preferring crevice-like roosts or attics which often become very warm. [464]

Evidence for bed bugs being vectors of disease is "equivocal". Half of these insects fed hepatitis B virus-infected blood contained the virus when tested 2 weeks later. [462]

Patients may be asymptomatic with only a small puncture. Reactions develop in 30% of patients (based on volunteer studies under controlled conditions). These include 2-5 mm pruritic maculopapular, erythematous lesions, one at each feeding site, and increasing size and pruritis at bite site for individuals experiencing multiple bites. Less frequent reactions include pruritic wheals (local urticaria) around a central punctum, papular urticaria, and diffuse urticaria at bite sites. Subsequent bites may illicit a bullous rash, and some rashes evolve into pruritic papules or nodules. Impetigo has occurred with scratching. Rare systemic reactions include asthma, generalized urticaria, anaphylaxis, and possibly erythema multiforme. [462]

Species identification is important as some bat-infesting insects may be the cause. *C. pilosellus* develop in bat roosts which is especially a problem when a colony of bats move into an attic or behind a wall. Bat bugs may move into human living areas and bite people when bats migrate or are excluded from a dwelling. But without the bat, they can't reproduce. [462].

Crevice-dwelling bats (*Pipistrellus pygmaeus*) engaged in roost-switching. Allogrooming was infrequent among infested bats. When bugs were added to the roost, bats left the infested roost and the resulting roost-switching resulted in lower parasite burdens. [465] Roost-switching behavior has also been observed in Bechstein's bats (*Myotis bechsteinii*) in order to decrease their contact with infective stages of the bat fly (*Basilia nana*). [463]

Allergic Disease

Asthmatic patients in the Sudan tend to associate their symptoms with exposure to bat guano. Six patients with asthma and allergic rhinitis, and 1 patient with asthma alone were tested using extracts from yellow hairy bats, black bats, and bat guano. All patients had positive skin prick tests and specific IgE antibodies (RAST) to bat guano. Three patients had positive RAST to yellow hairy bats and black bats, and 1 patient was positive only for the yellow hairy bat. [466]

Amyotrophic Lateral Sclerosis/Parkinsonism Dementia Complex of Guam

In 1945, H. Zimmerman noted a high prevalence of motor neuron disease among the Chamorro people of Guam. Similar disease was found in western New Guinea, the Kii peninsula of Japan. Approximately 5% exhibited predominantly amyotrophic lateral sclerosis (ALS) and 38% had this in combination with Parkinsonism and dementia (ALS/PDC syndrome). [467, 468] Cyad seed in the diet in the form of flour, and consumption of the cytotoxin β-methylamino-L-alanine (BMAA), and consumption of flying foxes (*Pteropus* spp.) that eat cycad seeds and concentrate BMAA in their muscle may be the environmental factor responsible for ALS/PDC syndrome. [467, 469-472]· Others have cast doubt on this theory. Although ALS/PDC syndrome is in decline, as is consumption of cyad flour and flying foxes, the disease is still present. [468]

> "There's comfort yet; they are assailable;
> Then be thou jocund: ere the bat hath flown
> His cloister'd flight, ere to black Hecate's summons
> The shard-borne beetle with his drowsy hums
> Hath rung night's yawning peal, there shall be done
> A deed of dreadful note."
> [William Shakespeare: Macbeth, Act III, Scene 2]

Disclaimer

The authors have indicated there is no funding in support of this document. Additionally, there are no other relationships that might pose a conflict of interest by any of the authors. None of the authors has any professional or financial relationships relevant to the subject matter in this paper. This manuscript and any tables or figures therein, have not been submitted to or are under consideration by any other publisher or publication.

Support

There are no sources of support that require acknowledgment.
This manuscript was prepared by Microsoft Word on a PC platform.

References

Introduction

[1] Whitman J. Bats – Wildlife Notebook Series: Alaska Department of Fish and Game. Available at: *www.adfg.state.ak.us/pubs/notebook/smgame/bats.pkkhp*. (Downloaded July 20, 2010).

[2] Castrodale L. Bats in Alaska. State of Alaska, Epidemiology Bulletin 16November 7, 2001. Available at: *www.state.ak.us/local/akpages/FISH.GAME/notebook/smgame/bats.htm*. (Downloaded July 20, 2010).

[3] Rydell J, Strann K-B, Speakman JR. First record of breeding bats above the Arctic Circle: northern bats at 68-70 "N in Norway. Communications from the Mammal Society. -No 68, p. 335, 1994. *www.abdn.ac.uk/ibes/speakman/pdf_docs/54.pdf,* downloaded June 20, 2010.

[4] Wong S, Lau S, Woo P, Yuen KY. Bats as a continuing source of emerging infections in humans. *Rev Med Virol*. 2007;17:67-91.

[5] Munshi-South J, Wilkinson GS. Bats and birds: exceptional longevity despite high metabolic rates. *Ageing Res Rev*. 2010;9:12-19.

[6] Brunet-Rossinni AK, Austad SN. Ageing studies on bats: a review. *Biogerontology*. 2004;5:211-222.

[7] Salmon AB, Leonard S, Masamsetti V, et al. The long lifespan of two bat species is correlated with resistance to protein oxidation and enhanced protein homeostasis. *FASEB J*. 2009;23:2317-2326.

[8] Turmelle AS, Olival KJ. Correlates of viral richness in bats (order chiroptera). *Ecohealth*. 2009;6:522-539.

[9] Constantine DG. Geographic translocation of bats: known and potential problems. *Emerg Infect Dis*. 2003;9:17-21.

[10] Breed AC, Field HE, Smith CS, et al. Bats without borders: long-distance movements and implications for disease risk management. *Ecohealth*. 2010 July 20. [Epub ahead of print].

[11] Fortuna MA, Popa-Lisseanu AG, Ibanez C, Bascompte J. The roosting spatial network of a bird-predator bat. *Ecology*. 2009;90:934-944.

[12] Dimitrov DT, King AA. Modeling evolution and persistence of neurological viral diseases in wild population. *Math Biosci Eng*. 2008;5:729-741.

[13] Baker ML, Tachedjian M, Wang LF. Immunoglobulin heavy chain diversity in Pteropid bats: evidence for a diverse and highly specific antigen binding repertoire. *Immunogenetics*. 2010;62:173-184.

[14] Calisher CH, Childs JE, Field HE, et al. Bats: important reservoir hosts of emerging viruses. *Clin Microbiol Rev*. 2006;19:531-545.

[15] Halpin K, Hyatt AD, Plowright RK, et al. Emerging viruses: coming in on a wrinkled wing and a prayer. *Clin Infect Dis*. 2007;44:711-717.

[16] Bausch DG, Nichol ST, Muyembe-Tamfum JJ, et al. Marburg Hemorrhagic Fever associated with multiple genetic lineages of virus. *N Engl J Med*. 2006;355:909-919.

[17] Chua KB, Crameri G, Hyatt A, et al. A previously unknown reovirus of bat origin is associated with an acute respiratory disease in humans. *Proc Natl Acad Sci U S A*. 2007;104:11424-11429.

[18] Reeves WK, Loftis AD, Gore JA, Dasch GA. Molecular evidence for novel *Bartonella* species in *Trichobius major* (Diptera: Streblidae) and *Cimex adjunctus* (Hemiptera: Cimicidae) from two southeastern bat caves, U.S.A. *J Vector Ecol*. 2005;30:339-341.

[19] Vayssier-Taussat M, Le Rhun D, Bonnet S, Cotte V. Insights in *Bartonella* specificity. *Ann N Y Acad Sci*. 2009;1166:127-132.

[20] De Lima H, Rodriguez N, Barrios MA, et al. Isolation and molecular identification of *Leishmania chagasi* from a bat (*Carollia perspicillata*) in northeastern Venezuela. *Mem Inst Oswaldo Cruz*. 2008;103:412-414.

[21] Evans NJ, Bown K, Timofte D, et al. Fatal borreliosis in bat caused by relapsing fever spirochete, United Kingdom. *Emerg Infect Dis*. 2009;15:1331-1333.

[22] Nascentes GA, Meira WS, Lages-silva E, Ramirez LE. Absence of experimental cross-protection induced by a *Trypanosoma cruzi*-like strain isolated from bats. *Rev Soc Bras Med Trop*. 2008;41:152-157.

[23] Childs-Sanford SE, Kollias GV, Abou-Madi N, et al. *Yersinia pseudotuberculosis* in a closed colony of Egyptian fruit bats (*Rousettus aegyptiacus). J Zoo Wildl Med.* 2009;40:8-14.

[24] Monahan AM, Miller IS, Nally JE. Leptospirosis: risks during recreational activities. *J Appl Microbiol*. 2009;107:717-716.

[25] Muhldorfer K, Wibbelt G, Haensel J, et al. *Yersinia* species isolated from bats, Germany. *Emerg Infect Dis*. 2010;16:578-580.

[26] Reeves WK, Streicker DG, Loftis AD, Dasch GA. Serologic survey of *Eptesicus fuscus* from Georgia, U.S.A. for *Rickettsia* and *Borrelia* and laboratory transmission of a *Rickettsia* by bat ticks. *J Vector Ecol*. 2006;31:386-389.

[27] Julg B, Elias J, Zahn A, et al. Bat-associated histoplasmosis can be transmitted at entrances of bat caves and not only inside the caves. *J Travel Med*. 2008;15:133-136.

[28] Krutzsch PH, Watson RH. Isolation of *Coccidioides immitis* from bat guano and preliminary findings on laboratory infectivity of bats with *Coccidioides imm*itis. *Life Sci*. 1978;22:679-684.

[29] Randhawa HS, Chaturvedi VP, Kini S, Khan ZU. *Blastomyces dermatitidis* in bats: first report of its isolation from the liver of *Rhinopoma hardwickei hardwickei* Gray. *Sabouraudia*. 1985;23:69-76.

[30] Sugita T, Kikuchi K, Makimura K, et al. *Trichosporon* species isolated from guano samples obtained from bat-inhabited caves in Japan. *Appl Environ Microbiol*. 2005;71:7626-7629.

[31] Banack SA, Cox PA. Biomagnification of cycad neurotoxins in flying foxes – Implications for ALS-PDC in Guam. *Neurology*. 2003;61-387-389.

[32] Armon C. Western Pacific ALS/PDC and flying foxes – What's next?. *Neurology*. 2003; 291-292.

Rabies and Lyssaviruses

[33] Cleri DJ, Hamid NS, Panesar M, et al. Rabies (Part I and Part II). *Infectious Disease Practice for Clinicians*. 2004;28 (May/June):303-310 and (July):315-318.

[34] Bourhy H, Reynes JM, Dunham EJ, et al. The origin and phylogeography of dog rabies virus. *J Gen Virol*. 2008 89;2673-2681.

[35] Talbi C, Holmes EC, de Benedictis P, et al. Evolutionary history of dynamics of dog rabies virus in western and central Africa. *J Gen Virol.* 2009;90:783-791.

[36] Staeheli P, Rinder M, Kaspers B. Avian bornavirus associated with fatal disease in psittacine birds. *J Virol*. 2010;84:6269-6275.

[37] Ludwig H, Bode L. Borna disease virus: new aspects on infection, disease, diagnosis and epidemiology. *Rev Sci Tech*. 2000;19:259-288.

[38] Kamhieh S, flower RL. Borna disease virus (BDV) infection in cats. A concise review based on current knowledge. *Vet Q*. 2006;28:66-73.

[39] Solbrig MV. Animal models of CNS viral disease: examples from borna disease viral moels. *Interdiscip Perspect Infect Dis*. 2010;2010:709791.

[40] Richt JA, Pfeuffer I, Christ M, et al. Borna disease virus infection in animals and humans. *Emerg Infect Dis*. 1997;3:343-352.

[41] Ma Y, Wang ZH, Kong FY, Zhang Y. Research on the association between Borna disease virus infection and the viral encephalitis. *Zhonghua Liu Xing Bing Xue Za Zhi*. 2009;30:1284-1287.

[42] Na KS, Tae SH, Song JW, Kim YK. Failure to detect borna disease virus antibody and RNA from peripheral blood mononuclear cells of psychiatric patients. *Psychiatry Investig*. 2009;6:306-312.

[43] Bassin SL, Rupprecht CE, Bleck TP. Rhabdoviruses. In: Mandell GL, Bennett JE, Dolin, editors. *Principles and practice of infectious diseases*, Volume 2, 7th edition. Churchill Livingston Elsevier, Philadelphia. p.2249-2258, 2009.

[44] Chadha MS, Arankalle VA, Jadi RS, et al. An outbreak of Chandipura virus encephalitis in the eastern districts of Gujarat state, India. *Am J Trop Med Hyg*. 2005;73:566-570.

[45] Hoffmann B, Beer M, Schutze H, Mettenleiter TC. Fish rhabdoviruses: molecular epidemiology and evolution. *Curr Top Microbiol Immunol*. 2005;292:81-117.

[46] Dietzgen RG, Callaghan B, Wetzel T, Dale JL. Completion of the genome sequence of lettuce necrotic yellows virus, type species of the genus Cytorhabdovirus. *Virus Res*. 2006;118:16-22.

[47] Bandyopadhyay A, Kopperud K, Anderson G, et al. An integrated protein localization and interaction map for potato yellow dwarf virus, type species of the genus Nucleorhabdovirus. *Virology*. 2010;402:61-71.

[48] Kondo H, Maeda T, Shirako Y, Tamada T. Orchid fleck virus is a rhabdovirus with an unusual bipartite genome. *J Gen Virol*. 2006;87:2413-2421.

[49] Tao JJ, Zhou GZ, Gui JF, Zhang QY. Genomic sequence of mandarin fish rhabdovirus with an unusual small non-transcriptional ORF. *Virus Res*. 2008;132:86-96.

[50] Duesund H, Nylund S, Watanabe K, et al. Characterization of a VHS virus genotype III isolated from rainbow trout (*Oncorhychus mykiss*) at a marine site on the west coast of Norway. *Virol J*. 2010;7:19

[51] Ammayappan A, LaPatra SE, Vakharia VN. Molecular characterization of the virulent infectious hematopoietic necrosis virus (IHNV) strain 220-90. *Virol J*. 2010;7:10.

[52] Kuzmin IV, Niezgoda M, Franka R, et al. Lagos bat virus in Kenya. *J Clin Microbiol*. 2008;46:1451-1461

[53] Johnson N, Fooks A, McColl K. Human rabies case with long incubation, Australia [letter]. *Emerg Infect Dis*. [serial on the Internet]. 2008 Dec [August 3, 2010]. Available from *http://www.cdc.gov/EID/content/14/12/1950.htm* .

[54] Centers for Disease Control. Imported human rabies – Australia, 1987. *MMWR Morb Mortal Wkly Rep*. 1988;37:351-353.

[55] Kuzmin IV, Hughes GJ, Rupprecht CE. Phylogenetic relationships of seven previously unclassified viruses within the family Rhabdoviridae using partial nucloprotein gene sequences. *J Gen Virol*. 2006;87:2323-2331.

[56] Shope RE. Rabies-related viruses. *Yale J Biol Med*. 1982;55:271-275.

[57] Quan PL, Junglen S, Tashmukhamedova A, et al. Moussa virus: a new membeer of the Rhabdoviridae family isolated from *Culex decens* mosquitoes in Cote d'Ivoire. *Virus Res*. 2010;147:17-24.

[58] Bourhy H, Cowley JA, Larrous F, et al. Phylogenetic relationships among rhabdoviruses inferred using the L polymerase gene. *J Gen Virol*. 2005;86:2849-2858.

[59] Trumelle A, Jackson F, Green D, et al. Host immunity to repeated virus infection in big brown bats. *J Gen Virol*. 2010 June 2. [Epub ahead of print].

[60] Centers for Disease Control and Prevention (CDC). Wildlife reservoirs for rabies. *www.cdc.gov/rabies/exposure/animals/wildlife_reservoirs.html.* (Downloaded August 4, 2010).

[61] Sodre MM, da Gama AR, de Almeida MF. Updated list of bat species positive for rabies in Brazil. *Rev Inst Med Trop Sao Paulo*. 2010;52:75-81.

[62] Wilkerson JA, III. Rabies. In: Auerbach PS, ed. *Wilderness medicine, 5th edition*. Mosby, Elsevier, Philadelphia. 2007.

[63] Centers for Disease Control and Prevention (CDC). Human rabies. *www.cdc.gov/rabies/location/usa/surveillance/human_rabies.html.* (Downloaded August 4, 2010).

[64] Centers for Disease Control and Prevention (CDC). Presumptive abortive human rabies – Texas, 2009. *MMWR Morb Mortal Wkly Rep.* 2010;59:185-190.

[65] Blanton JD, Robertson K, Palmer D, Rupprecht CE. Public veterinary medicine: public health – Rabies surveillance in the United States during 2008. *JAVMA*. 2009;235:676-689.

[66] Centers for Disease Control and Prevention (CDC). Notice to readers: final 2009 reports of nationally notifiable infectious diseases. *MMWR Morb Mortal Wkly Rep.* 2010;59:1025-1027-1029

[67] Centers for Disease Control and Prevention (CDC). Imported human rabies --- California, 2008. *MMWR Morb Mortal Wkly Rep*. 2009;58:713-716.

[68] Jackson FR, Turmelle AS, Farino DM, et al. Experimental rabies virus infection of big brown bats (*Eptesicus fuscus*). *J Wildl Dis*. 2008;44:612-621.

[69] Messenger SL, Smith JS, Orciari LA, et al. Emerging pattern of rabies deaths and increased viral infectivity. *Emerg Infect Dis*. 2003;9:151-154.

[70] Bourhy H, Kissi B, Lafon M, et al. Antigenic and molecular characterization of bat rabies virus in Europe. *J Clin Microbiol*. 1992;30:2419-2426.

[71] Roy A, Phares TW, Koprowski H, Hooper DC. Failure to open the blood-brain barrier and deliver immune effectors to central nervous system tissues leads to the lethal outcome of silver-haired bat rabies virus infection. *J Virol*. 2007;81:1110-1118.

[72] Helesic J, Bartonicka T, Krbkova L. Bat rabies in Europe and the Czech Republic. *Klin Mikrobiol Infekc Lek*. 2007;13:93-98.

[73] Centers for Disease Control and Prevention (CDC). Human rabies --- Alberta, Canada, 2007. *MMWR Morb Mortal Wkly Rep*. 2008;57:197-200.

[74] Velasco-Villa A, Orciari LA, Juarez-Juarez-Islas V, et al. Molecular diversity of rabies viruses associated with bats in Mexico and other countries of the Americas. *J Clin Microbiol*. 2006;44:1697-1710.

[75] Lucas CH, Pino FV, Baer G, et al. Rabies control in Mexico. *Dev Biol* (Basel). 2008;131:167-175.

[76] Kobayashi Y, Sato G, Mochizuki N, et al. Molecular and geographic analyses of vampire bat-transmitted cattle rabies in central Brazil. *BMC Vet Res*. 2008;4:44.

[77] Rabies Watch. Rabies – Mexico: Jalisco. Tuesday, June 9, 2009. *rabieswatch.blogspot.com/2009/06/rabies-mexico-jalisco.html.* (Downloaded August 4, 2010).

[78] Aguilar-Setien A, Loza-Rubio E, Salas-Rojas M, et al. Salivary excretion of rabies virus by healthy vampire bats. *Epidemiol Infect*. 2005;133:517-22.

[79] Pan American Health Organization. Elimination of Dog-Transmitted Human Rabies in Latin America: Situation Analysis. ©Organización Panamericana de la Salud, 2005. *http://www.who.int/rabies/rabnet/en/.* (Downloaded August 5, 2010).

[80] Belotto A, Vargas R, Schneider MC, et al. Regional program for the elimination of rabies transmitted by dogs in the Americas. Presented at the 28th World Congress of the World Small Animal Veterinary Association, October 24-27, 2003. *www.vin.com/ proceedings/Proceedings.plx CID=WSAVA2003andCategory=andPID=6491andO=Generic*. (Downloaded August 5, 2010).

[81] Pan American Health Organization. Boletin: Vigilancia epidemiologica de la rabia en las Americas -2004, vol XXXV, 2004. *http://bvs1.panaftosa.org.br/local/file/textoc/bolvera2004.pdf.* (Downloaded August 5, 2010).

[82] Schneider MC, Romijn PC, Uieda W, et al. Rabies transmitted by vampire bats to humans: an emerging zoonotic disease in Latin America?. *Rev Panam Salud Publica.* 2009;25:260-269.

[83] Mendes Wda S, Silva AA, Neiva RF, et al. An outbreak of bat-transmitted human rabies in a village in the Brazilian Amazon. *Rev Saude Publica*. 2009;43:1075-1077.

[84] Mayen F. Haematophagous bats in Brazil, their role in rabies transmission, impact on public health, livestock industry and alternatives to an indiscriminate reduction of bat population. *J Vet Med B Infect Dis Vet Public Health*. 2003;50:469-472.

[85] Salmon-Mulanovich G, Vasquez A, Albujar C, et al. Human rabies and rabies in vampire and nonvampire bat species, southeastern Peru, 2007. *Emerg Infect Dis*. 2009;15:1308-1310.

[86] Lord RD. Seasonal reproduction of vampire bats and its relation to seasonality of bovine rabies. *J Wildl Dis*. 1992;28:2892-294.

[87] Cunha EM, Silva LH, Lara Mdo C, et al. Bat rabies in the north-northwestern regions of the state of Sao Paulo, Brazil: 1997-2002. *Rev Saude Publica.* 2006;40:1082-1086.

[88] Carneiro NF, Caldeira AP, Antunes LA, et al. Rabies in *Artibeus lituratus* bats in Montes Claros, State of Minas Gerais. *Rev Soc Bras Med Trop*. 2009;42:449-451.

[89] Queiroz LH, de Carvalho C, Buso DS, et al. Epidemiological profile of rabies in the northwestern region of Sao Paulo State, from 1993-2007. *Rev Soc Bras Med Trop.* 2009;42:9-14.

[90] de Mattos CA, Favi M, Yung V, et al. Bat rabies in urban centers in Chile. *J Wildl Dis.* 2000;36:231-240.

[91] Favi M, de Mattos CA, Yung V, et al. First case of human rabies in Chile caused by an insectivorous bat virus variant. *Emerg Infect Dis*. 2002;8:79-81.

[92] World Health Organization. Rabies-Bulletin-Europe. The Journal. 2009;33:October-December 2009. *www.who-rabies-bulletin.org.* (Downloaded August 9, 2010).

[93] Bourhy H, Dacheux L, Strady C, Mailles A. Rabies in Europe in 2005. Euro Surveill. 2005;10(11):pii=575. Available online: *http://www.eurosurveillance.org/ViewArticle.aspx?ArticleId=575.* (Downloaded August 9, 2010).

[94] Davis PL, Holmes EC, Larrous F, et al. Phylogeography, population dynamics, and molecular evolution of European bat lyssaviruses. *J Virol.* 2005;79:10487-10497.

[95] Hughes GJ. A reassessment of the emergence time of European bat lyssavirus type 1. *Infect Genet Evol*. 2008;8:820-824.

[96] Franka R, Johnson N, Muller T, et al. Susceptibility of North American big brown bats (*Eptesicus fuscus*) to infection with European bat lyssavirus type 1. *J Gen Virol*. 2008;89:1998-2010.

[97] Stantic-Pavlinic M. Public health concerns in bat rabies across Europe. Euro Surveill. 2005;10(11):pii=576. Available online:*http://www.eurosurveillance.org/ViewArticle.aspx?ArticleId=576*. (Downloaded August 9, 2010).

[98] Bayard AC, Hartley M, Fooks AR. Reassessing the risk of rabies: a continuing threat to the UK?. *Virus Res*. 2010;152:79-84.

[99] Freuling C, Grossmann E, Conraths FJ, et al. First isolation of EBLV-2 in Germany. *Vet Microbiol*. 2008;131:26-34.

[100] Jakava-Viljanen M, Lilley T, Kyheroinen EM, Huovilainen A. First encounter of European bat lyssavirus type 2 (EBLV-2) in a bat in Finland. *Epidemiol Infect*. 2010;3:1-5.

[101] Johnson N, Selden D, Parsons G, et al. Isolation of European bat lyssavirus type 2 from a Daubenton's bat in the United Kingdom. *Vet Rec*. 2003;152:383-387

[102] Johnson N, Vos A, Neubert L, et al. Experimental study of European bat lyssavirus type 2 infection in Daubenton's bats (*Myotis daubentonii*). *J Gen Virol*. 2008;89:2662-2672.

[103] Wright E, Temperton NJ, Marston DA, et al. Investigating antibody neutralization of lyssaviruses using lentiviral pseudotypes: a cross-species comparison. *J Gen Virol*. 2008;89:2204-2213.

[104] Dodet B, Goswami A, Gunaseker A, et al. Rabies awareness in eight Asian countries. *Vaccine*. 2008;26:6344-6348.

[105] Knobel DL, Cleaveland S, Coleman PG, et al. *Bull World Health Organ*. 2005;83;360-368.

[106] Hampson K, Dobson a, Kaare M, et al. Rabies exposures, post-exposure prophylaxis and deaths in a region of endemic canine rabies. *PLoS Negl Trop Dis*. 2008;2:e339.

[107] Si H, Guo ZM, Hao YT, et al. Rabies trend in China (1990-2007) and post-exposure prophylaxis in the Guangdong province. *BMC Infect Dis*. 2008;8:113.

[108] Chatterjee P. India's ongoing war against rabies. *Bull World Health Organ*. 2009;87:890-891.

[109] Ichhpujani RL, Mala C, Veena M, et al. Epidemiology of animal bites and rabies cases in India. A multicentric study. *J Commun Dis*. 2008;40:27-36.

[110] Singh MP, Goyal K, Majumdar M, Ratho RK. Prevalence of rabies antibodies in street and household dogs in Chandigarh, India. *Trop Anim Health Prod*. 2010Jul 31. [Epub ahead of print].

[111] Wu X, Hu R, Zhang Y, et al. Reemerging rabies and lack of systemic surveillance in People's Republic of China. *Emerg Infect Dis*. 2009;15:1159-1164.

[112] Jiang Y, Wang L, Lu Z, et al. Seroprevalence of rabies virus antibodies in bats from southern China. *Vector Borne Zoonotic Dis*. 2010;10:177-181.

[113] Kuzmin IV, Botvinkin AD, Poleschuk EM, et al. Bat rabies surveillance in the former *Soviet Union. Dev Biol* (Basel). 2006;125:273-282.

[114] Kuzmin IV, Orciari LA, Arai YT, et al. Bat lyssaviruses (Aravan and Khujand) from central Asia: phylogenetic relationships according to N, P and G gene sequences. *Virus Res*. 2003; 97:65-79.

[115] Arai YT, Kuzmin IV, Kameoka Y, Botvinkin AD. New lyssavirus genotype from the lesser mouse-eared bat (*Myotis blythii*), Kyrgyzstan. *Emerg Infect Dis*. 2003;9:333-337.

[116] Kuzmin IV, Niezgoda M, Carroll DS, et al. Lyssavirus surveillance in bats, Bangladesh. *Emerg Infect Dis*. 2006;12:486-488.

[117] Reynes JM, Molia S, Audry L, et al. Serologic evidence of lyssavirus infection in bats, Cambodia. *Emerg Infect Dis*. 2004;10:2231-2234.

[118] Hanlon CA, Kuzmin IV, Blanton JD, et al. Efficacy of rabies biologics against new lyssaviruses from Eurasia. *Virus Res*. 2005;111:44-54.

[119] Lumiertdacha B, Boongird K, Wanghongsa S, et al. Survey for bat lyssaviruses, Thailand. *Emerg Infect Dis*. 2005;11:232-236.

[120] Kuzmin IV, Franka R, Rupprecht CE. Experimental infection of big brown bats (*Eptesicus fuscus*) with West Caucasian bat virus (WCBV). *Dev Biol* (Basel). 2008;131:327-337.

[121] Hughes GJ, Kuzmin IV, Schmitz A, et al. Experimental infection of big brown bats (*Eptesicus fuscus*) with Eurasian bat lyssaviruses Aravan, Khujand, and Irkut virus. *Arch Virol*. 2006;151:2021-2035.

[122] Kuzmin IV, Niezgoda M, Franka R, et al. Possible emergence of West Caucasian bat virus in Africa. *Emerg Infect Dis*. 2008; 14:1887-1889.

[123] Australian Quarantine and Inspection Service. Rabies. Australian Government. Last updated: 13 April, 2007. *www.daff.gov.au/aqis/quarantine/pests-diseases/livestock/rabies*. (Downloaded August 10, 2010).

[124] Fraser GC, Hooper PT, Lunt RA, et al. Encephalitis caused by a lyssavirus in fruit bats in Australia. *Emerg Infect Dis*. 1996;2:327-331.

[125] Centers for Disease Control and Prevention (CDC). Imported human rabies – Australia, 1987. *MMWR Morb Mortal Wkly Rep*. 1988;37:351-353.

[126] Johnson N, Fooks A, McCollK. Human rabies case with long incubation, Australia. *Emerg Infect Dis*. 2008;14:1950-1951.

[127] Grattan-Smith PJ, O'Regan WJ, Ellis PS, et al. Rabies. A second Australian case, with a long incubation period. *Med J Aust*. 1992;156:651-654.

[128] Communicable Disease Control, Public Health Branch, Rural and Regional Health and Aged Care Services Division of the Victorian State Government. Rabies and Australian bat lyssavirus. Victoria State Government, Department of Health, Australia. Last updated: 29 December, 2008. *www.health.vic.gov.au/ideas/bluebook/rabies*. (Downloaded August 10, 2010).

[129] Mackenzie JS, Chua KB, Daniels PW, et al. Emerging viral diseases of Southeast Asia and the Western Pacific. *Emerg Infect Dis*. 2001;7:497-504

[130] Hanna JN, Carney IK, Smith GA, et al. Australian bat lyssavirus infection: a second human case, with a long incubation period. *Med J Aust*. 2000;172:597-599.

[131] McCall BJ, Epstein JH, Neill As, et al. Potential exposure to Australian bat lyssavirus, Queensland, 1996-1999. *Emerg Infect Dis*. 2000;6:259-264.

[132] Field H, McCall B, Barrett J. Australian bat lyssavirus infection in a captive juvenile black flying fox. *Emerg Infect Dis*. 1999;5:438-440.

[133] Dendle C, Looke D. Review article: animal bites: an update for management with focus on infections. *Emerg Med Australas*. 2008;20:458-467.

[134] Torvaldsen S, Watson T. Rabies prophylaxis in western Australia: the impact of Australian bat lyssavirus. *Commun Dis Intell.* 1998;22:149-152.

[135] Gummow B, Roefs YA, de Klerk G. Rabies in south Africa between 1993 and 2005--what has been achieved? *J S Afr Vet Assoc.* 2010;81:16-21.

[136] Dzikwi AA, Kuzmin II, Umoh JU, et al. Evidence of Lagos bat virus circulation among Nigerian fruit bats. *J Wildl Dis.* 2010;46:267-271.

[137] Hayman DT, Fooks AR, Horton D, et al. Antibodies against Lagos bat virus in megachiroptera from West Africa. *Emerg Infect Dis*. 2008;14:926-928.

[138] Butenko AM. Arbovirus circulation in the Republic of Guinea. *Med Parazitol* (Mosk). 1996;(2):40-45.

[139] Konstantinov OK, Diallo SM, Inapogi AP, et al. The mammals of Guinea as reservoirs and carriers of arboviruses. *Med Parazitol* (Mosk). 2006;(1):34-9

[140] Sabeta CT, Markotter W, Mohale DK, et al. Mokola virus in domestic mammals, South Africa. *Emerg Infect Dis*. 2007;13:1371-1373.

[141] Nel LH, Rupprecht CE. Emergence of lyssaviruses in the Old World: the case of Africa. *Curr Top Microbiol Immunol*. 2007;315:161-193.

[142] Sabeta C, Blumberg L, Miyen J, et al. Mokola virus involved in a human contact (South Africa). *FEMS Immunol Med Microbiol*. 2010;58:85-90.

[143] Bingham J, Javangwe S, Sabeta CT, et al. Report of isolations of unusual lyssaviruses (rabies and Mokola virus) identified retrospectively from Zimbabwe. *J S Afr Vet Assoc*. 2001;72:92-94.

[144] Botvinkin AD, Poleschuk EM, Kuzmin IV, et al. Novel lyssaviruses isolated from bats in Russia. *Emerg Infect Dis*. 2003;9:1623-1625.

[145] Kuzmin IV, Orciari LA, Arai YT, et al. Bat lyssaviruses (Aravan and Khujand) from Central Asia: phylogenetic relationships according to N, P, and G gene sequences. *Virus Res*. 2003;97:65-79..

[146] Kuzmin IV, Mayer AE, Niezgoda M, et al. Shimoni bat virus a new representative of the lyssavirus genus. *Virus Res*. 2010;149:197-210..

[147] Vaidya SA, Manning SE, Dhankhar P, et al. Estimating the risk of rabies transmission to humans in the U.S.: a Delphi analysis. *BMC Public Health.* 2010;10:278.

[148] Gibbons RV. Cryptogenic rabies, bats, and the question of aerosol transmission. *Ann Emerg Med.* 2002;39:528-536,

[149] Johnson N, Phillpotts R, Fooks AR. Airborne transmission of lyssaviruses. *J Med Microbiol.* 2006;55:785-790.

[150] Srinivasan A, Burton EC, Kuehnert MJ, et al. Transmission of rabies virus from an organ donor to four transplant recipients. *N Engl J Med*. 2005;352;1103-1111.

[151] Kusne S, Smilack J. Transmission of rabies virus from an organ donor to four transplant recipients. *Liver Transpl*. 2005;1295-1297.

[152] Maier T, Schwarting A, Mauer D, et al. Management and outcomes after multiple corneal and solid organ transplantations from a donor infected with rabies virus. *Clin Infect Dis*. 2010;50:1112-1119.

[153] Fekadu M, Endeshaw T, Alemu W, et al. Possible human-to-human transmission of rabies in Ethiopia. *Ethiop Med J*. 1996;34:123-127.

[154] Ramsden RO, Johnston DH. Studies on the oral infectivity of rabies virus in carnivora. *J Wildl Dis*. 1975;11:318-324.

[155] Arizona Game and Fish Department. Living with bats. *www.azgfd.gov/w_c/urban_bats.shtml.* (Downloaded August 6, 2010)

[156] Amengual B, Bourhy H, Lopez-Roig M, Serra Cobo J. Active monitoring of EBLV infection in natural colonies of mouse-eared bat (*Myotis myotis*). *Dev Biol* (Basel). 2008;131:547-553.

[157] Amengual B, Bourhy H, Lopez-Roig M, Serra-Cobo J. Temporal dynamics of European bat lyssavirus type 1 and survival of Myotis myotis bats in natural colonies. *PLoS One*. 2007;2:e566.

[158] Shankar V, Bowen RA, Davis Ad, et al. Rabies in a captive colony of big brown bats (*Eptesicus fuscus*). *J Wildl Dis*. 2004;40:403-413.

[159] Almeida MF, Martorelli LF, Aires CC, et al. Experimental rabies infection in haematophagous bats *Desmodus rotundus*. *Epidemiol Infect*. 2005;133:523-527.

[160] Fooks AR, Johnson N, Muller T, et al. Detection of high levels of European bat lyssavirus type-1 viral RNA in the thyroid gland of experimentally-infected *Eptesicus fuscus* bats. *Zoonoses Public Health*. 2009;56:270-277.

[161] deDiaz AM, Fuenzalida E, Bell JF. Non-fatal rabies in dogs and cats. *Ann Microbiol* (Paris). 1975;126:503-509.

[162] Fekadu M, Shaddock JH, Baer GM. Intermittent excretion of rabies virus in the saliva of a dog two and six months after it had recovered from experimental rabies. *Am J Trop Med Hyg*. 1981;30:1113-1115.

[163] Fekadu M, Shaddock JH, Chandler FW, Baer GM. Rabies virus in the tonsils of a carrier dog. *Arch Virol* 1983;781:37-47.

[164] Fekadu M, Shaddock JH, Baer GM. Excretion of rabies virus in saliva of dogs. *J Infect Dis*. 1982;145:715-719.

[165] Minor R. Rabies in the dog. *Vet Rec*. 1977.101:516-520.

[166] Fuller JE. A case of equine rabies. *N Engl J Med*. 1984;310:525-526.

[167] Gillespie JH, Timoney JF. The rhabdoviridae. In: JH Gillespie, JF Timoney (authors). *Hagan and Bruner's Infectious Diseases of Domestic Animals. 7th ed.* Ithaca and London: Comstock Publishing Associates/Cornell University Press. 758-780, 1981.

[168] Hudson LC, Weinstock D, Jordan T, Bold-Fletcher NO. Clinical features of experimentally induced rabies in cattle and sheep. *Zentralbl Veterinarmed B*. 1996;43:85-95.

[169] Stoltenow CL, Solemasass K, Niezgoda M, et al. Rabies in an American bison from North Dakota. *J Wildl Dis*. 2000;36:169-171.

[170] Daszak P. Epidemiology. bats, in black and white. *Science*. 2010;329:634-635.

[171] Streicker DG, Turmelle AS, Vonhof MJ, et al. Host phylogeny constrains cross-species emergence and establishment of rabies virus in bats. *Science*. 2010;329:676-679.

[172] Jackson AC. Why does the prognosis remain so poor in human rabies? *Expert Rev Anti Infect Ther*. 2010;8:623-625.

[173] Gavrila I, Iurasog G, Luca E. Rabies in man. personal observations of seroprophylaxis, prolonged incubation and therapeutic trials. *Ann Inst Pasteur* (Paris). 1967;112:504-515.

[174] Warrell MJ, Warrell DA. Rabies. In: J Cohen, WG Powderly (eds). *Infectious Diseases*. Volume 2. 2nd ed. Edinburgh, London, New York. Oxford, Philadelphia: Mosby. pp. 2061-2066, 2004.

[175] Huppatz C, Kelly PM, Levi C, et al. Encephalitis in Australia, 1979-2006: trends and aetiologies. *Commun Dis Intell.* 2009;33:192-197.

[176] Gadre G, Satishchandra P, Mahadevan A, et al. Rabies viral encephalitis: clinical determinants in the diagnosis with special reference to paralytic form. *J Neurol Neurosurg Psychiatry*. 2010;81:812-820.

[177] Ghosh S. A case of rabies in England. *Br Med J*. 1964;2:167-168.

[178] Nejad AG. Hydrophobia as a rare presentation of Cotard's syndrome: a case report. *Acta Psychiatr Scand.* 2002;106:156-158.

[179] Hemachudha T, Wacharapluesadee S, Lumlertdaecha B, et al. Sequence analysis of rabies virus in humans exhibiting encephalitic or paralytic rabies. *J Infect Dis.* 2003;188:960-966.

[180] Hemachudha T, Wacharapluesadee S, Mitrabhakdi E, et al. Pathophysiology of human paralytic rabies. *J Neurovirol*. 2005;11:93-100.

[181] Mitrabhakdi E, Shuangshoti S, Wannakrairot P, et al. Difference in neuropathogenetic mechanisms in human furious and paralytic rabies. *J Neurol Sci.* 2005;238:3-10.

[182] Willoughby RE Jr, Tieves KS, Hoffman GM, et al. Survival after treatment of rabies with induction of coma. *N Engl J Med*. 2005;352:2508-2514.

[183] Hu WT, Willoughby RE Jr, Dhonau H, Mack KJ. Long-term follow-up after treatment of rabies by induction of coma. *N Engl J Med*. 2007;357:945-946.

[184] Centers for Disease Control and Prevention. Presumptive abortive human rabies – Texas, 2009. *MMWR Morb Mortal Wkly Rep*. 2010.59:185-190.

[185] Jackson AC. Therapy of rabies encephalitis. *Biomedica*. 2009;29:169-175.

[186] Jackson AC. Why does the prognosis remain so poor in human rabies? *Expert Rev Anti Infect Ther*. 2010;8:623-625.

[187] Roy A, Hooper DC. Lethal silver-haired bat rabies virus infection can be prevented by opening the blood-brain barrier. *J Virol*. 2007;81:7993-7998.

[188] Wang ZW, Sarmento L, Wang Y, et al. Attenuated rabies virus activates, while pathogenic rabies virus evades, the host innate immune responses in the central nervous system. *J Virol*. 2005;79:12554-12565.

[189] Huynh W, Cordato DJ, Kehdi E, et al. Post-vaccination encephalomyelitis: literature review and illustrative case. *J Clin Neurosci*. 2008;15:1315-1322.

[190] Kulkarni V, Nadgir D, Tapiawala S, et al. Biphasic demyelination of the nervous system following anti-rabies vaccination. *Neurol India*. 2004;52:106-108.

[191] Zaidman GW, Billingsley A. Corneal impression test for the diagnosis of acute rabies encephalitis. *Ophthalmology*. 1998;105:249-251.

[192] Johnson N, Cunningham AF, Fooks AR. The immune response to rabies virus infection and vaccination. Vaccine. 2010;28:3896-3901.

[193] Wacharapluesadee S, Hemachudha T. Ante- and post-mortem diagnosis of rabies using nucleic acid-amplification tests. *Expert Rev Mol Diagn*. 2010;10:207-218.

[194] Panning M, Baumgarte S, Pfefferie S, et al. Comparative analysis of rabies virus reverse transcription-PCR and virus isolation using samples from a patient infected with rabies virus. *J Clin Microbiol*. 2010;48:2960-2962.

[195] Jiang Y, Luo Y, Michel F, et al. Characterization of conformation-specific monoclonal antibodies against rabies virus nucleoprotein. *Arch Virol*. 2010;155:1187-1192.

[196] Awasthi M, Parmar H, Patankar T, Castillo M. Imaging findings in rabies encephalitis. *AJMR Am J Neuroradiol*. 2001;22:677-680.

[197] Pleasure SJ, Fischbein NJ. Correlation of clinical and neuroimaging findings in a case of rabies encephalitis. *Arch Neurol*. 2000;57:1765-1769.

[198] Sing TM, Soo MY. Imaging findings in rabies. *Australas Radiol*. 1996;40:338-341.

[199] Piyaphanee W, Shantavasinkul P, Phumratanaprapin W, et al. Rabies exposure risk among foreign backpackers in Southeast Asia. *Am J Trop Med Hyg*. 2010;82:1168-1171.

[200] Phanuphak P, Ubolyam S, Sirivichayakul S. Should travellers in rabies endemic areas receive pre-exposure rabies immunization? *Ann Med Interne* (Paris). 1994;145:409-411.

[201] Ertl HC. Novel vaccines to human rabies. *PLos Negl Trop Dis*. 2009;3:e515.

[202] Boz C, Velioglu S, Ozmenoglu M. Acute disseminated encephalomyelitis after be sting. *Neurol Sci*. 2003;23:313-315.

[203] Mihai C, Jubelt B. Post-infectious encephalomyelitis. *Curr Neurol Neurosci Rep*. 2005;5:440-445.

[204] Garg RK. Acute disseminated encephalomyelitis. *Postgrad Med J*. 2003;79:11-17.

[205] Manning SE, Rupprecht CE, Fishbein D, et al. Human rabies prevention—United States, 2008: recommendations of the Advisory Committee on Immunization Practices. *MMWR Recomm Rep*. 2008:23;57(RR-3):1-28.

Severe Acute Respiratory Syndrome (SARS)

[206] Cleri DJ, Ricketti AJ, Vernaleo JR. Severe acute respiratory syndrome (SARS). *Infect Dis Clin North Am*. 2010;24:175-202.

[207] Hui DSC, Chan PKS. Severe acute respiratory syndrome and coronavirus. *Infect Dis Clin North Am*. 2010:24:619-638.

[208] Jiang S, He Y, Liu S. SARS vaccine development. *Emerg Infect Dis*. 2005;11:1016-1020.

[209] McIntosh K, Perlman S. Coronaviruses, including severe acute respiratory syndrome (SARS)-associated coronavirus. In: Mandell GL, Bennett JE, Dolin, editors. *Principles and practice of infectious diseases, Volume 2, 7th edition*. Churchill Livingston Elsevier, Philadelphia. p.2187-2194, 2009.

[210] Misra V, Dumonceaux T, Dubois J, et al. Detection of polyoma and corona viruses in bats of Canada. *J Gen Virol*. 2009;90:2015-2022.

[211] Abdul-Rasool S, Fielding BC. Understanding human coronavirus HCoV-NL63. *Open Virol J*. 2010;4:76-84.

[212] Chiu SS, Chan KH, Chu KW, et al. Human coronavirus NL63 infection and other coronavirus infections in children hospitalized with acute respiratory disease in Hong Kong, China. *Clin Infect Dis*. 2005;40:1721-1729.

[213] Vabret A, Dina J, Gouarin S, et al. Human (non-severe acute respiratory syndrome) coronavirus infections in hospitalized children in France. *J Paediatr Child Health*. 2008;44:176-181.

[214] Vijaykrishna D, Smith GJ, Zhang JX, et al. Evolutionary insights into the ecology of coronaviruses. *J Virol.* 2007;81:4012-4020.

[215] Stockman LJ, Haynes LM, Miao C, et al. Coronavirus antibodies in bat biologists. *Emerg Infect Dis*. 2008;14:999-1000.

[216] Lau SK, Li KS, Huang Y, et al. Ecoepidemiology and complete genome comparison of different strains of severe acute respiratory syndrome-related *Rhinolophus* bat coronavirus in china reveal bats as a reservoir for acute, self-limiting infection that allows recombination events. *J Virol*. 2010;84:2808-2819.

[217] Watanabe S, Masangkay JS, Nagata N, et al. Bat coronaviruses and experimental infection of bats, the Philippines. *Emerg Infect Dis.* 2010;16:1217-1223.

[218] Dominguez SR, O'Shea TJ, Oko LM, Holmes KV. Detection of group 1 coronaviruses in bats in North America. *Emerg Infect Dis.* 2007;13:1295-1300.

[219] Pfefferle S, Oppong S, Drexler JF, et al. Distant relatives of severe acute respiratory syndrome coronavirus and close relatives of human coronavirus 229E in bats, Ghana. *Emerg Infect Dis*. 2009;15:1377-1384.

[220] Tong S, Conrardy C, Ruone S, et al. Detection of novel SARS-like and other coronaviruses in bats from Kenya. *Emerg Infect Dis*. 2009;15:482-485.

[221] Rihtaric D, Hostnik P, Steyer A, et al. Identification of SARS-like coronaviruses in horseshoe bats (*Rhinolophus hipposideros*) in Slovenia. *Arch Virol*. 2010;155:507-514.

[222] Wang LF, Shi Z, Zhang S, et al. Review of bats and SARS. *Emerg Infect Dis.* 2006;12:1834-1840.

[223] Hou Y, Peng C, Yu M, et al. Angiotensin-converting enzyme 2 (ACE2) proteins of different bat species confer variable susceptibility to SARS-CoV entry. *Arch Virol.* 2010 Jun 22 [Epub ahead of print].

[224] Sheahan T, Rockx B, Donaldson E, et al. Pathways of cross-species transmission of synthetically reconstructed zoonotic severe acute respiratory syndrome coronavirus. *J Virol.* 2008;82:8721-8732.

[225] Ren W, Qu X, Li W, et al. Difference in receptor usage between severe acute respiratory syndrome (SARS) coronavirus and SARS-like coronavirus of bat origin. *J Virol.*2008;82:1899-1907.

[226] Zhou P, Han Z, Wang LF, ShiZ. Immunogenicity difference between the SARS coronavirus and the bat SARS-like coronavirus (S) proteins. *Biochem Biophys Res Commun.* 2009;387:326-329.

[227] Li F. Structural analysis of major species barriers between humans and palm civets for severe acute respiratory syndrome coronavirus infections. *J Virol*. 2008;82:6984-6991.

[228] Ling D, Menzies D. Occupation-related respiratory infections revisited. *Infect Dis Clin North Am.* 2010;24:655-680.

[229] Christian MD, Poutanen SM, Loutfy MR, et al. Severe acute respiratory syndrome. *Clin Infect Dis*. 2004;38:1420-1427.

[230] Kwan BC, Leung CB, Szeto CC, et al. Severe acute respiratory syndrome in dialysis patients. *J Am Soc Nephrol*. 2004;15:1883-1888.

[231] Wang XW, Li JS, Guo TK, et al. Excretion and detection of SARS coronavirus and its nucleic acid from digestive system. *World J Gastroenterol*. 2005;11;4390- 395.

[232] Concentration and detection of SARS coronavirus in sewage from Xiao Tang Shan Hospital and the 309th Hospital of the Chinese People's Liberation Army. *Water Sci Technol.* 2005;52:213-221.

[233] Li Y, Duan S, Yu IT, Wong TW. Multi-zone modeling of probable SARS virus transmission by airflow between flats in block E, Amoy Gardens. *Indoor Air.* 2005;15:96-111
[234] McKinney KR, Gong YY, Lewis TG. Environmental transmission of SARS at Amoy Gardens. *J Environ Health*. 2006;68:26-30.
[235] Yip C, Chang WL, Yeung KH, Yu IT. Possible meteorological influence on the severe acute respiratory syndrome (SARS) community outbreak at Amoy Gardens, Hong Kong. *J Environ Health*. 2007;70:39-46.
[236] Chu CM, Cheng VC, Hung IF, et al. Viral load distribution in SARS outbreak. *Emerg Infect Dis*. 2005;11:1882-1886.
[237] Pitzer VE, Leung GM, Lipsitch M. Estimating variability in the transmission of severe acute respiratory syndrome to household contacts in Hong Kong, *China. Am J Epidemiol*. 2007;166:355-363.
[238] Guo Y, Korteweg C, McNutt MA, Gu J. Pathogenetic mechanisms of severe acute respiratory syndrome. *Virus Res*. 2008;133:4-12
[239] Chan VS, Chan KY, Chen Y, et al. Homozygous L-SIGN (CLEC4M) plays a protective role in SARS coronavirus infection. *Nat Genet*. 2006;38:38-46.
[240] Lee HK, Lee BH, Seok SH, et al. Production of specific antibodies against SARS-coronavirus nucleocapsid protein without cross reactivity with human coronaviruses 229E and OC43. *J Vet Sci*. 2010;11:165-167.
[241] Zhu Z, Chakraborti S, He Y, et al. Potent cross-reactive neutralization of SARS coronavirus isolates by human monoclonal antibodies. *PNAS*. 2007;104:12123-12128.
[242] Stockman LJ, Bellamy R, Garner P. SARS: systematic review of treatment effects. *PLoS Med*. 2006;3:e343.
[243] Raboud J, Shigayeva A, McGeer A, et al. Risk factors for SARS transmission from patients requiring intubation: a multicentre investigation in Toronto, Canada. *PLoS One*. 2010;19:e10717.

Paramyxoviridae

– Henipaviruses Genus: Hendra Virus and Nipah Virus

[244] Hanlon CA. *Health precautions for bat workers*. In: Bernard SM (ed.), Bats in captivity. Volume 1: biological and medical aspects, 1st edition. Logos Press, Washington, DC, p.1-24, 2009.
[245] McCormack JG. Hendra and Nipah viruses: new zoonotically-acquired human pathogens. *Respir Care Clin*. 2005;11:59-66.
[246] Khabbaz RF, Ostroff SM, LeDuc JW, et al. Emerging and reemerging infectius disease threats. In: Mandell GL, Bennett JE, Dolin, editors. *Principles and practice of infectious diseases*, Volume 1, 7th edition. Churchill Livingston Elsevier, Philadelphia. p.199-219, 2009.
[247] Field HE, Mackenzie JS, Daszak P. Henipaviruses: emerging paramyxoviruses associated with fruit bats. *Curr Top Microbiol Immunol*. 2007;315:133-159.
[248] Radostits OM, Gay CC, Hinchcliff, Constable PD. Diseases associated with viruses and Chlamydia-II. In: Radostits OM, Gay CC, Hinchcliff, Constable PD, *Veterinary medicine - a textbook of the diseases of cattle, horses, sheep, pigs and goats*, 10th

edition. Saunders Elsevier, Edinburg, London, New York, 1307-1438, 2007, reprinted, 2008.
[249] Weingartl HM, Berhane Y, Czub M. Animal models of henipavirus infection: a review. *Vet J.* 2009;181:211-220.
[250] Barker SC. The Australian paralysis tick may be the missing link in the transmission of Hendra virus from bats to horses to humans. *Med Hypotheses*. 2003;60:481-483.
[251] Thorner AR, Dolin R. Zoonotic paramyxoviruses: Nipah, Hendra, and Menangle viruses. In: Mandell GL, Bennett JE, Dolin, editors. *Principles and practice of infectious diseases*, Volume 1, 7th edition. Churchill Livingston Elsevier, Philadelphia. p.2237-2244, 2009.
[252] Reynes JM, Counor D, Ong S, et al. Nipah virus in Lyle's flying foxes, Cambodia. *Emerg Infect Dis*. 2005;11:1042-1047.
[253] Chua KB, Koh CL, Hooi PS, et al. Isolation of Nipah virus from Malaysian island flying foxes. *Microbes Infect*. 2002;4:145-151.
[254] Olson JG, Rupprecht C, Rollin PE, et al. Antibodies to Nipah-like virus in bats (Pteropus lylei), Cambodia. *Emerg Infect Dis*. 2002;8:987-988.
[255] Yob JM, Field H, Rashdi AM, et al. Nipah virus infection in bats (order Chiroptera) in peninsular Malaysia. *Emerg Infect Dis*. 2001;7:439-441.
[256] Johara MY, Field H, Rashdi AM, et al. Nipah virus infection in bats (order Chiroptera) in Peninsular Malaysia. *Emerg Infect Dis*. 2001;7:439-441.
[257] Montgomery JM, Hossain MJ, Gurley E, et al. Risk factors for Nipah virus encephalitis in Bangladesh. *Emerg Infect.* 2008;14:1526-1532.
[258] Nahar N, Sultana R, Gurley ES, et al. Date palm sap collection: exploring opportunities to prevent Nipah transmission. *Ecohealth*. 2010. Jul 9. [Epub ahead of print].

–Rubulavirus Genus: Menangle Virus, Tioman Virus

[259] Kirkland PD, Love RJ, Philbey AW, et al. Epidemiology and control of Menangle virus in pigs. *Aust Vet J.* 2001;79:199-206.
[260] Philbey AW, Kirkland PD, RossAD, et al. Infection with Menangle virus in flying foxes (Pteropus spp.) in Australia. *Aust Vet J.* 2008;86:449-454.
[261] Philbey AW, Ross AD, Kirkland PD, Love RJ. Skelatal and neurological malformations in pigs congenitally infected with Menangle virus. *Aust Vet J.* 2007;85:134-140.
[262] Chant K, Chan R, Smith M, et al. Probable human infection with a newly described virus in the family Paramyxoviridae. *Emerg Infect Dis.* 1998;4:273-275.
[263] Zvfirbliene A, Kucinskaite-Kodze I, Juozapaitis M, et al. Novel monoclonal antibodies against Menangle virus nucleocapsid protein. *Arch Virol*. 2010;155:13-18.
[264] Petraityte R, Tamosiunas PL, Juozapaitis M, et al. Generation of Tioman virus nucleocapsid-like particle in yeast *Saccharomyces cerevisiae*. *Virus Res.* 2009;145:92-96.
[265] Iehle C, Razafitrimo G, Razainirina J, et al. Henipavirus and Tioman virus antibodies in pteropodid bats, Madagascar. *Emerg Infect Dis*. 2007;13:159-161.
[266] Yaiw KC, Crameri G, Wang L, et al. Serological evidence of possible human infection with Tioman virus, a newly described paramyxovirus of bat origin. *J Infect Dis.* 2007;196:884-886.

[267] Yaiw KC, Bingham J, Crameri G, et al. Tioman virus, a paramyxovirus of bat origin, causes mild disease in pigs and has a predilection for lymphoid tissues. *J Virol.* 2008;82:565-568.

[268] Yaiw KC, Ong KC, Chua KB, et al. Tioman virus infection in experimentally infected mouse brain and its association with apoptosis. *J Virol Methods*. 2007;143;140-146.

[269] Yaiw KC, Hyatt A, Vandriel R, et al. Viral morphogenesis and morphological changes in human neuronal cells following Tioman and Menangle virus infection. *Arch Virol.* 2008;153:865-875.

[270] Drexler JF, Corman VM, Gloza-Rausch F, et al. Henipavirus RNA in African bats. *PLoS One*. 2009;4:e6367.

Other Paramyxoviridae Viruses

[271] Henderson GW, Laird C, Dermott E, Rima BK. Characterization of Mapuera virus: structure, proteins and nucleotide sequence of the gene encoding the nucleocapsid protein. *J Gen Virol.* 1995;76:2509-2518.

[272] Hagmaier K, Stock N, Precious B, et al. Mapuera virus, a rubulavirus that inhibits interferon signaling in a wide variety of mammalian cells without degrading STATs. *J Gen Virol.* 2007;88:956-966.

[273] Wang LF, Hansson E, Yu M, et al. Full-length genome sequence and genetic relationship to two paramyxoviruses isolated from bat and pigs in the Americas. *Arch Virol.* 2007;152:1259-1271.

[274] Lau SK, Woo PC, Wong BH, et al. Identification and complete genome analysis of three novel paramyxoviruses, Tuhoko virus 1, 2 and 3, in fruit bats from China. *Virology*. 2010;404:106-116.

Reoviridae

[275] Tee KK, Takebe Y, Kamarulzaman A. Emerging and re-emerging viruses in Malaysia, 1997-2007. *Int J Infect Dis*. 2009;13:307-318.

[276] DeBiasi RL, Tyler KL. Orthoreoviruses and orbiviruses. In: Mandell GL, Bennett JE, Dolin, editors. *Principles and practice of infectious diseases, Volume 2, 7th edition.* Churchill Livingston Elsevier, Philadelphia. p.2097-2100, 2009.

[277] DeBiasi RL, Tyler KL. Coltiviruses and seadornaviruses. In: Mandell GL, Bennett JE, Dolin, editors. *Principles and practice of infectious diseases, Volume 2, 7th edition.* Churchill Livingston Elsevier, Philadelphia. p.2101-2104, 2009.

[278] Dormitzer PR. Rotaviruses. In: Mandell GL, Bennett JE, Dolin, editors. *Principles and practice of infectious diseases, Volume 2, 7th edition*. Churchill Livingston Elsevier, Philadelphia. p. 2105-2116, 2009.

[279] Pritchard LI, Chua KB, Cummins D, et al. Palau virus; a new member of the Nelson Bay orthoreovirus species isolated from fruit bats in Malaysia. *Arch Virol.* 2006;151:229-239.

[280] Wellehan JF Jr, Childress AL, Marschang RE, et al. Consensus nested PCR amplification and sequencing of diverse reptilian, avian, and mammalian orthoreoviruses. *Vet Microbiol*. 2009;133:34-42.

[281] Furuse Y, Suzuki A, Kishi M, et al. Detection of novel respiratory viruses from influenza-like illness in the Philippines. *J Med Virol*. 2010;82:1071-1074.

[282] Chua KB, Voon K, Crameri G, et al. Identification and characterization of a new orthoreovirus from patients with acute respiratory infections. *PLoS One.* 2008;3:e3803.

Orbivirus Genus of the Family Reoviridae

[283] Kemp GE, Le Gonidec G, Karabatsos N, et al. IFE: a new African orbivirus isolated from *Eidolon helvum* bats captured in Nigeria, Cameroon, and the Central African Republic. *Bull Soc Pathol Exot Filiales.* 1988;81:40-48.

[284] Boiro I, Fidarov FM, Lomonossov NN, et al. Isolation of the Fomede virus from Chiroptera, *Nycteris nana*, in the Republic of Guinea. *Bull Soc Pathol Exot Filiales.* 1986:79:180-182.

Astroviruses

[285] Zhu HC, Chu DK, Liu W, et al. Detection of diverse astroviruses from bats in China. *J Gen Virol.* 2009;90:883-887.

[286] Dolin R, Treanor JJ. Astroviruses and Picobirnaviruses. In: Mandell GL, Bennett JE, Dolin, editors. *Principles and practice of infectious diseases, Volume 2, 7th edition.* Churchill Livingston Elsevier, Philadelphia. p.2407-2410, 2009.

[287] Chu DK, Poon LL, Guan Y, Peiris JS. Novel astroviruses in insectivorous bats. *J Virol.* 2008;82:9107-9114.

Papillomavirus

[288] Bonnez W, Reichman RC. Papillomaviruses. In: Mandell GL, Bennett JE, Dolin, editors. *Principles and practice of infectious diseases, Volume 2, 7th edition.* Churchill Livingston Elsevier, Philadelphia. p.2035-2049, 2009.

[289] Majewski S, Jablonska S. Skin autografts in epidermodysplasia verruciformis: human papillomavirus-associated cutaneous changes need over 20 years for malignant conversion. *Cancer Res.* 1997;57:4214-4216.

[290] Cultan JE, Rashid RM, Torres Cabala C, et al. Epidermodysplasia veruciformis after cutaneous T-cell lymphoma: periungual presentation. *Dermatol Online J.* 2010;16:12.

[291] McKnight CA, Wise AG, Maes RK, et al. Papillomavirus-associated basosquamous carcinoma in an Egyptian fruit bat (*Rousettus aegyptiacus*). *J Zoo Wildl Med.* 2006;37:193-196.

[292] Schaller J, Rohwedder A, Fuchs M, et al. HPV-5 typing with nested PCR and sequencing in epidermodysplasia verruciformis. *Hautarzt.* 1996;47:454-458.

[293] Boxman IL, Mulder LH, Russell A, et al. Human papillomavirus type 5 is commonly present in immunosuppressed and immunocompetent individuals. *Br J Dermatol.* 1999;141:246-249.

Polyoma Virus

[294] Tan CS, Koralnik IJ. JC, BK, and other polyomaviruses: progressive multifocal leukoencehalopathy. In: Mandell GL, Bennett JE, Dolin, editors. *Principles and practice of infectious diseases, Volume 2, 7th edition.* Churchill Livingston Elsevier, Philadelphia. p.2051-2058, 2009.

Other Viruses

Viral Hemorrhagic Fevers

[295] Cleri DJ, Ricketti AJ, Porwancher RB, et al. Viral hemorrhagic fevers: current status of endemic disease and strategies for control. *Infect Dis Clin N Am*. 2006;20:359-393.

[296] Abraham J, Kwong JA, Albarino CG, et al. Host-species transferrin receptor 1 orthologs are cellular receptors for nonpathogenic New World clade B arenaviruses. *PLoS Pathog*. 2009:5:e1000358.

[297] Hartman L, Towner JS, Nichol ST. Ebola and Marburg hemorrhagic fever. *Clin Lab Med*. 2010;30:161-177.

[298] Towner JS, Pourrut X, Albarino CG et al. Marburg virus infection detected in a common African bat. *PLoS One*. 2007;2:e764.

[299] Feldmann H. Marburg hemorrhagic fever—the forgotten cousin strikes. *N Engl J Med*. 2006;355:866-869.

[300] Towner JS, Amman BR, Sealy TK, et al. Isolation of genetically diverse Marburg viruses from Egyptian fruit bats. *PLoS Pathog*. 2009;5:e1000536.

[301] Kuzmin IV, Niezgoda M, Franka R, et al. Marburg virus in fruit bat, Kenya. *Emerg Infect Dis*. 2010. 16:352-354.

[302] Leroy EM, Kumulungui B, Pourrut X, et al. Fruit bats as reservoirs of Ebola virus. *Nature*. 2005;438:575-576.

[303] Groseth A, Feldmann H, Strong JE. The ecology of ebola virus. *Trends Microbiol*. 2007;15:408-416.

[304] Hayman DT, Emmerich P, Yu M, et al. Long-term survival of an urban fruit bat seropositive for ebola and Lagos bat viruses. *PLoS One*. 2010;5.pii:e11978.

[305] Leroy EM, Epelboin A, Mondonge V, et al. Human Ebola outbreak resulting from direct exposure to fruit bats, in Luebo, Democratic Republic of Congo, 2007. *Vector Borne Zoonotic Dis*. 2009;9:723-728.

[306] Pourrut X, Souris M, Towner JS, et al. Large serological survey showing cocirculation of Ebola and Marburg viruses in Gabonese bat populations, and a high seroprevalence of both viruses in *Rousettus aegyptiacus*. *BMC Infect Dis*. 2009;9:159.

[307] Bennett M. Bats and human emerging diseases. *Epidemiol Infect*. 2006;134:905-907.

[308] Biek R, Walsh PD, Leroy EM, Real LA. Recent common ancestry of ebola Zaire virus found in a bat reservoir. *PLoS Pathog*. 2006;2:e90.

[309] Pourrut X, Delicat A, Rolin PE, et al. Spatial and temporal patterns of Zaire ebolavirus antibody prevalence in the possible reservoir bat species. *J Infect Dis*. 2007;196 Suppl 2:S176-S183.

[310] Vaughn DW, Barrett A, Solomon T. Flaviviruses (Yellow fever, Dengue, Dengue hemorrhagic Fever, Japanese encephalitis, West Nile encephalitis, St. Louis encephalitis, tick-borne encephalitis) In: Mandell GL, Bennett JE, Dolin, editors. *Principles and practice of infectious diseases, Volume 2, 7th edition*. Churchill Livingston Elsevier, Philadelphia. p.2133-2156, 2009.

[311] Mehla R, Kumar SR, Yadav P, et al. Recent ancestry of Kyasanur Forest disease virus. *Emerg Infect Dis*. 2009;15:1431-1437.

[312] Mackenzie JS, Williams DT. The zoonotic Flaviviruses of Southern, South-Eastern and Eastern Asia, and Australasia: the potential for emergent viruses. *Zoonoses Public Health*. 2009; May 20. [Epub ahead of print].

[313] Bhat HR, Sreenivasan MA, Goverdhan MK, et al. Antibodies to Kyasanur forest disease virus in bats in the epizootic-epidemic area and neighbourhood. *Indian J Med Res*. 1978;68:387-392.

[314] Sreenivasan MA, Bhat HR, Naik SV, Banerjee K. Susceptibility of *Rousettus leschenaulti* to infection with Kyasanur forest disease virus—a note. *Indian J Med Res*. 1979;69:535-537.

[315] Gadkari DA, Banerjee K, Bhat HR. Critical evaluation of Kyasanur Forest disease virus neutralizing antibodies found in bats (a preliminary report). *Indian J Med Res*. 1976;64:64-67.

[316] Rajogopalan PK, Paul SD, Sreenivasan MA. Isolation of Kyasanur forest disease virus from the insectivorous bat, *Rhinolophus rouxi* and from *Ornithodoros* ticks. *Indian J Med Res*. 1969;57:805-808.

[317] Pavri KM, Singh KR. Kyasanur forest disease virus infection in the frugivorous bat, *Cynopterus sphinx*. *Indian J Med Res*. 1968;56:1202-1204.

[318] Parvi KM, singh KR. Demonstration of antibodies against the virus Kyasanur Forest disease (KFD) in the frugivorous bat *Rousettus leschenaulti*, near Poona, India. *Indian J Med Res*. 1965;53:956-961.

[319] Centers for Disease Control of Prevention. St. Louis encephalitis. Available at: *http://www.cdc.gov/ncidod/dvbid/sle/Sle_FactSheet.html* (Downloaded September 7, 2010).

[320] Ubico SR, McLean RG. Serologic survey of neotropical bats in Guatemala for virus antibodies. *J Wildl Dis*. 1995;31:1-9.

[321] Ray SC, Thomas DL. Hepatitis C. In: Mandell GL, Bennett JE, Dolin, editors.*Principles and practice of infectious diseases, Volume 2, 7th edition*. Churchill Livingston Elsevier, Philadelphia. p.2157-2185, 2009.

[322] Epstein JH, Quan PL, Briese T, et al. Identification of GBV-D, a novel GB-like flavivirus from old world frugivorous bats (*Pteropus giganteus*) in Bangladesh. *PLoS Pathog*. 2010;6:1000972.

[323] Charrel RN, de Lamballerie X. Zoonotic aspects of arenavirus infections. *Vet Microbiol*. 2010;140:213-220.

[324] Milazzo ML, Cajimat MN , Haynie ML, et al. Diversity among tacaribe serocomplex viruses (family Arenaviridae) naturally associated with the white-throated woodrat (*Neotoma albigula*) in the southwestern United States. *Vector Borne Zoonotic Dis*. 2008;8:523-540.

[325] Cajimat MN, Milazzo ML, Borchert JN, et al. Diversity among Tacaribe serocomplex viruses (family Arenaviridae) naturally associated with the Mexican woodrat (*Neotoma mexicana*). *Virus Res*. 2008;133:211-217.

[326] Inizan CC, Cajimat MN, Milazzo ML, et al. Genetic evidence for a tacaribe serocomplex virus, Mexico. *Emerg Infect dis*. 2010;16:1007-1010.

[327] Urata S, Yasuda J, de la Torre JC. The Z protein of the New World Arenavirus Tacaribe virus has bona fide budding activity that does not depend on known late domain motifs. *J Virol*. 2009;83:12651-12655.

Alphaviruses

Venezuelan Equine Encephalomyelitis, Eastern and Western Equine Encephalitis Viruses

[328] Markoff L. Alphaviruses. In: Mandell GL, Bennett JE, Dolin, editors. *Principles and practice of infectious diseases, Volume 2, 7th edition*. Churchill LivingstonElsevier, Philadelphia. p.2117-2125., 2009.

Chikungunya Fever\

[329] Bernard E, Solignat M, Gay B, et al. Endocytosis of chikungunya virus into mammalian cells: role of clathrin and early endosomal compartments. *PLoS one*. 2010;5:e11479
[330] Chhabra M, Mittal V, Bhattacharya D, et al. Chikungunya fever: a re-emerging viral infection. *Indian J Med Microbiol*. 2008;26:5-12.
[331] Diallo M, Thonnon J, Traore-lamizana M, Fontenille D. Vectors of chikungunya virus in Senegal: current data and transmission cycles. *Am J Trop Med Hyg*. 1999;60:281-286.

Vesicular Stomatitis Virus

[332] Fine SM. Vesicular stomatitis virus and related vesiculoviruses. In: Mandell GL, Bennett JE, Dolin, editors. *Principles and practice of infectious diseases*, Volume 2, 7th edition. Churchill Livingston Elsevier, Philadelphia. p. 2245-2247, 2009.

Adenoviruses

[333] Rhee EG, Barouch DH. Adenoviruses. In: Mandell GL, Bennett JE, Dolin, editors. *Principles and practice of infectious diseases, Volume 2, 7th edition*. Churchill Livingston Elsevier, Philadelphia. p. 2027-2033, 2009.
[334] Li Y, Ge X, Zhang H, et al. Host range, prevalence, and genetic diversity of adenoviruses in bats. *J Virol*. 2010;84:3889-3897.
[335] Sonntag M, Muhldorfer K, Speck S, et al. New adenovirus in bats, Germany. *Emerg Infect Dis*. 2009;15:2052-2055.
[336] Janoska M, Vidovszky M, Molnar V, et al. Novel adenoviruses and herpesviruses detected in bats. *Vet J*. 2010 Aug 31.[Epub ahead of print].

Herpesviruses

[337] Cohen JI. Introduction to Herpesviridae. In: Mandell GL, Bennett JE, Dolin, editors. *Principles and practice of infectious diseases, Volume 2, 7th edition*. Churchill Livingston Elsevier, Philadelphia. p.1937-1942, 2009.
[338] Schiffer JT, Corey L. Herpes simplex virus. In: Mandell GL, Bennett JE, Dolin, editors. *Principles and practice of infectious diseases, Volume 2, 7th edition*. Churchill Livingston Elsevier, Philadelphia. p.1943-1962, 2009.
[339] Whitley RJ. Varicella-zoster virus. In: Mandell GL, Bennett JE, Dolin, editors.*Principles and practice of infectious diseases, Volume 2, 7th edition*. Churchill Livingston Elsevier, Philadelphia. p.1963-1970, 2009.

[340] Cohen JI. Herpes B virus. In: Mandell GL, Bennett JE, Dolin, editors. *Principles and practice of infectious diseases, Volume 2, 7th edition*. Churchill Livingston Elsevier, Philadelphia. p.2023-2026., 2009.
[341] Crumpacker CS II, Zhang JL. Cytomegalovirus. In: Mandell GL, Bennett JE, Dolin, editors. *Principles and practice of infectious diseases, Volume 2, 7th edition*. Churchill Livingston Elsevier, Philadelphia. p.1971-2010, 2009.
[342] Cohen JL. Human herpesvirus types 6 and 7. In: Mandell GL, Bennett JE, Dolin, editors. *Principles and practice of infectious diseases, Volume 2, 7th edition*. Churchill Livingston Elsevier, Philadelphia. p.2011-2016, 2009.
[343] Johannsen EC, Kaye KM. Epstein-Barr virus (infectious mononucleosis, Epstein-Barr virus-associated malignant diseases, and other diseases. In: Mandell GL, Bennett JE, Dolin, editors. *Principles and practice of infectious diseases, Volume 2, 7th edition.* Churchill Livingston Elsevier, Philadelphia. p.1789-2010, 2009.
[344] Kay KM. Kaposi's sarcoma-associated herpesvirus (human herpesvirus type 8). In: Mandell GL, Bennett JE, Dolin, editors. *Principles and practice of infectious diseases, Volume 2, 7th edition.* Churchill Livingston Elsevier, Philadelphia. p.2017-2222, 2009.
[345] Tandler B. Cytomegalovirus in the principal submandibular gland of the little brown bat, *Myotis lucifugus*. *J Comp Pathol.* 1996;114:1-9.
[346] Wibbelt G, Kurth A, Yasmum N, et al. Discovery of herpesviruses in bats. *J Gen Virol.* 2007;88:2651-2655.
[347] Tsygankov AY. Cell transformation by Herpesvirus saimiri. *J Cell Physiol.* 2005;203:305-318,
[348] Springfeld C, Tidona CA, Kehm R, et al. Identification and characterization of the Tupaia herpesvirus DNA polymerase gene. *J Gen Virol*. 1998;79:3049-3053.
[349] Molnar V, Janoska M, Harrach B, et al. Detection of a novel bat gammaherpesvirus in Hungary. *Acta Vet Hung*. 2008;56:529-538.
[350] Watanabe S, Ueda N, Iha K, et al. Detection of a new bat gammaherpesvirus in the Philippines. *Virus Genes*. 2009;39:90-93.
[351] Watanabe S, Maeda K, Suzuki K, et al. Novel beta herpes virus in bats. *Emerg Infect Dis*. 2010;16:986-988.
[352] Razafindratsimandresy R, Jeanmaire EM, Counor D, et al. Partial molecular characterization of alphaherpesviruses isolated from tropical bats. *J Gen Virol.* 2009;90:44-47.
[353] Li L, Victoria JG, Wang C, et al. Bat guano virome: predominance of dietary viruses from insects and plants plus novel mammalian viruses. *J Virol*. 2010;84:6955-6965.

Bacteria

Enteric Pathogens

[354] Adesiyun AA, Stewart-Johnson A, Thompson NN. Isolation of enteric pathogens from bats in Trinidad. *J Wildl Dis*. 2009;45:952-961.

Kluyvera Species

[355] Carter JE, Evans TN. Clinically significant *Kluyvera* infections: a report of seven cases. *Am J Clin Pathol*. 2005;123:334-338.

[356] Moonah S, Deonarine K, Freeman C. Multidrug resistant *Kluyvera ascorbata* septicemia in an adult patient: a case report. *J Med Case Reports*. 2010;4:197.

[357] Altun Koroglu O, Yalaz, Ozalkaya E, et al. *Kluyvera cryocrescens* sepsis in a preterm infant. *Jpn J Infect Dis*. 2010;63:195-196.

[358] Rosso M, Rojas P, Garcia E, et al. *Kluyvera* meningitis in a newborn. *Pediatr Infect Dis J*. 2007;26:1070-1071.

[359] Dollberg S, Gandacu A, Klar A. Acute pyelonephritis due to a *Kluyvera* species in a child. *Eur J Clin Microbiol Infect Dis*. 1990;9:281-283.

[360] Sarria JC, vidal AM, Kimbrough RC 3rd. Infections caused by *Kluyvera* species in humans. *Clin Infect Dis*. 2001;33:E69-E74.

[361] Toprak D, Soysal A, Turel O, et al. Hickman catheter-related bacteremia with *Kluyvera cryocrescens*: a case report. *Jpn J Infect Dis*. 2008;61:229-230.

[362] Wong VK. Broviac catheter infection with *Kluyvera cryocrescens*: a case report. *J Clin Microbiol*. 1987;25:1115-1116.

[363] Sierra-Madero J, Pratt K, Hall GS, et al. *Kluyvera* mediastinitis following open-heart surgery: a case report. *J Clin Microbiol*. 1990;28:2848-2849.

[364] Cheruvattath R, Balan V, Stewart R, et al. *Kluyvera* co-infection in two solid organ transplant recipients: an emerging pathogen or a colonizer bystander? *Transpl Infect Dis*. 2007;9:83-86.

[365] Musgrove MT, Jones DR, Shaw JD, et al. Enterobacteriaceae and related organisms isolated from nest run cart shelves in commercial shell egg processing fcilities. *Poult Sci*. 2009;88:2113-2117.

[366] Han JE, Gomez DK, Kim JH, et al. Isolation of a zoonotic pathogen *Kluyvera ascorbata* from Egyptian fruit bat *Rousettus aegyptiacus*. *J Vet Med Sci*. 2010;72:85-87.

[367] Narvarrete P, Magne F, Mardones P, et al. Molecular analysis of intestinal microbiota of rainbow trout (*Oncorhynchus mykiss*). *FEMS Microbiol Ecol*. 2010;71:148-156.

[368] Lee HY, Stephen A, Sushela D, Mala M. Detection of protozoan and bacterial pathogens of public health importance in faeces of *Corvus* spp. (large-billed crow). *Trop Biomed*. 2008;25:134-139.

[369] Richard C, Vicens R, Coulanges P, et al. Isolation of *Kluyvera ascorbata* from Madagascan lemurs, healthy carriers of this unusual enterobacterium. *bull Soc Pathol Exot Filiales*. 1987;80:756-760.

Leptospirosis

[370] Adler B, de la Pena Moctezuma A. Leptospira and leptospirosis. *Vet Microbiol*. 2010;140:287-296.

[371] Levett PN. Leptospirosis. *Clin Microbiol Rev*. 2001;14:296-326.

[372] Levett PN, Haake DA. *Leptospira* species (Leptospirosis). In: Mandell GL, Bennett JE, Dolin, editors. *Principles and practice of infectious diseases, Volume 2, 7th edition*. Churchill Livingston Elsevier, Philadelphia. p.3059-3065, 2009.

[373] Vashi NA, Reddy P, Wayne DB, Sabin B. Bat-associated leptospirosis. *J Gen Intern Med*. 2010;25:162-164.

[374] Matthias MA, Diaz MM, Campos KJ, et al. Diversity of bat-associated Leptospira in the Peruvian Amazon inferred by bayesian phylogenetic analysis of 16S ribosomal DNA sequences. *Am J Trop Med Hyg.* 2005;73:964-974.

[375] Bunnell JE, Hice Cl, Watts DM, et al. Detection of pathogenic Leptospira spp. infections among mammals captured in the Peruvian Amazon basin region. *Am J Trop Med Hyg.* 2000;63:255-258.

[376] Everard CO, Fraser-Chanpong GM, Bhagwandin LJ, et al. Leptospires in wildlife from Trinidad and Grenada. *J Wildl Dis.* 1983;19:192-199.

[377] Fennestad KL, Borg-Petersen C. Leptospirosis in Danish wild mammals. *J Wildl Dis.* 1972;8:343-351

[378] Ganoza CA, Matthias MA, Collins-Richards D, et al. Determining risk for severe leptospirosis by molecular analysis of environmental surface waters for pathogenic Leptospira. *PLoS Med.* 2006;3:e308.

[379] Karaahmetoglu S, Ciffci A, Ozer I, et al. A case of acute pancreatitis due to Weil's disease. *Mikrobiyol Bul.* 2003;37:297-299.

[380] Sperber SJ, Schleupner CJ. Leptospirosis: a forgotten cause of aseptic meningitis and multisystem febrile illness. *South Med J.* 1989;82:1285-1288.

[381] Hill MK, Sanders CV. Leptospiral pneumonia. *Semin Respir Infect.* 1997;12:44-49.

[382] Bessa TA, Spichler A, Chapola EG, et al. The contribution of bats to leptospirosis transmission in Sao Paulo City, Brazil. *Am J Trop Med Hyg.* 2010;82:315-317.

[383] Cox TE, Smythe LD, Leung LK. Flying foxes as carriers of pathogenic *Leptospira* species. *J Wildl Dis.* 2005;41:753-757.

[384] Smythe LD, Field HE, Barnett LJ, et al. Leptospiral antibodies in flying foxes in Australia. *J Wildl Dis.* 2002;38:182-186.

[385] Acevedo-Whitehouse K, de la Cueva H, Gulland FM, et al. Evidence of *Leptospira interrogans* infection in California sea lion pups from the Gulf of California. *J Wildl Dis.* 2003;39:145-151.

Bartonella Species

[386] Slater LN, Welch DF. *Bartonella* including cat-scratch disease. In: Mandell GL, Bennett JE, Dolin, editors. *Principles and practice of infectious diseases, Volume 2, 7th edition.* Churchill Livingston Elsevier, Philadelphia. p.2995-3009, 2009.

[387] Reeves WK, Rogers TE, Durden LA, Dasch GA. Association of *Bartonella* with fleas (*Siphonaptera*) of rodents and bats using molecular techniques. *J Vector Ecol.* 2007;32:118-122.

Q Fever, Rickettsia, and Borrelia

[388] Marrie TJ, Raoult D. *Coxiella burnetii* (Q fever). In: Mandell GL, Bennett JE, Dolin, editors. *Principles and practice of infectious diseases, Volume 2, 7th edition.* Churchill Livingston Elsevier, Philadelphia. p.2511-2519, 2009.

[389] Gardon J, Heraud J, Laventure S, et al. Suburban transmission of Q fever in French Guiana: evidence of a wild reservoir. *J Infect Dis.* 2001;1843:278-284

[390] D'Auria SR, Camargo MC, Pacheco RC, et al. Serologic survey for rickettsiosis in bats from Sao Paulo city, Brazil. *Vector Borne Zoonotic Dis.* 2010;10:459-463.

[391] Gill JS, Ullmann AJ, Loftis AD, et al. Novel relapsing fever spirochete in bat tick. *Emerg Infect Dis.* 2008;14:522-523.

[392] Schwan TG, Raffel SJ, Schrumpf ME, et al. Characterization of a novel relapsing fever spirochete in the midgut, coxal fluid, and salivary glands of the bat tick *Carios kelleyi*. *Vector Brone Zoonotic Dis*. 2009;9:643-647.

[393] Loftis AD, Gill JS, Schriefer ME, et al. Detection of Rickettsia, Borrelia, and Bartonella in *Carios kelleyi* (Acari: Argasidae). *J Med Entomol*. 2005;42:473-480.

[394] Evans NJ, Brown K, Timofte D, et al. Fatal borreliosis in bat caused by relapsing fever spirochete, United Kingdom. *Emerg Infect Dis*. 2009;15:1331-1333.

[395] Radostits OM, Gay CC, Hinchcliff, Constable PD. Diseases associated with Rickettsiales. In: Radostits OM, Gay CC, Hinchcliff, Constable PD, *Veterinary medicine - a textbook of the diseases of cattle, horses, sheep, pigs and goats,* 10th edition. Saunders Elsevier, Edinburg, London, New York, 1455-1470, 2007, reprinted, 2008.

[396] Gibson KE, Rikihisa Y, Zhang C, Martin C. *Neorickettsia risticii* is vertically transmitted in the trematode *Acanthatrium oregonense* and horizontally transmitted to bats. *Environ Microbiol*. 2005;7:203-212.

Fungi

Histoplasmosis

[397] Deepe GS Jr. *Histoplasma capsulatum*. In: Mandell GL, Bennett JE, Dolin, editors. *Principles and practice of infectious diseases, Volume 2, 7th edition.* Churchill Livingston Elsevier, Philadelphia. p.3305-3318, 2009.

[398] Ferreira MS, Borges AS. Histoplasmosis. *Rev Soc Bras Med Trop*. 2009;42:192-198.

[399] Hunt PJ, Harden TJ, Hibbins M, et al. *Histoplasma capsulatum*. Isolation from an Australian cave environment and from a patient. *Med J Aust*. 1984;141:280-283.

[400] Reyes-Montes MR, Rodriguez-Arellanes G, Perez-Torres A, et al. Identification of the source of histoplasmosis infection in two captive maras (*Dolichotis patagonum*) from the same colony by using molecular and immunologic assays. *Rev Argent Microbiol.* 2009;41:102-104.

[401] Canteros CE, Ianchini RH, Rivas MC, et al. First isolation of *Histoplasma capsulatum* from the urban bat *Eumops bonariensis*. *Rev Argent Microbiol*. 2005;37:46-56.

[402] Gaur PK, Lichtwardt RW. Comparative study of a new *Chrysosporium* species with *Histoplasma capsulatum*. *Sabouraudia*. 1980;18:105-114.

[403] Huhn GD, Austin C, Carr M, et al. Two outbreaks of occupationally acquired histoplasmosis: more than workers at risk. *Environ Health Perspect.* 2005;113L585-589.

[404] Sorley DL, Levin ML, Warren JW, et al. Bat-associated histoplasmosis in Maryland bridge workers. *Am J Med*. 1979;67:623-626.

[405] Taylor ML, Perez-Mejia A, Yamamoto-Furusho JK, Granados J. Immunologic, genetic and social human risk factors associated to histoplasmosis: studies in the State of Guerrero, Mexico. *Mycopathologia*. 1997:138:137-142

[406] Suzaki A, Kimura M, Kimura S, et al. An outbreak of acute pulmonary histoplasmosis among travelers to a bat-inhabited cave in Brazil. *Kansenshogaku Zasshi.* 1995;69:444-449.

[407] Salomon J, Flament Saillour , De Truchis P, et al. An outbreak of acute pulmonary histoplasmosis in members of a trekking trip in Martinque, French West Indies. *J Travel Med*. 2003;10:87-93.

[408] de Vries PJ, Koolen MG, Mulder MM, Kortbeek LM. Acute pulmonary histoplasmosis from Ghana. *Travel Med Infect Dis*. 2006;4:286-289.

[409] Kikuchi K, Sugita T, Makimura K, et al. Is *Histoplasma capsulatum* a native inhabitant of Japan? *Microbiol Immunol*. 2008;52:455-459.

[410] Santos JW, Michel GT, Lazzarotto M, et al. Chronic cavitary pulmonary histoplasmosis. *J Bras Pneumol*. 2009;35:1161-1164.

[411] Reyes-Montes MR, Bobadilla-Del Valle MA, Rodriguez-Arellanes G, et al. Relatedness analysis of *Histoplasma capsulatum* isolates from Mexican patients with AIDS-associated histoplasmosis by using histoplasmin electrophoretic profiles and randomly amplified polymorphic DNA patterns. *J Clin Microbiol*. 1999;37:1404-1408.

[412] Greer DL, McMurray DN. Pathogensis of experimental histoplasmosis in the bat, *Artibeus lituratus*. *Am J Trop Med Hyg*. 1981;30:653-659.

[413] McMurray DN, Greer DL. Immune responses in bats following intranasal infection with Histoplasma capsulatum. *Am J Trop Med Hyg*. 1979:1036-1039.

[414] Estrada-Barcenas DA, Palacios-Vargas JG, Estrada-Venegas E, et al. Biological activity of the mite *Sancassania* sp. (Acari: Acaridae) from bat guano associated with the pathogenic fungus *Histoplasma capsulatum*. *Mem Inst Oswaldo Cruz*. 2010;105:127-131.

Pneumocystis Species in Bats

[415] Walzer PD, Smullian AG. *Pneumocystis* species. In: Mandell GL, Bennett JE, Dolin, editors. *Principles and practice of infectious diseases, Volume 2, 7th edition*. Churchill Livingston Elsevier, Philadelphia. p.3377-3390, 2009.

[416] Derouiche S, Deville M, Taylor ML, et al. *Pneumocystis* diversity as a phylogeographic tool. *Mem Inst Oswaldo Cruz*. 2009;104:112-117.

[417] Guillot J, Demanche C, Hugot JP, et al. Parallel phylogenies of *Pneumocystis* species and their mammalian hosts. *J Eukaryot Microbiol*. 2001;Suppl1:113S-115S.

Blastomycosis

[418] Chapman SW, Sullivan DC. *Blastomyces dermatitidis*. In: Mandell GL, Bennett JE, Dolin, editors. *Principles and practice of infectious diseases, Volume 2, 7th edition*. Churchill Livingston Elsevier, Philadelphia. p.3319-3332, 2009.

[419] Raymond JT, White MR, Kilbane TP, Janovitz EB. Pulmonary blastomycosis in an Indian fruit bat (*Pteropus giganteus*). *J Vet Diagn Invest*. 1997;9:85-87.

[420] Randhawa HS, Chaturvedi VP, Kini S, Khan ZU. *Blastomyces dermatitidis* in bats: first report of its isolation from the liver of *Rhinopoma hardwickei hardwickei*, gray. *Sabouraudia*. 1985;23:69-76.

[421] Khan ZU, Randhawa HS, Lulla M. Isolation of *Blastomyces dermatitidis* from the lungs of a bat, *Rhinopoma hardwickei hardwickei*, gray, in Delhi. *Sabouraudia*. 1982;20:137-144.

[422] Chaturvedi VP, Randhawa HS, Kini S, Khan ZU. Survival of *Blastomyces dermatitidis* in the gastrointestinal tract of an orally infected bat, *Rhinopoma hardwickei hardwickei* gray. *J Med Vet Mycol*. 1986;24:349-352.

Paracoccidioides

[423] Restrepo A, Tobon AM. *Paracoccidioides brasiliensis*. In: Mandell GL, Bennett JE, Dolin, editors. *Principles and practice of infectious diseases, Volume 2, 7th edition*. Churchill Livingston Elsevier, Philadelphia. p.3357-3363, 2009.

[424] Greer DL, McMurray DN. Pathogenesis and immune response to *Paracoccidioides brasiliensis* in the fructivorous bat, *Artibeus lituratus*. *Sabouraudia*. 1981;19:165-178.

[425] Greer DL, Molanos B. Role of bats in the ecology of *Paracoccidioides brasiliensis*: the survival of *Paracoccidioides brasiliensis* in the intestinal tract of frugivorous bat, *Artibeus lituratus*. *Sabouraudia*. 1977;15:273-282.

[426] Grose E, Tamsitt JR. *Paracoccidioides brasiliensis* recovered from the intestinal tract of three bats (*Artibeus lituatus*) in Columbia, S.A.. *Sabouraudia*. 1964;4:124-125.

Coccidioidomycosis

[427] Galgiani JN. *Coccidioides* species. In: Mandell GL, Bennett JE, Dolin, editors. *Principles and practice of infectious diseases, Volume 2, 7th edition*. Churchill Livingston Elsevier, Philadelphia. p.3333-3344, 2009.

[428] Kajihiro ES. Occurrence of dermatophytes in fresh bat guano. *Appl Microbiol*. 1965;13:720-724.

[429] Wanke B, Lazera M, Monteiro PC, et al. Investigation of an outbreak of endemic coccidioidomycosis in Brazil's northeastern state of Piaui with a review of the occurrence and distribution of *Coccidioides immitis* in three other Brazilian states. *Mycopathologia*. 1999;148:57-67.

Basidiobolus Ranarum Eidam

[430] van den Berk GE, Noorduyn LA, van Ketel RJ, et al. A fatal pseudo-tumour: disseminated basidiobolomycosis. *BMC Infect Dis*. 2006:6:140.

[431] Singh R, Xess I, Ramavat AS, Arora R. Basidiobolomycosis: a rare case report. *Indian J Med Microbiol*. 2008;26:265-267.

[432] Chaturvedi VP, Randhawa HS, Khan ZU, et al. Prevalence of *Basidiobolus ranarum* Eidam in the instestinal tract of an insectivorous bat, *Rhinopoma hardwickei hardwickei* gray, in Delhi. *Sabouraudia*. 1984;22:185-189.

[433] Nemenquani D, Yaqoob N, Khoja H, et al. Gastrointestinal basidiobolomycosis: an unusual fungal infection mimicking colon cancer. *Arch Pathol Lab Med*. 2009;133:1938-1942.

[434] Goyal A, Gupta N, Das S, Jain S. Basidiobolomycosis of the nose and face: a case report and a minireview of unusual cases of basidiobolomycosis. *Mycopathologia*. 2010;170:165-168.

White-Nose Syndrome

[435] Frick WF, Pollock JF, Hicks, AC, et al. An emerging disease causes regional population collapse of a common North American bat species. *Science*. 2010;329:679-682.

[436] Wibbelt G, Kurth A, Hellmann D, et al. White-nose fungus (*Geomyces destructans*) in bats Europe. *Emerg Infect Dis*. 2010;16:1237-1243.

[437] Blehert DS, Hicks AC, Behr M, et al. Bat white-nose syndrome: an emerging fungal pathogen? *Science*. 2009;323;227.
[438] Daszak P. Epidemiology. Bats in black and white. *Science*. 2010;329:634-635.
[439] Puechmaille SJ, Verdeyroux P, Fuller H, et al. White-nose syndrome fungus (*Geomyces destructans*) in bat, France. *Emerg Infect Dis*. 2010;16:290-293.
[440] Meteyer CU, Buckles EL, Blehert DS, et al. Histopathologic criteria to confirm white-nose syndrome in bats. *J Vet Diagn Invest.* 2009;21:411-414.
[441] Bouma HR, Carey HV, Kroese FG. Hibernation: the immune system at rest? *J Leukoc Biol.* 2010 Jun 2 [Epub ahead of print]
[442] Della Franca P, Caretta G. Keratinophilic fungi isolated from the air at Pavia. *Mycopathologia*. 1984;85:65-68.
[443] Christen-Zaech S, Patel S, Mancini AJ. Recurrent cutaneous *Geomyces pannorum* infection in three brothers with ichthyosis. *J Am Acad Dermatol*. 2008;58 (5 Suppl 1):S112-S113.
[444] Cosgrove L, McGeechan PL, Robson GD, Handley PS. Fungal communities associated with degradation of polyester polyurethane in soil. *Appl Environ Microbiol.* 2007;73:5817-5824.

Parasites

Trypanosoma Cruzi

[445] Kirchhoff LV. *Trypanosma* species (American trypanosomiasis, Chagas' disease): biology of trypanosomes. In: Mandell GL, Bennett JE, Dolin, editors. *Principles and practice of infectious diseases, Volume 2, 7th edition*. Churchill Livingston Elsevier, Philadelphia. p. 3481-3488, 2009.
[446] Noireau F, Diosque P, Jansen AM. *Trypanosoma cruzi*: adaption to its vectors and its hosts. *Vet Res*. 2009;40:26.
[447] Poinar G Jr. *Triatoma dominicana* sp. n. (Hemiptera: Reduviidae: Triatominae), and *Trypanosoma antiquus* sp. n. (Stercoraria: Trypanosomatidae), the first fossil evidence of a triatomine-trypanosomatid vector association. *Vector Borne Zoonotic Dis.* 2005;5:72-81.
[448] Lisboa CV, Pinho AP, Herrera HM, et al. *Trypanosoma cruzi* (Kinetoplastida, Trypanosomatidae) genotypes in neotropical bats in Brazil. *Vet Parasitol.* 2008;156:314-318.
[449] Lisboa CV, Xavier SC, Herrera HM, Jansen AM. The ecology of the *Trypanosoma cruzi* transmission cycle: dispersion of zymodeme 3 (Z3) in wild hosts from Brazilian biomes. *Vet Parasitol.* 2009;165:19-24.
[450] Cavazzana M Jr, Marcilli A, Lima L, et al. Phylogeographical, ecological and biological patterns shown by nuclear (ssr RNA and gGAPDH) and mitochondrial (Cyt B) genes of trypanosomes of the subgenus *Schizotrypanum* parasitic in Brazilian bats. *Int J Parasitol*. 2010;40:345-355.
[451] Marcili A, Limma L, Cavazzana M, et al. A new genotype of *Trypansoma cruzi* associated with bats evidenced by phylogenetic analyses using SSU RDNA,

cytochrome b and histone H2B genes and genotyping based on ITS1 rDNA. *Parasitology*.; 2009;136:641-655.

Microsporidiosis

[452] Weiss LM. Microsporidiosis. In: Mandell GL, Bennett JE, Dolin, editors. *Principles and practice of infectious diseases, Volume 2, 7th edition*. Churchill Livingston Elsevier, Philadelphia. p.3391-3407, 2009.

[453] Furuya K. Spore-forming *Microsporidian encephalitozoon*: current understanding of infection and prevention in Japan. *Jpn J Infect Dis*. 2009;62:413-422.

[454] Kasickova D, Sak B, Kvac M, Ditrich O. Sources of potentially infectious human microsporidia: molecular characterization of microsporidia isolates from exotic birds in the Czech Republic, prevalence study and importance of birds in epidemiology of the human microsporidial infections. *Vet Parasitol*. 2009;165:125-130.

[455] Slodkowicz-Kowalska A. Animal reservoirs of human virulent *Microsporidian* species. *Wiad Parazytol*. 2009;55:63-65.

[456] Slodkowicz-Kowalska A, Graczyk TK, Tamang L, et al. *Microsporidian* species known to infect humans are present in aquatic birds: implications for transmission via water?. *Appl Environ Microbiol*. 2006;72:4540-4544.

[457] Bart A. Wentink-Bonnema EM, Heddema ER, et al. Frequent occurrence of human-associated microsporidia in fecal droppings of urban pigeons in Amsterdam, the Netherlands. *Appl Environ Microbiol*. 2008;74:7056-7-58.

[458] Childs-Sanford SE, Garner MM, Raymond JT, et al. Disseminated microsporidiosis due to *Encephalitozoon hellem* in an Egyptian fruit bat (*Rousettus aegyptiacus*). *J Comp Pathol*. 2006;134:370-373.

[459] De Bosschere H, Wang Z, Orlandi PA. First diagnosis of *Encephalitozoon intestinalis* and *E. hellem* in a European brown hare (*Lepus europaeus*) with kidney lesions. *Zoonoses Public Health*. 2007;54:131-134.

Angiostrongylosis

[460] Nash TE. Visceral larva migrans and other unusual helminth infections. In: Mandell GL, Bennett JE, Dolin, editors. *Principles and practice of infectious diseases, Volume 2, 7th edition*. Churchill Livingston Elsevier, Philadelphia. p.3391-3407, 2009

[461] Barrett JL, Carlisle MS, Prociv P. Neuro-angiostrongylosis in wild black and grey-headed flying foxes (*Pteropus* spp.). *Aust Vet J*. 2002;80:554-558.

Bed Bugs And Bats

[462] Goddard J, deShazo R. Bed bugs (*Cimex lectularius*) and clinical consequences of their bites. *JAMA*. 2009;301;1358-1366.

[463] Reinhardt K, Siva-Jothy MT. Biology of the bed bugs (Cimicidae). *Ann Rev Entomol*. 2007;52:351-374.

[464] Bartonicka T. Survival rate of bat bugs (*Cimex pipistrelli*, Heroptera) under different microclimatic conditions. *Parasitol Res*. 2010;107:827-833.

[465] Bartonicka T. *Cimex pipistrelli* (Heteroptera, Cimicidae) and the dispersal propensity of bats: an experimental study. *Parasitol Res*. 2008;104:163-168.

Allergic Disease

[466] el-Ansary EH, Tee RD, Gordon DJ, Taylor AJ. Respiratory allergy to inhaled bat guano. *Lancet*. 1987;1:318-318.

ALS-PDC Neurodegenerative Disease in Guam

[467] Caban-Hol A, Mattingly M, Cooper G, Schmitt FA. Neurodegenerative memory disorders: a potential role of environmental toxins. *Neurol Clin*. 2005;23:485-521.

[468] Steele JC, McGreer PL. The ALS/PDC syndrome of Guam and cycad hypothesis. *Neurology*. 2008;70:1984-1990.

[469] Cox PA, Banack SA, Murch SJ. Biomagnification of cyanobacterial neurotoxins and neurodegenerative disease among the Chamorro people of Guam. *Proc Natl Acad Sci U S.A*. 2003;100:13380-13383.

[470] Cox PA, Sacks OW. Cycad neurotoxins, consumption of flying foxes, and ALS-PDC disease in Guam. *Neurology*. 2002;26:956-959.

[471] Banack SA, Murch SJ, Cox PA. Neurotoxic flying foxes as dietary itrems for the Chamorro people, Marianas Islands. *J Ethnopharmacol*. 2006;106:97-104.

[472] Stone R. Toxicology. Fruit bats linked to mystery disease. *Science*. 2002;296:241.

In: Bats: Biology, Behavior and Conservation
Editors: J. L. Zupan and S. L. Mlakar, pp. 91-134
ISBN 978-1-61209-749-7

Chapter 2

FORAGING BEHAVIOUR OF THE SHORT-NOSED FRUIT BAT, *CYNOPTERUS SPHINX*

Vadamalai Elangovan[1], Yuvana Satya Priya[1] and Ganapathy Marimuthu[2]

[1]Department of Applied Animal Sciences, Babasaheb Bhimrao Ambedkar University, Vidya Vihar, Uttar Pradesh, India

[2]Department of Animal Behaviour and Physiology, School of Biological Science, Madurai Kamaraj University, Tamil Nadu, India

ABSTRACT

Frugivorous bats are important members of tropical ecosystems and they play a key role in pollination and seed dispersal in tropical forests throughout the World. They evolved in a mutualistic exploitation system with plants in which they obtain food in the form of nectar, pollen or fruit while providing mobility of pollen grains or seeds. The interactions have been profoundly influenced the co-evolution between bats and plants. This Chapter deals about the foraging behaviour includes individual and group foraging, folivory and nectarivory behaviour, temporal variations in the foraging behaviour and influence of moonlight on the foraging behaviour of the short-nosed fruit bat, *Cynopterus sphinx* (Vahl, 1797). *Cynopterus sphinx* is a medium-sized fruit bat, common in many parts of India. It lives in small colonies, generally of three to four individuals. Its diurnal roosts include the underside of the dry fronds of palm trees *Borassus flabellifer*, the flower/fruit clusters of kitul palms *Caryota urens,* the stems of mast tree *Polyalthia longifolia* as well as under the roof of abandoned buildings.

Cynopterus sphinx feeds upon 23 species of fruits, 8 species of leaves, flowers / petals of *Coccinia indica*, and nectar from *Musa paradisiaca, Bassia latifolia* and *Ceiba pentandra*. *Cynopterus sphinx* begins to visit fruit-bearing trees about 30 min after sunset and often hovers nearer to food sources or lands on fruits or on nearby branches to remove whole or part of fruit by its mouth. These bats seldom remain in the fruit-bearing trees to feed, instead carry fruits to feeding roosts, repeating this behaviour several times throughout the night. Temporal distribution of the foraging behaviour of *C. sphinx* revealed two peaks of activity, one in the pre-midnight hours when bats fed mostly on "steady state" fruits and another during the post-midnight hours when bats fed on "big bang" fruits. Only solitary bats visited and fed on species with steady-state fruiting phenologies, whereas groups of bats regularly visited and fed on species with big bang fruiting phenologies.

Cynopterus sphinx fed the leaves mainly during the postmidnight hours. They snatched a leaf on flight without landing and hovering. The folivory in *C. sphinx* has been substantiated by collection of discarded bolus of the leaves of *Cassia fistula*, *Mimusops elengi* and *Coccinia indica* as fibrous pellets beneath the day roosts of *C. sphinx* and a mist-netted adult male bat had a fresh leaf of *C. fistula* in its mouth. *Cynopterus sphinx* exhibits folivory by ingesting only the protein-rich soluble contents of leaves and expelling the undigestible fibrous portions. Folivory in *C. sphinx* helps to meet their daily protein requirements.

Temporal variations have been observed on nectar feeding behaviour of *C. sphinx* when bats fed on flowers of a big bang tree *Bassia latifolia* and a steady state tree *Musa paradisiaca*. The bats foraged in-groups and solitarily upon the flowers of *B. latifolia* and *M. paradisiacal*, respectively. Group foraging facilitates an increased protection from predators, an increased sensory area and an increased knowledge on food patches. On the other hand, reduced feeding interference and reduced competition for food are the advantages with solitary foraging. The peak of bat-visits to the flowers of *M. paradisiaca* occurred at 2000 h and to flowers of *B. latifolia* it was at 2100 h. The peak visit coincided with the time of occurrence of maximum amount of nectar secretion and the relatively high concentration of sugar. The bats might acquire more energy by feeding upon the nectar, in addition to fruits. At the same time, the bats offer pollination service to the plants. These interactions have profoundly influenced the coevolution of bats and plants.

Moonlight plays an important role on the foraging behaviour of *C. sphinx*. Observations on bats were made when they foraged on the fruits of *Calophyllum inophyllum* and on the leaves of *Mimusops elengi* and *Cassia fistula*. The foraging activity gets modulated with different phases of the moon. During bright moonlit nights *C. sphinx* exhibited less activity as against that during new moon nights. The total number of feeding bouts was negatively correlated with percent moonlight. The number of feeding bouts decreased during the waxing phases of the moon. During the period of new moon the mean number of feeding bouts was significantly higher compared to the periods of other phases. During the first quarter and the last quarter phases, feeding occurred more during the dark part of the night, especially when the bats foraged upon *M. elengi* and *C. fistula*. While cloud cover enhanced the foraging activity, lightning and thunder suppressed it.

The radio-telemetry studies on the foraging behaviour of *C. sphinx* suggest that the mean commuting distance (0.22 ± 0.19 km) and mean foraging area (0.75 ± 0.27 km^2) of males comparatively lesser than the mean commuting distance (2.1 ± 1.0 km) and mean foraging area (0.83 ± 0.12 km^2) of females. Both males and females exhibit a high level of activity during early hours of the night soon after emergence and another activity peak during pre-dawn hours.

INTRODUCTION

Almost all megachiropteran bats are phytophagous (feed upon fruits, blossoms, nectar, pollen and leaves) and they belong to the Palaeotropical family Pteropodidae. Megachiropterans are mostly tree-dwellers, and some species roost in caves and temples. Old World fruit bats (Megachiroptera: Pteropodidae) are highly mobile (Webb and Tidemann, 1996) and at least 300 plant species of nearly 200 genera are known to rely on them for either pollination and or seed dispersal (Marshall, 1983; Fleming, 1988; Banack, 1998; Knudsen and Kiltgaard, 1998). Frugivorous bats are important pollen and seed dispersers in both the Paleotropics and Neotropics, and at least 300 plant species are known to rely on Old World fruit bats (Megachiroptera: Pteropodidae) for their propagation. However, rapid food transit times (generally less than 30 minutes) in frugivorous bats have been thought to limit their ability to disperse seeds to just a few tens of kilometers (Shilton *et al.*, 1999).

The causes for food choice and foraging behaviour of bats could be constrained by both extrinsic and intrinsic factors (Fleming, 1988). The potential extrinsic factors include abundance, diversity and seasonality of different food items; the relative energy costs of searching, handling and assimilation of different food items; location and renewal rate of food; feeding behaviour of competitors and risk of predation. Intrinsic factors include the individuals' age, sex, reproductive condition and social status (Fleming, 1988). For instance, plant-visiting bats have four major concerns regarding the nature of their food supply: what kinds of plant foods are available; how much is there of each kind; where is it located and for how long it is available? Gentry (1974) recognized four basic flowering and fruiting patterns, ranging from 'big bang' (production of a large number of flowers over a short period) to 'steady state' (production of a small number of flowers over an extended time period). The remaining two strategies include 'cornucopia' (large number of flowers or fruits over a month's time) and 'multiple bang' (several flushes in a year).

The foraging behaviour and group size of plant-visiting bats appear to be significantly influenced by the resource abundance and distribution patterns of a foraging habitat (Fleming, 1982, 1988; Krebs and McCleery, 1984). In an environment where food is more or less homogeneously distributed around a roost, 'refuge' theory predicts that feeding sites at different distances from the refuge should be visited at random by the roost inhabitants (Fleming, 1988). The optimal diet breadth has also been considered as a central issue in many discussions of foraging behaviour and it is influenced by the abundance, diversity and seasonality of different kinds of food (Fleming, 1982). This is well documented by the behaviour of *Phyllostomus discolor*, which consumes a variety of fruits and flowers in Costa Rica and Brazil (Heithaus *et al.*, 1974; Sazima and Sazima, 1977). Fleming (1982, 1988) suggested that *Carollia perspicillata* is a solitary forager and probably preferred food plants that produce only a small amount of fruits in a night, probably to reduce competition when food resources are limited. Fleming (1988) suggested that bats preferred to feed on scarce fruits early in the evening before switched over to more abundant and energy rich fruits (e.g. *Piper amalago* and *Cecrcopia peltata*). In addition, *C. perspicillata* food choice is influenced by fruit density, length of fruiting season and nutritional composition of fruits. Heithaus *et al.* (1975) suggested that competition among bats for food has influenced the structure of bat communities.

Group foraging provides protection from predators while commuting and searching for food, and perhaps functions in information transfer about which food patches have previously been visited by conspecifics (Wilson, 1975). Group foraging was observed in *Leptonycteris sanborni* (at Adansonia flowers), *Eonycteris spelaea* (at various big bang flowers) and probably Australian *Pteropus* (at *Eucalyptus* flowers) (Nelson, 1965; Ayensu, 1974; Start and Marshall, 1976; Howell, 1979). Flock foraging probably increases the rate of patch discovery and decreases the chance that individuals will waste time and visiting already depleted plants. In addition, flock foraging also increases the potential for feeding interference among individuals, but solitary foraging avoids this. Flock foraging for fruit is probably common in pteropodids that live in large colonies (e.g. *Eidolon helvum, Rousettus aegyptiacus* and

Pteropus sp.). Ayensu (1974) and Kingdon (1974) deduced that feeding interference could occur while bats cling on to flowers and fruits for lapping nectar and ingesting fruit pulp. *Cynopterus sphinx* forages solitarily and in groups on fruit and flower bearing trees which have either steady state or big bang phenological patterns. The steady state fruiting trees produce only in small numbers of fruits for a long period while big bang fruiting trees produce a large number of fruits for short period.

Fruits are not exclusive food items for most of the plant visiting bats. In addition to fruits, the Neo- and Paleotropical bats feed upon flowers, pollen, nectar and leaves (Gardner, 1977; Marshall, 1985; Wiles and Fujita, 1992). These additional food items must be beneficial to them. For example, nectar provides energy in the form of sugars (Petit and Freeman, 1997) and pollen (Law, 1992) and leaves (Fleming, 1982) provide an important protein source for bats. While there are relatively more number of reports available on bats feeding upon nectar (Heithaus *et al.*, 1975; Sazima and Sazima, 1977; Fleming, 1982; Heithaus, 1982; Fischer, 1992; Calley *et al.*, 1993; Petit, 1997; Petit and Freeman, 1997), reports on bats feeding upon leaves (folivory) are rare (Zortea and Mendes, 1993; Kunz and Diaz, 1995; Tan *et al.*, 1998). This might presumably due to the rarity of observing the bats visually when they snatch leaves from trees under natural conditions. Observations of folivory by bats mainly relied on the collection of discarded leaves and leaf pellets at their feeding as well as day roosts (Kunz and Diaz, 1995).

The activity patterns of nocturnal animals have been influenced by moonlight. Animals often show moonlight avoidance behaviours such as reducing the use of open space, restricting foraging activity and movements, switching the activity to dark periods and shifting the activity to areas of heavy cover (Erkert, 1974, 1976, 1989; Trent *et al.*, 1977; Morrison, 1978; Wright, 1981; Reith, 1982; Erkert and Groeber, 1986; Julien-Laferrier, 1997). The bannertail kangaroo rat *Dipodomys spectabilis* restricts the feeding activity to the dark part of the moonlight hours (Lockard and Owings, 1974). The influence of moonlight on the activity pattern of a few species of bats has been well documented both under field (Crespo *et al.*, 1972; Erkert, 1974; Fenton *et al.*, 1977; Morrison, 1978; Fleming and Heithaus, 1986) and laboratory conditions (Haeussler and Erkert, 1978).

Apart from bats, moonlight also influences the activity pattern of other nocturnal mammals such as the marsupial *Caluromys philander* (Julien-Laferriere, 1997), North American desert rodents (Lockard and Owings, 1974; Bowers, 1990; Longhand and Price, 1991), temperate rodents (Jahoda, 1973; Wolfe and Summerlin, 1989), desert and arctic lagomorphs (Butynski, 1984; Gilbert and Boutin, 1991), badgers (Cresswell and Harris, 1988) and primates (Erkert, 1976; Trent *et al.*, 1977). The moonlight avoidance behaviour is exhibited mainly to avoid visually orienting predators, because neither resource abundance nor social activity in bats is correlated with lunar illuminations (Morrison, 1978). There may be a chance for higher risk of predation by visually orienting predators like owl. During moonlight periods, terrestrial animals reduce activity or prefer to occupy more cluttered habitats, which presumably provide more cover (in birds: Imber, 1975; Storey and Grimmer, 1986; Watanuki, 1986; Nelson, 1989, in rodents: Kotler, 1984; Price *et al.*, 1984; Longhand and Price, 1991). Reith (1982) suggested that if bats exhibit 'lunar phobia' in response to increased predation risk then activity levels should decline with increasing levels of moonlight. Fleming and Heithaus (1986) reported that the foraging behaviour of the frugivorous bat *C. perspicillata* is sensitive to moonlight conditions in the dry season. All these studies show that when there is an increase in the intensity of moonlight, animals reduce the use of open space and restrict their foraging activity in the periods of darkness. In most of these field studies, observations on the animals have been made mainly on the particular days where the four moon phases occurred.

The foraging behaviour of frugivorous bats, such as utilization of resource patch, changes of the feeding location and dietary differences between males and females are well documented in the temperate zone species (Fleming, 1982; Tidemann *et al.*, 1985; Audet, 1990). Such systematic studies on the foraging behaviour of bats in the tropics are far less numerous. The foraging behaviour of tropical bats is known to change seasonally thereby affecting diet breadth, food choice, patterns of habitat use, defence of feeding areas, foraging group size and migratory behaviour (Heithaus *et al.*, 1975; Bradbury and Vehrencamp, 1976; Start and Marshal, 1976; Vaughan, 1976; Whitaker and Black, 1976; Bradbury, 1977; Fenton *et al.*, 1977; Bonaccorso, 1979; Thomas, 1983; Lemke, 1984; Barclay, 1985; Fenton *et al.*, 1985).

The short-nosed fruit bat *Cynopterus sphinx* (Vahl, 1797) ranges from Pakistan, India and Sri Lanka to southern China, Hong Kong, Malaysia and Java (Bates and Harrison, 1997). This is a medium-sized fruit bat which is common in many parts of India. *Cynopterus sphinx* lives in small colonies, generally of three to four individuals, although up to 25 have been found roosting together. Its diurnal roosts include the underside of the dry fronds of palm trees *Borassus flabellifer*, the flower/fruit clusters of kitul palms *Caryota urens,* and the stems of mast tree *Polyalthia longifolia* (Balasingh *et al.*, 1995; Bhat and Kunz, 1995). *Cynopterus sphinx* feeds upon 23 species of fruits, 8 species of leaves (Bhat, 1994), flowers of *Coccinia indica* (Elangovan and Marimuthu, 2001), and nectar from *Musa paradisiaca, Bassia latifolia* (Elangovan *et al.*, 2000) and *Ceiba pentandra* (Singaravelan and Marimuthu, 2004). Several individuals of *C. sphinx* feed together on the same fruit tree (Elangovan *et al.*, 1999) and share fruit trees with other species such as *Pteropus giganteus* and *Rousettus leschenaulti* (Singaravelan and Marimuthu, 2004; Nathan *et al.*, 2005). Generally, *C. sphinx* plucks smaller fruits and carries them away from the original tree to feeding perches where it chews the soft parts, swallows the juice and drops seeds and chewed fibers. In this way they are active seed dispersers.

Several studies have been conducted on biology and behaviour of *C. sphinx* (e.g. Subramanya and Radhamani, 1993; Balasingh *et al.*, 1993; Gopukumar *et al.*, 1998, 1999, 2003; Elangovan, 2000; Elangovan and Marimuthu, 2001; Elangovan *et al.*, 1999, 2000, 2001; Storz *et al.*, 2000a,b, 2001; Nathan, 2001; Nathan *et al.*, 2005). Data collected so far on the foraging behaviour of *C. sphinx* are more biased to male individuals since tent roosting harem males exhibit several behaviours that make them easier to study (Balasingh *et al.*, 1995; Marimuthu *et al.*, 1998; Nair *et al.*, 1999). However, little is known about nectarivory and folivory in fruit bats, temporal variations on foraging behaviour, effect of moonlit on foraging behaviour, group and solitary foraging behaviour of *C. sphinx*. In addition, there was a lacuna on the effect of intrinsic factors include individuals' age, sex, reproductive conditions and social status, and extrinsic factors such as location and renewal rate of food, feeding behaviour of competitors and risk of predation on foraging behaviour of *C. sphinx*. Therefore, the authors have conducted a series of studies to fulfill the above lacunae and presented the outcome in this Chapter.

METHODS

Foraging Behaviour: Study Area, Food Plants, Phenology and Visual Observations

Observations on the foraging behaviour of *C. sphinx* were carried out on ten species of fruit bearing trees, two species of flowering trees. Folivory was studied on three tree species. Observations were made near the Madurai Kamaraj University campus (9° 58' N, 78° 10' E) between May 1996 and January 1998. The fruit foraging behaviour of *C. sphinx* was observed on each three species of plants from Family Moraceae and Annonaceae, and on one species from Sapotaceae, Cucurbitaceae and Combretaceae. Nectar feeding behaviour was observed on each one species of Family Sapotaceae and Musaceae. Observations on folivory by bats were carried out on Indian Laburnum *Cassia fistula* (Caesalpiniaceae), *Mimusops elengi* (Sapotaceae) and Ivy gourd *Coccinia indica* (Cucurbitaceae) for 28, 27 and 14 nights, respectively.

The fruit and flower bearing species that were observed included Custard apple *Annona squamosa* (9 nights), Mast tree *Polyalthia longifolia* (18 nights), *Polyalthia pendula* (16 nights), Sapota *Achras sapota* (6 nights), Alexandrian Laurel *Calophyllum inophyllum* (24 nights), Ivy gourd *Coccinia indica* (14 nights), Indian Almond-tree *Terminalia catappa* (11 nights), Peepal tree *Ficus religiosa* (5 nights), *Ficus benjamina* (11 nights), and Banyan tree *Ficus bengalensis* (7 nights). Visual observations on bats visiting the flowers were made continuously on a Butter tree *Bassia latifolia* for 10 nights and eight trees of Banana *Musa paradisiaca* for 83 nights which added up to a total of 1116 hours of observations. Three species of *Ficus* and one species of *Bassia* have a big bang fruiting and flowering pattern, respectively, whereas the remaining six plant species exhibit steady state fruiting/flowering pattern (Gentry 1974). The foraging behaviour and number of feeding bouts were monitored continuously from a vantage point by using a red filtered torch (> 640 nm) and a night vision sniperscope (FJW Optical Systems INC, Palatine, Illinois). The number of feeding bouts of *C. sphinx* was counted by using a tally-counter and recorded continuously between 1800 h and 0500 h. In a typical feeding bout a bat collects a part or a whole fruit in its mouth. Food abundance was estimated by counting mean number of fruits in a twig or a branch and multiplying this figure by total number of branches in a tree (August, 1981).

The time duration spent by bats feeding upon the nectar during each visit was noted by using a stopwatch. The amount of nectar secreted by the entire inflorescence of *M. paradisiaca* was collected for every hour between 1800 h and 0600 h on five trees for a total of 46 collection-nights, producing a total of 552 hour-samples. During this quantification of nectar production, feeding by bats was avoided by covering the flowers with nylon-meshed polythene bags. The nectar samples were drawn from each flower by using a 1 ml syringe (needleless), collected in 1.5 ml eppendorf tubes and quantified by using a pipetman (Gilson 200 μl). To estimate the total sugar present in the nectar, 500 μl from each hour sample collected from 2-3 trees was used. Total sugar was estimated by following the method of Dubois *et al.* (1956). Since the flowers of *B. latifolia* began to drop from the tree at 2100 h, nectar collection after 2100 h was not continued. However at 2100 h, the volume of nectar from ten fresh flowers was measured. The number of flowers present in the tree at the beginning of observation was estimated by the method suggested by August (1981).

Bats were captured in mist nets, set near fruit and flower bearing trees and were fitted with aluminum collars (mean weight 0.9 g and mean width 0.4 mm) covered with reflective tape. In addition, chemiluminescent tags (Cormoran Mini Knicklicht, 3.0 mm x 2.5 mm) which glow for about eight hours were attached with the collar. A collar together with a chemiluminescent tag weighed less than 5% of the body mass of the bat. The reflective tape allowed us to locate bats in the fruit trees with a dim light source in the event that the brightness of the chemiluminescent tag deteriorated during the course of time. Data are presented as mean ± SD. The mean number of feeding bouts by bats near food resources was plotted for every hour, One-way Analysis of Variance and Student Newman-Keul's tests were used to evaluate differences in peak hours of activity.

Influence of Moonlight on Foraging Behaviour: Study Area and Visual Observations

Visual observations were carried out when bats were foraging on the fruits of *Calophyllum inophyllum* and on the leaves of *Mimusops elengi* and *Cassia fistula* for 31, 27 and 28 consecutive nights respectively, at the botanical garden of Madurai Kamaraj University campus (9°58' N, 78°10'E). There was no light source nearer to the food sources from which we made observations. The number of feeding bouts was recorded every hour for the whole night between 1800 h and 0500 h for a total of 946 observation hours. Visual observations were carried out from a vantage point with the help of natural light and by using a night vision sniperscope (Litton precision product, Germany, M - 972). The number of feeding bouts was counted by using of a tally counter, and total number of feeding bouts was recorded. A single feeding bout means removal of a part (or whole) of a fruit or a leaf from the parent tree by bats. Timings of moonrise and moonset were obtained from the Nautical Almanac and adjusted to Indian Standard Time (IST). The period available between onset of moon and offset of moon known as duration of moonlight in a night and this duration was converted into percentage. The number of feeding bouts was related to the percentage (duration) of moonlight. Paired t-tests and One Way Repeated Measures ANOVA using SIGMASTAT were carried out to compare the number of feeding bouts observed for each day with the percentage (duration) of moonlight available for the corresponding day.

Radio-telemetry studies were carried out to substantiate the observations on the influence of moonlight on the foraging behaviour of *C. sphinx*. For radio-telemetry studies, two male bats were captured within the Madurai Kamaraj University campus and they were fitted with radio transmitters (range covering 400 – 500 m) mounted over an aluminium collar which covered with light reflective tape. For the purpose of monitoring, two sets of Merlin receiver and collapsible 5 – element Yagi antenna (Customs electronics, Urbana, Illinois, USA) were used. One of the bats radio-tagged five days prior to new moon and other bat five days prior to full moon and their activities were monitored for seven days continuously. The tracking of radio-tagged bats was carried out continuously between emergence and return to the roost the next morning. Signals that fluctuated in amplitude were considered as activity, and the constant unfluctuated signals are considered as rest.

Radio-Telemetry

The radio-telemetry studies were carried out between May 1997 and June 1998 with 16 individuals of *C. sphinx* (five adult males and five adult females; three subadult males and three subadult females). The radio-tracking studies on each individual carried out for a period ranged from seven to 15 nights. The study area, Madurai Kamaraj University campus (9° 58' N, 78° 10' E) and adjoining areas, is surrounded by trees like *Azadiracta indica, Bassia latifolia, Borassus flabellifer, Caryota urens, Cocos nucifera, Ficus bengalensis, F. benjamina, F. religiosa, Mimusops elengi, Polyalthia pendula, P. longifolia,* etc and intercepted by hill ridges with a little vegetation at some places. Male *C. sphinx* (mean 47 $\pm$ 3.2 g) chewed and clipped the twigs of the interior of the foliage of *P. longifolia* and *B. flabellifer* to make tent roosts. Bats were captured by erecting mist nets (Avinet - Dryden, NY. 13053-1103, U S A) around these tents. The length of forearm was measured to the nearest 0.1 mm using vernier calipers and body mass was weighed nearest to the 1.0 g using a spring balance (Avinet - Dryden, NY. 13053-1103, U S A). Bats were fitted with radio-transmitters (mean = 2.6 g) that were mounted over aluminium collar which was covered by light reflective tape. The transmitter along with collar weighed about 5.5% (adults) and 8.2% (subadult) of the average body mass of *C. sphinx*, well below the upper limit 10% suggested for terrestrial animals (Brander and Cochran, 1969). Two sets of receivers and collapsible 5 - element Yagi antennae (Customs Electronics, Urbana, Illinois, USA) were used. The techniques, triangulation and estimation of the foraging area were carried out by the following method suggested by Kunz (1987).

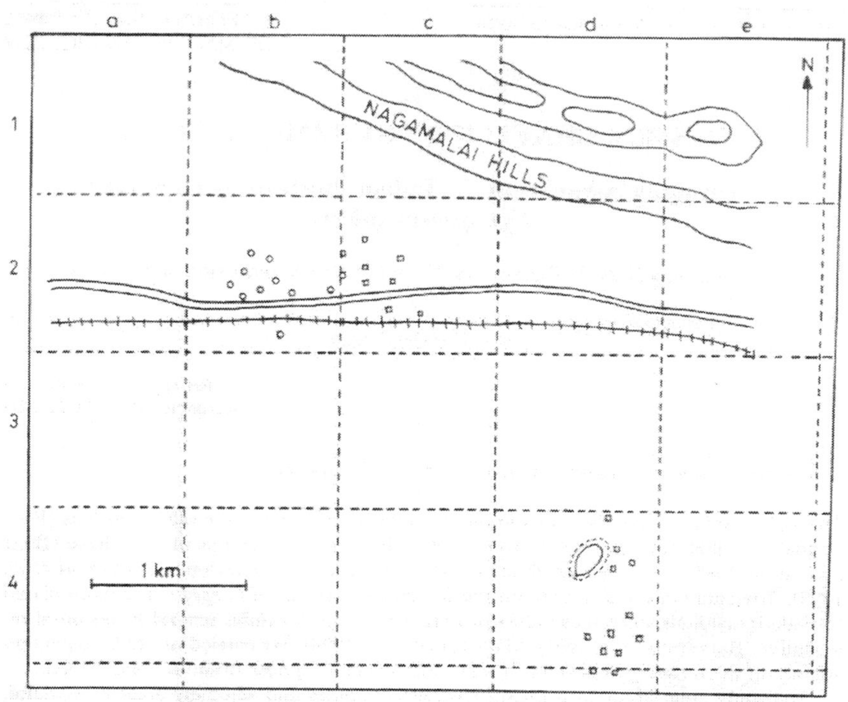

Plate 1. Typical representative triangulated foraging areas of a male, M4 (O) and female, F1 (□) *C. sphinx*, in a grid map of the study area in Madurai. Southern India, Each grid covers 1 km^2.

The centres of these foraging areas were estimated and the distance traveled by bats was calculated by measuring the distance between the center of foraging area and the day roost. The study area map (Plate 1) has been divided into 20 grids of 1 km^2 area each. It was marked 'a' to 'e' horizontally and '1' to '4' vertically. This facilitated the naming of specific foraging areas and day roost of each bat. The activity budget of the bats was calculated by the fluctuation of 'beep' pulses from the receiver and their behaviour was also monitored with a night vision sniperscope (Litton Precision Product, Germany, M - 972). Constant unfluctuated signals were considered as 'rest' whereas fluctuating signals were considered as 'activity', which may cover both flying as well as moving in the night roosts or in the fruit trees. Five males and five females were radio-tagged and monitored for 81 nights consisting of 640 night hours. The radio-tracking studies on adult bats were carried out on two seasons dry (June - July) and wet (September - November).

RESULTS

Fruiting Pattern of Food Plants and Frugivory Behaviour of C. Sphinx

All the three species of *Ficus* trees and one species of *Bassia* produced numerous fruits and flowers, respectively for a short period (big bang). However, the ficus trees flushed many times in a year, and are called 'multiple bang' or 'freak fruiting'. Species such as *T. catappa, P. longifolia, P. pendula* and *C. inophyllum* produced the minimum number of fruits over about two months, whereas *A. squamosa, A. sapota* and *M. paradisiaca* produced the least number of fruits and flowers for a month. *C. indica* produced fruits almost over the year. The fruit mass and daily availability of fruits recorded during the study period are summarized in Table 1.

Table 1. Fruit mass and estimated number of daily fruit availability.

Tree species	Fruit mass (g)	No. of fruits available / tree
Steady state species		
Annona squamosa	126.0 ± 5.8	3
Polyalthia pendula	1.5 ± 0.5	62
Polyalthia longifolia	1.8 ± 0.2	*
Coccinia indica	3.7 ± 1.3	5
Achras sapota	83.8 ± 9.8	7
Terminalia catappa	40.8 ± 11.9	18
Calophyllum inophyllum	11.1 ± 2.1	20
Big bang species		
Ficus religiosa	1.1 ± 0.2	36,480
Ficus benjamina	1.2 ± 0.3	19,636
Ficus bengalensis	2.1 ± 0.3	8,470

* Data are not collected.

Plate 2.(From top to bottom) Left panel: 1. *C. sphinx* hovering to pluck fruit of *F. religiosa*. 2. *C. sphinx* feeding upon fruit of *F. benjamina*. 3. *C. sphinx* hovering to pluck flower of *B. latifolia*. Right panel: 1. *C. sphinx* feeding upon fruit of *A. squamosa*. 2. *C. sphinx* snatching a leaf of *M. elengi*.

C. sphinx began visiting fruit trees about 30 min after sunset. Individuals briefly hovered or landed on the fruit or on nearby branches (Plate 2) and plucked off whole (*P. longifolia*, *P. pendula*, *T. catappa*, *C. indica*, *C. inophyllum*, *F. religiosa*, *F. benjamina, F. bengalensis* and *Bassia latifolia* - flowers) or parts of fruits (*A. squamosa*, *T. catappa*, *C. indica*, *C. inophyllum* and *A. sapota*). Circling flights around the trees preceded the removal of fruits by bats. These bats seldom remained in the fruit-bearing trees to feed, but instead carried fruits to feeding roosts, repeating this behaviour several times throughout the night. While transporting whole fruits of *T. catappa*, which weighed 84.8 % of *C. sphinx* body mass, individual bats sometimes dropped fruits at an average distance of 21.9 $\pm$ 11.8 m (n = 77) from fruit trees, before reaching a feeding roost. Similarly, bats often dropped whole fruits of *C. inophyllum*, which weighed 23.1 % of *C. sphinx* body mass, at an average distance of 33.0 $\pm$ 13.8 m (n = 168) from fruit trees.

In a few instances, bats briefly (3 to 5 min) fed on fruits in the fruit bearing trees, especially in *A. squamosa*, *T. catappa* and *A. sapota*. *C. sphinx* fed only on ripe fruits and routinely ignored unripe fruits. The temporal distributions of feeding bouts of *C. sphinx*, which fed on steady state and big bang fruits showed that there were two peaks of feeding bouts occurred in a night. The first peak occurred during the pre-midnight hours when bats fed mostly on steady state fruits, and the second peak of feeding activity occurred during the post-midnight hours when they fed upon the big bang fruits. However, there were temporal differences at which the feeding peaks occurred.

The peak of nightly visits occurred at 2100 h when bats fed on *A. squamosa* (Figure 1; $F_{10,\ 88} = 9.1$, $p < 0.01$). The tagged bats plucked a piece of fruit and carried to their feeding roost which was located about 15 – 20 m away from the food source. On rare occasions bats landed over the fruits or nearby branches and fed for a few minutes (3 – 5 min) and then flew away. Foraging conflicts were observed rarely when several bats commuted to feed upon fruits of *A. squamosa* (A tagged bat was observed to commute 33 times to the same fruiting tree in a night). The peak times for foraging on *P. pendula* (Figure 2; $F_{10,\ 165} = 15.2$, $p < 0.001$) and *P. longifolia* (Figure 3; $F_{10,\ 187} = 7.5$, $p < 0.01$) also occurred before midnight.

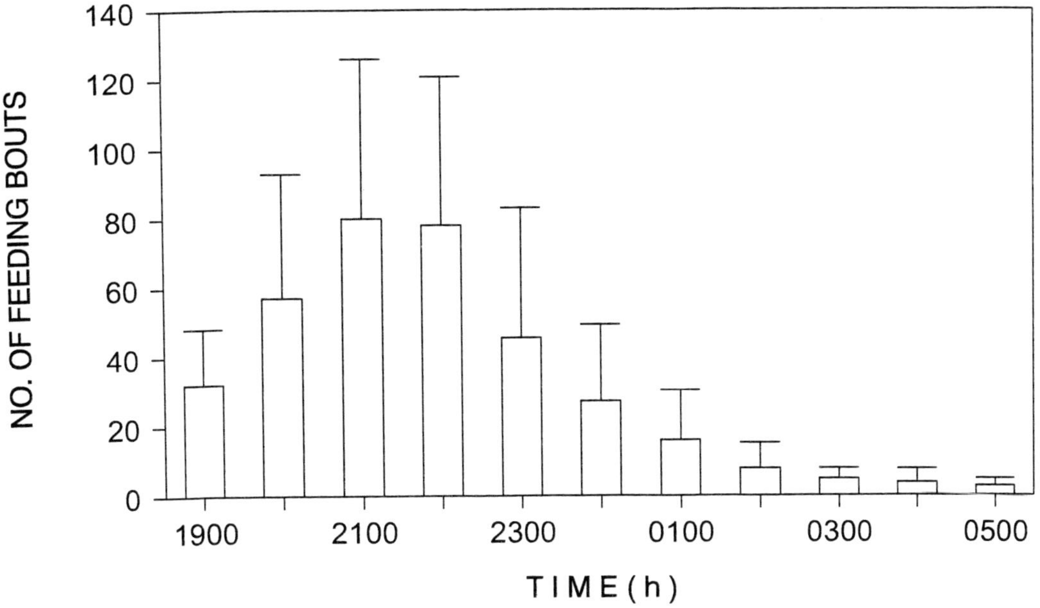

Figure 1. Temporal pattern of nightly foraging activity of *C. sphinx* feeding on *A. squamosa*. Data are given as mean and SD (n = 9 for each hour).

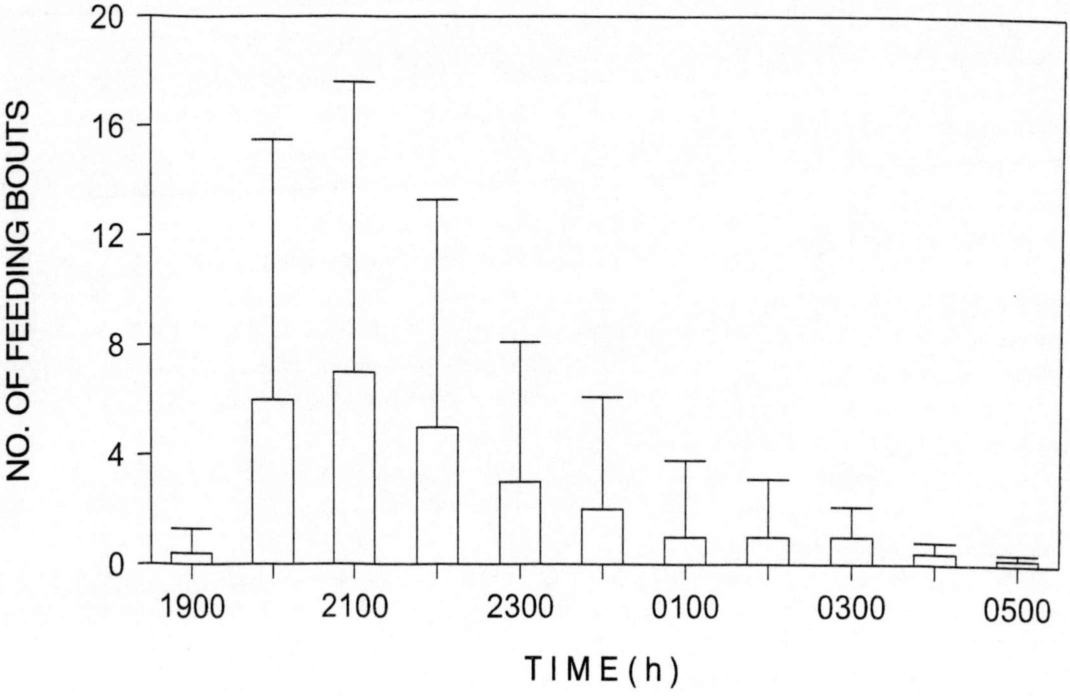

Figure 2. Temporal pattern of nightly foraging activity of *C. sphinx* feeding on *P. pendula.* Data are given as mean and SD (n = 16 for each hour).

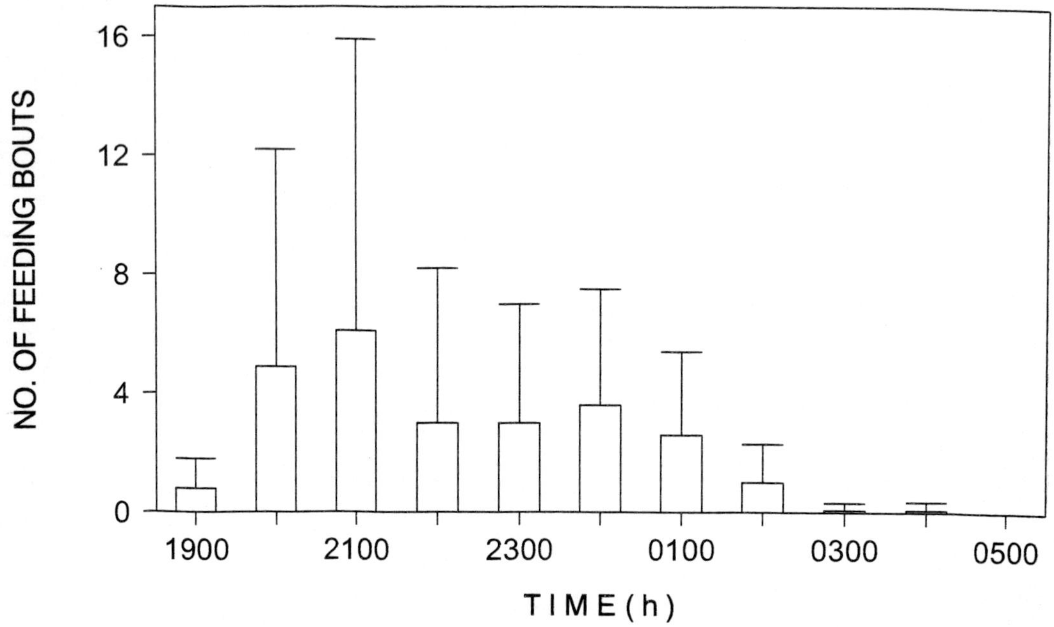

Figure 3. Temporal pattern of nightly foraging activity of *C. sphinx* feeding on *P. longifolia.* Data are given as mean and SD (n = 18 for each hour).

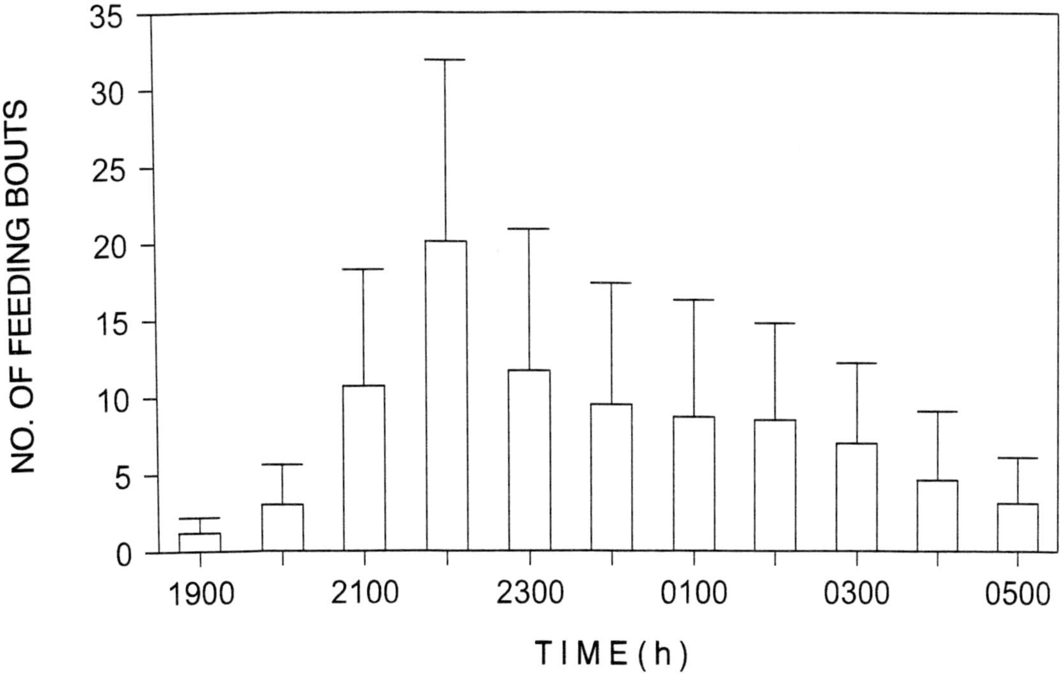

Figure 4. Temporal pattern of nightly foraging activity of *C. sphinx* feeding on *C. indica*. Data are given as mean and SD (n = 14 for each hour).

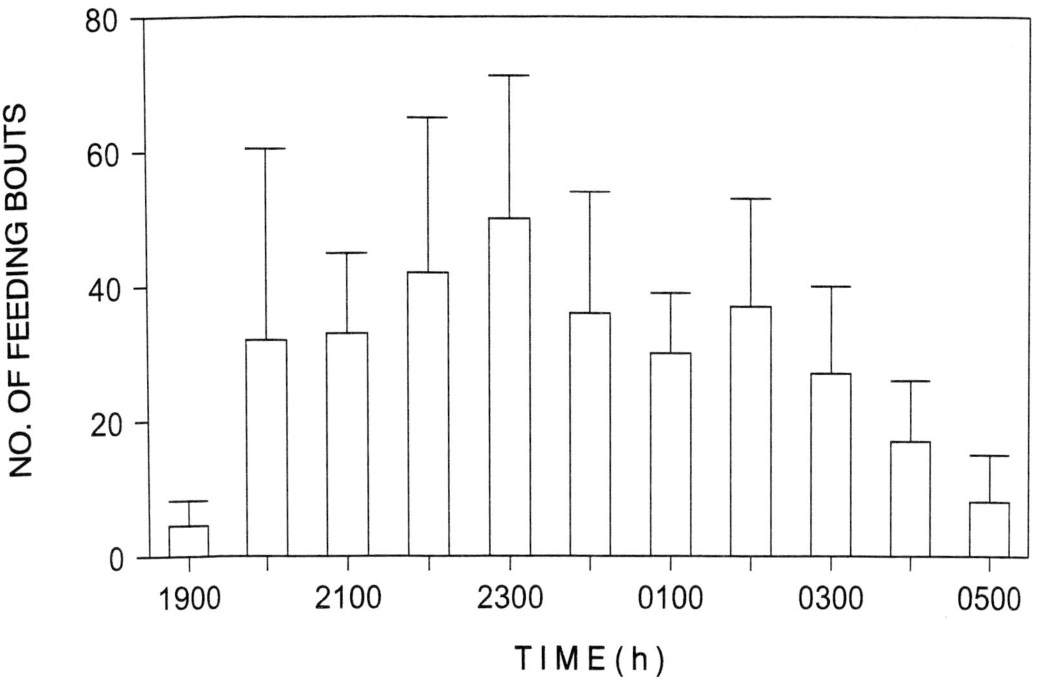

Figure 5. Temporal pattern of nightly foraging activity of *C. sphinx* feeding on *A. sapota*. Data are given as mean and SD (n = 6 for each hour).

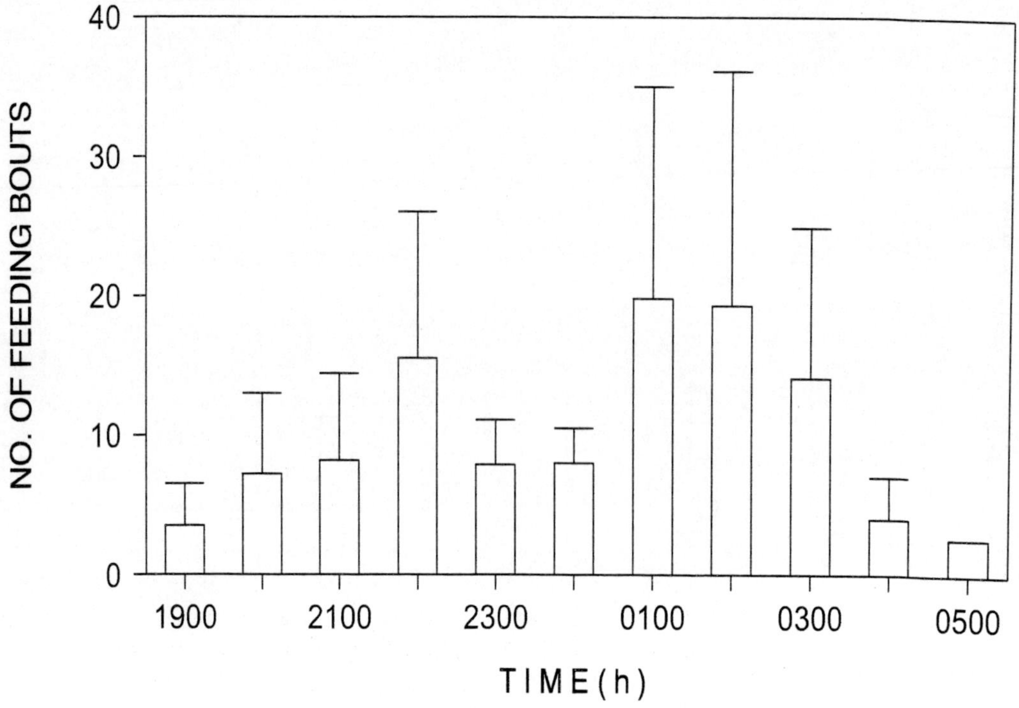

Figure 6. Temporal pattern of nightly foraging activity of *C. sphinx* feeding on *T. catappa*. Data are given as mean and SD (n = 11 for each hour).

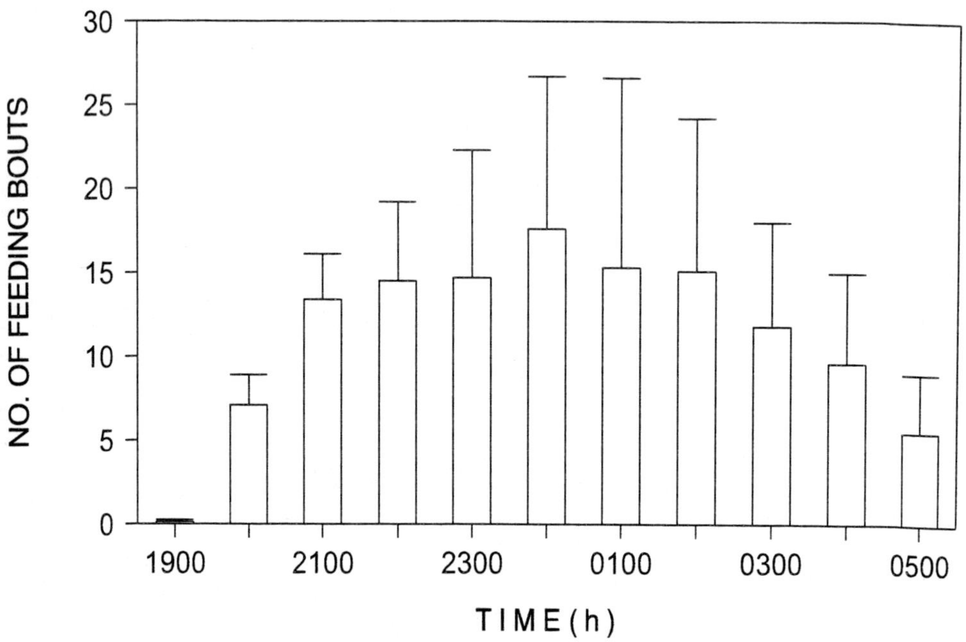

Figure 7. Temporal pattern of nightly foraging activity of *C. sphinx* feeding on *C. inophyllum*. Data are given as mean and SD (n = 24 for each hour).

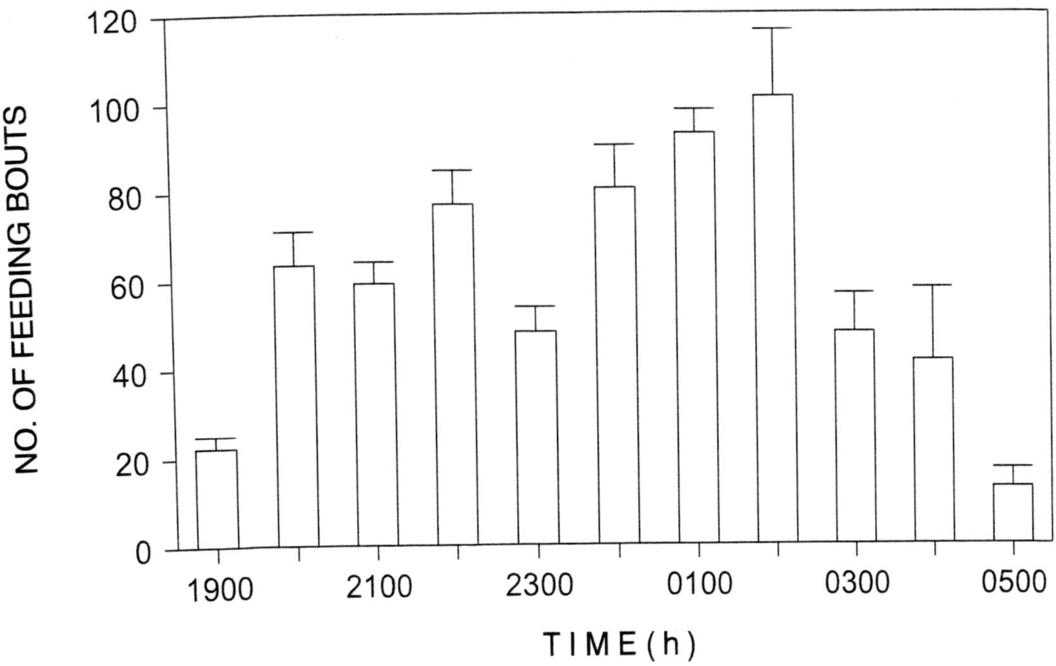

Figure 8. Temporal pattern of nightly foraging activity of *C. sphinx* feeding on *F. religiosa*. Data are given as mean and SD (n = 5 for each hour).

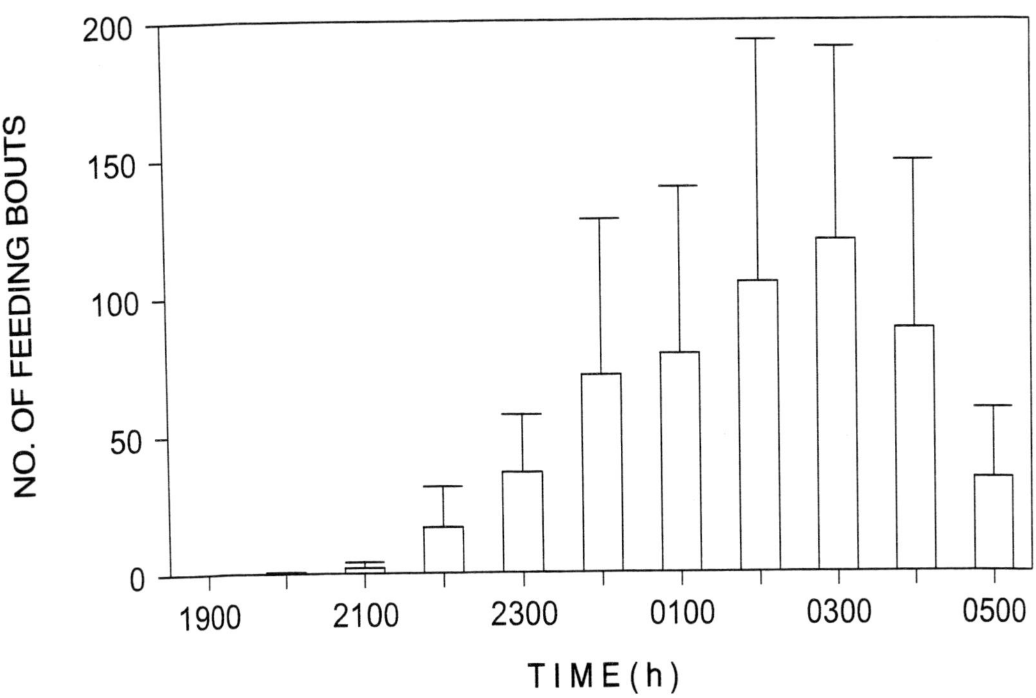

Figure 9. Temporal pattern of nightly foraging activity of *C. sphinx* feeding on *F. benjamina*. Data are given as mean and SD (n = 11 for each hour).

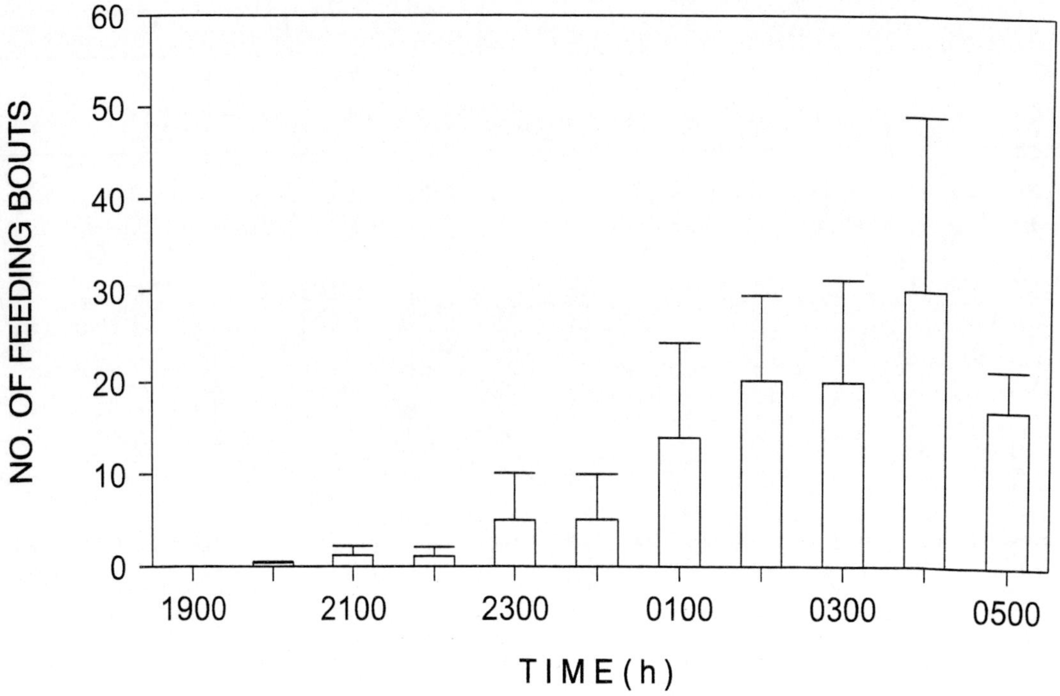

Figure 10. Temporal pattern of nightly foraging activity of *C. sphinx* feeding on *F. bengalensis*. Data are given as mean and SD (n = 7 for each hour).

Pteropus giganteus was the only species, seen foraging with *C. sphinx* when the latter fed on *P. pendula* and *P. longifolia.* However, there was no interspecific conflict observed, although intraspecific conflicts were observed on rare occasions (n = 11). The peak visits occurred at 2200 h when bats fed on *C. indica* (Figure 4; $F_{10,\ 143} = 7.6$, $p < 0.01$) and the foraging peak was observed at 2300 h when bats fed on *A. sapota* (Figure 5; $F_{10,\ 55} = 4.5$, $p < 0.01$). Although a low amplitude-feeding peak occurred at 2200 h in *T. catappa*, but a high amplitude feeding peak was observed at 0100 h (Figure 6; $F_{10,\ 110} = 4.8$, $p < 0.01$).

As bats visited *C. inophyllum* trees, the number of feeding bouts steadily increased from 1900 h, reached a peak at 2400 h, and declined thereafter (Figure 7; $F_{10,\ 253} = 2.6$, $p < 0.05$). When bats fed on fruits of *F. religiosa,* the peak of foraging occurred at 0200 (Figure 8; $F_{10,\ 44} = 34.5$, $p < 0.001$) in addition a small peak occurred at pre-midnight hours. Foraging activity on *F. benjamina* (Figure 9; $F_{10,\ 110} = 8.5$, $p < 0.01$) and *F. bengalensis* (Figure 10; $F_{10,\ 66} = 9.1$, $p < 0.01$) also showed post-midnight foraging peak at 0300 h and 0400 h, respectively.

In all the three species of *Ficus*, bats picked up a single fruit during each feeding bout and carried to their feeding roosts. The frequency of feeding bouts was very high in *F. religiosa* and *F. benjamina* compared with *F. bengalensis*. *Pteropus giganteus* was observed to hunt along with *C. sphinx* when foraging on the three species of *Ficus* trees and on *B. latifolia* (flower). *Pteropus giganteus* fed off the upper part of trees and *C. sphinx* foraged among the lower part of trees. However, there were no interspecific conflicts were observed between *P. giganteus* and *C. sphinx*, but intraspecific conflicts were observed very frequently.

Flowering Pattern of Food Plants and Nectarivory Behaviour of C. Sphinx

The number of flowers of *B. latifolia* available for bats for each night was approximately 10,000. The flowers were white, tiny (mean mass of a flower was 240 μg) and available at a height of about 4 m from the ground. The bats made circling flights around the tree, hovered over the flowers for a second, briefly landed on a branch with wing spread, picked up a single flower, sucked the nectar in flight and dropped the flowers. The bats visited the tree in groups of 5 – 25 individuals. The peak of bat-visits occurred at 2100 h (Figure 11; $F_{10, 99} = 19.8$, $p < 0.001$). As the flowers of *B. latifolia* started shedding at 2100 h, and the bat-visits also gradually declined then. The mean amount of nectar available per flower was 170 μl (n = 20).

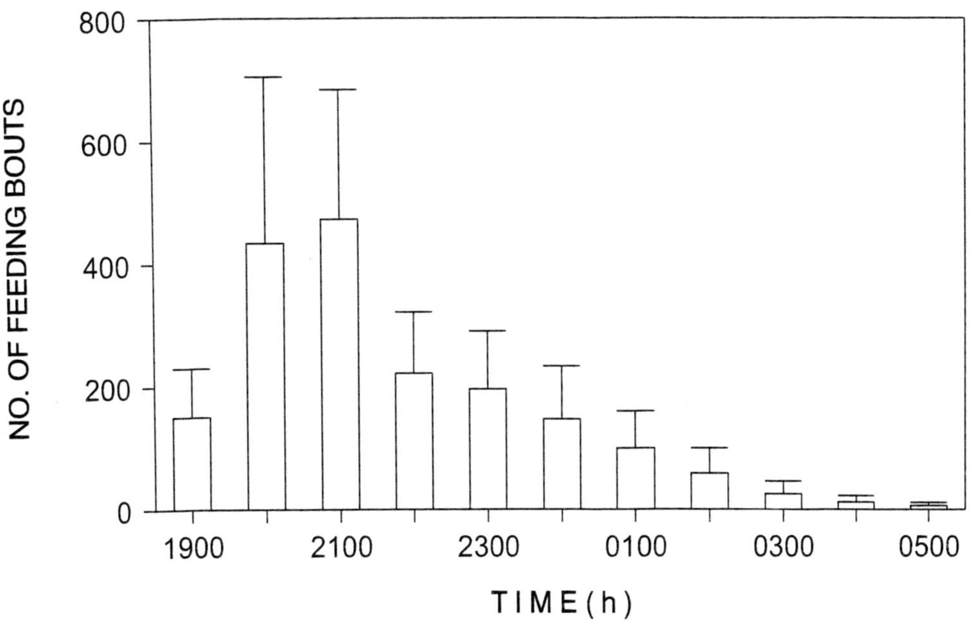

Figure 11. Temporal pattern in the foraging activity of *C. sphinx* feeding on *B. latifolia*. Data are given as mean and SD (n = 10 for each hour).

The bracts, covering the inflorescence of *M. paradisiaca*, opened about 30 min after sunset. The number of creamy white flowers varied from 10 to 11. In the boat shaped flowers, a curved petal carried the nectar. The inflorescence withers and gets dropped within one night. *C. sphinx* started visiting the banana flowers at the same timings when the bracts were open. Only solitary bats visited the flowers of *M. paradisiaca*. The bats hovered over the flowers, landed on the cone of the unopened bracts and immediately moved towards the inflorescence. The bats then inserted their noses into flowers and started lapping the nectar (Plate 2). They moved around the inflorescence, lapped on two to three flowers and flew away. The duration of time taken for lapping the nectar during each visit was 46.4 $\pm$ 40.9 sec (range 15 to 180 sec, n = 688). Before landing on the bract the bats made circling flights two to three times around the tree. A significant peak of bat visits occurred at 2000 h (Figure 12; $F_{10, 903} = 3.5$, $p < 0.01$). The nectar secretion showed a steady increase from 1900 h and reached a significant peak at 2100 h ($F_{10, 542} = 2.94$, $p < 0.05$). The minimum amount of nectar secreted in a single flower was 24.5 ± 0.3 µl (n = 46) at 0500 h and the maximum amount of nectar secreted was 361.5 ± 223.9 µl (n = 46) at 2100 h (Figure 13). The mean amount of nectar secretion per flower, per night was 2520 ± 817.6 µl (range 1550 µl to 4315 µl, n = 46). The total sugar production showed an apparent constancy between 224.0 mg/ml and 235.3 mg/ml (mean) for the nectar samples collected between 1900 h and 2400 h. The mean minimum value, 157.0 mg/ml, was obtained for the samples collected at 0500 h (Figure 14).

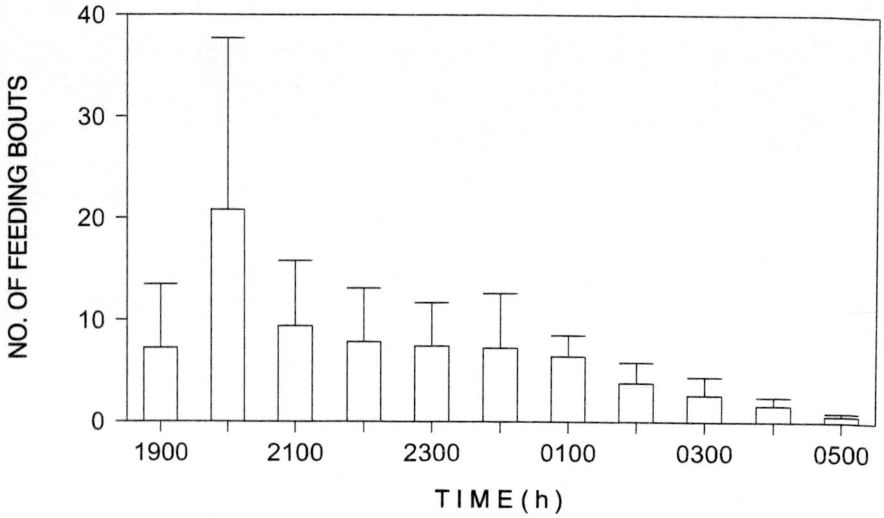

Figure 12. Temporal pattern of foraging activity of *C. sphinx* feeding on nectar of *M. paradisiaca*. Data are given as mean and SD (n = 83 for each hour).

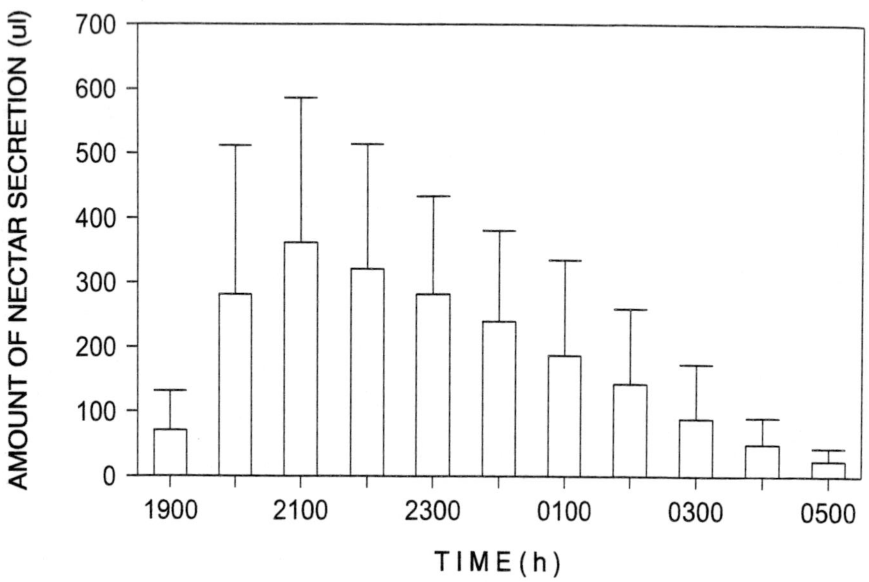

Figure 13. Temporal pattern of nectar secretion in *M. paradisiaca*. Data are given as mean and SD (n = 46 for each hour).

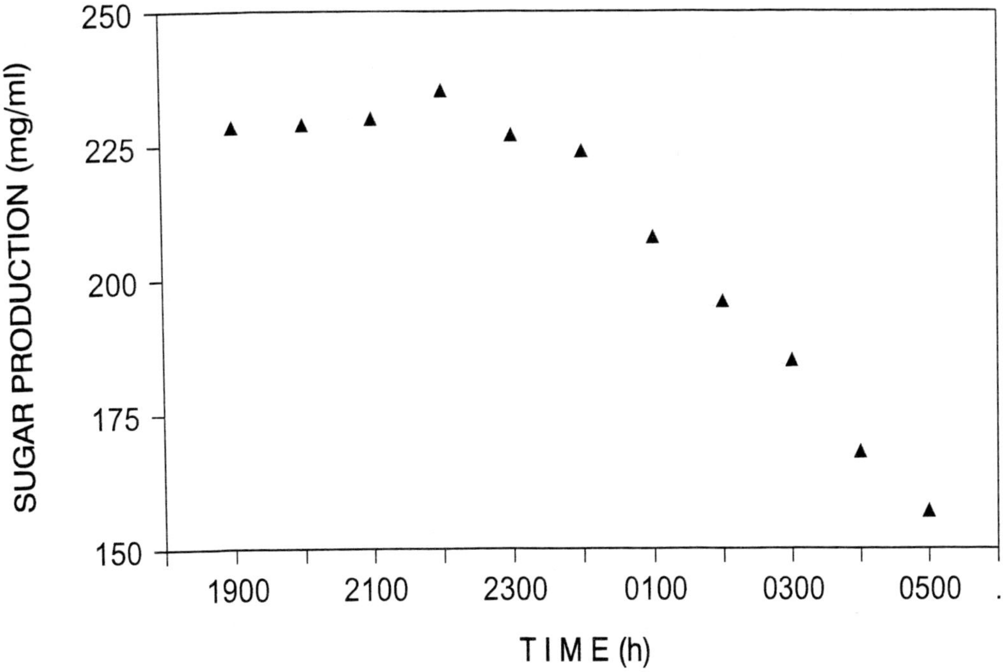

Figure 14. Temporal pattern of average total sugar production of *M. paradisiaca*.

Folivory Behaviour of C. Sphinx

Generally bats landed on the fruit after briefly hovering over it and removed a part (*A. squamosa*) or whole (*C. indica*) of it in the mouth, but folivory is by means of snatching leaves while flying without landing and hovering on leaves (Plate 2). But circling flights around the trees preceded removal of both the fruits and leaves. Bats were not observed to perch on the trees to feed on leaves, instead they carried the leaves to their feeding roost and repeated the feeding bouts after certain time intervals which varied over the nights. Folivory began after 1930 h and the peak of foraging bouts occurred always at postmidnight hours. The bats fed upon the leaves of *C. indica* the peak occurred at 0200 h (Figure 15; $F_{10, 55} = 11.6$, $p < 0.001$).

Similarly the peak of activity occurred at 0100 h and 0200 h when they fed upon the leaves of *C. fistula* (Figure 16; $F_{10, 55} = 3.3$, $p < 0.05$) and *M. elengi* respectively, although the peak for the latter was not significant (Figure 17; $F_{10, 33} = 1.5$, $p = 0.19$). In addition to the visual observations the phenomena of folivory by *C. sphinx* was confirmed by collecting 148 discarded fibrous pellets of partially chewed leaves of *C. fistula*, *M. elengi* and *C. indica*, beneath 20 day roosts of *C. sphinx*. In one occasion a mist netted male *C. sphinx* had a fresh leaf of *C. fistula* in its mouth.

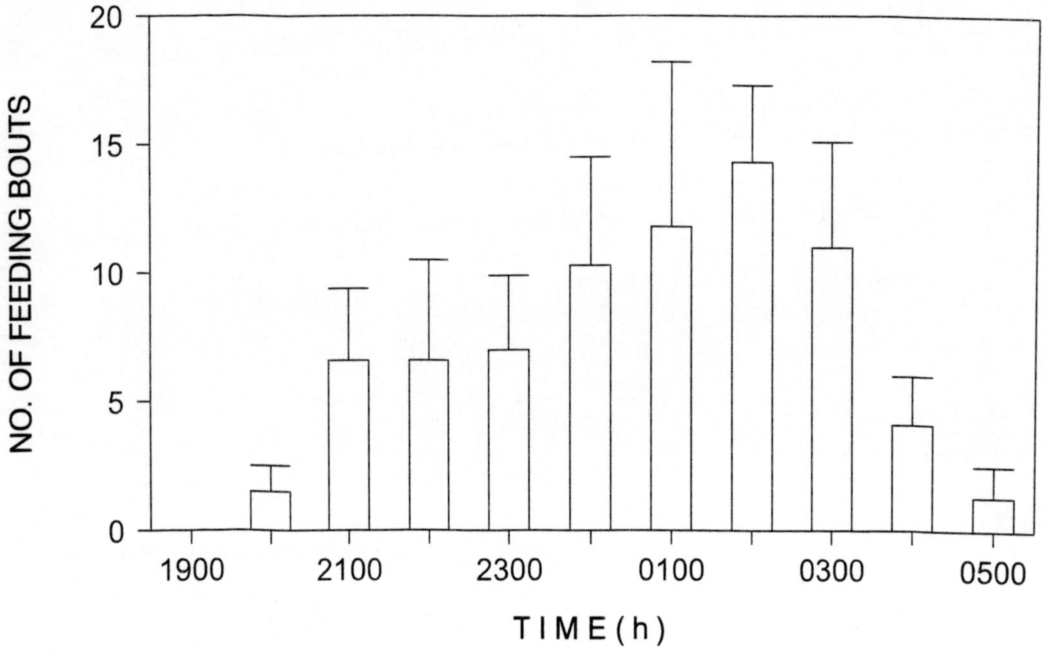

Figure 15. Temporal pattern of the foraging activity of *C. sphinx* feeding on leaves of *C. indica*. Data are given as mean and SD (n = 14 for each hour).

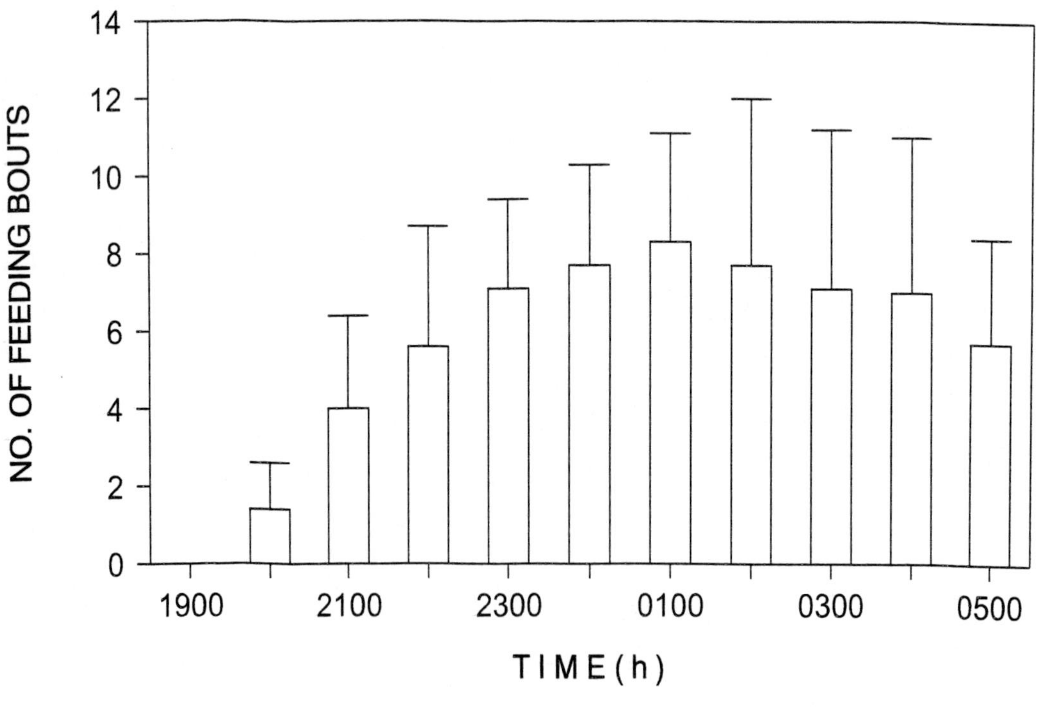

Figure 16. Temporal pattern of the foraging activity of *C. sphinx* feeding on the leaves of *C. fistula*. Data are given as mean and SD (n = 28 for each hour).

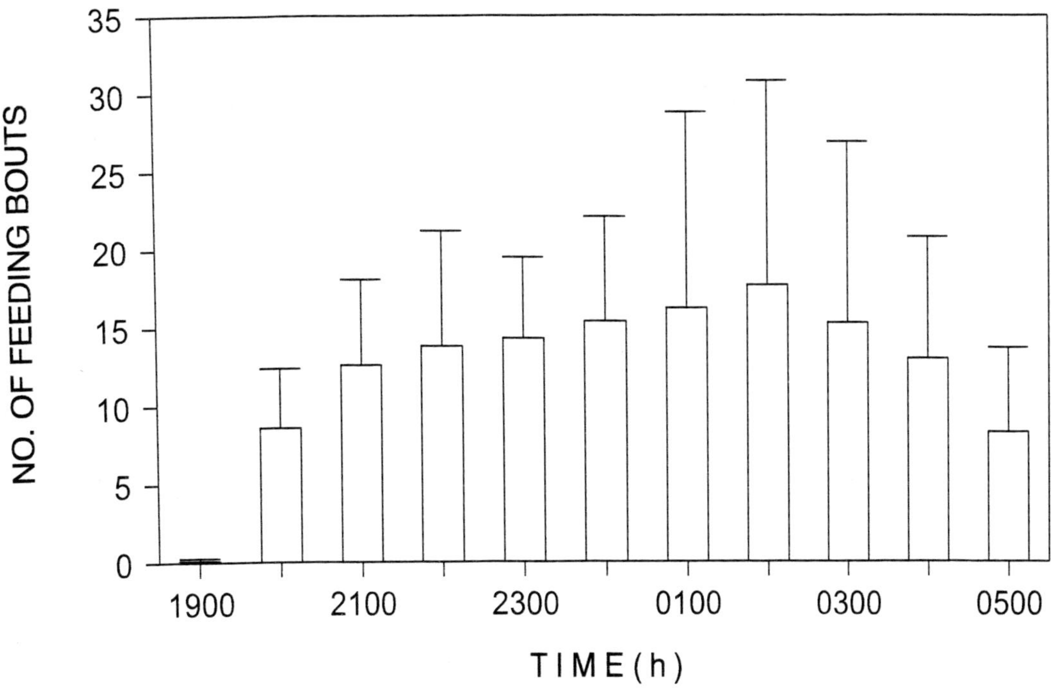

Figure 17. Temporal pattern of foraging activity of *C. sphinx* feeding on the leaves of *M. elengi*. Data are given as mean and SD (n = 27 for each hour).

ACTIVITY PATTERNS OF BATS AND DIFFERENT PHASES OF MOONLIGHT

Bats started visiting the trees individually to pluck a fruit of *C. inophyllum* or a leaf of *M. elengi* and *C. fistula* about one hour after sunset. They landed on the fruit after briefly hovering over it and removed a part or whole of the fruit in the mouth, whereas they snatched a leaf on flight without hovering. Circling flights around the trees preceded removal of both the fruits and leaves. Bats never stayed on the trees to feed, instead they carried fruits and leaves to a feeding roost where they masticate the food and expel the fibrous portions. Bats repeated the feeding bouts after an interval that varied over the nights and which depended upon the phase of moon.

Nightly variations in the bouts of feeding activity of *C. sphinx* observed for a complete lunar cycle showed that the activity decreased linearly with the increased duration or percentage of moonlight. The foraging activity of bats was significantly higher during the phase of the new moon nights compared with the phase of the full moon nights. The consequent nightly foraging activity of *C. sphinx* on *C. inophyllum* shows the direct influence of moonlight (Figure 18). The bats exhibit the least numbers of feeding bouts during the long moonlight nights. Clouds enhanced the activity of *C. sphinx* even during the moonlight phase, whereas thunder, lightning, drizzling and rain obstruct the foraging activity of *C. sphinx* (Figure 18).

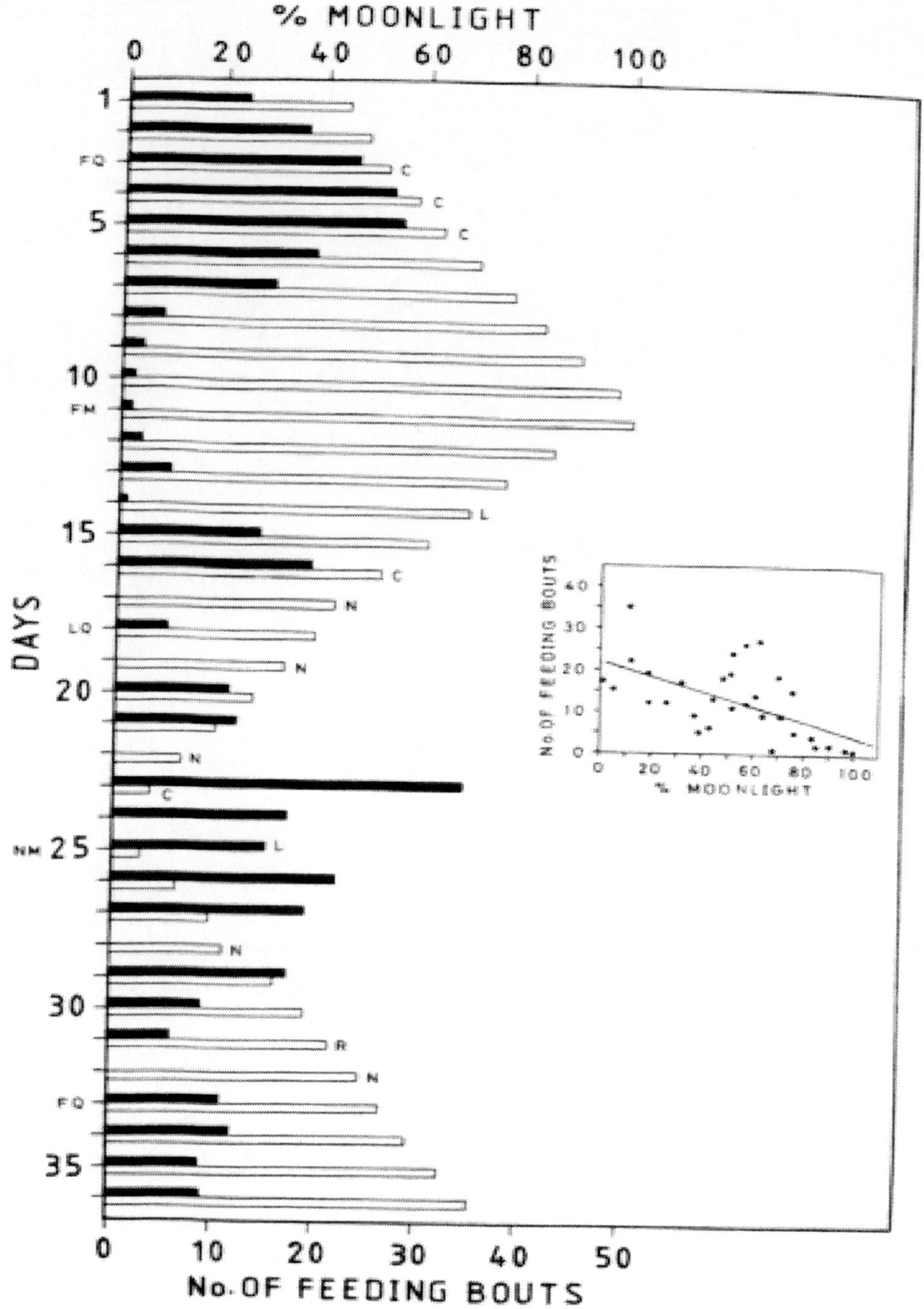

Figure 18. Night-to-night foraging activity patterns of *C. sphinx* while feeding upon the fruit *C. inophyllum*; black bars indicate number of feeding bouts, open bars indicate duration (per cent) of moonlight. The inset shows the linear decrease in the number of feeding bouts with the increase in the percentage of moonlight, $r = -0.6$, $p < 0.01$; C – cloudy nights, L lightning, N – no observation was made, R – rainy night, FQ – first quarter, FM – full moon, LQ – last quarter and NM – new moon.

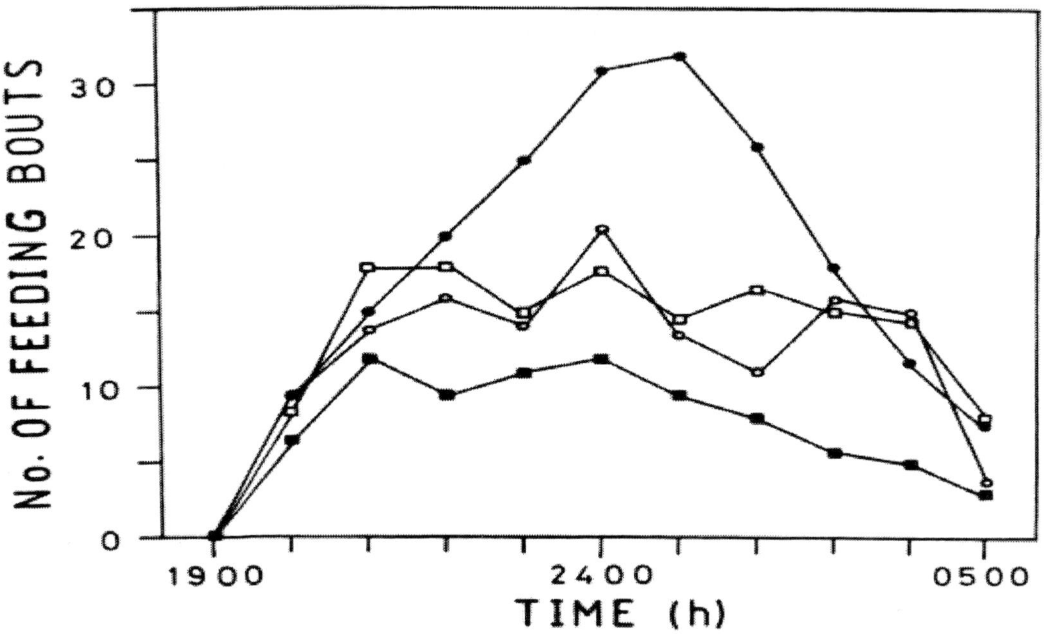

Figure 19. Mean number of feeding bouts of *C. sphinx* observed over the nights, while feeding upon *C. inophyllum* at different moon phases; solid circles: new moon phase (n = 5 nights), open circles: full moon phase (n = 6 nights), open squares: first quarter phase (n = 6 nights), solid squares: last quarter phase (n = 6 nights).

The foraging activity of *C. sphinx* shows a linear decrease in the number of feeding bouts with an increase in the duration (%) of moonlight (Figure 18, inset: $r = -0.6$, $p = 0.01$). During the new moon phase the foraging peak occurred between 2400 h and 0100 h whereas the full moon phase showed early foraging peak at 2400 h. The foraging activities of two moon phases differed significantly (Figure 19: $t = 2.8$, $df = 10$, $p < 0.05$). Feeding activity was also higher during the new moon phase compared with the phase of the first quarter ($t = 3.8$, $df = 10$, $p < 0.01$) and the last quarter ($t = 4.5$, $df = 10$, $p < 0.01$).

The foraging activities of *C. sphinx* on the leaves of *M. elengi* are also influenced by the occurrence of moonlight (Figure 20). Nightly foraging activity of *C. sphinx* while feeding upon leaves of *M. elengi* showed a linear decrease in the number of feeding bouts with an increasing duration (per cent) of moonlight (Figure 20, inset: $r = -0.5$, $p < 0.01$). Although the peak of foraging activity occurred at 0200 h both during new moon and full moon phases, the amount of activity is significantly higher during the new moon nights (Figure 21; $t = 7.7$, $df = 10$, $p < 0.001$).

The feeding activities of *C. sphinx* were not significantly higher during the new moon phase compared to the phase of the first quarter. Though, the number of feeding bouts which was observed on new moon did not show considerable difference when compared with first quarter (Figure 7.4; $t = -2.0$, $df = 10$, $p < 0.07$). However, when comparing activity during the light and dark halves of the nights within the phase of first quarter, the bats were significantly more active during the dark period while foraging on *M. elengi* (Figure 24; $t = -3.0$, $df = 4$, $p < 0.05$). The feeding activities of bats were significantly higher during the new moon phase compared to the phase of the last quarter (Figure 24; $t = 4.1$, $df = 10$, $p < 0.01$). However, when comparing the activities between dark and light periods of the last quarter, a significant difference occurred (t = 1.7, df = 4, p < 0.01) and the bats were more active during the dark period.

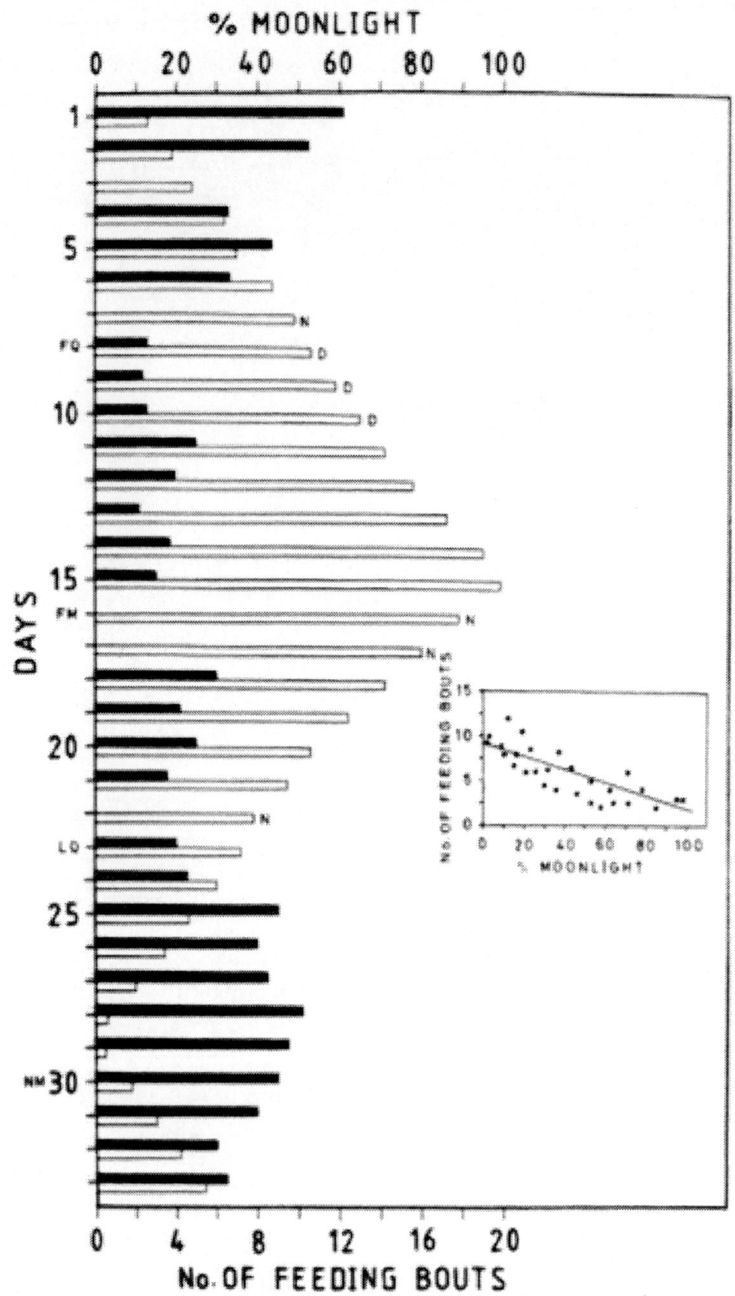

Figure 20. Daily foraging activity pattern of *C. sphinx* while feeding upon the leaves of *M. elengi*; black bars indicate number of feeding bouts, open bars indicate duration (per cent) of moonlight. The inset shows the linear decrease in the number of feeding bouts with the increase in the percentage of moonlight, r = - 0.5, p < 0.01; C – cloudy nights, L - lightning, N – no observation was made, R – rainy night, FQ – first quarter, FM – full moon, LQ – last quarter and NM – new moon.

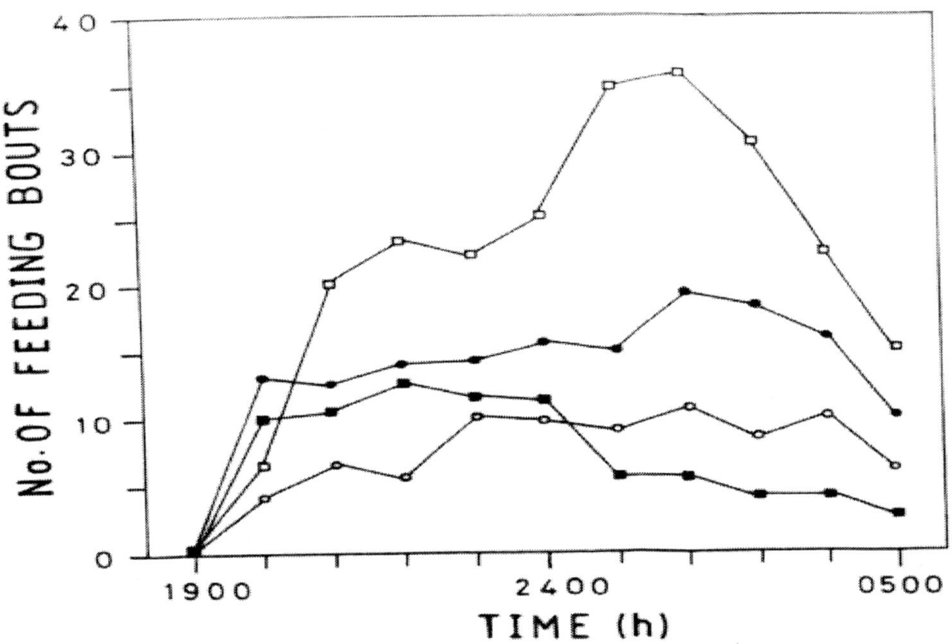

Figure 21. Mean number of feeding bouts of *C. sphinx* observed over the nights, while feeding upon the leaves of *M. elengi* at different moon phases; solid circles: new moon phase (n = 6 nights), open circles: full moon phase (n = 6 nights), open squares: first quarter phase (n = 7 nights), solid squares: last quarter phase (n = 6 nights).

Similarly, the nightly foraging activity of *C. sphinx* was also influenced by moonlight when the bats fed upon the leaves of *C. fistula* (Figure 22). The number of feeding bouts decreased with increasing percent of moonlight duration (Figure 22, inset: $r = -0.8$, $p < 0.001$). The peak foraging activity of *C. sphinx* during new moon phase showed between 2400 h and 0100 h. Similarly, the full moon phase showed peak foraging activity between 2400 h and 0100 h. However, the activities of the two moon phases showed a significant difference on the number of feeding bouts (Figure 23; $t = 6.9$, $df = 10$, $p < 0.001$). The feeding activity was not significantly higher during the new moon phase compared to the phase of the first quarter when bats fed on leaves of *C. fistula* (Figure 23; $t = 1.3$, $df = 10$, $p = 0.2$). However, when comparing the activities of light and dark halves of the nights within the phase of first quarter, a significant difference occurred ($t = -8.0$, $df = 4$, $p < 0.01$) and the bats were significantly more active during the dark period. The foraging activity of bats on *C. fistula* was not significantly higher during the new moon phase compared to the phase of the last quarter also (Figure 23; $t = 1.7$, $df = 10$, $p = 0.1$). However, when comparing the activity between dark and light periods of the last quarter significant difference occurred ($t = 1.7$, $df = 4$, $p < 0.01$) and the bats were significantly more active during the dark period.

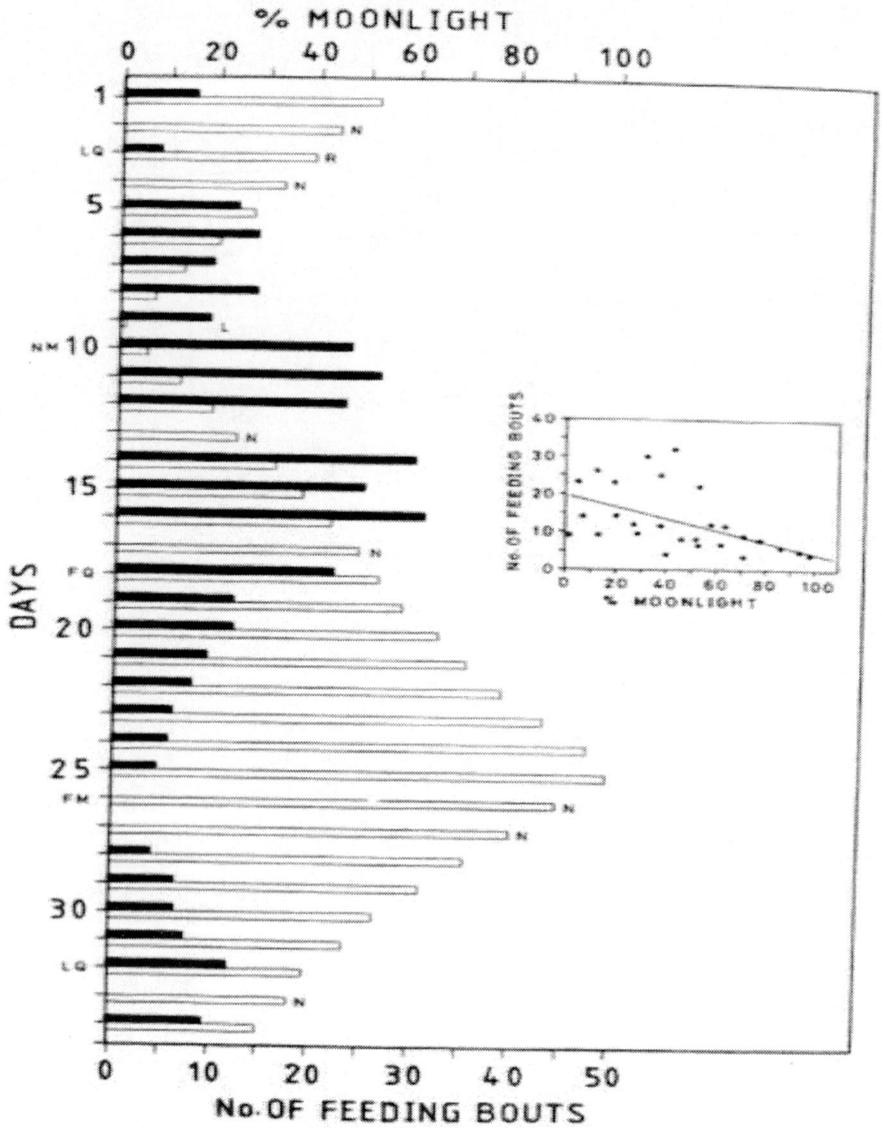

Figure 22. Daily foraging activity pattern of *C. sphinx* while feeding upon the leaves of *C. fistula*; black bars indicate number of feeding bouts, open bars indicate duration (per cent) of moonlight. The inset shows the linear decrease in the number of feeding bouts with the increase in the percentage of moonlight, $r = -0.8$, $p < 0.001$; D – drizzling, C – cloudy nights, L - lightning, N – no observation was made, R – rainy night, FQ – first quarter, FM – full moon, LQ – last quarter and NM – new moon.

The feeding activity was significantly higher during the hours of cloudiness compared to the period with clear sky (ANOVA, $F_{1,65}$=44.8, $P < 0.001$). Bats were less active during the hours of drizzling rain and completely stopped foraging whenever drizzling occurred with heavy lightning and thunder compared to the period with clear weather (ANOVA, $F_{1,165}$ = 41.8, $P < 0.001$).

It is clear that *C. sphinx* fed more upon fruits than leaves. One Way Repeated Measures ANOVA indicates that the bats showed significantly more feeding bouts on *C. inophyllum* than on *M. elengi* and *C. fistula* during the new moon (F = 14.5, df = 2,20 p < 0.001). Similarly, Friedman Repeated Measures ANOVA showed that *C. sphinx* exhibited significantly more feeding bouts on *C. inophyllum* than *M. elengi* and *C. fistula* during full moon phase ($X^2 = 15.3, p < 0.01$).

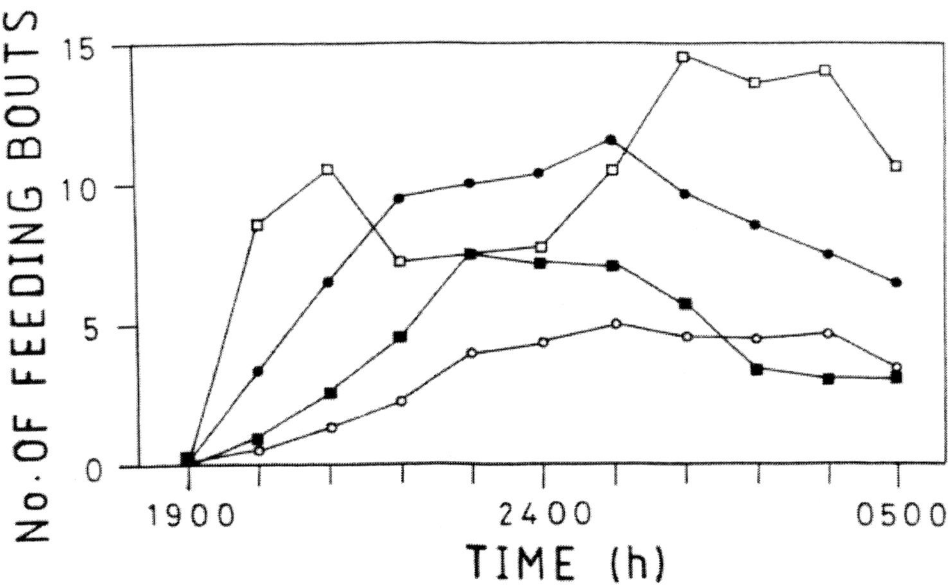

Figure 23. Mean number of feeding bouts of *C. sphinx* observed over the nights, while feeding upon the leaves of *C. fistula* at different moon phases; solid circles: new moon phase (n = 8 nights), open circles: full moon phase (n = 6 nights), open squares: first quarter phase (n = 7 nights), solid squares: last quarter phase (n = 4 nights).

The radio-telemetry studies also showed the influence of moonlight on the foraging activity of *C. sphinx*. The radio-tagged bats emerged from the day roost shortly after sunset and visited their foraging area. The radio-tagged bats removed food sources (fruits, flowers and leaves) from the resource tree and carried to their feeding roost, which was found 15 m to 30 m away form the resource trees. Bats spent most of their night hours for rest during the full moon phase whereas they spent fewest hours during the new moon phase. Remarkably, bats were inactive during the post midnight hours of the full moon phase. In the first day of observation during the full moon phase the tagged bat was active in the feeding area until 2310 h and then returned to its night roost (place used to ingest transported food from nearby feeding areas, resting place for bats to one or more feeding bouts) *Guettarda speciosa.* During the waxing of moon phases, the foraging patterns reduced to a lesser extent and during the full moon night the radio-tagged bat returned to the night roost at 2030 h showing a mere 11% of time in activity within a home range of 0.542 km^2 from a total of eight hours observation. Similarly the second bat which was radio-tagged 5 days prior to new moon showed increased activity (52%) with a home range of 0.832 km^2. These results showed that bats avoid active foraging bouts during full moon nights.

Radio-Telemetry: Influence of Sex on Foraging Behaviour

The radio-tagged bats of both sexes belong to five different colonies and each colony consists of 6 - 9 individuals. All the radio-tagged adult female bats roosted in *P. longifolia, C. nucifera* and *B. flabellifer*. Three of five male bats roosted solitarily in the leaflets of *C. nucifera* and the remaining two bats roosted under the dry fronds of *B. flabellifer*. During dry (June - July) and wet (September - November) seasons bats mainly fed upon seasonally available food sources (Table 2). The timings of emergence varied over the seasons (Table 3). During the wet season the radio-tagged bats emerged during the early hours of dusk (mean 1827.5 $\pm$ 0006.9 h), earlier compared with emergence during the dry season (mean 1915.5 $\pm$ 0004.4 h) and showed significant difference ($F_{1,8} = 5.3$ $p < 0.05$). The male and female bats exhibited a bimodal activity pattern with peaks during pre and post-midnight hours (Figs. 24 and 25).

Table 2. Food sources available for C. sphinx in the study area during the dry and wet seasons.

Tree species	**Food sources**	**Season**
Achras sapota	fruits	wet
Annona squamosa		
Azadiracta indica	fruits	wet
Bassia latifolia		
Calophyllum inophyllum	fruits	dry
Carica papaya	flowers and fruits	dry and wet
Cassia fistula	fruits	dry
Coccinia indica	fruits	dry and wet
Ficus bengalensis	leaves	dry and wet
F. benjamina	fruits and leaves	dry and wet
F. religiosa	fruits	dry and wet
Mimusops elengi	fruits	dry and wet
Pithecellobium dulce	fruits	dry and wet
Polyalthia pendula	fruits and leaves	dry and wet
P. longifolia	fruits	dry and wet
Psidium guajava	fruits	wet
Terminalia catappa	fruits	wet
	fruits	dry and wet
	fruits	dry and wet

Table 3. Forearm length, body mass, season and time of emergence of both male and female *C. sphinx*.

Bat No.	Forearm length (mm)	Body mass (g)	Mean time of emergence (h)	**Season**
Male (M)				
M1	71.0	47.0	1918 ± 0024	Dry
M2	70.6	45.0	1917 ± 0026	Dry
M3	71.8	49.0	1918 ± 0026	Dry
M4	69.7	40.0	1909 ± 0054	Dry
M5	71.2	46.0	1839 ± 0028	Wet
Female (F)				
F1	70.0	44.0	1832 ± 0054	Wet
F2	71.9	48.0	1826 ± 0038	Wet
F3	70.0	43.0	1820 ± 0050	Wet
F4	70.5	44.0	1823 ± 0047	Wet
F5	71.0	45.0	1825 ± 0043	Wet

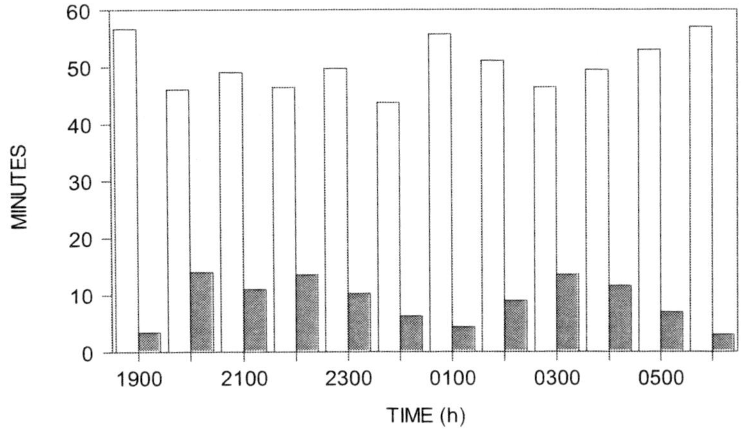

Figure 24. Activity budget of male *C. sphinx*. Open bars indicate rest and solid bars indicate activity.

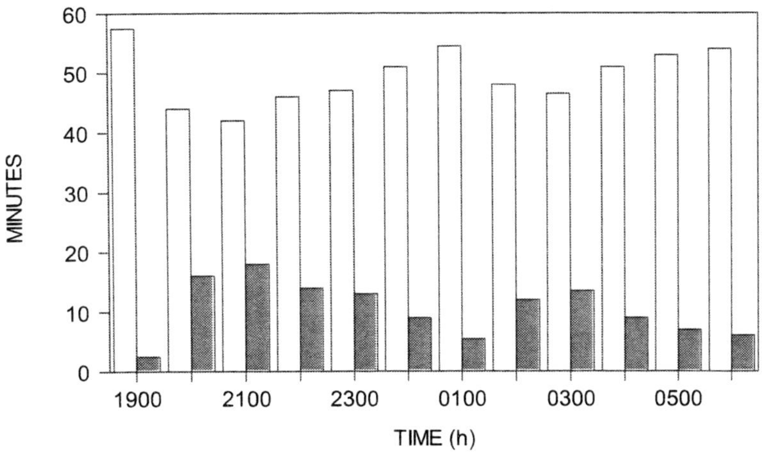

Figure 25. Activity budget of female *C. sphinx*. Open bars indicate rest and solid bars indicate activity.

Of the five radio-tagged male bats, four (M1, M2, M3 and M4) roosted close together and utilized only one foraging area (FA 1 - Foraging Area 1) throughout the tracking period whereas the other tagged male (M5) utilized two foraging areas (FA 1 and FA 2). The day roost and foraging areas of male and female bats are given in Table 4.

Table 4. Location of day roost and foraging areas of five males and five females of *C. sphinx* in the study area (Refer Plate. 1)

Bat No.	Day roost	Foraging area coordinates
M1	2b	2b
M2	2b	2b
M3	2b	2b
M4	2b	2b
M5	2c	2b and 3c
F1	4d	2c and 4d
F2	2b	3b, 2b and 2c
F3	2b	2a, 2b and 2c
F4	2b	2b and 4b
F5	2b	2b

The mean commuting distance of all five males was 0.2 $\pm$ 0.2 km and the mean size of feeding area of males was 0.7 $\pm$ 0.3 km^2. The foraging area overlapped very often among M1, M2, M3 and M4 due to the occurrence of their day roosts within the food resource patch. On one occasion the male bat (M5) spent 152 min (21.1% from the total observation hours) in the day roost during the night hours.

Generally females commuted farther than males and the females spent more time in the foraging ground than the males, whereas the males spent more time in day roosts and commuted very close to their day roosts. Among the five radio-tagged females, we could find all foraging areas for only three females (F1, F2 and F3). The mean commuting distance of all the three females was 2.0 $\pm$ 1.3 km and the mean size of foraging area was 0.8 $\pm$ 0.1 km^2. Among the five radio-tagged female bats, three (F1, F2 and F3) commuted to three different foraging areas. For instance, F1 whose day roost located in grid 4d foraged at three different areas (Plate 1; grids 2c and two sites at 4d). The first foraging area (FA1) of this female located at a distance of 2.1 km North West of the day roost and FA2 and FA3 were at south of its day roost at a distance of 0.2 km and 0.8 km, respectively. It (F1) spent more time (60%) in the farthest foraging area (FA 1) compared the other two foraging areas, which were located closest to its day roost.

Similarly F2 and F3 utilized three foraging grounds (Plate 1; grids 2b, 2c and 3b; 2a, 2b and 2c, respectively). The fourth female (F4) used two foraging areas in one night and the first foraging area (Plate 1; 4b) was farthest than the second foraging area (Plate 1; 2b) which located closest to the day roost (Table 5). The fifth female's (F5) foraging area was at a distance of 0.2 km towards north from the day roost and it spent 20% of the nightly foraging hours there. However it visited FA2 and FA3 regularly before reaching day roost at dawn. Apparently these two foraging areas were found beyond 4 km distance and the tracking unit could not locate them. The subadult females (body mass: mean 18.6 $\pm$ 1.1 g) commuted for two kilometers (mean 1.79 $\pm$ 0.42 km) whereas the subadult males (body mass: mean 19.6 $\pm$

2.07 g) preferred to forage at shorter distances of 0.5 km (mean 0.39 $\pm$ 0.10 km). The subadult females used two foraging areas, one close to the day roost and the other further away (Table 6). The commuting distance of both subadult and adult females did not show significant difference ($F_{1,8} = 0.36$, $p = 0.564$). Similar results were obtained for subadult and adult males ($F_{1,8} = 1.0$, $p = 0.346$). However, a significant difference was observed on the commuting distance between subadult males and subadult females ($F_{1,8} = 36$, $p = 0.001$).

Table 5. The commuting distance between centres of foraging area (FA) and day roost for five adult males and five adult females of *C. sphinx*.

Bat No.	Distance between the day roost and FAs (km)		
	FA1	FA2	FA3
M1	0.12	nil	nil
M2	0.10	nil	nil
M3	0.09	nil	nil
M4	0.20	nil	nil
M5	0.10	0.50	nil
F1	2.10	0.20	0.75
F2	1.20	0.50	0.20
F3	1.80	0.20	0.70
F4	2.20	0.40	nil
F5	0.20	NF	NF

FA1 – First Foraging Area, FA2 – Second Foraging Area, FA3 – Third Foraging Area and NF – Not Found.

Table 6. Distance between the day roost and foraging areas of subadult males and females.

Sex	**Distance between the day roost and FAs (km)**	
	FA1	**FA2**
Subadult Male	0.50	nil
,,	0.50	nil
,,	0.50	nil
Subadult Female	1.90	0.20
,,	2.10	0.30
,,	0.95	0.20

FA1 – First Foraging Area and FA2 – Second Foraging Area

The results show that the temporal distribution of feeding behaviour of *C. sphinx* constrained by food abundance, fruit size, intraspecific competition and by the temporal availability of food sources. Fleming (1982) reported that optimal diet breadth has been considered as a central issue in many discussions of foraging behaviour. The foraging behaviour of *P. discolor*, which consumes a variety of fruits and flowers in Costa Rica and

Brazil has been influenced by abundance, diversity and seasonality of different kinds of food items (Heithaus *et al.*, 1974; Sazima and Sazima, 1977). Temporal differences in foraging might reduce competition among some frugivorous bat species and such differences would serve to reduce direct interference during foraging (Heithaus *et al.*, 1975). There were two peaks of activity, one in the pre-midnight hours when bats fed mostly on steady state fruiting trees, which produce low number of fruits in a night. Such strategies have also been reported in *C. perspicillata* when they fed on *Piper amalago* (Fleming, 1982). These foraging patterns might promote scramble competition among conspecifics in a given foraging area. Fruits such as *A. squamosa, A. sapota, C. indica, P. pendula* and *P. longifolia* are removed by bats at early night hours and early feeding would appear to be advantageous to exploit the food available at this time, rather than promoting scramble competition. A later foraging peak occurred in post-midnight hours when bats fed on big bang fruiting trees such as *F. religiosa, F. benjamina* and *F. bengalensis*, plants which produce a large number of fruits in a night. Apparently, the big bang fruiting trees might allow more number of bats to forage without competition in a feeding area.

Wilson (1975) suggested that the group foraging provide protection from predators while commuting and searching for food. Group foraging has been observed in *L. sanborni* (at Adansonia flowers), *E. spelaea* (at various big bang flowers) and probably Australian *Pteropus* (at Eucalyptus flowers) (Nelson, 1965; Ayensu, 1974; Start and Marshall, 1976; Howell, 1979). Flock foraging probably increases the rate of patch discovery and decreases the chance that individuals will waste time and visiting already depleted plants but solitary foraging avoids this. Kunz (1982) suggested that flock foraging for fruit is probably common in pteropodids that live in large colonies (e.g. *E. helvum, R. aegyptiacus* and *Pteropus sp.*).

Only solitary bats visited the steady state fruiting trees which have a low abundance of fruits, whereas groups of bats regularly visited the big bang fruiting trees, which produce large number of fruits and appear to promote group foraging during the latter hours of the night. Even though big bang fruits are patchy, when they occur in such abundance competition for them becomes insignificant. Group foraging might facilitate the location of patchy food that is unpredictable in its distribution. We deduce that the temporal use of available fruit in the feeding area makes it possible for *C. sphinx* to successfully exploit available fruits, and thereby reduce feeding interference and competition with conspecifics.

Modulation in the patterns of nightly visits by *C. sphinx* to fruit-bearing trees suggests that the abundance of fruits influence the foraging activity of these bats. Similar patterns of nightly activity were reported for some Neotropical species that are influenced by either food quality or quantity (Fleming, 1988). Temporal differences in foraging were also reported for *C. perspicillata* based on the quality of available fruits. In the early evening hours individuals fed on protein-rich but scarce fruits and during later hours they switched to energy-rich fruits (Fleming, 1988).

The results suggest that trees with less abundant fruits promote early foraging. This behaviour presumably makes it possible for bats to exploit available resources at early hours of the night and thereby, bats could avoid feeding interference and competition from conspecifics. The Neotropical short-tailed fruit bat, *C. perspicillata* is mostly a solitary forager and it preferred food plants that produced ripe fruits only in small quantities (Fleming, 1982). However, another Neotropical frugivore, *A. jamaicensis*, forages in groups, and feeds mostly on *Ficus* fruits which are usually available in large numbers over a short period

(Handley and Morrison, 1991). Heithaus *et al.* (1974) found that larger phyllostomids foraged in groups and smaller species foraged solitarily.

Similarly, *C. sphinx* foraged upon *B. latifolia* in groups since the flowers were available large in numbers, and solitarily on *M. paradisiaca* since the flowers were less abundant. The flowers of *B. latifolia* started to shed at early hours of the night, and the bats therefore exploited the flowers when they are available. Thus the temporal foraging pattern of the nectarivorous bats is related to the timing of availability of food (Sazima and Sazima, 1977; Fleming, 1982; Heithaus, 1982).

Cynopterus sphinx feeds primarily on fruits (Bhat, 1994; Marimuthu *et al.*, 1998). In addition, it feeds upon nectar because the latter is an important source of energy and water (Howell, 1974; Petit and Freeman, 1997). The time at which the occurrence of peak of floral visitation by our study bats corresponded to periods of high nectar production and sugar concentration as reported for columnar cacti visited by *L. curasoae* and *Glossophaga longirostris* (Nassar *et al.*, 1997). While lapping nectar *C. sphinx* provides cross-pollination of the flowers (Subramanya and Radhamani, 1993). Plant characteristics that are helpful in defining pollination syndromes include time of anthesis, nectar secretion, colour, odour, morphology and positioning of flowers and the volume and concentration of nectar (Faegri and van der Pijl, 1979). In addition to chiropterophily, bat-visiting flowers are positioned outside the foliage. Examples include *Ravenala madagascariensis* pollinated by *P. alecto* (Calley *et al.*, 1993), *Agave schottii* pollinated by *L. nivalis*, *Parkia clappertoniana* pollinated by *Epomophorus gambianus* (Faegri and van der Pijl, 1979) and *Heliconia solomonensis* pollinated by *Melonycteris woodfordi* (Kress, 1985). The timings of bract opening, nectar secretion, the faintly musty sweet odour of *M. paradisiaca* coincided with the emergence of *C. sphinx*. The tiny flowers of *B. latifolia* facilitate *C. sphinx* to carry them while flying. The flowers of *M. paradisiaca* and *B. latifolia* shed within a night and such features promote the bats as pollinators rather than birds.

Cynopterus sphinx forages on fruit early in the night and concentrating on leaf-eating later. At least 16 species of plant-visiting bats are known to feed on leaves, although none appears to feed exclusively on leaves (Kunz and Diaz, 1995). Cunningham van Someren (1972) reported that *R. aegyptiacus* carries the leaves of *Erythrina* when returning to its day roosts, probably to consume the leaves. Morton (1973) suggested that many frugivorous vertebrates rely on leaves as alternative food sources to satisfy their protein requirements. Leaves of several tree species offer potentially rich source of dietary protein that would compensate for the low protein content in many fruits (Thomas, 1984; Stellar, 1986). Folivory has also been observed in *A. jamaicensis* and *P. longanus*, which feed on *Erythrina* leaves (Harris and Baker, 1959). Earlier reports on folivory by Old World plant-visiting bats have been anecdotal and all observations were conducted based on the expelled fibrous leaf pellets (Tedman and Hall, 1985; Lowry, 1989; Bhat, 1994; Kunz and Diaz, 1995). Extensive observations on folivory were reported for *C. brachyotis* (Tan *et al.*, 1998). *Cynopterus sphinx* removes fresh leaves from trees, carries them in its mouth, expels partially chewed and rejected leaf pellets beneath both day and night roosts. The above evidence provides additional support that this species relies extensively on leaves as a part of their diet.

Herrera (1987) suggested that most fruits eaten by bats are low in protein content, so bats should either select alternate food items that are rich in protein or modify their feeding tactics to increase protein consumption. *Cynopterus sphinx* feeds upon fruits, nectar and flowers during early hours before switching to leaves. The consumption of soluble protein rich leaf fraction fulfils the daily protein requirements of *A. jamaicensis* (Kunz and Diaz, 1995). Although leaves are potentially a super abundant resource for phytophagous bats, leaf digestion is generally very slow (Eisenberg, 1978; Waterman, 1984). These above criteria are consistent for *C. sphinx*, therefore it exploits the super abundant food sources (leaves) at later hours of the night. Also it carries the leaves to the day roosts when returning from the foraging areas. There it can slowly masticate the leaves, ingest the juice and expel the fibrous portion of the leaves. Bats that feed on leaves appear to overcome the problem associated with consuming a fiber rich diet, by extracting leaf juice rather than ingesting whole leaf (Kunz and Diaz, 1995). The food handling and folivory behaviour of *C. sphinx* are consistent with the foraging behaviour of *A. jamaicensis* (Morrison, 1978; Handley and Morrison, 1991) and *R. aegyptiacus* (Cunningham van Someren, 1972). Perhaps, leaf fraction also provides an important source of dietary calcium and other nutrients which are essential for maintenance and reproduction in bats (Barclay, 1994). *Cynopterus sphinx* fed both young and mature leaves, however mature leaves may contain a higher concentration of secondary compounds than developing leaves (Telek and Martin, 1983; Kunz and Diaz, 1995).

Temporal patterns of *C. sphinx* feeding on fruits and leaves support the prediction that individuals of this species predominantly feed on fruits during the early hours of the night and on the leaves later in the night. Ripe fruits of many tropical species are typically patchy in both space and time (Howe and Westley, 1988; Fleming, 1982), but leaves are ubiquitous and usually available in large numbers compared to fruits. Ripe fruits are considered to be limited resource for bats, therefore they compete for these food sources immediately after the onset of nightly foraging activity (Fleming, 1979). Consumption of figs by some plant visiting bats (Handley and Morrison, 1991; Handley *et al.*, 1991; Bhat, 1994; Tan *et al.*, 1998) is not surprising, since the *Ficus* fruits give high concentration of important minerals and protein (O'Brien *et al.*, 1998). Phytoestrogens, plant-derived estrogenic substances are consumed by many species of mammals, both in wild and in captivity. Numerous green leafy plants have estrogenic properties (Labov, 1977). While the estrogenic activity of these plants is not directly due to the presence of steroidal estrogens such as those produced by mammals, these plants contain isoflavones and are related to substances with extra hydroxyl groups (Moule *et al.*, 1963). Flavenoids mimic the actions of estrogen when they contain extra hydroxyl group. Naturally occurring phytoestrogens (include genistein, luteolin and coumestrol) that show steroidogenic activity in mammals (Baker, 1992). Gynomastia (enlarged mammary glands) is reported in males of *C. sphinx* (Ryberg, 1947; Dobson, 1978) and evidence for male lactation in the Dayak fruit bat *Dyacopterus spadecius* (Francis *et al.*, 1994), may result from the ingestion of leaves which contain these compounds. This study indicates that *C. sphinx* employs both solitary and group foraging behaviours, early fruit feeding and late leaf feeding. These behaviours could be largely influenced by the diet breadth, intraspecies competition and temporal availability of fruits.

The foraging behaviour of *C. sphinx* appears to be sensitive to moonlight. However, bats left their day roosts after sunset even on nights when a bright moon was already present. At sunset hunger may be an overriding factor driving the bats to emerge for a short bout of feeding despite the full moon. Bats were more active during the periods of new moon than during the full moon phase partly explaining the inverse linear correlation between the foraging activity of bats and the period of lunar illumination. The restriction of activity to the dark half of the nights during the first and last quarter moon phases and an enhancement of activity when the moon was obscured by clouds further support our observations on the influence of moonlight on the activity of bats. Cloud cover also enhances the foraging activity of the African insectivorous bats (Fenton *et al.*, 1977). Similarly the number of bat passes in the foraging area significantly increased during a lunar eclipse (Usman *et al.*, 1980) in a field study, a pattern also observed in neotropical phyllostomid bats in laboratory experiments which simulate the moon light conditions including the eclipse (Haeussler and Erkert, 1978).

The reduction in the activity pattern during full moon nights as against new moon night may be considered as an antipredatory adaptation against visually active predators such as snakes and owls. In order to avoid bright moonlight *C. sphinx* spent most of the night either in the night roost or in the day roost (Nair *et al.*, 1998), as observed in other species of bats such as *Megaderma lyra* (Balasingh, 1990) and *Artibeus jamaicensis* (Handley and Morrison, 1991). In contrast, African insectivorous bats confine their foraging flights within the canopy or even gave up the second feeding period under bright moonlight conditions (Fenton *et al.*, 1977). Exceptionally, moonlight did not alter the activity pattern of the insectivorous bat *Myotis lucifugus* (Negraef and Brigham, 1995). They monitored bat activity only for one hour at the beginning and one hour at the end of the night. However, Anthony *et al.* (1981) evaluated the effects of moonlight on the same species and came to a different conclusion. The nocturnal monkey *Aotus trivirgatus* also shows increased activity during full moon under natural conditions (Wright, 1981) and in captivity (Erkert and Groeber, 1986).

Restrictions of feeding activity during the nights of bright moonlight are generally construed as an adaptation to avoid nocturnal predators (Lockard and Owings, 1974; Fenton *et al.*, 1977; Fleming and Heithaus, 1986). Isaac and Marimuthu (1993) observed that the black-winged kite *Elanus caeruleus* caught Indian pygmy bats *Pipistrellus mimus* while they flew out of a tunnel during the evening hours. We have noted the presence a barn owl *Tyto alba* and an Indian great horned owl *Bubo bubo* in the vicinity of the trees on which *C. sphinx* foraged. However, we have not observed predation on them at the time of emergence or in fruit trees while foraging. Red fig-eating bats *Stenoderma rufum* did not alter their activity or movement in response to levels of moonlight since the study site lacks most of the visually oriented bat predators (Gannon and Willig, 1997). It might also be possible to elude interference from conspecifics in the bright moonlight. We have observed on earlier occasions that an individual *C. sphinx* chased away a conspecific while the latter was feeding on fruits in *Psidium guajava* and *Annona squamosa* trees. Thus, pressure from predation and conspecifics seems to be the likely explanation for the evolving of 'lunar phobia' (Morrison, 1978) which limits foraging flights to nights with minimal moonlight and which make bats carry fruits or leaves to a feeding roost.

The observations on the foraging pattern of *C. sphinx* is similar to that of *Artibeus jamaicensis, Phyllostomus hastatus* and *Carollia perspicillata* (Morrison, 1978; August, 1981; McCracken and Bradbury, 1981); bats leave the day roost shortly after sun set and commute to the feeding areas while they begin to search for ripe fruits. The harvested fruits are transported to the night roosts for consumption. Such processes in the night roosts might promote digestion and energy conservation, offer retreat from predators, serve as centers of information transfer about the location of fruit patches and facilitate social interaction (Kunz, 1982). Throughout the study only one male bat was found to be faithful to a single night roost (*Guetterda speciosa* was used as night roost constantly). A regular path exhibited by this bat between day roost and feeding area may be attributed to the constancy of resource availability, with the bat showing "trap-lining" behaviour to minimize commuting distance and energy expenditure. But other tagged bats of both sexes used more than one night roost such as *A. sapota, A. indica, Areca catechu, C. nucifera, P. longifolia,* etc. A high risk of predatory pressure may be attributed for the usage of more night roost.

It is apparent that the males restrict their foraging area closer to the day roost whereas the females commute further and utilized several foraging areas. Since males are involved in tent construction, harem formation and protection of tent might promote harem defense strategies near the day roost (Fleming, 1988; Marimuthu *et al.*, 1998). Males remained in their territories in the absence of females as well as in the presence to defend the territories against the intruders. Males of *C. perspicillata* stay close to their territories both day and night (Fleming, 1988). The observations of short distance foraging flights of males are consistent with the males of *A. jamaicensis, P. hastatus* and *C. perspicillata*. They were also foraged predominantly in the vicinity of their day roosts (Morrison, 1979; Morrison and Morrison, 1981; Fleming, 1988; Handley and Morrison, 1991; McCracken and Bradbury, 1991). The reasons for farther commuting distances in females may be to utilize several foraging areas, or to select potential males. Pregnant female *Myotis myotis* increased foraging time (Audet, 1990), probably to meet the increased energy requirements. The female bats use several foraging areas and change their primary foraging area in an unpredictable fashion as observed in *C. perspicillata* (Kunz, 1982). Every foraging area may not contain the same potential food sources and therefore, the unpredictable "shuttles" might cause the dietary diversity. The foraging areas of females are isolated whereas the males foraging area showed overlap due to the occurrence of day roosts within a rich food patch.

Cynopterus sphinx exhibits two peaks of foraging activity, one in the pre-midnight hours when bats fed mostly on steady state fruiting trees, which produce low number of fruits in a night. Such strategies have also been reported in *C. perspicillata* when they fed on *Piper amalago* (Fleming, 1982). These foraging patterns might promote scramble type of competition among their conspecifics in a given foraging area. Fruits such as *A. squamosa, A. sapota, C. indica, P. pendula* and *P. longifolia* are removed by bats early in the night and early feeding bouts would appear the advantageous for exploitation of the food resources available at that time. Another foraging peak occurred during post-midnight hours when bats fed on big bang fruiting trees such as *F. religiosa, F. benjamina* and *F. bengalensis*, which produce a large number of fruits in a night. Apparently, the big bang fruiting trees might allow more number of bats to forage without competition in a feeding area (Elangovan *et al.*, 1999).

The observations on the commuting distance and number of foraging areas utilized by subadult males and females showed clear differences. These differences in the above behavioural repertories appeared to reflect the behaviour of adults. However, the earlier study on the food preference of *C. sphinx* corresponds with that of the availability of fruits thereby exhibiting opportunistic feeding behaviour (Rajan *et al.*, 1999). The actual expectation is the subadult males and females would prefer to exploit the resources available nearer to their day roost. But the commuting behaviour of subadult females turned out against the expectation. It suggests the inheritance of the mother's foraging behaviour. Another possibility is that the subadult bats spent considerable amount of time with their mother and the consequent learning of foraging technique before taking independent foraging flights. Perhaps only the subadult females exhibit the long distance foraging flights like their mother whereas the subadult males prefer to forage at a shorter distance. Unlike the adult females, the adult males forage at shorter distances and this can be attributed to their social system.

REFERENCES

Anthony, E. L. P., Stack, M. H. and Kunz, T. H. 1981. Night roosting and the nocturnal time budget of the little brown bat, *Myotis lucifugus*: effects of reproductive status, prey density and environmental conditions. *Oecologia* 51: 151 – 156.

Audet, D. 1990. Foraging behavior and habitat use by a gleaning bat, *Myotis myotis* (Chiroptera: Vespertilionidae). *Journal of Mammalogy* 71: 420 – 427.

August, P. V. 1981. Fig consumption and seed dispersal by *Artibeus jamaicensis* in the Llanos of Venezuela. *Reprod. Botany, Supplement to Biotropica* 13: 70 - 76.

Ayensu, E. S. 1974. Plant and bat interactions in West Africa. *Annals of Botany* 61: 702 – 727.

Baker, M. E. 1992. Evolution of regulation of steroid-mediated intercellular communication in vertebrates: insights from flavenoids, signals that mediate plant-rhizobia symbiosis. *Journal of Steroid Biochemistry and Molecular Biology* 41: 301 – 308.

Balasingh, J. 1990. *The behaviour of the Indian false vampire* bats, *Megaderma lyra* Geoffroy 1810: field ethological studies. Ph.D. Thesis, Madurai Kamaraj University.

Balasingh, J., Isaac, S. S and Subbaraj, R. 1993. Tent-roosting by the frugivorous bat *Cynopterus sphinx* (Vahl 1797) in Southern India. *Current Science* 65: 418.

Balasingh, J., John Koilraj, A and Kunz, T. H. 1995. Tent construction by the short-nosed fruit bat, *Cynopterus sphinx* (Chiroptera: Pteropodidae) in Southern India. *Ethology* 100: 210 – 229.

Banack, S. A. 1998. Diet selection and resource use by flying foxes (Genus *Pteropus*). *Ecology* 79: 1949 – 1967.

Barclay, R. M. R. 1985. Foraging behavior of the African insectivorous bat, *Scotophilus leucogaster*. *Biotropica* 17: 65 – 70.

Barclay, R. M. R. 1994. Constraints on reproduction by flying vertebrates: energy and calcium. *American Naturalist* 144: 1021 – 1031.

Bates, P. J. J and Harrison, D. L. 1997. *Bats of the Indian subcontinent*. Harrison Zoological Museum, Sevenoaks.

Bhat, H. R. 1994. Observations on the food and feeding behaviour of *Cynopterus sphinx* Vahl (Chiroptera, Pteropodidae) at Pune, India. *Mammalia* 58: 363 – 370.

Bhat, H. R. and Kunz, T. H. (1995). Altered flower/fruit clusters of the kitul palm used as roosts by the short-nosed fruit bat, *Cynopterus sphinx* (Chiroptera: Pteropodidae). *Journal of Zoology* (London) 234: 597 – 604.

Bowers, M. A. 1990. Exploitation of seed aggregates by Merriam's kangaroo rat: harvesting rates and predatory risk. *Ecology* 71: 2334-2344.

Bradbury, J. W. and Vehrencamp, S. L. (1976). Social organization and foraging in emballonurid bats. *Behavioral Ecology and Sociobiology* 1: 337 – 404.

Brander, R. B. and Cochran, W. W. 1969. Radio-location telemetry. In: *Wildlife Management Techniques*, Giles, R. H. (Ed.). 95 – 103. Washington: Wildlife Society.

Butynski, T.M. 1984. Nocturnal ecology of the springhare, *Pedetes capensis* in Botswana. *African Journal of Ecology* 22: 7 – 22.

Calley, M., Braithwaite, R. W. and Ladd, P. G. 1993. Reproductive biology of *Ravenala madagascariensis* Gmel. as an alien species. *Biotropica* 25: 61-72.

Crespo, R.F., Linhart, S.B., Burns, R.J. and Mitchell, G.C. 1972. Foraging behavior of the common vampire bat related to moonlight. *Journal of Mammalogy* 53: 366 – 368.

Cresswell, W. J. and Harris, S. 1988. The effects of weather conditions on the movement and activity of badgers, *Meles meles* in a suburban environment. *Journal of Zoology* (London) 216: 187 – 194.

Cunningham van Someren, G. R. 1972. Some fruit bats eat leaves. *East African Natural History Society Bulletin* 2: 24 – 25.

Dobson, G. E. 1978. *Catalogue of the Chiroptera in the collection of the British Museum.* London.

Dubois, M., Gilles, K. A., Roberts, P. A and Smith, F. 1956. Calorimetric method for determination of sugars and related substances. *Annals of Chemistry* 25: 350 – 356.

Eisenberg, J. F. 1978. The evolution of arboreal folivores in the class Mammalia. In: *The ecology of arboreal folivores*, Montgomery, G. (Ed.). 135 - 152. Washington: Smithsonian Institution Press.

Elangovan, V and G. Marimuthu. 2001. Comparative analysis on food, energy and nitrogen intake of a megachiropteran bat, *Cynopterus shinx*. *Proceedings of the National Academy of Sciences B* (India), 71(2): 115–120.

Elangovan, V and Marimuthu, G. 2001. Effect of moonlight on the foraging behaviour of a megachiropteran bat *Cynopterus sphinx*. Journal of Zoology (London) 253: 347 – 350.

Elangovan, V. 2000. *Foraging behaviour of a frugivorous bat Cynopterus sphinx*. Ph. D. dissertation, Madurai Kamaraj University, Madurai, India.

Elangovan, V. Marimuthu, G. and Kunz, T. H. 1999. Temporal patterns of individual and group foraging behaviour in the short-nosed fruit bat, *Cynopterus sphinx*, in South India. *Journal of Tropical Ecology* 15: 681 – 687.

Elangovan, V., Marimuthu, G and Kunz, T. H. 2000. Nectar feeding behavior in the short-nosed fruit bat *Cynopterus sphinx* (Pteropodidae). *Acta Chiropterologica* 2: 1 – 5.

Elangovan, V., Marimuthu, G and T. H. Kunz. 2001. Temporal patterns of resource use by the short-nosed fruit bat *Cynopterus sphinx* (Megachiroptera: Pteropodidae). *Journal of Mammalogy*, 82: 161–165.

Elangovan, V., Raghuram, H., Yuvana Satya Priya, E and Marimuthu, G. 2002. Postnatal growth, age estimation and development of foraging behaviour in the fulvous fruit bat *Rousettus leschenaulti*. *Journal of Biosciences* 27: 695 – 702.

Elangovan, V., Raghuram, H., Yuvana Satya Priya, E and Marimuthu, G. 2004. Wing morphology and flight performance in *Rousettus leschenaulti*. *Journal of Mammalogy* 85: 806 – 812.

Elangovan, V., Yuvana Satya Priya, E., Raghuram, H and Marimuthu, G. 2003. Postnatal development in the Indian short-nosed fruit bat *Cynopterus sphinx*: growth rate and age estimation. *Acta Chiropterologica* 5: 107 – 116.

Erkert, H. G. 1989. Lighting requirements of nocturnal primates in captivity: a chronobiological approach. *Zoo Biology* 8: 179 – 191.

Erkert, H.G and Groeber, 1986. Direct modulation of activity and body temperature of owl monkeys (*Aotus lemurinus grisemembra*) by low light intensities. *Folia Primatology* 47: 171 – 188.

Erkert, H.G. 1974. Der Einfluss des Mondlichtes auf die Aktivitaetsperiodik nachtaktiver Saeugetiere. *Oecologia* 14: 269 – 287.

Erkert, H.G. 1976. Beleuchtungsabhaengiges Atkivitaetsoptimum bei Nachtaffen (*Aotus trivirgatus*). *Folia Primatology* 25: 186 – 192.

Faegri, K. and van der Pijl, L. 1979. *The principles of pollination ecology*. Oxford: Pergamon Press.

Fenton, M. B., Brigham, R. M., Mills, A. M. and Rautenbach, I. L. 1985. The roosting and foraging areas of *Epomophorus wahlbergi* (Pteropodidae) and *Scotophilus viridis* and (Vespertilionidae) in Kruger National Park, South Africa. *Journal of Mammalogy* 66: 461 – 468.

Fenton, M.B., Boyle, N.G.H., Harrison, T.M. and Oxley, D.J. 1977. Activity patterns, habitat use, and prey selection by some African insectivorous bats. *Biotropica* 9: 73 – 85.

Fischer, E.A. 1992. Foraging of nectarivorous bats on *Bauhinia ungulata*. *Biotropica* 24: 579 – 582.

Fleming, T. H. 1982. Foraging strategies of plant visiting bats. In: *Ecology of bats,* Kunz, T. H. (Ed.). 287 - 325. New York: Plenum Press.

Fleming, T. H. 1988. *The short-tailed fruit bat*. Chicago: The University of Chicago Press.

Fleming, T. H. and Heithaus, E. R. 1986. Seasonal foraging behavior of the frugivorous bat *Carollia perspicillata*. *Journal of Mammalogy* 67: 660 – 671.

Fleming, T.H. 1979. Tropical frugivores compete for food? *American zoologist*. 19: 1157 – 1172.

Francis, C. E. L. P., Anthony, J. B and Kunz, T. H. 1994. Lactation in male fruit bats. *Nature* 367: 691 - 692.

Gannon, M. R., Willig, M.R. 1997. The effect of lunar illumination on movement and activity of the red fig-eating bat (*Stenoderma rufum*). *Biotropica* 29: 525 – 529.

Gardner, A. L. 1977. Feeding habits. In: *Biology of bats of the New World family Phyllostomatidae*, Baker, R.J., Jones, J.K., Jr. and Carter, D.C. (Eds.). 293 - 350. Lubbock: Texas Tech. Univ., Part II, Special Publication.

Gentry, A. H. 1974. Flowering phenology and diversity in tropical Bignoniaceae. *Biotropica* 6: 64 - 68.

Gilbert, B.S. and Boutin, S. 1991. Effect of moonlight on winter activity of snowshoe hares. *Archives Alp. Research* 23: 61 – 65.

Gopukumar, N., Elangovan, V and Subbaraj, R. 1998. Influence of moonlight on the foraging behaviour of Indian short-nosed fruit bat *Cynopterus sphinx*. *Current Science* 74: 688 – 689.

Haeussler, U. and Erkert, H.G. 1978. Different direct effects of light intensity on the entrained activity rhythm in neotropical bats (Chiroptera: Phyllostomidae). *Behav. Processes* 3: 223 – 239.

Handley, C. O. Jr. and Morrison, D. W. 1991. Foraging behavior. In: *Demography and natural history of the common fruit bat, Artibeus jamaicensis, on Barro Colorado Island, Panama,* Handley, C.O., Jr., Wilson, D. E. and Gardner, A.L. (Eds.). 137 - 140. Washington: Smithsonian Contributions in Zoology.

Handley, C. O. Jr., Gardner, A. L. and Wilson, D. E. 1991. Food habits. In: *Demography and natural history of the common fruit bat, Artibeus jamaicensis, on Barro Colorado Island, Panama*, Handley, C. O. Jr., Wilson, D. E. and Gardner, A. L. (Eds.). 141 - 146. Washington: Smithsonian Contributions in Zoology.

Harris, B. J. and Baker, H. G. 1959. *Pollination of flowers by bats in Ghana*. Niger. Fld. 24: 151 – 159.

Heithaus, E. R. 1982. Coevolution between bats and plants. In: *Ecology of bats*, Kunz, T.H. (Ed.). 327 - 367. New York: Plenum Press.

Heithaus, E. R., Fleming, T. H. and Opler, P. A. 1975. Foraging pattern and resource utilisation by eight species of bats in a seasonal tropical forest. *Ecology* 56: 841 - 854.

Heithaus, E. R., Fleming, T. H. and Opler, P. A. 1975. Foraging pattern and resource utilisation by eight species of bats in a seasonal tropical forest. *Ecology* 56: 841 - 854.

Heithaus, E. R., Opler, P. A. and Baker, H. G. 1974. Bat activity and pollination of *Bauhinia puletia*: plant-pollinator coevolution. *Ecology* 55: 412 – 419.

Herrera, C. M. 1987. Vertebrate dispersed plants of the Iberian Peninsula: a study of fruit characteristics. *Ecological Monograph* 57: 305 – 331.

Howe, H. F and Westley, L. C. 1988. *Ecological relationships of plants and animals*. New York: Oxford University Press.

Howell, D. J. 1974. Bats and pollen: physiological aspects of the syndrome of chiropterophily. *Comparative Biochemistry and Physiology* 111: 617 – 832.

Howell, D. J. 1979. Flock foraging in nectar-feeding bats: advantages to the bats and to the host plants. *American Naturalist* 114: 23 – 49.

Imber, M. J. 1975. Behaviour of petrels in relation to the moon and artificial lights. *Notornis* 22: 302 – 306.

Isaac, S. S and Marimuthu, G. 1993. Infant's calls attract mother *Pipistrellus mimus*. *Bat Research News* 34: 57.

Jahoda, J. C. 1973. The effect of the lunar cycle on the activity pattern of *Onychomys leucogaster brevianritus*. *Journal of Mammalogy* 54: 544 – 549.

Julien-Laferriere, D. 1997. The influence of moonlight on activity of woolly opossums, *Caluromys philander*. *Journal of Mammalogy* 78: 251 – 255.

Kingdon, J. 1974. *East African mammals*. London: Academic Press.

Knudsen, J. T and Kiltgaard, B. B. 1998. Floral scent and pollination in *Browneopsis disepala* (Leguminosae: Caesalpinioideae) in Western Ecuador. *Brittonia* 50: 174 - 182.

Kotler, B. P. 1984. Risk of predation and the structure of desert rodent communities. *Ecology* 65: 689 - 701.

Krebs, J. R and McCleery, R. H. 1984. Optimization in behavioural ecology. In: *Behavioural ecology*, Krebs, J. R. and Davies, N. B. (Eds.). 91 - 121. Sunderland Mass: Sinauer.

Kress, W. 1985. Bat pollination of an Old World Heliconia. *Biotropica* 17: 302 - 308.

Kunz, T. H and Diaz, C. A. 1995. Folivory in fruit-eating bats, with new evidence from *Artibeus jamaicensis* (Chiroptera: Phyllostomidae). *Biotropica* 27: 106 – 120.

Kunz, T. H. 1982. Roosting ecology of bats. In: *Ecology of bats*, Kunz, T. H. (Ed.). 1 - 55. New York: Plenum Press.

Kunz, T. H. 1987. Post-natal growth and energetics of suckling bats. In: *Recent advances in the study of bats*, Fenton, M. B., Racey, P. A. and Rayner, J. M. V. (Eds.). 395 - 420. Cambridge: Cambridge University Press.

Labov, J. B. 1977. Mini review: phytoestrogens and mammalian reproduction. *Comparative Biochemistry and Physiology* A 57: 3 – 9.

Law, B. S. 1992. Physiological factors affecting pollen use by Queensland blossom bats, *Syconycteris australis*. *Functional Ecology* 6: 257 - 264.

Lemke, T. O. 1984. Foraging ecology of the long-nosed bat, *Glossophaga soricina*, with respect to resource availability. *Ecology* 65: 538 - 548.

Lockard, R. B and Owings, D. M. 1974. Moon related surface activity of bannertail, *Dipodomys spectabilis* and Fresno (*D. nitratoides*) kangaroo rats. *Animal Behaviour* 22: 262 – 273.

Longhand, W. S and Price, M. V. 1991. Direct observations of owls and heteromyid rodents: can predation risk explain microhabitat use? *Ecology* 72: 2261 – 2273.

Lowry, J. B. 1989. Green-leaf fractionation by fruit bats: is this feeding behaviour a unique nutritional strategy for herbivores? *Australian Wildlife Research* 16: 203 - 206.

Marimuthu, G., Rajan, K. E., John Koilraj, A., Isaac, S. S. and Balasingh, J. 1998. Observations on the foraging behaviour of a tent roosting megachiropteran bat *Cynopterus sphinx*. *Biotropica* 30: 321 – 324.

Marshall, A. G. 1983. Bats, flowers, and fruits: evolutionary relationship in the Old World. *Biological Journal of Linnaean Society* 20: 115 – 135.

Marshall, A. F. 1985. Old World phytophagous bats (Megachiroptera) and their food plants: a survey. *Zoological Journal of Linnaean Society* 83: 351 – 369.

McCracken, G. F and Bradbury, J. W. 1981. Social organization and kinship in the polygynous bat *Phyllostomus hastatus*. *Behavioral Ecology and Sociobiology* 8: 11 – 34.

Morrison, D. W. 1978. Lunar phobia in a neotropical fruit bat *Artibeus jamaicensis* (Chiroptera: Phyllostomidae). *Animal Behaviour* 26: 852 – 855.

Morrison, D. W. 1979. Apparent male defense of tree hollows in the fruit bat, *Artibeus jamaicensis*. *Journal of Mammalogy* 60: 11 – 15.

Morrison, D. W and Morrison, S. H. 1981. Economics of harem maintenance by a Neotropical bat. *Ecology* 62: 864 - 866.

Morton, E. S. 1973. On the evolutionary advantages and disadvantages of fruit eating tropical birds. *American Naturalist* 107: 8 - 22.

Moule, G. R., Braden, A. W. H and Lamond, D. R. 1963. The significance of oestrogens in pasture plants in relation to animal production. *Animal Breeding Abstract* 31: 139 – 157.

Nair, N. G., Elangovan, V., Sripathi, K., Marimuthu, G and Subbaraj, R. 1999. Foraging behavior of a frugivorous bat, *Cynopterus sphinx*. *International Journal of Mammalian Biology* 64: 187 – 191.

Nair, N.G., Elangovan, V and Subbaraj, R. 1998. Influence of moonlight on the foraging behaviour of the Indian short-nosed fruit bat *Cynopterus sphinx*: radio-telemetry studies. *Current Science* 74: 688 – 689.

Nassar, J. M., Ramirez, N and Linares, O. 1997. Comparative pollination biology of Venezuelan columnar cacti and the role of nectar-feeding bats in their sexual reproduction. *American Journal of Botany* 84: 918 – 927.

Nathan, P. T. 2001. *Behaviour of the Indian short-nosed fruit bat Cynopterus sphinx (Vahl, 1797): field and seminaturalistic ethological studies*. Ph. D. dissertation, Manonmaniam Sundaranar University, Tirunelveli, India.

Nathan, P. T., Raghuram, H., Elangovan, V., Karuppudurai, T and Marimuthu, G. 2005. Bat pollination of kapok tree, *Ceiba pentandra*. *Current Science* 88(10): 1679 – 1681.

Nathan, P. T., Swami Doss, D. P., Isaac, S. S., Balasingh, J., Emmanuvel Rajan, K., Gopukumar, N and Subbaraj, R. 2001. Mist net capture and field observation on the short-nosed fruit bat (Chiroptera: Pteropodidae) *Cynopterus sphinx*. *Journal of Bombay Natural History Society* 98: 373 – 378.

Negraeff, E and Brigham, R. M. 1995. The influence of moonlight on the activity of little brown bats, *Myotis lucifugus. International Journal of Mammalian Biology* 60: 330 – 336.

Nelson, D. A. 1989. Gull predation on Cassin's Auklet varies with the lunar cycle. *Auk* 106: 495 – 497.

Nelson, J. E. 1965. Behaviour of Australian Pteropodidae (Megachiroptera). *Animal Behaviour* 13: 544 – 557.

O'Brien, T. G., Kinnard, M. F., Diernfeld, E. S., Conklin-Brittain, N. L., Wrangham, R. W and Silver, S. C. 1998. What's so special about figs? *Nature* 292: 668.

Petit, S. 1997. The diet and reproductive schedules of *Leptonycteris curasoae curasoae* and *Glossophaga longirostris elongata* (Chiroptera: Glossophaginae) on Curacao. *Biotropica* 29: 214 – 223.

Petit, S. and Freeman, C. E. 1997. Nectar production of two sympatric species of columnar cacti. *Biotropica* 29: 175 – 183.

Price, M. V., Waser, N. M. and Bass, R. A. (1984). Effects of moonlight on microhabitat use by desert rodents. *Journal of Mammalogy* 65: 353 – 362.

Rajan, K. E., Nair, N. G and Subbaraj, R. 1999. Seasonal food preference of the Indian short-nosed fruit bat *Cynopterus sphinx* (Vahl) (Chiroptera: Pteropodidae). *Journal of Bombay Natural History Society* 96: 24 – 27.

Reith, C. C. 1982. Insectivorous bats fly in shadows to avoid moonlight. *Journal of Mammalogy* 63: 685 – 688.

Reith, C. C. 1982. Insectivorous bats fly in shadows to avoid moonlight. *Journal of Mammalogy* 63: 685 – 688.

Ryberg, O. 1947. *Studies on bats and bat parasites, especially with regard to Sweden and other neighboring countries of the north. Stockholm* Bokforlaget: Svensk Natur.

Sazima, I and Sazima, M. 1977. Solitary and group foraging: two flower-visiting patterns of the lesser spear-nosed bat, *Phyllostomus discolor. Biotropica* 9: 213 – 215.

Shilton, L. A., Altringham, J. D., Compton, S. G and Whittaker, R. J. 1999. Old World fruit bats can be long-distance seed dispersers through extended retention of viable seeds in the gut. *Proceedings of the Royal Society of London, Biological Series* 266: 219 – 223.

Singaravelan, N and Marimuthu, G. 2004. Nectar feeding and pollen carrying from *Ceiba pentandra* by pteropodid bats. *Journal of Mammalogy* 85: 1 – 7.

Start, A. N and Marshall, A. G. 1976. Nectarivorous bats as pollinators of trees in West Malaysia. In: *Tropical trees: variation, breeding and conservation*, Buraly, J. and Styles, B. T. (Eds.). 243. London: Linnaean Society Symposium Series No. 2, Academic Press.

Start, A. N and Marshall, A. G. 1976. Nectarivorous bats as pollinators of trees in West Malaysia. In: *Tropical trees: variation, breeding and conservation*, Buraly, J. and Styles, B. T. (Eds.). 243. London: Linnaean Society Symposium Series No. 2, Academic Press.

Stellar, D. D. 1986. The dietary energy and nitrogen requirements of the grey-headed flying fox, *Pteropus poliocephalus* (Temminck) (Megachiroptera). *Australian Journal of Zoology* 34: 339 – 349.

Storey, A. E and Grimmer, B. L. 1986. Effect of illumination on the nocturnal activities of Manx Shear-water: Colony avoidance or inconspicuous behavior? *Bird Behavior* 6: 85 – 89.

Storz, J. F., Balasingh, J., Nathan, P. T., Immanuel, J. K and Kunz, T. H. 2000b. Dispersion and site fidelity in a tent-roosting population of the short-nosed fruit bat (*Cynopterus sphinx*) in Southern India. *Journal of Tropical Ecology* 16: 117 – 131.

Storz, J. F., Bhat, H. R and Kunz, T. H. 2000a. Social structure of a polygynous tent-making bat, *Cynopterus sphinx* (Megachiroptera). *Journal of Zoology* (London) 251: 151 – 165.

Storz, J. F., Bhat, H. R and Kunz, T. H. 2001. Genetic consequences of polygyny and social structure in an Indian fruit bat, *Cynopterus sphinx*. II. Variance in male mating success and effective population size. *Evolution* 55: 1224 – 1232.

Subramanya, S and Radhamani, T. R. 1993. Pollination by birds and bats. *Current Science* 65: 201 – 209.

Tan, K. H., Zubaid, A and Kunz, T. H. 1998. Food habits of *Cynopterus brachyotis* (Myullar) (Chiroptera: Pteropodidae) in peninsular Malaysia. *Journal of Tropical Ecology* 14: 299 – 307.

Tedman, R. A and Hall, L. S. 1985. The morphology of the gastrointestinal tract and food transit time in the fruit bats, *Pteropus alecto* and *P. poliocephalus* (Megachiroptera). *Australian Journal of Zoology* 33: 625 – 640.

Telek, L and Martin, W. 1983. Tropical plants for leaf protein concentrates. In: *Leaf protein concentrates*, Telek, L. and Graham, H. D. (Eds.). 81 - 116. Westport, Connecticut: AVI Publishing Company, Inc.

Thomas, D. W. 1983. The annual migration of three species of West African fruit bats (Chiroptera: Pteropodidae). *Canadian Journal of Zoology* 61: 2266 – 2272.

Tidemann, C. R., Priddeel, D. M., Nelson, J. E and Pettigrew, J. D. 1985. Foraging behavior of the Australian ghost bat, *Macroderma gigas* (Microchiroptera: Megadermatidae). *Australian Journal of Zoology* 33: 705 – 713.

Trent, B. K., Tucker, M. E and Lockard, J. S. 1977. Activity changes with illumination in slow loris *Nycticebus coucang*. *Applied Animal Ethology* 3: 281 – 286.

Usman, K., Habersetzer, J., Subbaraj, R., Gopalakrishnaswamy, G and Paramanandam, K. 1980. Behaviour of bats during a lunar eclipse. *Behavioral Ecology and Sociobiology* 7: 79 – 81.

Vaughan, T. A. 1976. Nocturnal behavior of the African false vampire bat, *Cardioderma cor*. *Journal of Mammalogy* 57: 227 – 248.

Watanuki, Y. 1986. *Moonlight avoidance behavior in Leach's Storm-Petrels as a defense against Slaty-backed Gulls.* Auk 103: 14 – 22.

Waterman, P. G. 1984. Food acquisition and processing as a function of plant chemistry. In: *Food acquisition and processing in primates*, Chivers, D. J., Wood, B. A. and Bilsborough, A. (Eds.). 177 - 211. New York: Plenum Press.

Webb, N. J and Tidemann, C. R. 1996. Mobility of Australian flying-foxes, *Pteropus spp.* (Megachiroptera) evidence from genetic variation. *Proceedings of the Royal Society of London, Biological Series* 263: 497 – 502.

Wiles, G. J and Fujita, M. S. 1992. The food plants and economic importance of flying foxes on Pacific Islands. In: *Pacific Island flying foxes: Proceedings of an International Conference*: Wilson, D. W. and Graham, G. (Eds.). 90(23): 24 - 35. Washington D C.

Wilson, E. O. 1975. *Sociobiology*. Cambridge: Belknap Press.

Wolfe, J. L and Summerlin, C. T. 1989. The influence of lunar light on nocturnal activity of the old field mouse. *Animal Behaviour* 37: 410 – 414.

Wright, P. C. 1981. The night monkeys, genus *Aotus*. In: *Ethology and behaviour of neotropical primates*, Coimbra-Filho, A. F. and Mittermeir, R. A. (Eds.). 1: 1 - 496. Brazil: Academia Brasileira de Ciencias, Rio de Janeiro.

Zortea, M and Mendes, S. L. 1993. Folivory in the big fruit-eating bat, *Artibeus lituratus* (Chiroptera: Phyllostomidae) in eastern Brazil. *Journal of Tropical Ecology* 9: 117 – 120.

In: Bats: Biology, Behavior and Conservation
Editors: J. L. Zupan and S. L. Mlakar, pp. 135-175
ISBN 978-1-61209-749-7

Chapter 3

REPRODUCTIVE BIOLOGY OF MALE BATS: ANATOMY, PHYSIOLOGY AND ENDOCRINOLOGY

Regiana Lúcia Oliveira and Cleida Aparecida Oliveira
Department of Morphology, Universidade Federal de Minas Gerais,
Belo Horizonte, MG - Brazil

ABSTRACT

The chiropterans are the only mammals capable of true flight. They represent one of the largest and most diversified orders of mammals, inhabiting every continent, except Antarctica. This diversity is reflected in the anatomy and physiology of the reproductive organs that have been adapted for flight and geographic distribution, as seen in hibernating species inhabiting temperate zones and non-hibernating species generally inhabiting the tropics. Here we summarize the current knowledge of male bat reproductive strategies and genital organs morphophysiology. Similar to other mammals, the male genital system of Chiroptera is composed of the testis, efferent ductules, epididymis, vas deferens, urethra, penis and accessory sex glands, however a variety of peculiar features are observed in the morphophysiology of these organs. The bat testis presents a large variation in position, according to the presence or absence of scrotum, being scrotal, subcutaneous or abdominal. The testicular position may be permanent or transitory depending on the species considered. According to the acrossomic system method, ten to eleven stages of the cycle of seminiferous epithelium are recognized in bat seminiferous tubules. During regression, seminiferous tubules shelter only Sertoli cells and spermatogonia. Curiously, the Sertoli cells are located above the spermatogonia, forming a protective barrier for these cells that are responsible for re-colonization of the seminiferous tubules. The efferent ductules connect the rete testis to the epididymis. A description of *Artibeus lituratus* revealed 12 to 15 flexuous ductules arising separately from the *rete testis* and running parallel towards the epididymis, then anastomozing and forming multiple entries into the epididymal duct. This pattern of disposition is similar to that found for large eutherian mammals and men. The epididymis of bats may be anatomically divided into initial segment, caput, corpus and cauda. As shown in *A. lituratus*, the epididymal epithelium is composed of principal, basal, apical, narrow, clear and halo cells, with different distribution along the duct. In several hibernating species, the epididymis is responsible for storing viable sperm during long periods. Information about the male accessory sex glands of bats is scarce. The prostate and bulbourethral glands have been described in all species studied, whereas the ampullary gland and seminal vesicles are present in some species and absent in others. The bat penis is prominent and pendulous, presenting a well-developed glans. In some species, the penis presents a baculum (*os penis*), extending throughout the glans. In seasonal animals, environmental factors as photoperiod play an important role in gonadal activity. In these animals, the exposure to short days leads to decreased hypothalamic GnRH and pituitary gonadotropins (FSH and LH) releases, leading to a decrease in gonadal activity. Testosterone levels vary along the reproductive cycle of hibernating and non-hibernating bats. Besides androgens, it is now recognized that estrogens also play an important role in male reproduction. However, information about estrogen in male bats has been limited to one report about *A. lituratus*, a seasonal non-hibernating Neotropical species, in which both estrogen receptors ERα and ERβ were detected in the testis. During testicular regression, levels of estrogen receptors, especially ERβ, are increased, indicating that estrogens may be directly involved in the regulation of the bat male reproductive cycle.

I. Introduction

The chiropterans or bats are the only mammals capable of true flight. They represent the second largest and most diversified order of mammals in terms of geographical distribution as well as number of species, losing only for rodents. Bats are found in all continents, except Antarctica, and in some oceanic islands, they constitute the only native mammal species (Brunet-Rossinni and Austad, 2004). The order Chiroptera is comprised of two suborders, the Megachiroptera and Microchiroptera, which are distributed in 18 families, 202 genera and 1120 species, comprising approximately 22% of the total species of mammals (Singh and Krishna, 1996; Reis et al., 2007). The diversity of Chiroptera are reflected in the anatomy and physiology of the reproductive organs that have been adapted for flight and diverse geographical distribution, as seen in hibernating species inhabiting temperate zones and non-hibernating species mostly inhabiting the Tropics.

Despite the numerous reproductive peculiarities, very little is known about the anatomy and physiology of the genital system of male bats. The available information is often incomplete and fragmented, especially when dealing with species that inhabit tropical regions. Here, we summarize the current knowledge of male bat genital tract morphophysiology and reproductive strategies, in addition to some of our own findings on the tropical non-hibernating, big fruit-eating bat *Artibeus lituratus* (Olfers, 1818).

II. Reproductive System

The reproductive system of bats consists of paired testes, efferent ductules, epididymides, vas deferens and bulbourethral or Cowper's glands, unpaired urethra and penis, as well as variable accessory glands surrounding the proximal urethra, such as prostate, ampullary gland and seminal vesicle.

1. TESTES

The testes are pair organs, symmetrical and responsible for sperm production and sex steroid synthesis, both essential for reproductive functions. The bat testes are variable in shape depending on species and reproductive phase. In some species, the gonad is ovoid or elliptical (Matthews, 1941 - *Hipposideros caffer*; Krutzsch et al., 1976 - *Macrotus waterhousii*; Krutzsch and Crichton, 1986 - *Pipistrellus subflavus*;

Oliveira et al., 2009 - *Artibeus lituratus*), whereas in others, the testis is globular (Matthews, 1941 - *Nycteris luteola* and *N. hispida*; Vamburkar, 1958 - *Cynopterus sphinx*), becoming elliptical during spermatogenic activity (Vamburkar, 1958). Testis size also undergoes remarkable seasonal variation, as the greatest size is achieved at maximal spermatogenic activity. Therefore, even though the timing of spermatogenesis varies between species, the testis weight has been a reliable indicator of the organ activity (Hosken et al., 1998).

It is noteworthy that the bat testis presents large variations in position among species, according to the absence or presence of scrotum, being permanently abdominal, partially abdominal, subcutaneous or scrotal. Moreover, in some species, the testes may undergo migration into the abdominal cavity at certain periods of the year.

The testes lie permanently in the abdominal cavity without migration through the inguinal canal in members of Rhinopomatidae (e.g. *Rhinopoma kinneari)* and Pteropodidae (e.g. *Cynopterus sphinx gangeticus)* families (Vamburkar, 1958; Anand Kumar, 1965; Krutzsch, 1979; Singwi and Lall, 1983). In these species, the scrotum is absent and the spermatogenesis occurs only when the bats enter into winter torpor, possibly because at this period, the metabolism, including body temperature, is reduced (Anand Kumar, 1965).

Migration of the testes between the scrotum and abdominal cavity via the inguinal canal occurs in several species of the Emballonuridae (e.g. *Taphozous sp* and *Thapozous georgianus*) (Matthews, 1941; Jolly and Blackshaw, 1988) and Pteropodidae families, in which the testis are scrotal during the breeding season (usually in summer) and intra-abdominal at other seasons (Krutzsch, 1979). In *Taphozous georgianus,* the testicular descent is closely influenced by high temperature but not daylength (Jolly and Blackshaw, 1988). The fact that in this species there is no pampiniform plexus and the testicular arteries are not coiled is interesting (Jolly and Blackshaw, 1988). Differing from testis, in *T. georgianus,* the cauda epididymis is permanently in the scrotum, thus providing cooler temperatures adequate for sperm storage. In some members of the Molossidae family, as *Mormopterus planiceps* and *Tadarida (=Chaerephon) hindei*, the testes are inguinal in position and undergo migration into the abdominal cavity when spermatogenesis is ceased (Krutzsch, 1979; Krutzsch and Crichton, 1987).

The testes are subcutaneous and do not undergo migration into the abdominal cavity in some Hipposideridae (Matthews, 1941 - *Hipposideros caffer, Triaenops afer*), Vespertilionidae (Krutzsch, 1975 - *Pipistrellus hesperus*), and Phyllostomidae (Krutzsch et al., 1976 - *Macrotus waterhousii*; Beguelini et al., 2009 - *Artibeus lituratus, Artibeus planirostris, Carollia perspicillata, Platyrrhinus lineatus* and *Sturnira lilium*; Oliveira et al., 2009 - *Artibeus lituratus*). These species present the testes located in a subcutaneous pouch at a parapenial position, without forming a true scrotum (Krutzsch, 1975; Krutzsch et al., 1976;

Beguelini et al., 2009; Oliveira et al., 2009). The pouch is non-pigmented and richly vascular at the breeding season (Krutzsch et al., 1976).

Testes located in a true scrotum, without migration to the abdominal cavity, are observed in several species of the families Nycteridae, Megadermidae, Emballonuridae and Vespertilionidae (Matthews, 1941 - *Nycteris luteola, N. hispida, Cardioderma cor*; *Miniopterus minor, M. dasythrix*; Krutzsch and Crichton, 1986 - *Pipistrellus subflavus*; Beguelini et al., 2009 - *Myotis nigricans*). In *Myotis nigricans*, the testis is maintained into the scrotum by a suspensory ligament that attaches the cauda epididymis to the tail skeleton (Beguelini et al., 2009). Pigmentation of the scrotum is frequent in Vespertilionidae bats that store sperm (Krutzsch and Crichton, 1986; Krutzsch and Crichton, 1987).

1.1. Testicular Structure

The structural and functional characteristics of bat testes are similar to other mammals. They are surrounded by a thick albuginea, which emits septa of connective tissue in the mediastinum direction, dividing the parenchyma in lobules (Oliveira et al., 2009). The testicular parenchyma is formed by the seminiferous tubules interspersed by the interstitium, thus characterizing the tubular and intertubular compartments, respectively.

1.1.1. Intertubular Compartment

The intertubular compartment or interstitium is distributed between the seminiferous tubules and contains Leydig cells, fibroblasts, macrophages, mast cells, as well as blood and lymphatic vessels (Gustafson, 1987; Oliveira et al., 2009). The lymphatic vessels present in the interstitium are more prominent during the spermatogenic period (Gustafson, 1979; Gustafson, 1987). In hibernating bats, macrophages are frequently found in the interstitium at hibernation and periarousal periods, usually in close association with Leydig cells. Macrophages appear involved in endocytosis of residues of involuting Leydig cells. As a result of the endocytotic activity, these macrophages accumulate numerous cytoplasmic lipofuscin granules (Gustafson, 1987).

Leydig cells responsible for androgen production are usually organized in small clusters, close to blood vessels or to lymphatic spaces, lined by a thin endothelium, as seen in *Artibeus lituratus* (Oliveira et al., 2009). The Leydig cells are usually rounded or polyhedral (Richardson, 1977). The abundance of Leydig cells in bat testis is variable, as in some species, they are abundant whereas in others, they are scarce (Racey, 1974; Richardson, 1977; Oliveira et al., 2009). The number of cells also varies depending on the reproductive period considered, as Leydig cells appear more abundant in involuted testis, possibly due to the decreasing in seminiferous tubules size, thus providing more interstitial space (Krutzsch, 1975; Krutzsch et al., 1976; Richardson, 1977; Krutzsch and Crichton, 1987; Oliveira, nonpublished data).

The cytoplasm of active Leydig cells is characterized by the presence of numerous lipid droplets and well-developed smooth endoplasmic reticulum which can form concentric layers around the lipid droplets (Gustafson, 1979; Bernard, 1986; Jolly and Blackshaw, 1989; Bernard et al., 1991; Aoki, 1997). Mitochondria are numerous and possess primarily tubular cristae, even though mitochondria with lamellar cristae may be rarely found. Lysosomes, sparse free ribosomes and profiles of rough endoplasmic reticulum and Golgi apparatus are also observed (Gustafson, 1979; Jolly and Blackshaw, 1989; Krutzsch and Crichton, 1990). The nucleus of Leydig cells is usually euchromatic, central, spherical or round with one or two prominent nucleoli (Richardson, 1977; Jolly and Blackshaw, 1989; Krutzsch and Crichton, 1990; Beguelini et al., 2009). The cell surface is characterized by the presence of intricate interdigitations (Jolly and Blackshaw, 1989; Aoki, 1997).

In the seasonal species, during the testicular regression, Leydig cells undergo atrophy characterized by the smaller size, decreased nuclear diameter, heterochromatic nuclei and scarce cytoplasm, as observed in *Pipistrellus hesperus* (Krutzsch, 1975), *Pipistrellus subflavus* (Krutzsch and Crichton, 1986), *Rhinolophus capensis* (Bernard, 1986), *Mormopterus planiceps* (Krutzsch and Crichton, 1987), *Myotis lucifugus* (Gustafson, 1979; Gustafson, 1987), *Miniopterus schreibersii* (Krutzsch and Crichton, 1990; Bernard et al., 1991), *Tadarida brasiliensis* (Aoki, 1997), *Nyctophyllus geoffroyi* (Hosken et al., 1998) and *Artibeus lituratus* (Oliveira et al., 2009). In *Artibeus lituratus*, Leydig cell apoptosis can be observed during testicular regression (Figure 1) (Oliveira et al., 2009). The regressed cells possess few lipid droplets, little or no smooth endoplasmic reticulum, mitochondria with predominantly lamellar cristae, many free rybosomes or polysomes and some lipofuscin granules and dense bodies (probably lysosomes) (Gustafson, 1979; Bernard, 1986; Krutzsch and Crichton, 1987; Krutzsch and Crichton , 1990; Bernard et al., 1991; Aoki, 1997). These cells can undergo seasonal changes in volume according to the period of the reproductive cycle (Gustafson, 1987; Krutzsch and Crichton, 1990; Hosken et al., 1998), even though in some species such as *Nyctalus noctula, Pipistrellus pipistrellus* and *Taphozous georgianus*, no apparent change in the volume of Leydig cells is observed during the seasonal reproductive cycle (Racey, 1974; Racey and Tam, 1974; Jolly and Blackshaw, 1987).

The existence of Leydig cell functional phases in hibernating bat species is still a matter of debate, as ultrastructural features, histochemistry detection of spermatogenic enzymes and endocrine evaluations has been controversial (Gustafson, 1979; Bernard, 1986; Hosken et al., 1998). Nevertheless, in every species studied to date, these cells are active during spermatogenesis, when high levels of androgens are required. During mating and hibernation, the Leydig cells remain active in some species, whereas in others, they can be less active or regressed at all (Gustafson, 1979; Bernard, 1986; Gustafson, 1987; Hosken et al., 1998).

1.1.2. Tubular Compartment

The tubular compartment of the testis is comprised of the seminiferous tubules, formed by the seminiferous epithelium, composed by the Sertoli cells and spermatogenic cells, which are surrounded by a thin tunica propria, containing the peritubular myoid cells.

Sertoli Cells

Sertoli cells are present in the seminiferous epithelium of bats, resting on the basement membrane and extending to the lumen (Beguelini et al., 2009). These cells involve the developing germ cells, being responsible for essential functions such as nutrition and physical support of spermatogenic cells, as well as phagocytosis of residual cytoplasm released during spermiogenesis. As in other mammals, adjacent Sertoli cells of bats are joined by tight junctions and desmosomes, which in conjunction with the tunica propria, form the blood-testis barrier (Bernard and Hodgson, 1989; Crichton et al., 1993; Beguelini et al., 2009). This barrier creates a specialized and protective environment for the spermatogenic cell development. The tight junctions are formed by many parallel or anastomosing strands of ridges and complimentary grooves that cover large areas of the basal Sertoli cell membranes (Crichton et al., 1993). These specialized tight junctions between Sertoli cells demarcate two intraepithelial compartments, denominated basal and adluminal compartments (Bernard and Hodgson, 1989; Russell et al., 1990).

Despite the recognized importance of Sertoli cells, few studies have been found for bat species. The most complete study of Sertoli cell structures were performed in *Rhinolophus capensis* and *Miniopterus schreibersii* (Bernard and Hodgson, 1989), in which species the description below is based. The Sertoli cells are columnar in shape and present an irregular surface, where the spermatogenic cells are located. The nucleus is oval or elongated, highly euchromatic, with well-defined nucleolus (Bernard and Hodgson, 1989; Beguelini et al., 2009). The cytoplasm is rich in mitochondria, however with morphological differences depending on the apicobasal position, as mitochondria are usually round to oval in the basal cytoplasm and elongated or tubular when located at the apical cytoplasm. Morphology of the smooth endoplasmic reticulum is also variable depending on the germ cell to which they are associated. In this sense, the endoplasmic reticulum is sparse and forms platelike cisternae, when associated with spermatogonia, spermatocytes and early spermatids at the basal and middle portion of Sertoli cells, respectively. Conversely, the apical portion of the Sertoli cells, which contacts the late spermatids, presents tubular, smooth endoplasmic reticulum. Besides the nucleus, the basal cytoplasm contains lysosomes, lipofuscin granules, multivesicular bodies, Golgi apparatus, glycogen granules, as well as microtubules. Lipid droplets may be found in the cytoplasm of active Sertoli cells, however they are more conspicuous in the regressed cells.

During testicular regression, the diameter of the inactive seminiferous tubules is considerably reduced (Kurohmaru et al., 2002; Oliveira et al., 2009). The regressive Sertoli cells are characterized by the presence of pyknotic, ovoid to oblong nuclei (Krutzsch et al., 1976), and large lipid droplets distributed throughout the cytoplasm (Bernard and Hodgson, 1989). Curiously, the Sertoli cells cover the spermatogonial cells, thus protecting these testicular cells responsible for tubular re-colonization (Figure 1) (Racey, 1974; Racey and Tam, 1974; Gustafson, 1979; Gustafson, 1987; Krutzsch and Crichton, 1987; Bernard and Hodgson, 1989; Hosken et al., 1998; Kurohmaru et al., 2002; Oliveira et al., 2009).

Spermatogenic Cells

The seminiferous epithelium is composed by Sertoli cells in association with the spermatogenic cells, namely spermatogonia, spermatocytes and spermatids. The spermatogonia are cells restricted to the basal compartment (Bernard and Hodgson, 1989), which undergoes several mitoses and originates a large population of spermatocytes. The spermatocytes undergo meiotic division originating spermatids which differentiate into sperm. The spermatocytes and spermatids are located in the adluminal compartment closest to the lumen, where the sperm are delivered.

Bat species show three classes of spermatogonia, namely A, intermediary and B types, identified based on the shape, size and nuclear morphology (Singwi and Lall, 1983 - *Rhinopoma kinneari*; Saidapur and Patil, 1992 - *Rousettus leschenaultia*; Morigaki et al., 2001 - *Pteropus vampyrus* and *Rhinolophus cornutus*; Oliveira et al., 2009 - *Artibeus lituratus*). The A spermatogonia have large ovoid nuclei containing homogenously distributed chromatin and one or two nucleoli. According to the chromatin condensation, these cells may be subdivided in A1, A2 and A3 (Saidapur and Patil, 1992) or dark A spermatogonia and pale

A spermatogonia (Beguelini et al., 2009 - *Artibeus lituratus, Artibeus planirostris, Carollia perspicillata, Platyrrhinus lineatus* and *Sturnira lilium*). The chromatin is homogenous and diffuses in A1 spermatogonia, but gradually become more compact, being darkest at A3 stage. Intermediate spermatogonia are smaller than A type and present round and darker nuclei with heterochromatin disposed below the nuclear envelope (Saidapur and Patil, 1992; Oliveira et al., 2009). The B spermatogonia are round in shape and present dark nuclei with one or two nucleoli (Saidapur and Patil, 1992; Beguelini et al., 2009; Oliveira et al., 2009).

The primary spermatocytes in leptotene and zygotene may be found in the base of the seminiferous epithelium close to the basement membrane or at the first layer of the adluminal compartment. The pachytene presents a more condensed chromatin and occupies the second and third layers of cells in the seminiferous epithelium (Singwi and Lall, 1983; Beguelini et al., 2009; Oliveira et al., 2009). They represent the largest of all spermatogenic cells (Saidapur and Patil, 1992). The diplotene nuclei have condensed chromatin and are placed in superior layers, where they originate the round spermatids (Bernard and Hodgson, 1989; Beguelini et al., 2009). The primary and secondary spermatocytes are in close association with the Sertoli cells (Bernard and Hodgson, 1989).

Round spermatids are located adjacent to the seminiferous tubules lumen. Their nuclei are round, with evident nucleoli and homogenous chromatin. As the spermatids differentiate, the nuclei become elongated and the chromatin more compact, characterizing the elongated spermatids (Saidapur and Patil, 1992; Beguelini et al., 2009). Developing spermatids are associated with Sertoli cells by junctional complexes restricted to the region in contact with the developing acrosome. At this portion of the Sertoli cell cytoplasm, there are several cisterns of smooth endoplasmic reticulum placed in parallel to the cellular membrane of spermatids. Patches of microfilaments connect the Sertoli cell plasma membrane and the first cisternae of the smooth endoplasmic reticulum (Bernard and Hodgson, 1989). Late spermatids and Sertoli cells present justaposed membranes. At this point, the Sertoli cell cytoplasm presents bundles of microfilaments disposed at right angles to the head of the spermatids as seen in *Rhinolophus capensis* (Bernard and Hodgson, 1989). In other species, such as *Miniopterus schreibersii*, the microfilaments form a layer around the head of late spermatids, instead of perpendicular bundles. Subjacent to the microfilaments, there are one or two cisternae of smooth endoplasmic reticulum, whereas the remaining cytoplasm around the head of spermatid is filled with tubular smooth endoplasmic reticulum (Bernard and Hodgson, 1989).

At spermiation, the association between Sertoli cells and spermatids are changed, as the Sertoli cell separates from the spermatid heads forming a cavity, where the head of spermatids are located. Mature sperm are then released into the lumen of the seminiferous tubules and the residual cytoplasm of spermatids is phagocytosed by Sertoli cells (Bernard and Hodgson, 1989). Sperm are dimorphic with conical or blunt heads in *Rhinopoma kinneari* (Singwi and Lall, 1983), whereas the heads of sperm are scoop-like in the *Pteropus vampyrus* and spatula-like in the *Rhinolophus cornutus* (Morigaki et al., 2001). In *Myotis lucifugus*, the sperm head is oval in the frontal section and triangular in the sagittal section. The acrosome covers the anterior two-thirds of the nucleus. A pronounced constriction is located at the connecting piece segment. Posterior to this constriction, the tail thickens abruptly with the appearance of the mitochondrial sheath of the midpiece region. At the junction between the midpiece and principal piece, the annulus appears as a triangular wedge of electrondense material (Hoffman et al., 1987).

The arrangement of spermatogenic cells and Sertoli cells within the tubule vary depending on the spermatogenic activity (Bernard and Hodgson, 1989). The seminiferous tubules of seasonal bats have two distinct phases characterizing the testicular activity and testicular regression. During the testicular activity, the tubular pattern follows that of other mammals, with more immature spermatogonial cells positioned in the basal epithelium. The spermatocyte cells are found in the middle position and the more mature spermatids are located near the lumen. During the regression period, the seminiferous epithelium contains only inactive Sertoli cells and basally-located spermatogonia (Racey, 1974; Racey and Tam, 1974; Gustafson, 1979; Gustafson, 1987; Krutzsch and Crichton, 1987; Bernard and Hodgson, 1989; Hosken et al., 1998; Kurohmaru et al., 2002; Oliveira et al., 2009). The spermatogonia are characterized by voluminous nuclei and prominent nucleolus found in close contact to the basement membrane (Racey, 1974; Oliveira et al., 2009). Degenerating spermatocytes and spermatids are rarely observed in the regressive epithelium. There are no sperm in the regressed seminiferous tubules.

1.2. Spermatogenesis

Spermatogenesis is a cyclic process that occurs in seminiferous tubules, in which spermatogonia undergo sequential transformations culminating in the formation of sperm (Russell et al., 1990; França and Russell, 1998; Hess and Franca, 2007). This phenomenon is species-specific, as each species presents distinct durations of spermatogenesis, type, size, shape and population of germ cells, number of cells and associated mechanism for renewal of stem cells. Within the seminiferous epithelium, the differentiating cells do not have random distribution, but specific cell types are associated, forming the stages. The sequences of events that occur in a particular area of the seminiferous epithelium between two successive appearances of the same pattern of cell association or stage define the cycle of the seminiferous epithelium (Leblond and Clermont, 1952). The different stages of the seminiferous tubules have been determined by two methodologies: the acrosomic system and tubular morphology method. The method of the acrosomic system is based in the identification of the stages according to the acrosomic development. This method determines a variable number of stages for each species. The methodology of tubular morphology utilizes the meiosis occurrence, alterations in spermatid nuclei and the position of spermatogenic cells in the seminiferous epithelium. This method determines eight stages for all mammalian species (Berndtson, 1977).

According to the acrossomic system method, ten to eleven stages of the cycle of seminiferous epithelium are recognized in bat seminiferous tubules. Eleven stages of the seminiferous epithelial cycle are identified in *Rhinopoma kinneari* (Singwi and Lall, 1983) and *Pteropus vampyrus* (Morigaki et al., 2001), whereas *Rhinolophus cornutus* (Morigaki et al., 2001) and *Rousettus leschenaultia* (Saidapur and Patil, 1992) have a cycle of ten stages. Concerning the spermiogenic process, the developing spermatids present 13 (*Pteropus vampyrus* and *Rhinolophus cornutus*), 14 (*Rousettus leschenaulti*) or 16 (*Rhinopoma kinneari*) distinct steps of acrosomal system formation, including Golgi, cap, acrosome and maturation phases, culminating in the formation of sperm (Singwi and Lall, 1983; Saidapur and Patil, 1992; Morigaki et al., 2001).

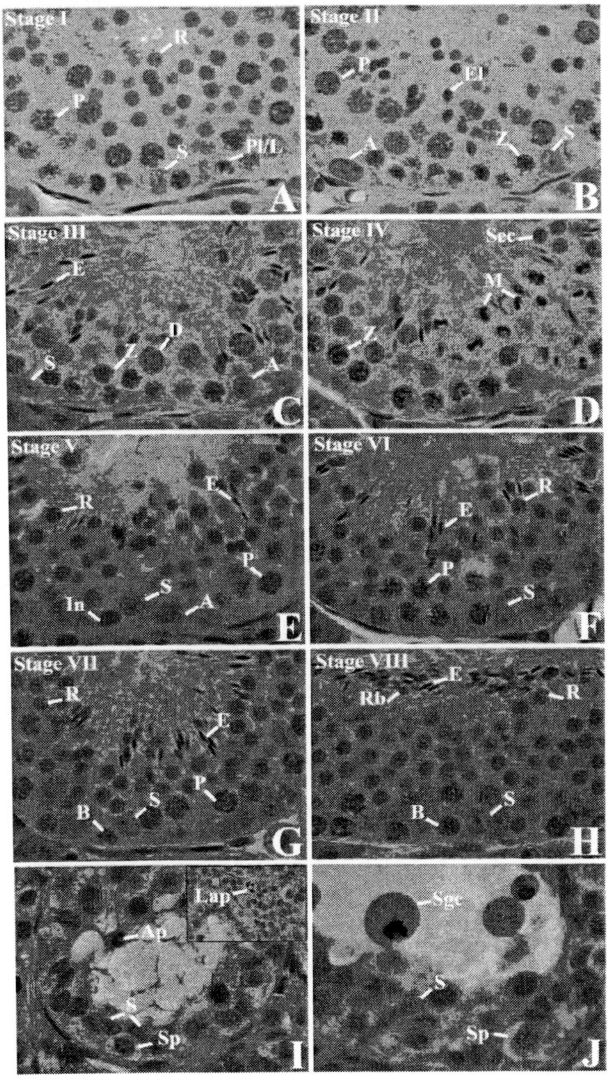

Figure 1. Morphology of the testis of *Artibeus lituratus* during reproductive and regressive periods. (A-H) Seminiferous epithelium at different stages of the spermatogenic cycle classified according to the tubular morphology method. Stage I - characterized by the presence of several layers of round spermatids close to the lume. Stage II - The spermatid nuclei begin to elongate. Stage III - The elongated spermatids are arranged in bundles deeply embedded in the seminiferous epithelium. Stage IV - The occurrence of metaphase and anaphase of meiosis in primary and secondary spermatocytes can be visualized in this stage. Stage V - At this stage, there are both round and elongated spermatids. Stage VI - bundles of elongated spermatids are found in the middle of the epithelium. Stage VII - The elongated spermatids are closer to the luminal surface. The B spermagonia are present in the base of the epithelium. Stage VIII - The elongated spermatids are located at the luminal surface ready for spermiation. The residual bodies are large and numerous. (I-J) During the regressive period, the seminiferous tubules are atrophic, formed mostly by Sertoli cells whose nuclei are located above the nuclei of spermatogonia cells. Insert in I = Leydig cells in apoptosis. Sloughed germ cells are common into the lumen (Sgc) (J). A = spermatogonia A; Ap = apoptotic cells; B = spermatogonia B; D = diplotene; E = elongated spermatid; El = elongating spermatid; In = intermediate spermatogonia; L = leptotene; Lap = Leydig cells in apoptosis; M = metaphase; P = paquitene; Pl = prelelptotene; R = round spermatid; Rb = residual bodies; S = Sertoli cells; Sec = secondary spermatocyte; Sp = spermatogonia; Z = zigotene. Bar in (A) = 30 μm.

Based on the tubular morphology method, eight stages of the seminiferous tubule cycle were identified in *Artibeus lituratus* (Figure 1) (Oliveira et al., 2009) and several other species of Neotropical Phyllostomidae bats (Beguelini et al., 2009). As seen in most mammals, each seminiferous tubule section of the Phyllostomidae species analyzed presents a single stage of the cycle (Beguelini et al., 2009; Oliveira et al., 2009). In contrast, there is overlapping of stages of seminiferous tubule of the Tropical Vespertilionid *Myotis nigricans* (Beguelini et al., 2009).

During the annual reproductive cycle, the testis undergoes remarkable changes in the equilibrium between cell proliferation and cell death or apoptosis, at the recrudescence and involution phases, respectively. Cell proliferation determines testicular growth, whereas apoptosis is an important event leading to testicular regression. It is known that oxidative stress is a key factor stimulating apoptosis in the testis and the occurrence of antioxidant scavenging systems is crucial for preventing apoptosis. Therefore, the balance between reactive oxygen species (ROS) generation and ROS scavenging is essential to determine the rate of cell death and growth. The testicular mechanism of antioxidant defense includes the activity of several enzymes, such as superoxide dismutase (SOD), catalase and glutathione peroxidase (GPX). There is evidence that in the Vespertilionidae *Corynorhinus mexicanus*, specific activity of these scavenger enzymes changes during the annual reproductive cycle in a manner closely related to the testis activity, as they peak during or slightly before the regressive period and decrease at the period of testis growth (Arenas-Rios et al., 2007).

2. Excurrent Ducts

The bat excurrent ducts is comprised of the *rete testis*, efferent ductules, epididymal ducts, vas deferens and urethra.

2.1. Rete Testis

The seminiferous tubules converge to the mediastinum of the testis, where they gradually lose the germ cells, persisting only Sertoli cells, which modifies in a simple cubic epithelium lining the straight tubules. In the transition zone from seminiferous tubules to straight tubules, myoid cells forms one or two continuous layers below the epithelium (Oliveira et al., 2009). Following the straight tubules, the intratesticular *rete testis* is lined by a simple cubic epithelium, sustained by the connective tissue of the mediastinum.

2.2. Efferent Ductules

The efferent ductules consist of several tiny tubules that connect the *rete testis* to the epididymal duct (Ilio and Hess, 1994; Hess, 2002). Studies in several mammal species indicate that the efferent ductules are responsible for the reabsorption of more than 90% of the testicular fluid released from the testes, thus increasing the concentration of sperm in the epididymis and facilitating their maturation and storage (Clulow et al., 1994; Ilio and Hess,

1994). Due to the small size and difficulty in distinguishing this segment, the efferent ductules have been largely neglected in most studies concerning the male genital tract (Ilio and Hess, 1994). Therefore, information about efferent ductules was found only for the Italian species Vesperugo (=Pipistrellus) savii and *Vesperugo piccolo* (Azzali, 1983) and the Brazilian *Artibeus lituratus* (Oliveira, non-published data).

In *Artibeus lituratus,* the efferent ductules are composed by 12 to 15 flexuous ductules, which arise separately from the *rete testis* and run parallel towards the epididymis, then anastomosing and forming about six terminal ductules that enter the epididymal duct separately (Figure 2). This pattern of disposition is similar to that found for man and large eutherian mammals (Ilio and Hess, 1994). Also similar to humans (Yeung et al., 1991), the efferent ductules of *Artibeus lituratus* are composed of most of the head of the epididymis.

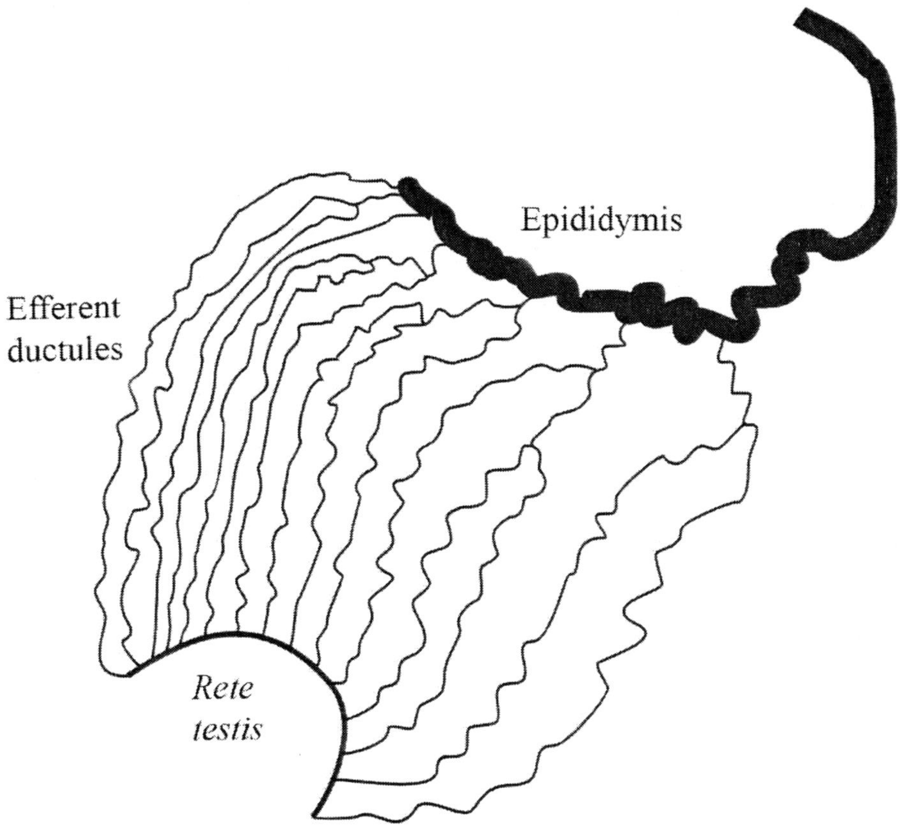

Figure 2. Squematic drawing showing the anatomical disposition of the efferent ductules of *Artibeus lituratus*.

The efferent ductules are lined by a columnar epithelium, composed of ciliated and non-ciliated cells randomly interspersed (Figure 3) (Azzali, 1983); Oliveira, non-published data). The non-ciliated cells are characterized by the presence of abundant microvillus in the apical surface, an oval nucleus placed in the basal cytoplasm and numerous PAS-positive granules. The PAS-positive granules are primarily located in the apical cytoplasm, even though they may also be found in the perinuclear region. The ciliated cells are apparently less numerous than non-ciliated cells. They are goblet-like and present numerous and long cilia. The nucleus of the ciliated cells is oval, but unlike the non-ciliated cells, it is more apical in location. Eventually, the cytoplasm of these cells presents few PAS-positive granules.

Externally, the efferent ductules are surrounded by one or two layers of smooth muscle cells. The tubular lumen is filled with fluid, but sperm are rarely seen, except in the terminal ductules. The intertubular connective tissue contains blood and lymphatic vessels as well as numerous mast cells.

During testicular regression, the efferent ductules present several morphological changes, including significant decrease in epithelial height and tubular diameter (Figure 3). The PAS-positive granules are decreased in non-ciliated cells and barely detectable in the ciliated cells.

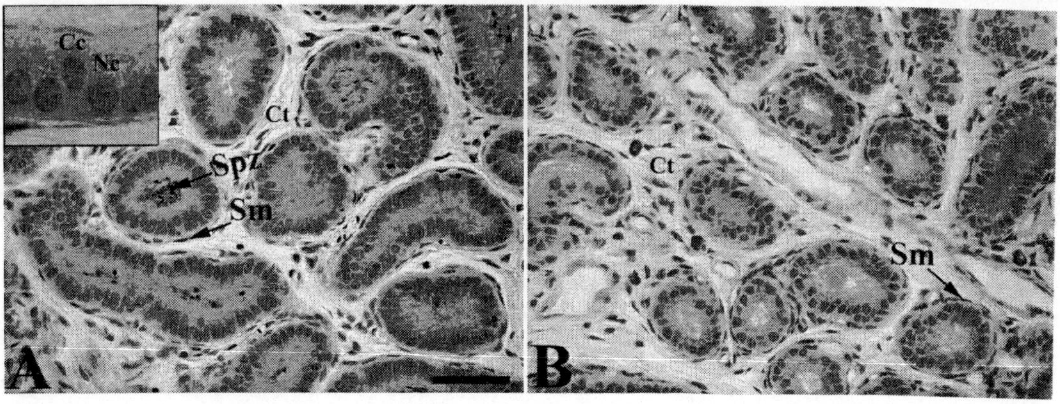

Figure 3. Efferent ductules of *Artibeus lituratus* during the reproductive (A) and regressive (B) periods. The epithelium of efferent ductules is composed of ciliated (Cc) and non-ciliated cells (Nc) (insert). Externally, the efferent ductules are surrounded by two or three layers of smooth muscle cells (Sm). Sperm (Spz) may be seen in the lumen of terminal ductules. During the regressive period, there is an expressive reduction in the epithelial height and tubular diameter. The intertubular connective tissue (Ct) became more conspicuous. Bar in A = B = 50μm.

2.3. Epididymis

The epididymis of bats consists of a single flexuous duct positioned laterally to the testis, continuing as the vas deferens on each side (Vamburkar, 1958). Anatomically the epididymis may be divided into caput, corpus and cauda (Matthews, 1941; Vamburkar, 1958; Azzali, 1983), even though an initial segment has also been easily distinguished in the epididymis of *Artibeus lituratus* (Oliveira, non-published data). In some species, the head and body of the epididymis are closely applied to the testis but the cauda lies caudal to the testis and is not closely applied to it (Matthews, 1941 - *Cardioderma cor*). Variation in the proportion of each epididymal region has been described for several bat species. In *Taphozous sp, Miniopterus minor* and *M. dasythrix,* the caput of epididymis is small and the cauda is very long, which usually projects caudally, beyond the posterior border of the testis (Matthews, 1941). The caput and cauda epididymis have the same size and the convolution of the duct is not well-defined in *Cynopterus sphinx gangeticus* (Vamburkar, 1958). In the *Nycteris luteola* and *N. hispida,* the caput, corpus and cauda are all well-developed, whereas *Triaenops afer* has a slender body connecting the moderately expanded head to a very large tail (Matthews, 1941). In *Hiposideros caffer,* the epididymis is small when compared with many other bat species and the cauda is not very extended caudally, suggesting that this segment is not involved in the reservoir of sperm (Matthews, 1941).

The epididymal duct is lined by a pseudostratified columnar epithelium (Krutzsch et al., 1976), which in some species, changes from columnar to cuboidal from the caput to the cauda (Richardson, 1977). As shown in *Artibeus lituratus*, the epididymal epithelium is composed by principal, basal, apical, narrow, clear and halo cells, which presents different distribution along the duct (Figure 4) (Oliveira, non-published data).

The principal cells are present in all segments of the epididymal duct. These cells are usually very tall columnar in the initial segment and become low columnar or cubic in the cauda region. The principal cell nucleus is round or oval, located in the basal cytoplasm and presenting one to three evident nucleoli (Oliveira, nonpublished data). These cells are characterized by the presence of many long microvilli (stereocilia) in the luminal surface (Krutzsch, 1975; Crichton et al., 1993). The cytoplasm of principal cells shows numerous PAS-positive granules whose distribution, amount and size vary along the epididymis. Usually, fewer clear PAS-positive granules are scattered in the supranuclear cytoplasm of the principal cells at the initial segment. The number and staining intensity of the granules increases in the caput region. In the corpus, the number of PAS-positive granules is even higher and they occupy both supranuclear and subnuclear regions (Oliveira, nonpublished data). In the cauda epididymis, the epithelium is lower and the microvillus are not as evident as in the caput (Krutzsch, 1975); Oliveira, nonpublished data}. Scarce PAS-positive granules are detected in the principal cells (Oliveira, nonpublished data).

Ultrastructural data concerning bat epididymis is scarce. One study in *Antrozous pallidus* revealed that the lateral plasma membranes of adjoining principal cells present junctional complexes at the cell apices, whereas basally, the lateral plasma membranes are characterized by the presence of many folds and interdigitation (Crichton et al., 1993). Tight junctions at caput and corpus epididymis appear as a series of branching interlacing strands, approximately 12 in number, whose horizontal meshwork run parallel to the lumen. These tight junctions form the blood-epididymis barrier to maintain a specialized luminal microenvironment for the maturation and maintenance of the sperm in transit. In the cauda epididymis, tight junction architecture is essentially similar to that of caput and corpus epididymis, during early sperm storage prior to winter torpor (named "ordinary" tight junctions). As the sperm storage season progresses, an additional tight junction type appears in the cauda. These tight junctions are composed of multiple, long parallel rows of strands, resembling those of the testis Sertoli cells. The strands often exceed 70 in number and mostly run tangential to the long axis of the cell. The presence of this highly specialized "Sertoli cell-like" tight junction may explain the ability of the cauda epididymis to withstand the excessively hyperosmolar epididymal luminal environment that is essential for long-term sperm storage in the hibernating bats (Crichton et al., 1993).

The basal cells are also present in all segments of the epididymis. These cells are small but numerous, and located at the base of the epithelium, apparently, without communication with the tubular lumen (Oliveira, nonpublished data). They present reduced cytoplasm without granules or vacuoles. The nucleus is small, round, ovoid or elongated, with one or two evident nucleoli.

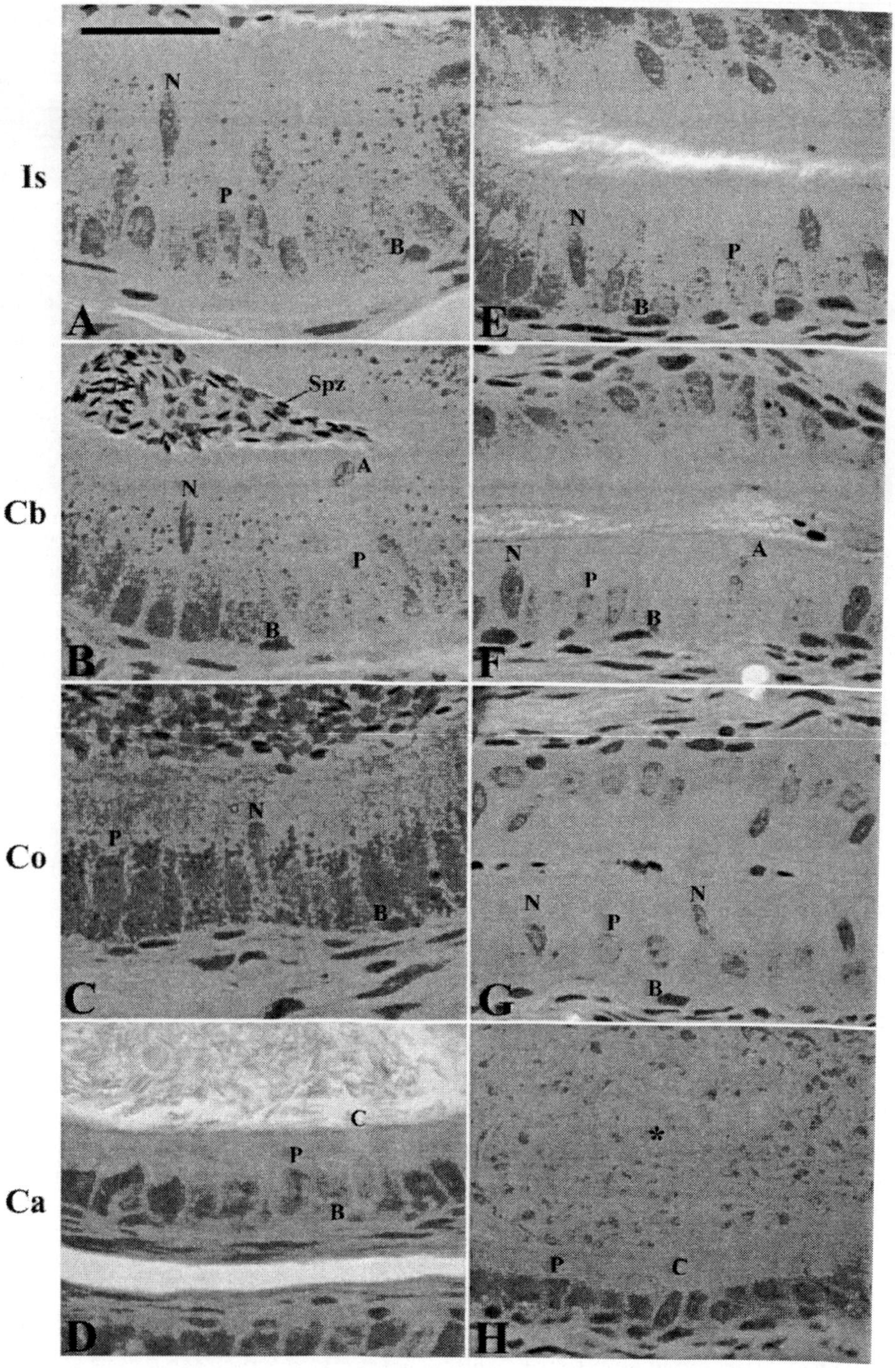

Figure 4. The epididymis epithelium of *Artibeus lituratus* in the reproductive (A-D) and regressive (E-H) periods. The epithelium presents principal (P), basal (B), narrow (N), apical (A) and clear (C) cells, evident in both reproductive and regressive phases. During the reproductive period, the epithelium presents PAS-positive granules that increase gradually from initial segment (IS) to caput (Ca) and corpus (Co). In cauda (Ca), the PAS-positive granules are nearly absent, as well as in all segments during the regressive period. During the regressive period, the epithelial height is greately reduced compared with active period. Spz = luminal sperm; * = luminal degenerating cells; Bar in (A) = 30μm.

The apical and narrow cells are present in the initial segment, caput and corpus regions. The apical cells are located between the principal cells and present a round or ovoid nucleus at the apical cytoplasm. Narrow cells are intercalated with the principal and/or apical cells. The nucleus of the narrow cells is vertically elongated and positioned in the medial to apical cytoplasm (Oliveira, non-published data). Scarce clear cells are found restricted to the cauda region. The cytoplasm has no evident vesicles or granules (Oliveira, non-published data). The halo cells are present in all regions of the epididymis. Located at the base of the epithelium near the basal cells, the halo cells are hardly visualized. Halo cells are characterized by the small spherical and intensely basophilic nucleus. The cytoplasm does not stain with routinely used stains, thus forming a lighter halo around the nucleus (Oliveira, non-published data).

Externally, the epididymal duct presents three to four layers of peritubular smooth muscle cells in initial segment, caput and corpus regions. In the caudal region, the number of layers of smooth muscle cells is increased. The intertubular connective tissue is scant, containing evident fibroblasts, mast cells, blood and lymphatic vessels.

2.3.1. Seazonal Variation

The annual reproductive cycle of bats is marked by relevant morphological and physiological changes in the epididymis of both hibernating and non-hibernating seasonal species. During the regression period of *Artibeus lituratus*, a tropical, non-hibernating species, all segments of the epididymis show a significant reduction in epithelial height (Figure 4) and tubular diameter (Oliveira, non-published data). The tubular diameter is reduced by approximately 50% and become free of sperm, thus differing from most hibernating species. In contrast, numerous sloughed cells are found in the epididymal lumen (Figure 4). Even with considerable atrophy of the epididymal duct, it is possible to distinguish all of the cell types found in the epithelium during reproductive activity. Regarding principal cells, the amount and size of PAS-positive granules decrease in the regressive epididymis. Externally, there is an evident increase in the thickness of muscle layer, especially in the cauda epididymis.

Reduction in tubular diameter and absence of luminal sperm have also been described for other species (Krutzsch, 1975 - *Pipistrellus hesperus*; Krutzsch et al., 1976 - *Macrotus waterhousii*; Krutzsch and Crichton, 1987 - *Mormopterus planiceps*; Crichton et al., 1993 - *Antrozous pallidus*). During the reproductive inactivity, the epithelium becomes cubic and microvilli are more evident with the gradual reduction in the amount of luminal sperm. When a new spermatogenic cycle starts, this situation is reversed (Krutzsch, 1975). Upon involution, the intertubular connective tissue is more conspicuous (Krutzsch et al., 1976; Oliveira, non-published data). In addition, a large amount of adipose tissue lay down around the testis and epididymis during the winter, as seen in *Miniopterus australis* and *Miniopterus schreibersii* (Richardson, 1977).

In some hibernating species, the head and cauda of epididymis reach a maximum weight and epithelial height in the course of the summer associated with theengorgement with sperm. Thereafter, involution takes place and the tubules regress in the head epididymis, whereas the cauda remains distended and filled with sperm throughout the winter until middle spring (Racey and Tam, 1974 - *Pipistrellus pipistrellus*; Krutzsch and Crichton, 1986 - *Pipistrellus subflavus*). At the end of the sperm storage period, the cauda epididymis presents principal cells with irregular-shaped basal nucleus and prominent supra-nuclear cytoplasm. Conspicuous Golgi apparatus, mitochondria, rough endoplasmic reticulum, free ribosomes, and vacuoles of varying sizes, including apical coated pits, as well as lysosomes are observed. The cell surface facing the lumen presents many microvilli (Crichton et al., 1993).

2.3.2. Physiology

In several hibernating and some non-hibernating bats, the cauda epididymis is responsible for storing viable sperm during long periods. Therefore, studies on epididymis histology and physiology have been mostly focused on the cauda region.

Stored sperm form an extensive, densely packed mass within the caudal lumen, showing no specific orientation with respect to the epithelial lining (Hosken et al., 1998). The storage of sperm in the cauda epididymis usually occurs after ceasing spermatogenesis and may occur for a short period for species that mate before and during hibernation, or can extend over several months as observed in species that mate after hibernation (Anand Kumar, 1965; Racey, 1974; Bernard, 1984; Krutzsch and Crichton, 1986; Gustafson, 1987; Jolly and Blackshaw, 1987; Krutzsch and Crichton, 1987; Jolly and Blackshaw, 1988; Encarnação et al., 2004). In the last case, the time to store viable sperm in the epididymis exceeding the period observed in other mammals, epididymal storage may last about six months (Van der Merwe and Rautenbach, 1990 - South African *Pipistrellus rusticus*), seven months (Racey, 1974 - *Nyctalus Noctula*), eight months (Hosken et al., 1998 - Australian *Nyctophylus geoffroyi, N. major* and *Vespadelus regulus*) or ten months (Bernard, 1984 - African *Rhinolophus capensis*). In African and Australian Vespertilionidae of the genus *Miniopterus*, epididymal storage of sperm is retained for several months after the end of spermatogenesis, and the tip of the cauda filled with sperm is macroscopically enlarged projecting caudally from the testis limit (Matthews, 1941 - *Miniopterus minor* and *Miniopterus schreibersii*; (Richardson, 1977 - *Miniopterus australis* and *Miniopterus schreibersii*). It is not uncommon that aggregated sperm are recognized in the cauda epididymis throughout the year, even though the spermatogenesis is restricted to a short period of time (usually in the summer and early fall) (Anand Kumar, 1965).

The storage of sperm for long periods as observed in bats can be a natural consequence of hypothermia produced by hibernation. This ability is an important adaptation that allows the asynchronous reproductive cycle of males and females, important to reconcile the reproductive period with hibernation (Jolly and Blackshaw, 1987). The storage of sperm in the epididymis appears restricted to species of temperate zone belonging to three families of bats: *Emballonuridae,* Vespertilionidae and Rhinopomatidae (Gustafson, 1987; Jolly and Blackshaw, 1988; Van der Merwe and Rautenbach, 1990; Setchell et al., 1993; Hosken et al., 1998). However, it is now known that some species of non-hibernating tropical bats also store sperm in the epididymis (Cervantes et al, 2008). Prolonged sperm storage throughout the year is described for the Tropical Australian Emballonuridae *Taphozous georgianus* (Jolly and Blackshaw, 1987), the Tropical Indian Hipposideridae *Hipposideros speori* (Gopalakrishna and Bhatia, 1980) and Rhinopomatidae *Rhinopoma hardwickei* (Karim and Banerjee, 1985), as well as the Japanese *Rhinolophus cornutus*, which belongs to the Rhinolophidae family (Kurohmaru et al., 2002).

Nevertheless, it is unknown whether the prolonged survival of sperm in the epididymal cauda is possible due to characteristics of the luminal fluid, intrinsic factors of the sperm or both (Setchell et al., 1993). It is interesting that bat sperm do not differ from other mammals in terms of plasma membrane stability that could justify their resistance to long-term storage

(Crichton et al., 1993). Long-term epididymal sperm preservation appears to be controlled by change in osmolality of the cauda milieu, as seen in male hibernating vespertilionid bats - *Eptesicus fuscus*, *Myotis lucifugus* and *M. sptentrionalis* (Crichton et al., 1994). Osmolalities of epididymal fluid in hibernating species of bats rise during sperm storage periods to values as high as 1,523 mmol/kg H_2O (5 times that of plasma). Seasonal establishment of hyperosmotic conditions driven by the epithelial cells dehydrate the sperm and thereby minimize their metabolic needs, which conserves energy by the imposition of quiescence. Protection from cold-induced membrane damage is afforded by the reduced activation energy of water at elevated osmolality. Reduction of osmolality (to 500-600 mmol/kg H_2O) induces swelling of sperm and allows the re-initiation of motility and increased metabolic rate (Crichton et al., 1994).

The fate of excess sperm stored in the bat cauda epididymis is still a matter of debate. In African *Rhinolophus capensis*, the sperm cells that remain in the epididymis after copulation are phagocytosed by luminal macrophages (Bernard, 1984). Similar occurrence is described for *Myotis lucifugus* and *M. velifer* (Krutzsch et al., 1982). Little or no evidence of spermiophagy by the epididymal epithelial cells is observed (Krutzsch et al., 1982; Bernard, 1984). Destruction of luminal sperm by macrophages is not a common finding in mammals at natural conditions, thus revealing another particularity of bat reproductive physiology.

Frequently, cytoplasmic droplets are seen on sperm at the cauda epididymis; thus indicating that, besides storage and spermiophagy, the cauda may also be involved in the final maturation of the sperm, during the storage period (Hoffman et al., 1987; Cervantes et al., 2008). Sequential induction of capacitation and acrosome reaction in sperm obtained from different epididymal regions (caput, corpus, cauda) throughout the annual reproductive cycle of *Corynorhinus mexicanus*, also reinforce the involvement of the cauda epididymis in sperm maturation during the storage period (Cervantes et al., 2008).

2.4. Ductus Deferens

The cauda epididymis is continuous with the ductus or vas deferens, which enters the abdominal cavity through the vaginal canal as part of the spermatic cord. Within the abdominal cavity, the vas deferens leaves the spermatic cord, passes medially and loops dorsally to the ureter. In most bat species, the vas deferens dilates terminally forming the ampullary gland (ampulla of Henle) (Matthews, 1941; Krutzsch and Crichton, 1986; Oliveira, non-published data). In some species, the vas deferens does not open directly into the urethra; they enter the seminal vesicle and opens into the vesicle collecting duct, which opens into the urethra. In other species, the vas deferens and the seminal vesicle duct join to each other forming a common ejaculatory duct that opens into the urethra (Matthews, 1941; Vamburkar, 1958). In species without a seminal vesicle, the distal part of the ampulla becomes narrower close to the urethra, forming the ejaculatory duct (Matthews, 1941).

In some hibernating species, the storage of sperm is not restricted to the cauda epididymis, but may also occur in the ductus deferens, primarily at its distal dilation, the ampullary gland (Anand Kumar, 1965 - *Rhinopoma kinneari*; Krutzsch and Crichton, 1986 - *Pipistrellus subflavus*).

2.5. Urethra

The urethra emerges from the neck of the urinary bladder as a uniform tube, cylindrical in shape, which is surrounded, partially or completely, by the prostate, and in some species, also by the seminal vesicle (Vamburkar, 1958; Krutzsch et al., 1976). The urethra extends from the neck of the bladder to the tip of the penis. On this course, the urethra is called a prostatic urethra while passing through the prostate, penile urethra while passing through the penis or membranous urethra in the portion located between the prostatic and penile urethra. The length of all urethral segments is variable. Accordingly, the prostatic plus membranous urethra are about 5.0 to 6.0mm in *Thaphozous sp*, 7.0 mm in *Cardioderma cor*, 8.0 to 9.0mm in *Triaenops afer* (Matthews, 1941) and 10mm long during the breeding season in *Cynopterus sphinx gangeticus* (Vamburkar, 1958). The prostatic and membranous urethra are lined with a transitional epithelium, thinner than in the penile urethra (Krutzsch et al., 1976). In some species, the lamina propria of the membranous urethra encloses numerous small simple alveolar mucous glands (Matthews, 1941; Krutzsch, 1975; Krutzsch et al., 1976). Externally, the prostatic urethra is surrounded by a thick muscular sheath (*musculi compressor urethrae*) (Krutzsch et al., 1976).

The prostatic urethra presents a well-defined median crista urethralis, in which numerous excretory ducts of the prostate glands open (Matthews, 1941; Vamburkar, 1958; Krutzsch, 1975). The ejaculatory ducts, when present, also enter the urethra at the colliculus seminalis in the apice of the crista urethralis (Krutzsch, 1975; Krutzsch and Crichton, 1987). A conspicuous utriculus prostaticus has been described in several species of bats (*Cardioderma cor; Hipposideros caffer, Nycteris luteola* and *Taphozous sp* - Matthews, 1941), but it is not present in others (*Cynopterus sphinx, Miniopterus minor, Miniopterus schreibersii, Nycteris hispida* and *Triaenops afer,* - Matthews, 1941; *gangeticus* - Vamburkar, 1958; *Mormopterus planiceps* - Krutzsch and Crichton, 1987). When present, the utriculus prostaticus opens on the crista urethralis distal to the ejaculatory ducts. It consists of a tubular evagination with a very small lumen, lined by a stratified cuboid epithelium (Matthews, 1941; Krutzsch and Crichton, 1987).

The caudal end of the membranous urethra widens to form the bulb on the base of the penis, where it is enclosed in paired, thick bulbocavernosus muscles (Krutzsch, 1975). The bulbous urethra receives the ducts of the Cowper's glands (Krutzsch, 1975; Krutzsch et al., 1976). It continues in the penile urethra, which extends from the base of the penis to the urethral orifice (Vamburkar, 1958). The penile urethra dilates distally into a *sinus navicularis*, before the opening to the exterior through the external urethral meatus (Krutzsch et al., 1976). The external urethral meatus is subterminal and may open on the ventral *(Cynopterus sphinx gangeticus, Hipposideros caffer, Macrotus waterhousii, Miniopterus minor, Nycteris hispida, Taphozous sp* and several species of Molossidae) or dorsal (*Cardioderma cor; Miniopterus schreibersii*) surface of the glans, whereas in some species, it is terminal (*Nycteris luteola; Triaenops afer*) (Matthews, 1941; Vamburkar, 1958; Krutzsch et al., 1976; Ryan, 1991a).

The penile urethra is lined by a transitional (Krutzsch et al., 1976; Ryan, 1991b) or stratified columnar epithelium two to six- layers thick (Krutzsch, 1975). Distally, the urethral epithelium gradually changes to stratified squamous epithelium which lines the sinus and external meatus (Krutzsch, 1975; Krutzsch et al., 1976). This epithelium is continuous with the external epithelium, covering the glans and the internal surface of the prepuce (Krutzsch, 1975). In some species, urethral glands lie in the connective tissue of the mucosa, whereas in others, the penile urethra is devoid of intramural glands (Krutzsch, 1975; Krutzsch et al., 1976).

3. Penis

The remarkable interspecific variation in the morphology of the mammalian penis has usually served as an important phylogenetic tool in several orders, such as marsupials, rodents and primates. Lesser is known about interspecific diversity of the bat penis, therefore it is still premature to use phallic morphological variation as a systematic character for Chiroptera. The bat penis is prominent, highly variable in size, structure and position.

3.1. Anatomy

For descriptive purposes, the bat penis may be divided into three regions: the root or crura, the body or shaft, and the glans. The chiropterans penis is similar to the cavernosum penis of most mammals. It is formed by two symmetrical corpora cavernosa and the corpus spongiosum. The basis of these erectile structures in conjunction with tissues that surround them form the root of the penis. At the root, the membranous urethra expands to form the erectile bulb and the corpora cavernosa form the crura, which are attached to the isquiopubic bones. The bulb is surrounded by thick bulbospongiosum muscle. The corpora cavernosa form the bulk of the penis shaft. The glans of bats penile is defined as the part of the penis distal to the origin of the prepuce (Ryan, 1991a). The glans is well-developed in most bats and, in some species, a baculum (*os penis*) is found extending throughout the glans. The form of the glans and the baculum differs considerably among bats.

The bat penis may be pendulous (*Miniopterus minor* - Matthews, 1941; *Cynopterus sphinx gangeticus* - Vamburkar, 1958; *Mormopterus planiceps* - Krutzsch and Crichton, 1987), semipendulous (*Pipistrellus hesperus* - Krutzsch, 1975) or not pendulous (*Macrotus waterhousii* - Krutzsch et al., 1976; *Artibeus lituratus* - Oliveira, non-published data). The pendulous penis is directed caudally, whereas the non-pendulous penis is rostrally directed (Krutzsch et al, 1976). The organ is usually cylindrical and elongated (Matthews, 1941; Vamburkar, 1958), but varies in diameter (Matthews, 1941). The length of the penis is variable among species (4.5 mm in *Taphozous sp*; 5.0 to 7.0 mm in *Triaenops afer*; 5.5 to 7.5 mm in *Pipistrellus hesperus;* 6.0 mm in *Hipposideros caffer*; *Cardioderma cor;* 6.5 to 7.5 mm in *Macrotus waterhousii*; 7.0 mm in *Miniopterus minor;* 9.4 mm in *Mormopterus planiceps;* 14 mm in *Cynopterus sphinx gangeticus* - and 18 mm in *Artibeus lituratus*) (Matthews, 1941; Vamburkar, 1958; Krutzsch, 1975; Krutzsch et al., 1976; Krutzsch and Crichton, 1987); Oliveira, nonpublished data). Proportionally, the bat penis correspond to 7-10% of the body length in most species, except *Artibeus lituratus* in which the penis reaches up to 16% of the body length.

3.2. Shaft Structure

Similar to other mammals, the bat penis possesses three cylindrical columns of erectile tissue: two corpora cavernosa and the corpus spongiosum. The bulk of the penis shaft is composed by paired erectile, bilaterally symmetrical, corpora cavernosa, which lie above the urethra and are enclosed by a very thick tunica albuginea (Matthews, 1941; Vamburkar, 1958; Krutzsch, 1975; Krutzsch and Crichton, 1987; Ryan, 1991b). The corpora cavernosa are constituted by trabeculae of fibroeslastic and muscular tissues, which traverse the interior of the corpora, separating numerous vascular sinuses or lacunae that fill with blood during erection (Ryan, 1991b).

The corpora cavernosa originate in the form of two lateral crura, which are connected to the rami of the pubic arc (Vamburkar, 1958). The two corpora cavernosa are apparently fused in some species but remain internally separated by a median fibrous septum of tunica albuginea (Vamburkar, 1958; Krutzsch, 1975; Ryan, 1991a; Ryan, 1991b), which may be complete or incomplete in some species (Vamburkar, 1958; Ryan, 1991b). In other species, the corpora cavernosa is completely fused throughout the length of the penis to form a single erectile body (Matthews, 1941).

Hipposideros caffer, *Miniopterus minor*, *Miniopterus schreibersii*, *Nycteris hispida*, *Nycteris luteola* and *Taphozous sp*). In *Triaenops afer,* the corpora cavernosa are fused in the shaft of the penis but become separated in the base of the glans, where the tunica albuginea remain joined (Matthews, 1941). In most bats, the bat corpora cavernosa is not restricted to the shaft of the penis but extends well into the glans (Ryan, 1991a).

The corpus spongiosum of bat penis is usually poorly developed. It lies ventral to the corpora cavernosa and surrounds the urethra from the bulb to the base of the glans, as a very thin cylinder of erectile tissue consisting of a small venous plexus (Matthews, 1941 - *Cardioderma cor, Miniopterus minor, Miniopterus schreibersii, Nycteris hispida, Taphozous sp* and *Triaenops afer*). The tunica albuginea is thinner and more elastic than that of the corpora cavernosum (Ryan, 1991a; Ryan, 1991b). Usually, the corpus spongiosum does not extend into the glans (Matthews, 1941; Ryan, 1991a; Ryan, 1991b). In *Cynopterus sphinx* and *Nycteris luteola,* the corpus spongiosum is considerably more developed than in other species of bats, and extends at a short distance into the glans (Matthews, 1941; Vamburkar, 1958). Corpus spongiosum is also present in the proximal glans of some Molossidae species (Ryan, 1991a). Conversely, in the Hipposideridae *Hipposideros caffer,* a minute corpus spongiosum surrounds the urethra as far as the middle of the penis, distal to which it is absent (Matthews, 1941).

In addition to the corpora cavernosum and spongiosum, a specialized accessory erectile body, the accessory cavernosus tissue, is present in some chiropterans (Matthews, 1941; Ryan, 1991a; Ryan, 1991b), but absent in others (Matthews, 1941; Krutzsch et al., 1976; Krutzsch and Crichton, 1987). The accessory cavernosous tissue contains large trabecular sinus, rich in elastic fibers, which is not enclosed by the albuginea tunica, but by the penile deep fascia (Ryan, 1991b). This tissue may be restricted to the distal half of the glans in Molossidae, surrounding the corpus spongiosum and *os penis* (Ryan, 1991a; Ryan, 1991b). In *Molossus ater,* some accessory erectile tissue is found dorsal to the corpora cavernosa in the mid-glans and also in the bacular mound (Ryan, 1991a). They are more conspicuous in *Cardioderma cor* and *Nycteris luteola* (Matthews, 1941). The accessory cavernous body when engorged, forms a large, bulbous, hyperemic distention. These specializations may be important in the reproductive process of species whose solitary roosting habits presumably provide fewer copulatory opportunities (Krutzsch, 1975).

The tunica albuginea of the erectile corpora cavernosum and corpus spongiosum are surrounded by the deep penile fascia, followed by the superficial fascia or subcutaneous tissue and, externally, the skin (Ryan, 1991b). The penile skin is lightly pigmented, containing sparsely distributed hair, including the prepuce (Krutzsch, 1975; Krutzsch et al., 1976; Ryan, 1991b). In some species, the penis skin presents thin short hairs amongst which project a number of longer (2-3 mm) stiff hairs (Matthews, 1941).

3.3. Glans Structure

The glans is well-developed in most bat species, but can be minute in others (Table 1) (Matthews, 1941; Krutzsch and Crichton, 1987). In *A. lituratus* ,the glans represents about 15% of the penis length, in *Macrotus waterhousii* 20% (Krutzsch et al., 1976) and 40% in *Pipistrellus hesperus* (Krutzsch, 1975). On the other hand, a very well-developed glans, comprising 90% of the penis length, is found in *Mormopterus planiceps* (Krutzsch and Crichton, 1987). In this species, the glans presents two parts: a long, dorsal primary portion and a short ventral secondary portion. The primary portion contains a well-developed baculum and the urethra, which emerges from the ventral surface (Krutzsch and Crichton, 1987).

The form of the bat glans has been described as conical, ovoid, cylindrical or club-shaped (Matthews, 1941; Vamburkar, 1958; Ryan, 1991b). The glans of chiropterans presents several specializations that appears species-specific. In this sense, the glans of *Pipistrellus hesperus* terminates in a pair of blunt, rounded processes conspicuous upon erection as bright, red, rounded knobs. The surface of flaccid glans in this species is wrinkled, presenting a number of deep longitudinal furrows and folds that permit the glans to expand upon erection (Krutzsch, 1975). In *Cynopterus sphinx gangeticus,* the glans surface is covered with minute denticles, which point cranially and are arranged in longitudinal rows (Vamburkar, 1958). The glans of *Cardioderma cor* is mostly covered with minute spiniform tubercles (Matthews, 1941). Most Molossidae species present epidermal spines in the glans, which are proximally directed (Ryan, 1991a; Ryan, 1991b). The keratinized spines are large in *Mormopterus jugularis* (75-100μm), but varies from 25-50μm in *M. planirostris* and *M. teminckii* to 15-30μm in *Nyctinomops laticaudatus* and *N. luteola* (Ryan, 1991b). In *Tadarida brasiliensis* and *T. aegyptiaca,* the glans surface is covered by several rows of blunt epithelial papillae instead of spines (Ryan, 1991b). Also in *Mormopterus planiceps*, the surface of the glans is covered by keratinized spines, which project backward and originate from nests of epithelial basal cells (Krutzsch and Crichton, 1987).

Differing from other mammal glans that are mainly filled by spongiosum tissue (e.g. stallion, human), the bat glans may contain corpora cavernosa, corpus spongiosum and/or accessory cavernosus tissue as the main erectile tissue. Accordingly, the glans of several Molossidae and some Vespertilionidae species contains mostly corpora cavernosa (Matthews, 1941; Krutzsch, 1975; Ryan, 1991a; Ryan, 1991b). In the Pteropodidae *Cynopterus sphinx gangeticus,* the corpora cavernosa end at the proximal part of the *os penis*, but the corpus spongiosum that surrounds the urethra, enters the glans, where they became thicker and practically fill the glans space beyond the corpora cavernosa (Vamburkar, 1958). In the

Emballonuridae *Taphozous sp,* the corpora cavernosa also terminates at the proximal end of the *os penis*, however the bulk of the tissue within the glans is formed by accessory corpora cavernosa (Matthews, 1941). A similar situation is found in *Molossops planirostris* and *M. temminckii* (Ryan, 1991b). In the Nycteridae *Nycteris hispida,* there is practically no erectile tissue within the glans, so the urethra and the *os penis* fill most of the glans (Matthews, 1941).

The *os penis* is characteristic of a wide range of genera representing virtually all bat families, except Phyllostomidae and Noctilionidae (Table 1) (Brown et al., 1971; Krutzsch, 1975; Hosken et al., 2001). When present, the *os penis* lies above the urethra and is contained into a prominent bacular mound, which is found distal to the urinary meatus (Ryan, 1991a; Herdina et al., 2010). The form and size of the *os penis* is greatly variable. It can be minute and restrict to the proximal half of the glans (*Taphozous sp*), distal half of the glans (*Cynopterus sphinx gangeticus, Hipposideros caffer,* Mollosidae species), well-developed extending for the entire glans (*Mormopterus planiceps*, *Triaenops afer*) or even from the glans into the shaft of the penis (*Nycteris luteola)* (Matthews, 1941; Vamburkar, 1958; Krutzsch and Crichton, 1987; Ryan, 1991a; Ryan, 1991b). The length of the *os penis* may be greatly variable depending upon species: 0.2 mm in *Thaphozous sp*, 0.76 mm in *Pipistrellus subflavus*, 1.5 mm in *Nycteris hispida,* 2.0 mm in *Cardioderma cor,* 2.5 mm in *Cynopterus sphinx gangeticus*, 5.0 mm in *Nycteris luteola* and 7.9 mm in *Mormopterus planiceps* (Matthews, 1941; Vamburkar, 1958; Krutzsch and Crichton, 1986; Krutzsch and Crichton , 1987). The smallest baculum is found in members of the families, Emballonuridae (0.07 mm in *Balantiopterix io*) and Molossidae, whereas the greatest is found in Pteropodidae (12.9 mm in *Pteropus neohibernicus*) (Hosken et al., 2001). The large variation in baculum size across bats does not appear to be associated with sexual selection or risk of sperm competition (Hosken et al., 2001).

Concerning the baculum shape, it may be cylindrical *(Taphozous sp, Nycteris hispida* - Matthews, 1941), Y-shaped (Krutzsch and Crichton, 1986 - *Pipistrellus subflavus*; Herdina et al., 2010 - *Plecotus austriacus*), disc-shaped (Nwoha, 2000 - *Eidolon helvum*), elongated and strongly concave ventrally (Krutzsch, 1975 - *Pipistrellus hesperus*) or curved with a ventral convexity and a dorsal concavity (Vamburkar, 1958 - *Cynopterus sphinx gangeticus*). The baculum periosteum is continuous with the dense connective tissue of the thick tunica albuginea of the corpora cavernosa (Krutzsch, 1975; Krutzsch and Crichton, 1987).

The functional significance of this ossicle is not clear. In species presenting well-developed baculum, it has been proposed that it may support and stiffen the penis (Krutzsch, 1975). The elongated baculum and spiny surface of the glans, as seen in *Mormopterus planiceps*, are considered specializations important for penetration and retention of the penis into the vagina at copulation (Krutzsch and Crichton, 1987). However, acçording to Nwoha, (2000), in the Nigerian *Eidolon helvum*, the *os penis* may function to support erection of the penis more for micturition than for intromission. A recent study on *Plecotus austriacus* (Herdina et al., 2010) suggested that the microanatomy of baculum implies a mechanical function, as it forms a functional unit with the corpora cavernosa. The collagen fibers of the tunica albuginea of both corpora cavernosa insert via fibrocartilage into the woven bone of the baculum branches. This fibrocartilage may represent alternating shearing forces on this joint-like interface during erection and copulation. In addition, a function in keeping the orifice of the urethra opened during copulation has also been proposed for the concave surface of the distal tip of the baculum (Herdina et al., 2010).

Table 1. Characteristics of the glans penis of bats.

Suborder/ Families/ Species	Glans Form	Glans Length (mm)	Baculum	Urethral meatus	References
MEGACHIROPTERA					
Pteropodidae					
Cynopterus sphinx	conical	4	Present	Subterminal	Vamburkar, 1958
Emballonuridae					
MICROCHIROPTERA					
Taphozous sp	Conical	0.8	Present	Subterminal	Matthews, 1941
Megadermatidae					
Cardioderma cor	Ovoid	4.0	Present	Subterminal	Matthews, 1941
Molossidae					
Eumops auripendulus	Oval	ND	Present	Subterminal	Ryan, 1991a
Eumops bonariensis	Oval	1.5-2.5	Present	Subterminal	Ryan, 1991a
Eumops perotis	ND	ND	Absent	ND	Ryan, 1991a
Mormopterus jugularis	Oval	3.5	Present	Subterminal	Ryan, 1991b
Molossus ater	Cone-shaped	5.6	Present	Subterminal	Ryan, 1991a
Molossus bondae	ND	5.0	Present	ND	Ryan, 1991a
Molossus coibensis	ND	4.5	Present	ND	Ryan, 1991a
Molossus molossus	ND	3.4	Present	ND	Ryan, 1991a
Molossus sinaloae	ND	3.0	Present	ND	Ryan, 1991a
Molossops temminckii	Oval	2.0	Absent	Terminal	Ryan, 1991b
Molossops planirostris	Oval	1.2	Absent	Subterminal	Ryan, 1991b
Nyctinomops (=Tadarida) macrotis	Cylindrical	ND	Present	Subterminal	Ryan, 1991b
Nyctinomops laticaudatus	Cylindrical	2.0	Present	Subterminal	Ryan, 1991b
Tadarida brasiliensis	Cylindrical	2.5-3.0	Present	Subterminal	Ryan, 1991b
Promops centralis	Long	4.0	Absent	ND	Ryan, 1991a
Nycteridae					
Nycteris luteola	club-shaped	ND	Present	Terminal	Matthews, 1941
Nycteris hispida	ND	0.8	Present	Subterminal	Matthews, 1941
Phyllostomidae					
Macrotus waterhousii	ND	1.7 2.0	Absent	ND	Krutzsch et al, 1976
Rhinolophidae					
Hipposideros caffer	Long and narrow	3.0	Present	Subterminal	Matthews, 1941
Triaenops afer	large and flattened dorso-ventrally	2.5	Present	Terminal	Matthews, 1941
Vespertilionidae					
Miniopterus minor	conical	0.5	Absent	Subterminal	Matthews, 1941
Miniopterus schreibersii	conical	0.5	Absent	Subterminal	Matthews, 1941
Pipistrellus hesperus	elongated and broad	2.3 - 3.5	Present	ND	Krutzsch, 1975

ND = Not detected.

In most species of bats, the glans is covered by a thin prepuce, readily retractable (Matthews, 1941; Vamburkar, 1958; Krutzsch and Crichton, 1987; Ryan, 1991a). In some species, the prepuce is thick, but still retractable, as seen in *Pipistrellus hesperus* and *Macrotus waterhousii* (Krutzsch, 1975; Krutzsch et al., 1976). In *Miniopterus schreibersii,* the glans is covered by a long prepuce which extends about 1.0 mm beyond its end (Matthews, 1941). Also, in *Cynopterus sphinx gangeticus, Macrotus waterhousii* and *Pipistrellus Hesperus,* the prepuce extends beyond the glans tip (Vamburkar, 1958; Krutzsch, 1975; Krutzsch et al., 1976). The prepuce is lightly pigmented and presents sparse hairs (Krutzsch and Crichton, 1987). Short hairs also cover the outer sheath of the prepuce in other Molossidae species (Ryan, 1991a). In these species, there is also a tuft of long stiff hairs along the dorsal margin of the prepuce. Preputial glands may be present in some species but absent in others (Krutzsch, 1975).

4. Accessory Sex Glands

Information about the accessory sex glands of bats is scarce and mostly restricted to gross morphology. The available information revealed that bat accessory sex glands show great diversity and remarkable differences even when closely related species are considered (Mokkapati and Dominic, 1977). The complex of accessory gland of bats comprises the prostate, seminal vesicles, ampullary glands, bulbourethral glands (Cowper's glands) and urethral glands (Figure 5). The prostate and bulbourethral glands have been described in all species studied to date, whereas seminal vesicles, ampullary glands and urethral glands are present in some species but absent in others.

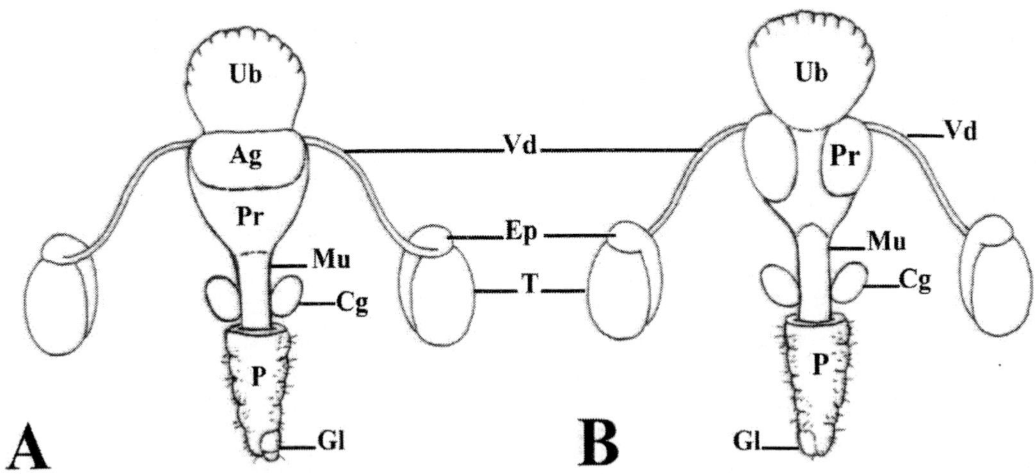

Figure 5. The accessory sex glands in *Artibeus lituratus*. (A) Dorsal and (B) ventral view. The accessory sex gland is comprised of the ampullary gland (Ag), prostate (Pr) and the Cowper's glands (Cg). Ub = urinary bladder; Mu = membranous urethra; P = penis; Gl = glans, T = testis; EP = epididymis; Vd = vas deferens.

4.1. Prostate

The prostate is an unpaired gland present in all bat species (Figure 5 and Table 2). It is located below the neck of the urinary bladder, however its position related to the prostatic urethra exhibit variation among species, being ventral to the urethra (Mokkapati and Dominic, 1977 - *Cynopterus sphinx, Pipistrellus mimus*), encircling dorsolaterally the urethra (Matthews, 1941 - *Nycteris luteola*; Krutzsch, 1975 - *Pipistrellus Hesperus*; Krutzsch, 1975 - *Macrotus waterhousii*; Mokkapati and Dominic, 1977 - *Scotophilus heathi, Scotophilus temmincki*; Krutzsch and Crichton, 1986 - *Pipistrellus subflavus;* Oliveira, non-published data - *A. lituratus*) or completely surrounding the urethra (Matthews, 1941 - *Taphozous sp; Hipposideros caffer, Triaenops afer, Miniopterus minor, M. schreibersii*; Mokkapati and Dominic, 1977 –

Rhinopoma hardwickei, Taphozous longimanus). Even in species that present the urethra surrounded by the prostate, the gland is larger in the dorsal portion than the ventral (Matthews, 1941).

The prostate is enclosed by a thick, connective tissue and smooth muscle capsule (Krutzsch, 1975; Krutzsch et al., 1976). The gland has no lobulations, except in *Pipistrellus hesperus* where the gland is described as ventrally bilobed (Krutzsch et al., 1976). The prostate consists of branched tubuloalveolar glands (Krutzsch, 1975; Mokkapati and Dominic, 1977). As seen in *Cynopterus sphinx*, it is composed of alveoli with 40 to 200 μm of diameter, which is lined by a tall simple columnar epithelium (Mokkapati and Dominic, 1977). However, the epithelium may vary from columnar to squamous, depending on the amount of secretory product accumulated in the alveoli lumen (Krutzsch, 1975; Krutzsch and Crichton, 1987). The nuclei of epithelial cells are located basally (Mokkapati and Dominic, 1977).

The prostate contains eosinophilic, PAS-positive, Alcian blue negative, diastase resistant secretion, suggesting a glycoprotein (Racey and Tam, 1974 - *Pipistrellus pipistrellus*; Krutzsch, 1975 - *Pipistrellus hesperus*; Bernard, 1986 - *Rhinolophus capensis*; Krutzsch and Crichton, 1987 - *Mormopterus planiceps*). The majority of the secretion is aggregated into clumps of ovoid to multi-sided globules into the lumen (Krutzsch and Crichton, 1990). This globular secretion is the result of a major breakdown of the glandular epithelium. Prostates secrete fructose in *Scotophilus beatbi, Taphozous longimanus, Macrotus waterhousii* and *Pipistrellus subflavus,* especially during breeding activity (Krutzsch et al., 1976; Mokkapati and Dominic, 1976; Krutzsch and Crichton, 1986). A great amount of citric acid is also produced by the prostate (Mokkapati and Dominic, 1976). The prostate secretion flows via multiple excretory ducts that open in the prostatic urethra in the neighborhood of the *crista urethralis* (Matthews, 1941; Krutzsch and Crichton, 1990).

During the inactive period, the gland alveoli are collapsed and interspersed by a large amount of connective tissue (Richardson, 1977; Krutzsch and Crichton, 1990).

4.2. Seminal Vesicles

The seminal vesicle consists of a pair of glands, usually pear-shaped, located laterally or proximal to the prostate on the outer border of the ampullary gland, when present (Matthews, 1941; Mokkapati and Dominic, 1976; Mokkapati and Dominic, 1977). The seminal vesicle is present in few species of bats (Figure 5 and Table 2) (Matthews, 1941; Krutzsch, 1975; Mokkapati and Dominic, 1977; Krutzsch and Crichton, 1990). As the structure of the bat seminal vesicles are similar to that of ampullary gland, it is possible that they were confounded in some studies. It is noteworthy that in all species examined to date, at least one of these glands is always present in bats. The occurrence of both ampullary glands and seminal vesicles is not a common finding (Matthews, 1941).

Mokkapati and Dominic (1977) described the accessory sex glands of seven species of Indian bats and found seminal vesicles in three species - *Cynopterus sphinx, Pipistrellus mimus* and *Taphozous longimanus*. The seminal vesicles of *Cynopterus sphinx* are large and coiled tubular structures, consisting of a rounded distal portion and a tubular proximal end opening into the urethra. The diverticula of the seminal vesicle are lined by low columnar epithelium (Matthews, 1941) or squamous secretory epithelium (Mokkapati and Dominic, 1976; Mokkapati and Dominic, 1977), as the height of the epithelium may vary depending on the amount of luminal secretion (Krutzsch, 1975).

The secretory product of the seminal vesicle is more amorphous, homogenous and mucinous in appearance than the prostate (Richardson, 1977 - *Miniopterus australis e Miniopterus schreibersii*; Krutzsch and Crichton, 1990 - *Miniopterus schreibersii*). Seminal vesicle secretion contains fructose and citric acid (Mokkapati and Dominic, 1976).

4.3. Ampullary Glands

The ampullary glands are present in most bat species (Figure 5 and Table 2). They consist of pear-shaped or elongated oval terminal enlargements of the distal extremity of the vas deferens, which become highly folded and undergo differentiation into a specialized secretory gland (Mokkapati and Dominic, 1977; Krutzsch and Crichton, 1990). The ampullary glands lie between the seminal vesicles when these are present. At the distal end of the ampulla, the ejaculatory ducts emerge, which pass through the prostatic tissue to join the urethra on its posterior surface, opening on the *crista urethralis* (Matthews, 1941). In some species, the vas deferens lies embedded in the center of the gland (Racey and Tam, 1974 - *Pipistrellus pipistrellus*; Mokkapati and Dominic, 1977 - *Rhinopoma hardwickei*; Krutzsch and Crichton, 1987 - *Mormopterus planiceps*). The vas deferens can be differentiated from the glandular diverticula because of the large lumen and the thicker muscular layer (Matthews, 1941).

The entire ampullary gland is ensheathed by a connective tissue and muscular capsule (Racey and Tam, 1974; Mokkapati and Dominic, 1977; Richardson, 1977; Krutzsch and Crichton, 1987). They consist of a mass of tubular diverticula, 30-60 μm in diameter, lined by a simple unfolded cuboidal secretory epithelium (Matthews, 1941; Racey and Tam, 1974; Bernard, 1986). The gland presents luminal secretions along the year, but the structure of the epithelium changes along the annual reproductive cycle. In the inactive condition, the epithelial cells present few and short microvilli, few secretory granules, but large autolytic vacuoles. Conversely, in the active gland, the microvilli are longer and abundant and the apical cytoplasm is filled with secretory granules (Bernard, 1986).

The ampullary gland produces fructose and citric acid in *Taphozous longimanus* but only fructose in *Scotophilus heathi* (Mokkapati and Dominic, 1976). Sperm are abundant in the lumen of ampullary gland collected in the breeding season (Richardson, 1977; Krutzsch and Crichton, 1987; Krutzsch and Crichton , 1990) or throughout the year (Anand Kumar, 1965).

Table 2. Occurrence of accessory sex glands in bat species.

Suborder/ Families/ Species	Prostate	Seminal vesicle	Ampullary gland	Cowper gland	Urethral glands	References
MEGACHIROPTERA						
Pteropodidae						
Cynopterus sphinx	Present	Present	Absent	Present	Present	Mokkapati and Dominic, 1977 Vamburkar, 1958
MICROCHIROPTERA						
Emballonuridae						
Taphozous georgianus	Present	Present	Present	ND	ND	Jolly and Blackshaw, 1987
Taphozous longimanus	Present	Present	Present	Present	Absent	Mokkapati and Dominic, 1977
Thaphozous sp	Present	Present	Present	Present	Present	Matthews, 1941
Megadermatidae						
Cardioderma cor	Present	Absent	Present	Present	Present	Matthews, 1941
Molossidae						
Mormopterus planiceps	Present	Present	Present	Present	Present	Krutzsch and Crichton, 1987
Nycteridae						
Nycteris luteola	Present	Absent	Present	Present	Absent	Matthews, 1941
Nycteris hispida	Present	Absent	Present	Present	Absent	Matthews, 1941
Phyllostomidae						
Artibeus lituratus	Present	ND	Present	Present	Present	Oliveira, nonpublished data
Macrotus waterhousii	Present	Present	Present	Present	Absent	Krutzsch et al 1976
Rhinolophidae						
Hipposideros speoris	Present	Absent	Present	Present	Present	Mokkapati and Dominic, 1977
Hipposideros caffer	Present	Absent	Present	Present	Present	Matthews, 1941
Rhinolophus capensis	Present	Absent	Present	Absent	Present	Bernard, 1986
Triaenops afer	Present	Present	Absent	Present	Present	Matthews, 1941
Rhinopomatidae						
Rhinopoma hardwickei	Present	Absent	Present	Present	Absent	Mokkapati and Dominic, 1977
Rhinopoma kinneari	Present	Absent	Present	Present	Absent	Anand Kumar, 1965
Vespertilionidae						
Miniopterus australis	Present	Present	Present	Present	ND	Richardson, 1977
Miniopterus minor	Present	Absent	Present	Present	Present	Matthews, 1941
Miniopterus schreibersii	Present	Present	Present	Present	Absent	Krutzsch and Crichton, 1990 Richardson, 1977
Myotis lucifugus	Present	Present	Present	Present	ND	Hunter et al, 1971
Nyctalus noctula	Present	Absent	Present	Present	Absent	Racey, 1974
Pipistrellus hesperus	Present	Present	Present	Present	Absent	Krutzsch, 1975
Pipistrellus mimus	Present	Present	Present	Present	Present	Mokkapati and Dominic, 1977
Pipistrellus pipistrellus	Present	Absent	Present	Present	Present	Racey and Tam, 1974
Pipistrellus subflavus	Present	Absent	Present	Present	Absent	Krutzsch and Crichton, 1986
Scotophilus heathi	Present	Absent	Present	Present	Present	Mokkapati and Dominic, 1977
Scotophilus temmincki	Present	Absent	Present	Present	Present	Mokkapati and Dominic, 1977

ND = Not detected.

4.4. Bulbourethral or Cowper's Glands

The bulbourethral glands are small, paired, bilaterally symmetrical glands present in all bat species (Figure 5 and Table 2). The gland size is variable depending on the species considered (0.6 mm in diameter in *Miniopterus minor*; 0.8 mm in *Nycteris luteola;* 1 mm in *Taphozous sp, Triaenops afer*; 1.2 mm in *Miniopterus schreibersii;* 1,7 mm in *A. lituratus*) (Matthews, 1941; Oliveira, non-published data). They are bean-shaped, spherical or ovoid in shape, lying cephalic to the bulbocavernosus muscles and opening into the penile urethra, cephalic to the bulb (Krutzsch, 1975; Mokkapati and Dominic, 1977; Richardson, 1977).

Histologically, they consist of compound tubuloalveolar glands, whose acini are lined by cubical cells (Krutzsch, 1975 - *Pipistrellus Hesperus*; Mokkapati and Dominic, 1976; Mokkapati and Dominic, 1977 - *Cynopterus sphinx*) or tall columnar cells, which produces a PAS-positive secretion (Richardson, 1977 - *Miniopterus schreibersii e Miniopterus australis*; Krutzsch and Crichton, 1987 - *Mormopterus planiceps*; Krutzsch and Crichton, 1990 - *Miniopterus schreibersii*). The cytoplasm of the epithelial cells is filled with secretory granules that flatten the nucleus in the basal cytoplasm (Richardson, 1977; Krutzsch and Crichton, 1987; Krutzsch and Crichton , 1990). The Cowper's glands are surrounded by a connective tissue capsule, which is covered by a thick tunica of striated muscle (*bulboglandularis*) (Krutzsch, 1975; Mokkapati and Dominic, 1977). Contraction of this muscle at ejaculation rapidly empties the gland secretion (Krutzsch, 1975). The glandular parenchyma is separated into lobules by thin septa of connective tissue originating from the capsule (Krutzsch et al., 1976).

4.5. Urethral Glands

The urethral glands are present in several species of bats, but is absent in others (Table 2). They consist of prominent tubuloalveolar glands embedded in the connective tissue surrounding the urethra (Matthews, 1941; Mokkapati and Dominic, 1977; Krutzsch and Crichton, 1987; Krutzsch and Crichton , 1990). The tubuloalveoli are lined with very tall, simple columnar epithelium, with nuclei at the base of the cells and cytoplasm filled with fine granules (Matthews, 1941; Krutzsch and Crichton, 1987). The lumen contains homogenous, fine, granular eosinophilic secretion (Krutzsch and Crichton, 1987).

The large urethral gland present in some species is apparently peculiar to chiropterans. Its presence is possibly correlated with the formation of a large vaginal plug in the female at insemination; however this function has not been proven (Matthews, 1941).

4.6. Seasonal Variation

The accessory complex gland of bats undergoes seasonal variation, which may be synchronous or asynchronous in relation to the testicular cycle. The accessory glands of *Macrotus waterhousii, Scotophilus heathi* and *Artibeus lituratus* undergo annual cycles in synchrony with the testis (Krutzsch, 1975; Krishna and Singh, 1998; Oliveira, non-published data). Also, the prostate, ampullary and Cowper's glands show seasonal variation, according to the testicular cycle of *Rhinopoma kinneari*, as they hypertrophy when Lẹydig cells are active in hormonal secretion, and become regressed when the testis are quiescent (Anand Kumar, 1965). Differing from other hibernating vespertilionid species, the accessory glands of Australian *Miniopterus schreibersii* regress during winter in synchrony with testis involution (Richardson, 1977; Krutzsch and Crichton, 1990).

However, in most hibernating species, there is asynchrony between testis and accessory gland activity. In this case, the glands remain maximally developed and secretorally active during mating and hibernation periods (through winter and early spring), after spermatogenesis has terminated, and Leydig cells are involuted (this occurs in later summer/early fall).

The glands regress during arousal (at spring) reaching minimum weight in late spring/early summer, when spermatogenesis occurs.

The gland weights are recovered in August/September reaching a maximum at autumnal copulation. (Racey, 1974 - *Nyctalus noctula*; Racey and Tam, 1974 - *Pipistrellus pipistrellus;* Gustafson, 1979; Richardson, 1977 - *Miniopterus australis*; Hosken et al., 1998 - *Nyctophylus geoffroyi; N. major* and *Vespadelus regulus*). The inactive glands are distinguished by the reduced size, empty alveoli and more conspicuous stroma, contrasting with active glands, which presents alveoli filled with secretion and the stroma is inconspicuous (Racey and Tam, 1974; Krutzsch, 1975; Krutzsch and Crichton, 1987; Krutzsch and Crichton , 1990).

The mechanism regulating the asynchrony between primary and secondary reproductive functions is intriguing and still not solved. There is evidence that a peak on the plasm testosterone levels occurs in the spermatogenic period which is the highest among mammal species (Gustafson and Shemesh, 1976). Therefore, even the lowered level of androgen occurring at hibernation would be enough for the maintenance of the accessory sex gland (Gustafson, 1979). It has also been proposed that the body metabolic reduction at hibernation may contribute for retarding the gland's physiological regression (Gustafson, 1979).

5. Hormonal Regulations

Chiroptera is an order of mammals with many unusual reproductive strategies, including some unique peculiarities, which depends fundamentally on endocrine regulation. Despite this fact, studies about the regulation of male reproductive function are scarce, outdated and mainly concentrated in hibernating species belonging to the Vespertilionidae and Rhinolophidae families. However, there are numerous critical pathways of hormonal regulation that are still not explored even in these animals. On the other hand, there are interesting reproductive strategies for non-hibernating bats that inhabit tropical regions, whose hormonal regulation remains unknown.

Concerning reproductive function, the hypothalamic-pituitary-gonadal axis is crucial in controlling each step of this complex process. The hypothalamus produces several neuropeptides, including the gonadotropin releasing hormone (GnRH), which stimulates the hypophysis to secrete gonadotropins that acts in the testis, modulating both spermatogenic and steroidogenic functions.

5.1. GnRH

The hypothalamic GnRH is a key regulator of reproductive function in all mammalian groups, including bats. This hormone is responsible for stimulating the release of follicle stimulating hormone (FSH) and luteinizing hormone (LH) from gonadotropes located in the *pars distalis* of the adenohypophysis. By acting through the GnRH receptors on the plasma membrane of gonadotropes, GnRH stimulates a cascade of intracellular events, such as an increase of intracellular calcium and activation of protein kinase C, responsible for the synthesis and secretion of the gonadotropins. This primaryhormonal control is required for the initiation and maintenance of spermatogenesis, which involves the action of FSH and testosterone acting in concert on Sertoli cells, whereas LH modulates steroidogenesis in Leydig cells (Gnessi et al., 1997).

In several Vespertilionidae species, the hypothalamic GnRH neurons are fusiform or bipolar, mostly located in the medial basal hypothalamus, especially in the arcuate nucleus, a feature that resembles primates, including humans (King et al., 1984; (Oelschlager and Northcutt, 1992; Kawamoto, 2003). Conversely, GnRH neurons in hibernating Rhinolophidae bats are concentrated in the medial preoptic area and medial basal hypothalamus (Kawamoto et al., 1998; Kawamoto, 2003). It is noteworthy that GnRH neurons of the medial basal region vary seasonally in number and immunoreactivity in both Vespertilionidae and Rhinolophidae hibernating bats, thus indicating that these neurons play a key role in the seasonal variation of gonadotropins secretion in these bats (Kawamoto, 2003).

Seasonal changes in the release of gonadotropin required for gametogenesis and gonadal steroidogenesis is observed in hibernating bats (Kawamoto et al., 2000). The percentages of LH- or FSH-secreting cells and the secretory activity of gonadotropin, as well as the responsiveness of gonadotropes to GnRH increase in the spermatogenic period (summer), as seen in *Rhinolophus ferrumequinum* (Kawamoto et al., 2000). During the hibernation, the secretory activity of gonadotropes decreases markedly, but the stimulatory effect of GnRH on the secretion of gonadotropins is relatively persistent. This data suggests that GnRH receptors and the post-receptor events that mediate GnRH action on basal gonadotropin secretion do not disappear under heterothermic conditions of bats (Kawamoto et al., 2000).

5.2. Gonadotropins

The testicular functions are dependent on the gonadotropins FSH and LH secreted from the *pars distalis* of the adenohypophysis, under stimulation of GnRH. Most individual gondotropes of bats produce both FSH and LH, even though some cells contain only one of these hormones within each cell (Richardson, 1979; Mikami et al., 1988; Bernard et al., 1991). The male gonadotropes are large, oval, rounded or irregular in shape, exhibiting cytoplasmic processes which surround other secretory cells of the gland (Anthony and Gustafson, 1984; Mikami et al., 1988).

In bats, the secretion of FSH in response to GnRH is increased during the spermatogenic period compared to hibernation (Hayashi et al., 2002; Kawamoto, 2003). FSH indirectly influences spermatogenesis by modulating the functions of Sertoli cells (Richardson, 1979; Francis et al., 1981). The response of Sertoli cells to FSH is mediated by cyclic adenosine 3', 5'-monophosphate (cAMP) and/or possibly Ca^{2+} as second messengers.

The testicular gonadotropin receptors are crucial for the gonadal responsiveness to gonadotropin. Information about LH receptors in the bat testis was not found in the current literature. The receptors for FSH are expressed in Sertoli cells, as observed in rhinolophid bats (Hayashi et al., 2002). The concentration of FSH receptors in the testis varies in accordance with the reproductive cycle. During the breeding season, high concentrations of a FSH receptor is detected in the Sertoli cells, whereas during hibernation, the receptor concentration is low (Hayashi et al., 2002). These seasonal fluctuations in the number of FSH receptors in the testes are accompanied by corresponding changes in the responsiveness of the testes to FSH stimulation. Accordingly, the production of cAMP in response to FSH is far less in the hibernation period than in the spermatogenic period. These findings suggest that during hibernation, the number of FSH-binding sites per Sertoli cell in the bat testis decrease significantly, parallel to a decrease in the activity of the adenylate cyclase system stimulated by the FSH-receptor complex (Hayashi et al., 2002).

LH is responsible for stimulating the synthesis of testosterone by Leydigs cells, thus indirectly regulating spermatogenesis by increasing intratesticular testosterone. In hibernating Vespertilionid and Rinolophid species, the LH level is closely associated with the reproductive cycle and the testosterone levels. Secretion of LH is higher during spermatogenic and mating periods, whereas its level decreases significantly during hibernation and post-arousal, when the testosterone levels are lower (Gustafson, 1979; Kawamoto, 2003). Similar patterns of synchrony between LH secretion and Leydig cell and seminiferous epithelium activity is described for seasonal non-hibernating species (O'Brien et al., 2003 - *Pteropus poliocephalus*). Contrasting with activity, gonadotropes producing LH is detected in the *pars distalis* of male bats throughout the year apparently with constant size and shape (Anthony and Gustafson, 1984; Mikami et al., 1988).

It is interesting that in some species of hibernating vespertilionid, gonadotropes reactive to LH and FSH are found in the *pars tuberalis* besides the *pars distalis* of the adenohypophysis (Anthony and Gustafson, 1984; Mikami et al., 1988; Azzali et al., 2003; Kawamoto, 2003). The LHβ-positive cells are present during recrudescence, spermatogenic and hibernation periods, but are not detected in the mating period, thus indicating that this gonadotropin secretion is increased in autumn (Kawamoto, 2003). The functional activity of the bat *pars tuberalis* gonadotropins is presently unknown.

5.3. Androgens

Androgens are steroid hormones essential for the maintenance of male reproductive functions, whose actions are mediated by the androgen receptors (AR). These receptors are widely distributed in the male genital system and have been described in several species of wild and domestic animals. Nevertheless, only recently AR were mapped in the testis of *Artibeus lituratus*, which is a tropical bat presenting marked reproductive seasonality, with a period of testicular activity from August to December and regression from December to April (Oliveira et al., 2009). During the reproductive period, the AR are restricted to Sertoli cells and their expression is stage-specific, being more prominent during stages VII-VIII, in which spermatogenic cells associated with Sertoli cells are in a more differentiated phase. In the regressive testis, the AR expression remains restricted to Sertoli cells, but immunostaining is intense in all seminiferous tubules, despite the low levels of plasma testosterone (Oliveira et al., 2009).

The main androgen produced by Leydig cells under LH stimulation is testosterone, which is responsible for the initiation and maintenance of spermatogenesis. Besides control of spermatogenesis, testosterone is also involved in the regulation of gonadotropin release, both directly or indirectly via aromatization to estrogens (Hayes et al., 2001). In all bat species analyzed to date, the testosterone levels are directly related to the spermatogenic function, being higher at this period (Gustafson and Shemesh, 1976; Gustafson, 1979; Gustafson and Damassa, 1985; Bernard, 1986; Singh and Krishna, 2000), except for the Tropical *Taphozous georgianus* in which spermatogenesis occurs in summer/ autumn but the peak of testosterone level is found during winter (Jolly and Blackshaw, 1989). In hibernating bats (Vespertilionidae and Rhinolophidae), testosterone concentrations are elevated in the spermatogenic period but rapidly decrease in the mating period, reaching minimal values during hibernation (Hayashi et al., 2002). The androgen level at the spermatogenic period of the hibernating bat is the highest known for mammals (Table 3) (Gustafson and Shemesh, 1976; Gustafson, 1979). It is curious that in these species there is a marked asynchrony between spermatogenesis, that occurs under high testosterone levels in the summer, and the activity of sex accessory glands, that occurs during winter, when the testosterone level is low (Gustafson and Shemesh, 1976; Gustafson, 1979, 1987; Hayashi et al., 2002). However, there is evidence that even in hibernating species, steroidogenic enzymes can be detected in the regressed Leydig cells, as seen for the key enzyme 3β-HSD. According to Kurohmaru et al (2002), the Leydig cells of *Rhinolophus cornutus* are positive for 3β-HSD throughout the year, being more intense just prior to the beginning of spermatogenesis and weaker after spermatogenesis and during hibernation. *Nyctalus noctula* also show little variation in Leydig cell morphology and 3β-HSD expression throughout the year (Racey, 1974).

Table 3. Blood testosterone levels in bat species

Suborder/ Families/ Species	Spermatogenic period	Dormance	Hibernation	After hibernation/ dormance	References
MICROCHIROPTERA					
Emballonuridae					
Taphozous longimanus	22.46 ng/ml	6.91 ng/ml	No	ND	Singh and Krishna, 2000
Rhinolophidae					
Rhinolophus capensis	6.2 ng/ml	No	0.2/3.2 ng/ml	3.5 ng/ml	Bernard, 1986
Vespertilionidae					
Myotis lucifugus	60 ng/ml	No	2.5 ± 0.50/ 9.4 ± 3.4 ng/ml	2.8 ±0.4 ng/ml	Gustafson and Shemesh, 1976
Myotis lucifugus	105.4 ±19.3 ng/ml	No	2.0±0.2 ng/ml	2.0±0.2 ng/ml	Gustafson and Damassa, 1985
Nyctalus noctula	134 ng/ml	No	19 ng/ml	73 ng/ml	Racey, 1974*
Scotophilus heathi	19.32±1.14 ng/ml	5.4±0.94 ng/ml	No	4.31±0.19 ng/ml	Singh and Krishna 1996

ND = Not determined; * = animals in captivity.

Data about other androgens, such as dihydrotestosterone (DHT) and androstenedione, is scarce. Available information for the Australian Tropical common sheated-tail bat *Taphozous georgianus* shows that DHT and androstenedione levels are high compared to testosterone, but no significant seasonal changes is observed for these testosterone metabolites (Jolly and Blackshaw, 1989). Androstenedione peaks are parallel to testosterone, before the onset of spermatogenesis, and show stronger correlation with body mass than with size of the reproductive organs (Racey and Tam, 1974; Singh and Krishna, 1996; Singh and Krishna, 2000).

5.4. Estrogens

Estrogens are now recognized as a hormone of great relevance for maintenance of the male genital system morphophysiology and fertility. In males, the testis is the main source of estrogens and their biosynthesis occurs from androgens, in a reaction catalyzed by the P450 aromatase enzyme (Simpson et al., 1997; Carreau et al., 2002). The estrogen actions are mediated by the estrogen receptors ERα and ERβ, which are widely distributed in the male genital system of several mammalian species (Hess, 2003). In bats, estrogen level is very low (<10pg/ml) or indetectable in males of *M. lucifugus*, during the breeding season (Damassa and Gustafson, 1984). There is no information about local estrogen concentration in the male tract. Nevertheless, one recent study described the occurrence and distribution of ERα and ERβ in the testes of *Artibeus lituratus* (Oliveira et al., 2009). In this species, immunoreactivy for ERα is restricted to some nuclei of Leydig cells, during both reproductive and regressive periods. Conversely, ERβ are widely distributed in the somatic and spermatogenic cells of the testis. In the reproductive period, the ERβ distribution in the seminiferous tubules is stage-specific. In the regressive period, more intense immunostaining is observed for ERβ in both Sertoli and Leydig cells (Oliveira et al., 2009). Variations in the estrogen receptor levels during the annual reproductive cycle of *A. lituratus* indicates that estrogens may indeed play an important role in the hormonal regulation of the reproductive cycle of bats.

Interestingly, there is information that only in bats and primates, including humans, the steroid hormone binding globulin (SHBG) bind to estrogens with a higher affinity than androgens (Damassa et al., 1982; Damassa et al., 1985; Kwiecinski et al., 1987). This is in contrast with most mammals in which SHBG presents higher affinity for testosterone and dihydrotestosterone. SHBG is responsible for the regulation of steroid-free concentration, protecting them from degradation and excretion, therefore the higher affinity of this glycoprotein for estrogen may explain the low levels of circulating hormones in bats. This intriguing feature of SHBG was studied in some species of bats and a specific SHBG was identified in some Phyllostomid and Vespertilionid (Kwiecinski et al., 1987). However, most of these studies were discontinued, so the precise role of SHBG in bats and its relationship with estrogen activity has not been fully elucidated.

6. Reproductive Patterns

The wide geographical distribution is in part a consequence of a high reproductive success of bats, even though most species produce just one litter per gestation. To ensure this reproductive success, bats adopt some interesting and often unique strategies only observed in this order. One strategy to adapt to different climates is the existence of hibernating and non-hibernating species usually inhabiting temperate and tropical regions, respectively. The difference in geographical distribution implies that some families or even genus of bats may present hibernating and non-hibernating representants, which may differ in the reproductive pattern.

Distinct reproductive patterns are important to reconcile the mating and the hibernation period with the reproductive cycle of females, ensuring that the litters are born in a period of greater availability of food, contributing to reproductive success (Gustafson, 1979; Oxberry, 1979; Kawamoto, 2003).The same diversity applies to non-hibernating species, which may be seasonal or non-seasonal in spite of the family or genus to which they belong.

The hibernating male bats belonging to Vespertilionidae and Rhinolophidae families present two different reproductive patterns, determined based on the duration of spermatogenesis in relation to mating and hibernating periods. In all patterns, the spermatogenesis occurs during summer, but the period of activity of the Leydig cells and male genital accessory gland is variable (Gustafson, 1979). In some species, the mating occurs shortly after spermatogenesis, before hibernation or torpor. These species usually store sperm in the female genital tract or delay the implantation. This pattern of reproduction, characterized by mating after spermatogenesis, is described in *Nyctalus noctula* (Racey, 1974), *Pipistrellus pipistrellus* and *P. hesperus* (Racey and Tam, 1974; Krutzsch, 1975), *Myotis lucifugus* (Gustafson, 1979; Gustafson, 1987), *Pipistrellus subflavus* (Krutzsch and Crichton, 1986), *Miniopterus schreibersii* (Richardson, 1977; Gustafson, 1979) and *Rhinolophus ferrumequim* (Hayashi et al., 2002). Other species mate after hibernation; therefore, they present prolonged sperm storage in the male epididymis, which will be used in mating occurring during or after arousal from hibernation, as seen in *Rhinolophus capensis* (Bernard, 1984; Bernard, 1986).

Although the bats belonging to the Rhinolophidae and Vespertilionidae families are mainly hibernating, they also present members that are non-hibernating with reproductive seasonal patterns like Rhinolophidae members: *Hipposideros caffer* (Nigeria), *H. commersoni* (Africa), *Rhinolophus landeri* (Nigeria) and *Triaenops afer* (East Africa); Vespertilionidae members: *Eptesicus furinalis* (Paraguai), *Lasiurus ega* (Paraguai), *Miniopterus australis* (Australia), *Myotis nigricans* (Panama), *Tylonycteris pachypus* (Malaysia) and *T. robustula* (Malaysia) (Richardson, 1977; Krutzsch, 1979).

Non-hibernating bats inhabiting regions where the seasons are poorly defined also present seasonality associated with various reproductive strategies. In these bats, the reproductive cycle is also directly linked to environmental conditions, especially in relation to rainfall (Fleming et al., 1972; Krutzsch, 1979). In some species of the family Pteropodidae (*Eonycteris spelaea, Epomorphorus labiatus, Myonyscteris torquata, Stenonycteris lanosus*), Nicteridae (*Nycteris hispida* and *N. luteola*) and Phyllostomidae (*Artibeus jamaicensis, Desmodus rotundus* and *Micronycteris megalotis*), the testis and accessory sex glands are active throughout the year. However, it is not known whether all individuals maintain the reproductive activity during the year (Krutzsch, 1979).

Seasonality in non-hibernating bats is observed in members of Pteropodidae (Eidolon helvum, Pteropus eotinus, P. geddei, P. giganteus, P. paliocephalus, P. scapulatus), Megadermatidae (Macroderma gigas, Megaderma lyra and Megaderma spasma), Natalidae (Natalus stramineus - Krutzsch, 1979), Molossidae (Otomops martiensseni, Tadarida (=Mops) condylurus, Tadarida (=Mops) longica, Tadarida (=Mops) nanulus, Tadarida (=Chaerephon) hindei, Tadarida (=Chaerephon) pumila and Tadarida brasiliensis - Aoki, 1997; Krutzsch, 1979) and Phyllostomidae (Anoura geoffroyi, Choeronycteris Mexicana, Leptonycteris curasoae, Macrotus californicus and Pteronotus parnellii - Krutzsch, 1979; Artibeus lituratus - Oliveira et al., 2009). In these species, testis and accessory sex glands activities are synchronous. In contrast, in the non-hibernating Emballonuridae Taphozous georgianus there is asynchrony between spermatogenesis and accessory sex gland activity (Jolly and Blackshaw, 1987; Jolly and Blackshaw, 1989).

Conclusion

The male bats present many reproductive particularities and strategies that enable them to have high reproductive success even when inhabiting regions inaccessible for other mammals. The peculiarities include testicular regression after completion of spermatogenesis, prolonged retention of viable sperm in the epididymis, asynchrony between spermatogenesis, genital glands activity and mating period. These reproductive adaptations raise several questions concerning physiology and neuroendocrine regulation of the male genital organs. On the other hand, bats present several similarities with primates, including men, suggesting that they may provide a useful model for study of some aspect of the reproductive functions. Additionally, the species that reproduces seasonally represent excellent models for the study of several aspects of the male reproduction, without interference of surgical, chemical or genetical manipulations that could interfere with other physiological events. Together, these advantages indicate that studies on the male reproductive system of bats may bring new perspectives for the reproductive biology of mammals, thus justifying an effort of investigators worldwide to go deeper on the studies of this peculiar order of mammals and insist on the possibility of maintaining breeding colonies in controlled environments that could ensure a more experimental than descriptive design of investigations.

Acknowledgments

We thank Dr. José Carlos Nogueira and Dr. Germán Arturo Bohórquez Mahecha for their critical reviewing of the manuscript; Myrian Morato Duarte for illustration of the male genital system of *A. lituratus* and André A. Oliveira, Lílian C. Praes, Rubem A. P. Dornas and Thiago S. Maia for assistance in the field captures of bats. The *Artibeus lituratus* bats were captured under permission of the Brazilian Institute of Natural Environment and Renewable Resources (IBAMA, Brazil) and studied in accordance with the Ethics Committee in Animal Experimentation of the Universidade Federal de Minas Gerais animal welfare policy. Cleida

A. Oliveira and Regiana L. Oliveira were recipient of a research and doctoral fellowship from the Conselho Nacional de Desenvolvimento Cientifico e Tecnológico - CNPq/Brazil, respectively.

REFERENCES

Anand Kumar, T. C. (1965). Reproduction in the rat-tailed bat Rhinopoma kinneari. *J. Zoology*, 147, 147–155.

Anthony, E. L. and GUSTAFSON, A. W. (1984). Seasonal variations in pituitary LH-gonadotropes of the hibernating bat Myotis lucifugus lucifugus: an immunohistochemical study. *Am J Anat,* 170, 101-115.

Aoki, A. (1997). Seasonal and experimental reactivation of Leydig cells of the bat Tadarida brasiliensis. Biocell, 21, 19-28.

Arenas-Rios, E.; Leon-Galvan, M. A.; Mercado, P. E.; Lopez-Wilchis, R.; Cervantes, D. L. and Rosado, A. (2007). Superoxide dismutase, catalase, and glutathione peroxidase in the testis of the Mexican big-eared bat (Corynorhinus mexicanus) during its annual reproductive cycle. *Comp Biochem Physiol A Mol Integr Physiol*, 148, 150-158.

Azzali. (1983). Fine struttura dei condottini efferenti e del dotto dell'epididimo di alcuni Chirotteri. *Acta Bio-Medica de Ateno Parmense*, 54, 67-83.

Azzali, G.; Arcari, M. L.; Spaggiari, B. and Romita, G. (2003). Ultrastructural aspects of the follicular cells of the pars tuberalis in bats related to the seasonal cycle. *Anat Rec A Discov Mol Cell Evol Biol,* 273, 763-771.

Beguelini, M. R.; Moreira, P. R.; Faria, K. C.; Marchesin, S. R. and Morielle-Versute, E. (2009). Morphological characterization of the testicular cells and seminiferous epithelium cycle in six species of Neotropical bats. *J Morphol*, 270, 943-953.

Bernard, R. T. (1984). The occurrence of spermiophagy under natural conditions in the cauda epididymidis of the Cape horseshoe bat (Rhinolophus capensis). *J Reprod Fertil,* 71, 539-543.

Bernard, R. T. (1986). Seasonal changes in plasma testosterone concentrations and Leydig cell and accessory gland activity in the Cape horseshoe bat (Rhinolophus capensis). *J Reprod Fertil*, 78, 413-422.

Bernard, R. T.; Bojarski, C. and Millar, R. P. (1991). Patterns of Leydig cell and LH gonadotroph activity, and plasma testosterone concentrations in the seasonally-reproducing Schreibers' long-fingered bat (Miniopterus schreibersii). *J Reprod Fertil*, 91, 479-492.

Bernard, R. T. and Hodgson, A. N. (1989). Ultrastructural changes in the seminiferous epithelium of two seasonally-reproducing bats (Mammalia: Chiroptera). *J Morphol,* 199, 249-258.

Berndtson, W. E. (1977). Methods for quantifying mammalian spermatogenesis: a review. *J. Anim. Sci,* 44, 818-833.

Brown, R. E.; Genoways, H. H. and Jones JR, J. K. (1971). Bacula of some neotropical bats. *Mammalian,* 35, 456 - 464.

BRUNET-ROSSINNI, A. K. and AUSTAD, S. N. (2004). Aging studies on bats: a review. *Biogerontology*, 5, 211-222.

Carreau, S.; Bourguiba, S.; Lambard, S.; Galeraud-Denis, I.; Genissel, C. and Levallet, J. (2002). Reproductive system: aromatase and estrogens. *Mol Cell Endocrinol,* 193, 137-143.

Cervantes, M. I.; Arenas-Rios, E.; Leon-Galvan, M. A.; Lopez-Wilchis, R.; Ambriz, D. and Rosado, A. (2008). Spermatozoa epididymal maturation in the Mexican big-eared bat (Corynorhinus mexicanus). *Syst Biol Reprod Med,* 54, 196-204.

Clulow, J.; Jones, R. C. and Hansen, L. A. (1994). Micropuncture and cannulation studies of fluid composition and transport in the ductuli efferentes testis of the rat: comparisons with the homologous metanephric proximal tubule. *Exp Physiol,* 79, 915-928.

Crichton, E. G.; Hinton, B. T.; Pallone, T. L. and Hammerstedt, R. H. (1994). Hyperosmolality and sperm storage in hibernating bats: prolongation of sperm life by dehydration. *Am J Physiol,* 267, R1363-1370.

Crichton, E. G.; Suzuki, F.; Krutzsch, P. H. and Hammerstedt, R. H. (1993). Unique features of the cauda epididymidal epithelium of hibernating bats may promote sperm longevity. *Anat Rec,* 237, 475-481.

Damassa, D. A. and Gustafson, A. W. (1984). Control of plasma sex steroid-binding protein (SBP) in the little brown bat: effects of photoperiod and orchiectomy on the induction of SBP in immature males. *Endocrinology,* 115, 2355-2361.

Damassa, D. A.; Gustafson, A. W. and King, J. C. (1982). Identification of a specific binding protein for sex steroids in the plasma of the male little brown bat, Myotis lucifugus lucifugus. *Gen Comp Endocrinol,* 47, 288-294.

Damassa, D. A.; Gustafson, A. W.; Kwiecinski, G. G. and Pratt, R. D. (1985). Control of plasma sex steroid-binding protein (SBP) in the little brown bat: effects of thyroidectomy and treatment with L- and D-thyroxine on the induction of SBP in adult males. *Biol Reprod,* 33, 1138-1146.

Encarnação, J. A.; Dietz, M. and Kierdorf, U. (2004). Reproductive condition and activity pattern of male Daubenton's bats (Myotis daubentonii) in the summer habitat. *Mammalian Biology,* 69, 163 -172.

Fleming, T. H.; Hooper, E. T. and Wilson, D. E. (1972). Three central American bat communities: structure, reproductive cycles, and movement patterns. *Ecology,* 53, 555-569.

França, L. R. and Russell, L. D. (1998). The testis of domestic mammals. In: Martínez-García, F. and Regadera, J. (eds.), *Male Reproduction* (1st edition,197-219). Madrid, Churchill Communications Europe España.

Francis, G. L.; Brown, T. J. and Bercu, B. B. (1981). Control of Sertoli cell response to FSH: Regulation by homologous hormone exposure. *Biology of Reproduction,* 24, 955-961.

Gnessi, L.; Fabbri, A. and Spera, G. (1997). Gonadal peptides as mediators of development and functional control of the testis: an integrated system with hormones and local environment. *Endocr Rev,* 18, 541-609.

Gopalakrishna, A. and Bhatia, D. (1980). Storage of spermatozoa in the epididymis of the bat, Hipposideros speoris (Schneider). *Current Science,* 49, 951-953.

Gustafson, A. W. (1979). Male reproductive patterns in hibernating bats. *J. Reprod. Fertil,* 56, 317-331.

Gustafson, A. W. (1987). Changes in Leydig cell activity during the annual testicular cycle of the bat Myotis lucifugus lucifugus: histology and lipid histochemistry. *Am J Anat,* 178, 312-325.

Gustafson, A. W. and Damassa, D. A. (1985). Annual variations in plasma sex steroid-binding protein and testosterone concentrations in the adult male little brown bat: relation to the asynchronous recrudescence of the testis and accessory reproductive organs. *Biol Reprod,* 33, 1126-1137.

Gustafson, A. W. and Shemesh, M. (1976). Changes in plasma testosterone levels during the annual reproductive cycle of the hibernating bat, Myotis lucifugus lucifugus with a survey of plasma testosterone levels in adult male vertebrates. *Biol Reprod*, 15, 9-24.

Hayashi, T.; Uchida, K. and Kawamoto, K. (2002). Basic properties and annual changes of follicle-stimulating hormone receptors in the testis of horseshoe bats, Rhinolophus ferrumequinum. *J Exp Zool,* 292, 304-313.

Hayes, F. J.; Decruz, S.; Seminara, S. B.; Boepple, P. A. and Crowley, W. F., JR. (2001). Differential regulation of gonadotropin secretion by testosterone in the human male: absence of a negative feedback effect of testosterone on follicle-stimulating hormone secretion. *J Clin Endocrinol Metab*, 86, 53-58.

Herdina, A. N.; Herzig-Straschil, B.; Hilgers, H.; Metscher, B. D. and Plenk, H., JR. (2010). Histomorphology of the penis bone (Baculum) in the gray long-eared bat Plecotus austriacus (Chiroptera, Vespertilionidae). *Anat Rec* (Hoboken), 293, 1248-1258.

Hess, R. A. (2002). The efferent ductules:structure and functions. In: Robaire, B. and Hinton, B. T. (eds.), T*he epididymis: from molecules to clinical practic*e (1st edition, 49 - 81). New York, Kluwer Academic/ Plenum Publishers.

Hess, R. A. (2003). Estrogen in the adult male reproductive tract: a review. *Reprod Biol Endocrinol*, 1, 52.

Hess, R. A. and Franca, L. R. (2007). Spermatogenesis and Cycle of the Seminiferous Epithelium. In: Cheng, C. Y. (ed.) *Molecular Mechanisms in Spermatogenesis* (1st edition,1-10). Texas, Springer Science Busines Media and Landes Bioscience.

Hoffman, L. H.; Wimsatt, W. A. and Olson, G. E. (1987). Plasma membrane structure of bat spermatozoa: observations on epididymal and uterine spermatozoa in Myotis lucifugus. *Am J Anat*, 178, 326-334.

Hosken, D. J.; Blackberry, M. A.; Stewart, T. B. and Stuck, A. F. (1998). The male reproductive cycle of three species of Australian vespertilionid bat. *J Zool Lond,* 245, 261-270.

Hosken, D. J.; Jones, K. E.; Chipperfield, K. and Dixson, A. (2001). Is the bat penis sexually selected? *Behav. Ecol Sociobiol*, 50, 450-460.

Ilio, K. Y. and Hess, R. A. (1994). Structure and function of the ductuli efferentes: a review. *Microsc Res Tech*, 29, 432-467.

Jolly, S. E. and Blackshaw, A. W. (1987). Prolonged epididymal sperm storage, and the temporal dissociation of testicular and accessory gland activity in the common sheath-tail bat, Taphozous georgianus, of tropical Australia. *J Reprod Fertil*, 81, 205-211.

Jolly, S. E. and Blackshaw, A. W. (1988). Testicular migration, spermatogenesis, temperature regulation and environment of the sheath-tail bat, Taphozous georgianus. *J Reprod Fertil,* 84, 447-455.

Jolly, S. E. and Blackshaw, A. W. (1989). Sex steroid levels and Leydig cell ultrastructure of the male common sheath-tail bat, Taphozous georgianus. *Reprod Fertil Dev*, 1, 47-53.

Karim, K. B. and Banerjee, S. (1985). Storage of spermatozoa in the epididymis of the tropical bat, Rhinopoma hardwickei hardwickei (Gray). *The Anatomical Record,* 211, 95A, abstract.

Kawamoto, K. (2003). Endocrine control of the reproductive activity in hibernating bats. *Zoolog Sci*, 20, 1057-1069.

Kawamoto, K.; Kurahashi, S. and Hayashi, T. (1998). Changes in the gonadotropin-relesing hormone (GnRH) neuronal system during the annual reproductive cycle of the horseshoe bat, Rhinolophus ferrumequinum. *Zoolog Sci*, 15, 779-786.

Kawamoto, K.; Tanaka, S. and Hayashi, T. (2000). Secretory activity of gonadotropin and the responsiveness of gonadotrophs to gonadotropin-releasing hormone during the annual reproductive cycle of male bats, Rhinolophus ferrumequinum: analysis by cell immunoblot assay. *J Exp Zool,* 287, 213-224.

King, J. C.; Anthony, E. P.; Gustafson, A. W. and Damassa, D. A. (1984). Luteinizing hormone-releasing hormone (LH-RH) cells and their projections in the forebrain of the bat Myotis lucifugus lucifugus. *Brain research,* 298, 289-301.

Krishna, A. and Singh, K. (1998). Changes in the thyroid gland during the reproductive cycle of the male vespertilionid bat, Scotophilus heathi. *Rev. Brasil. Biol.*, 58, 707-716.

Krutzsch, P. H. (1975). Reproduction of the canyon bat, Pipistrellus hesperus, in southwestern United States. *Am J Anat,* 143, 163-200.

Krutzsch, P. H. (1979). Male reproductive patterns in nonhibernating bats. *J Reprod Fertil,* 56, 333-344.

Krutzsch, P. H. and Crichton, E. G. (1986). Reproduction of the male eastern pipistrelle, Pipistrellus subflavus, in the north-eastern United States. *J Reprod Fertil*, 76, 91-104.

Krutzsch, P. H. and Crichton, E. G. (1987). Reproductive biology of the male little mastiff bat, Mormopterus planiceps (Chiroptera:Molossidae), in southeast Australia. *Am J Anat,* 178, 352-368.

Krutzsch, P. H. and Crichton, E. G. (1990). Reproductive biology of the male bent-winged bat, Miniopterus schreibersii (Vespertilionidae) in southeast South Australia. *Acta Anat* (Basel), 139, 109-125.

Krutzsch, P. H.; Crichton, E. G. and Nagle, R. B. (1982). Studies on prolonged spermatozoa survival in Chiroptera: a morphological examination of storage and clearance of intrauterine and cauda epididymal spermatozoa in the bats Myotis lucifugus and M. velifer. *Am J Anat,* 165, 421-434.

Krutzsch, P. H.; Watson, R. H. and Lox, C. D. (1976). Reproductive biology of the male leaf-nosed bat, Macrotus waterhousii in Southwestern United States. *Anat Rec,* 184, 611-635.

Kurohmaru, M.; Saruwatari, T.; Kimura, J.; Mukohyama, M.; Watanabe, G.; Taya, K. and Hayashi, Y. (2002). Seasonal changes in spermatogenesis of the Japanese lesser horseshoe bat, Rhinolophus cornutus from a morphological viewpoint. *Okajimas Folia Anat Jpn*, 79, 93-100.

Kwiecinski, G. G.; Damassa, D. A.; Gustafson, A. W. and Armao, M. E. (1987). Plasma sex steroid binding in Chiroptera. *Biol. Reprod,* 36, 628-635.

Leblond, C. P. and Clermont, Y. (1952). Definition of the stages of the cycle of the seminiferous epithelium in the rat. *Ann N Y Acad Sci,* 55, 548-573.

Matthews, L. H. (1941). Notes on the genitalia and reproduction of some african bats. *Proceedings of the Zoological Society of London,* 111B, 289-346.

Mikami, S.; Chiba, S.; Hojo, H.; Taniguchi, K.; Kubokawa, K. and Ishii, S. (1988). Immunocytochemical studies on the pituitary pars distalis of the Japanese long-fingered bat, *Miniopterus schreibersii fuliginosus. cell tissue res,* 251, 291-299.

Mokkapati, S. and Dominic, C. J. (1976). Sites of production of fructose and citric acid in the accessory reproductive glands of three species of male chiropterans. *Biol. Reprod,* 14, 627-629.

Mokkapati, S. and Dominic, S. J. (1977). Morphology of the accessory reproductive glands of some male Indian chiropterans. *Anatomischer Anzeiger*, 141, 391-397.

Morigaki, T.; Kurohmaru, M.; Kanai, Y.; Mukohyama, M.; Hondo, E.; Yamada, J.; Agungpriyono, S. and Hayashi, Y. (2001). Cycle of the seminiferous epithelium in the Java fruit bat (Pteropus vampyrus) and the Japanese lesser horseshoe bat (Rhinolophus cornutus). *J Vet Med Sci,* 63, 773-779.

Nwoha, P. U. (2000). Sex differences in the bony pelvis of the fruit-eating bat, Eidolon helvum. *Folia Morphol* (Warsz), 59, 291-295.

O'brien, G. M.; Mcfarlane, J. R. and Kearney, P. J. (2003). Pituitary content of luteinizing hormone reveals species differences in the reproductive synchrony between males and females in Australian flying-foxes (genus Pteropus). *Reprod Fertil Dev,* 15, 255-261.

Oelschlager, H. A. and Northcutt, R. G. (1992). Immunocytochemical localization of luteinizing hormone-releasing hormone (LHRH) in the nervus terminalis and brain of the big brown bat, Eptesicus fuscus. *J Comp Neurol*, 315, 344-363.

Oliveira, R. L.; Oliveira, A. G.; Mahecha, G. A.; Nogueira, J. C. and Oliveira, C. A. (2009). Distribution of estrogen receptors (ERalpha and ERbeta) and androgen receptors in the testis of big fruit-eating bat Artibeus lituratus is cell- and stage-specific and increases during gonadal regression. *Gen Comp Endocrinol,* 161, 283-292.

Oxberry, B. A. (1979). Female reproductive patterns in hibernating bats. *J Reprod Fertil,* 56, 359-367.

Racey, P. A. (1974). The reproductive cycle in male noctule bats, Nyctalus noctula. *J Reprod Fertil,* 41, 169-182.

Racey, P. A. and TAM, W. H. (1974). Reproduction in male Pipistrellus pipistrellus (Mammalia: Chiroptera). *J Zool,* 172, 101-121.

Reis, N. R.; Shibatta, O. A.; Peracchi, A. L.; Pedro, W. A. and Lima, I. P. (2007). *Sobre os Morcegos do Brasil.* In: Reis, N. R.;Peracchi, A. L.;Pedro, W. A. and Lima, I. P. (eds.), Morcegos do Brasil (1st edition, 17-25).Londrina, Universidade Estadual de Londrina.

Richardson, B. A. (1979). The anterior pituitary and reproduction in bats. *J Reprod Fertil,* 56, 379-389.

Richardson, E. G. (1977). The biology and evolution of the reproductive cycle of Miniopterus schreibersii and M. australis Chiroptera:Vespertilionidae). *Journal of Zoology*, London, 183, 353–375.

Russell, L. D.; Ettlin, R. A.; Sinha Hikim, A. P. and Clegg, E. D. (1990). Mammalian spermatogenesis. In: Bolesta (ed.) *Histological and histopatological evaluation of the testis* (1st edition, 1-40). Clearwater, Cache River Press.

Ryan, J. M. (1991a). Comparative Morphology of the Glans Penis in Molossus, Promops, and Eumops (Chiroptera: Molossidae). *Bulletin American Museum of Natural History,* 122-135.

Ryan, J. M. (1991b). Morphology of the glans penis in four genera of molossid bats (Chiroptera: Molossidae. *Journal of Mammalogy,* 72, 658-668.

Saidapur, S. K. and Patil, S. B. (1992). Kinetics of spermatogenesis in megachiropteran bat, Rousettus leschenaulti (Desmarset): seminiferous epithelial cycle, frequency of stages, spermatogonial renewal and germ cell degeneration. *Indian J Exp Biol*, 30, 1037-1044.

Setchell, B. P.; Sanchez-Partida, L. G. and Chairussyuhur, A. (1993). Epididymal constituents and related substances in the storage of spermatozoa: a review. *Reprod Fertil Dev,* 5, 601-612.

Simpson, E. R.; Michael, M. D.; Agarwal, V. R.; Hinshelwood, M. M.; Bulun, S. E. and Zhao, Y. (1997). Cytochromes P450 11: expression of the CYP19 (aromatase) gene: an unusual case of alternative promoter usage. *Faseb J,* 11, 29-36.

Singh, K. and Krishna, A. (1996). Seasonal changes in circulating serum concentration and in vitro testicular secretion of testosterone and androstenedione in the male vespertilionid bat (Scotophilus heathi). *J Exp Zool,* 276, 43-52.

Singh, U. P. and Krishna, A. (2000). Androstenedione concentration and their correlation with the anomalous reproductive pattern in the male Indian sheath-tailed bat, Taphozous longimanus. *Journal of experimental zoology,* 287, 54-61.

Singwi, M. S. and Lall, S. B. (1983). Spermatogenesis in the non-scrotal bat-Rhinopoma kinneari Wroughton (microchiroptera: mammalia). *Acta Anat* (Basel), 116, 136-145.

Vamburkar, S. A. (1958). The male genital tract of the indian Megachiropteran bat Cynopterus sphinx gangeticus and. *Proc. zool. Soc. Lond,* 130, 57-77.

Van Der Merwe, M. and Rautenbach, I. L. (1990). Reproduction in the rusty bat, Pipistrellus rusticus, in the northern Transvaal bushveld, South Africa. *Journal of Reproduction and Fertility,* 89, 537-542.

Yeung, C. H.; Cooper, T. G.; Bergmann, M. and Schulze, H. (1991). Organization of tubules in the human caput epididymidis and the ultrastructure of their epithelia. *Am J Anat,* 191, 261-279.

In: Bats: Biology, Behavior and Conservation
Editors: J. L. Zupan and S. L. Mlakar, pp. 177-188
ISBN 978-1-61209-749-7

Chapter 4

FORAGING ECOLOGY OF PTEROPODID BATS: POLLINATION AND SEED DISPERSAL

Hanumanthan Raghuram*[*1], Natarajan Singaravelan[2], Parthasarathy Thiruchenthil Nathan[3], Koilmani Emmanuvel Rajan[4,5] and Ganapathy Marimuthu[6]

[1]Department of Zoology, The American College, Madurai, India
[2]Past Affiliations: Department of Biology, University of Haifa-Oranim, Tivon, Israel
Institute of Evolution, University of Haifa, Mount Carmel, Haifa, Israel
[3]Mayiladuthurai (Taluk), Nagappattinam District, Tamil Nadu, India
[4]Department of Zoology, DDE, Alagappa University, Karaikudi, India
[5]Past Affiliation: Department of Animal Science, School of Life Sciences,
Bharathidasan University, Tiruchirappalli, India
[6]Department of Animal Behaviour and Physiology,
School of Biological Sciences, Madurai Kamaraj University, Madurai, India

ABSTRACT

Pteropodid bats are important aerial pollinators and seed dispersers in the Old World tropics and thus play a key role in forest dynamics and regeneration. In India, among the 13 species of pteropodid bats, the foraging ecology of four species *Cynopterus sphinx, C. brachyotis, Rousettus leschenaulti* and *Pteropus giganteus* were extensively studied. These species occur sympatrically and forage on resources like leaves, flowers and fruits, and inadvertently effect pollination and seed dispersal of the resource trees. These bats either consume fruits and flowers *in situ* or carry them to nearby roosts for feeding *ex situ* depend on the carrying capacity of bats and size and the nutritional quality of consuming resource. We suggest that small pteropodid bats aid short distance pollination and the dispersal of seeds of small sized fruits, whereas large pteropodid bats aid long distance pollination and the dispersal of seeds of large sized fruits with some exceptions that depended on the mass of the fruit. At least 300 plant species of nearly 200 genera rely mainly on these fruit bats for their propagation. Some of the plants are economically

* Corresponding Author: hraghuram@gmail.com (H. Raghuram).

important and they produce approximately 500 economically valuable products including fruits, dyes, tannins, timber, medicines, fiber and fire wood. In this review, we discuss the foraging ecology of pterpodid fruit bats in the Old world tropics particularly with reference to chiropterophilly and zoochory by fruit bats.

INTRODUCTION

Bats are the only mammals capable of powered flight and are expected to originate about 50 million years ago. They come under the Order Chiroptera, which is further divided into Megachiroptera with the only family Pteropodidae which is exclusively plant dependant and Microchiroptera with 17 families in which the only family Phyllostomidae visit plants for floral resources (Simmons, 2005). The Pteropodidae of Old World is an ancient and diverse group of bats with over 170 species distributed throughout the tropical regions of Africa, Asia and Indo-Australia (Hill and Smith, 1984). They adapted to live in humid, forested areas of the warm zones of the Old World, and not migrated to the New World, although certain areas of the tropical zones of Central and South America would provide good habitats in which the phyllostomid bats inhabit.

Pteropodid bats are quite larger in size with the forearm length ranges between 47 to 200 mm (Bates and Harrison, 1997). Their diet is primarily composed of fruits, leaves, flowers and flower products such as nectar (Marshall, 1983, 1985 and Vaughan et al., 2010). All the plant-visiting bats have large eyes compared to the insect eating counter parts. Since they do not have the ability to echolocate, like phyllostomid bats they don't have nose leaf and the ear is also simple without the tragus. Although species belonging to the genus *Rousettus* are capable of producing high frequency sounds on clicking its tongue and using them to find their way inside dark caves (Altringham, 1996). The ears are simple in form, the sides of each pinnae forming a complete ring at the base, there is no tragus (Bates and Harrison, 1997).

Table 1. Species of fruit bats recorded from India and their IUCN conservation status as determined by CAMP 2002

Bat species	Common name	IUCN status
Macroglossus sobrinus	Greater long-tongued fruit bat	NT
Sphaerias blanfordi	Blanford's fruit bat	NT
Megaerops niphanae	Ratanaworabhan's fruit bat	NT
Cynopterus brachyotis	Lesser short-nosed fruit bat	LC
Latidens salimalii	Salim Ali's fruit bat	EN
Cynopterus sphinx	Greater short-nosed fruit bat	LC
Eonycteris spelaea	Long-tongued dawn fruit bat	LC
Rousettus leschenaulti	Dog-faced fruit bat	LC
Pteropus faunulus	Nicobar flying fox	EN
Pteropus hypomelanus	Island flying fox	EN
Pteropus melanotus	Island flying fox	VU
Pteropus giganteus	Indian flying fox	LC
Pteropus vampyrus	Large or Malayan flying fox	EN

NT – near threatened, LC – least concerned, EN – endangered, VU – vulnerable

In India, among the 13 species of pteropodid bats (Table 1), the foraging ecology of four species *Cynopterus sphinx, C. brachyotis, Rousettus leschenaulti* and *Pteropus giganteus* were extensively studied. In this chapter we discuss certain general and ecological aspects of their diet, foraging, pollination and seed dispersal of pteropodid bats of India.

Table 2. Food plants of C. sphinx and R. leschenaulti (after Rajan et al., 1999; Tang et al., 2005; Chen et al., 2007). '+' indicate feeding on the food type and '–' for not feeding.

Family	Species	Food type	*Cynopterus sphinx*	*Rousettus leschenaultia*
Annonaceae	*Polyalthia longifolia*	Fruit	+	–
	Annona squamosa	Fruit	+	+
Moraceae	*Ficus bengalensis*	Fruit	+	+
	Ficus religiosa	Fruit	+	+
	Morus alba	Fruit	+	
Mimosaceae	*Enterolobium saman*	Fruit	+	
	Pithecellobium dulce	Fruit	+	
	Acacia nilotica	Fruit	+	
Sapotaceae	*Achras sapota*	Fruit	+	+
	Bassia latifolia	Fruit and Flowers	+	+
	Mimusops elengi	Fruit	+	–
Myrtaceae	*Psidium guajava*	Fruit	+	+
	Eugenia jambolana	Fruit	+	+
Combretaceae	*Terminalia catappa*	Fruit	+	+
Caesalpiniaceae	*Cassia fistula*	Leaves	+	
Anacardiaceae	*Mangifera indica*	Fruit and Flowers	+	+
Rutaceae	*Murray koenigii*	Fruit	+	
Clusiaceae	*Calophyllum inophyllum*	Fruit	+	
Punicaceae	*Punica granatum*	Fruit	+	
Cucurbitaceae	*Cephalandra indica*	Fruit	+	
Meliaceae	*Azadirachta indica*	Fruit	+	
Musaceae	*Musa acuminata*	Fruit	+	+
Solanaceae	*Solanum torvum*	Fruit	+	
Rhamnaceae	*Zizphus jujuba*	Fruit	+	
Spindaceae	*Nephelium lappaceum*	Fruit	+	+
	Dimocarpus longan	Fruit	+	+
	Sapindus rarak	Leaves	+	–
Euphorbiaceae	*Sapium baccatum*	Fruit	+	–
Rosaceae	*Pygeum latifolium*	Fruit	+	–
Rubiaceae	*Anthrocephalus chinensis*	Fruit	+	+
Rhizophoraceae	*Pellacalyx yunnanensis*	Fruit	+	–
Ebenaceae	*Diospyros nigrocart*	Fruit	+	+
Lauraceae	*Beilschmiedia purpurascens*	Leaves	+	+

Diet of Pteropodid Bats

Because of abundant harvest of fruit trees in the tropical and sub-tropical areas, the diet of the fruit bats is well provided for during all seasons of the year. The Old World fruit bats are obligate frugivores and nectivores. However, nitrogen rich diets such as insects and liquid fraction of leaves have also been reported in the diets of some obligate fruigivores (Courts, 1998). The short nosed fruit bat *Cynopterus sphinx* feed on about 25 plant species (Table 2). They mostly prefer the fruit of *Terminalia catappa*, which is available for a longer period about 7 months from December to June (Rajan et al., 1999). They also feed regularly on the leaves of *Cassia fistula* (Balasubramanian, 1988; Bhat, 1994; Rajan et al., 1999). Thus folivory may be energetically more advantageous for bats than the ingestion of large amounts of low protein fruits (Kunz and Ingalls, 1994; Elangovan and Marimuthu, 2001). Apart from fruits and leaves, C. *sphinx* is also identified as causing destructive foraging on flowers of *Parkia biglandulosa*, *Bassia latifolia* and *Mangifera indica* (Bhat, 1994). But recent studies on *Madhuca latifolia* indicate that they are not destructive forager rather than as effective pollinators (Nathan et al., 2009). The diet pattern of second predominant fruit bat species of India, *Rousettus leschenaulti* apparently overlaps with *C. sphinx* (Tang et al., 2005; Chen et al., 2007). Although both *C. sphinx* and *C. brachyotis* occur sympatrically in early successional forest mixed with old-growth forest, they can partition both food and foraging habitat (Bumrungsri et al., 2007). Being a forest dwelling species, the diet of *C. brachyotis* is vast and diverse than compared to *C. sphinx*. They feed on fruits of 54 plant species and leaves of 14 species and flower parts of four species (see Tan et al., 1998). The diet of Indian flying fox *Pteropus giganteus* includes both wild fruits like *Ficus sp., Madhuca, Polyalthia longifolia*, *Terminalia catappa* (Ezhilvendan, 2003), and commercial fruits like *Vitis vinifera, Achras sapota, Psidium guajava, Mangifera indica* and *Carica papaya* (Singaravelan, 2002).

Foraging and Feeding Behavior of Pteropodid Bats

Bat foraging behaviour varies depending upon the distribution and densities of resources. Normally group foraging occurs when the resources are rich but ephemeral such as "big bang" production of fruits and flowers (Gentry, 1974, Elangovan et al., 1999). A predator rich environment will reinforce the advantage of group foraging when resources are patchily distributed. But when the resources are predictable in space and time or in "steady state" flowering and fruiting pattern, solitary foraging occurs. Most of the phytophagous bats live gregariously in the day roost but each of them chooses to forage in different foraging areas. So visiting different foraging areas helps the bats to avoid competition and also roosting place may serve as "information centers" about the resources during lean period.

Normally larger bats like *Pteropus giganteus* spent longer time or *in situ* foraging (Figure 1) in the resource trees with fruits and flowers to maximize their resource utilization (Singaravelan, 2002) . Whereas small sized bat such as *C. sphinx*, usually pluck the fruit of *Muntingia calabura* (Singapore cherry) (Singaravelan and Marimuthu, 2006) and *Azadirachta indica* (Neam) (Bhat, 1994; Rajan et al., 1999) from the resource tree and taken it to the feeding roost for consumption. This bat mainly uses *ex situ* foraging strategy to avoid predators (Singaravelan, 2002).

Figure 1. In situ foraging - Indian flying fox lapping nectar from the inflorescence of *Ceiba pentandra* (Nathan et al., 2005).

The morphological, physiological and behavioural adaptations of the bats facilitate economical feeding by optimizing the time and energy from plant resources (Elangovan et al., 1999; Elangovan et al., 2001). The exclusive fruit diet of the fruit bats has resulted in particular adaptations of the facial structure, usually the fruit eating bats have comparatively short muzzle and the teeth and the palates are modified for chewing fruits and sucking juices. The teeth are well developed and relatively long, together with the small incisors to cut or open the skin of fruits (Dumont, 2003; Dumont and O'Neal, 2004). The flattened back molar teeth then crush the fruit pulp. When eating, the fruit bat often uses one foot to hold the fruit and bring it to its mouth (Singaravelan and Marimuthu, 2008). In most species, the mouth cannot be open wide. Thus, the back of the mouth cavity remains enclosed. This kind of cheek formation prevents the soft or juicy fruit pulps from slipping out of the mouth during chewing. There are many papillae on the upper surface of the long tongue, including the soft taste papillae and the horny papillae which are situated in the middle of the tongue, and which face the rear of the mouth cavity (Wimsatt, 1977). These papillae apparently help in the breaking down the hard fruit cover. Bat's hair has also modified to serve as pollen collectors. The hairs of flower-visiting bats have projecting scales, whereas insectivorous bats have smooth hair. The projecting scales increase the pollen-carrying capacity of the bat's pelage during flower visitation (Figure 2). Bats groom pollen from their bodies now and then and ingest it to gain protein and amino acids.

Many species do not eat the whole fruit. Instead, they crush it in order to release the fruit juice, which they drink. Gut analysis revealed that in the stomach of Indian flying fox *Pteropus giganteus*, filled with white, milky mucus or a sticky fluid. This proves that fruit bats do not eat the fibrous fruit pulp or kernels; rather they spit these out (Elangovan and Marimuthu, 2001). This may be the reason why natives in many areas believe that fruit bats do not have anal openings, but eat and defecate through their mouth. One can find the scattering of chewed leaves along with chewed fruits under fruit trees visited by fruit bats during the night (Rajan et al., 1999). It is obvious that fruits are commonly considered to be nutritionally poor and therefore, large fruit bats cannot achieve their energy requirement without over-ingestion. As a result, they have large, expandable stomach which increases on one side into a sac-like form in contrast to the long intestinal tract of insectivorous bats. The droppings of fruit bats are relatively liquid (Elangovan and Marimuthu, 2001; Raghuram, 2001).

Figure 2. The ventral body surface of the hovering Indian short-nosed fruit bat *Cynopterus sphinx* is covered with pollen (Nathan et al., 2005).

Except the bats like *Eonycteris spelaea* and – none of the pteropodid bats are specialized nectarivorous bats like phyllostomids. The phyllostomid flower visiting bats tend to have better hovering ability, which is facilitated by wing tips that are long relative to the rest of the wing, and are able to drink nectar from flowers without landing. This allows less robust plant species to use bats as pollen vectors. Although hovering is an expensive mode of flight, it allows bats, like humming birds, to visit larger number of flowers from time unit and therefore improves total foraging efficiency. However, the small sized pteropodid bat *Cynopterus sphinx* lands on the flower and laps the nectar or plucks the fruits and carries it to the feeding roost (Nathan et al., 2005). The ability of hovering is absent in most species of bats in Paleotropics, here they have short broad wings that aid in carrying loads and bat pollinated plants are adapted for landing pteropodid bats by having sturdy flowers which withstand the weight of the bats. The larger pteropodid bat *Pteropus giganteus* lands on the tree and simply crawls along the branches using forearm and legs to reach flowers and pull them close to its mouth for lapping nectar (Singaravelan and Marimuthu, 2004). In order to reach deep inside the corolla and suck the nectar, the nectarivorous bats have usually long tongue (Muchala, 2006). Pteropodid bats tongue length is shorter when compare to the nectarivorous phyllostomid bats. In nectarivorous bats the tip of the tongue has brush like filiform papillae that are used to collect nectar out of the flowers by capillary action. Fewer papillae are on the tongues of species that usually eat fruits but visit flowers seasonally; filiform papillae are absent on the tongues of insectivorous bats. Bat's hair has also modified to serve as pollen collectors. The hairs of flower-visiting bats have projecting scales, whereas insectivorous bats have smooth hair (Howell and Hodgkin, 1976). The projecting scales increase the pollen-carrying capacity of the bat's pelage during flower visitation. Bats groom pollen from their bodies now and then and ingest it to gain protein and amino acids.

POLLINATION BY PTEROPODID BATS

Flower-visiting bats provide two important benefits to plants: they deposit large amounts of pollen and a variety of pollen genotypes on plant stigmas and, compared with many other pollinators, they are long-distance pollen dispersers (Fleming et al., 2010). In India, pollination of economically important trees like *Ceiba pentandra, Ficus* sp. (fig), *Mangifera indica* (mango), *Musa* sp. (banana), *Careya arborea* by pteropodids like *C. sphinx, C. brachyotis, R. leschenaulti* and *P. giganteus* is well documented (Table 3). Studies on pollination biology of *Ceiba pentandra, Madhuca latifolia* and *Cullenia exarillata* in southern India have shown that apart from insects, fruit bats are major pollinators of these trees (Devy and Davidar, 2003; Nathan et al., 2005, Nathan et al., 2009). Floral characteristics of these plants like nocturnal blooming, strong odour, sturdy flowers with wide mouth, and copious amount of nectar probably evolved to aid nocturnal pollination especially by bats (Marshall, 1983). Pteropodid bats have keen senses of smell and appear to use olfaction or long-distance detection of fruits and flowers (Raghuram et al., 2009). Captive studies suggest that *C. sphinx* appeared to be more attracted to aromas containing ethyl acetate. In contrast, phyllostomids are attracted more to aromas of dimethyl sulphides (von Helversen, 1993; Elangovan et al., 2006).

Table 3. List of plant species pollinated by pteropodid bats

Plants	References
Ceiba pentandra	Singaravelan and Marimuthu, 2004, Nathan et al. 2005
Oroxylum indicum	Subramanya and Radhamani, 1993
Bauhinia sp.	Marshall, 1985; Corlett, 2004
Musa sp.	Corlett, 2004
Eugenia jambolana	Marshall, 1985
Madhuca indica	Subramanya and Radhamani, 1993; Nathan et al. 2009
Durio zibethinus	Gould, 1978
Cullenia exarillata	Devy and Davidar, 2003
Kigelia pinnata	Subramanya and Radhamani, 1993
Adansonia digitata	Subramanya and Radhamani, 1993

Fruiting and flowering pattern of the plants also very much influence the visitation rate of the bats (Gentry, 1974). For example "big bang" plants like *Bassia latifolia* and *Ceiba pentandra* produce more number of flowers for short duration and attract more number of bats (Nathan et al., 2005, Nathan et al., 2009) The "big bang" plants sacrifice the quality of animal service to maximize the quantity of visitations. Whereas "steady state" plants like *Cresentia cujete* and *Musa paradisiaca* (Figure 3) attract few numbers of bats over the long period of time. The steady state patterns promote trap-line foraging by bats and maximize the quality of the services their visitors provide while sacrificing the quantity of propagules dispersed.

Figure 3. Steady state flowering tree Musa paradiasica is visited by *Cynopterus sphinx*.

In addition to the seasonal patterns of flower availability, pteropodid bats face nightly interspecific variation in the timing and amounts of nectar production. Some flowers (e.g. *Durio zibethinus*, *Musa paradisiaca*) begin nectar production at or before sunset, whereas others (e.g. *Parkia biglandulosa*, *Kigelia pinnata*) do not begin secreting until well after sunset. The temporal secretion differences between sympatric species have evolved to reduce interspecific competition for the same pollinating agent. Among pollinating bats to avoid interspecific competition there is spatial as well as temporal variation in the foraging visitation. For example in *C. pentandra*, the small sized fruit bat *C. sphinx* comes early in the foraging ground and feeds on the nectar from flowers of the understorey inflorescences whereas the larger *P. giganteus* comes late in the flowering tree and feeds on the nectar from the flowers at the upper canopy (Singaravelan and Marimuthu, 2004; Nathan et al., 2005).

SEED DISPERSAL BY PTEROPODID BATS

The list of Indian plant species dispersed by *C. sphinx*, *C. brachyotis* and *Pteropus giganteus* is given in Table 4. On an average *C. sphinx* can expell nine seed pellets per day and disperse at a range of 200 m to 32 km (Tang et al., 2007; Shilton et al., 1999). Generally, the frugivorous bats have comparatively short and less complicated intestine than their insect eating microbats. This facilitates the bats to consume more fruits than their body mass to gain energy and defecating within 30 minutes. Since they void the food so fast, by looking at the colour and smell of the bat feces we can identify the type of fruits they are consuming. Due to this reason the egested seeds and even the pollen grains remain viable even after defecation. Even though defecation is fast, some viable seeds can be retained in the gut for more than 18 hours and dispersed to a longer distance (Shilton et al., 1999).

It is reported that bats are effective dispersers of seed than birds in tropical ecosystem and in forests near to the human habitats. Because most of the soft-coated seeds consumed by birds get grained inside the gizzard and they also perch and defecate on monuments and buildings and the emerging plants on them are nuisance to humans. But such a problem does not arise in bat dispersed seeds and even the bat defecated pollen grains are viable.

Table 4. List of seeds of plant species dispersed by pteropodid bats

Plants	References
Ficus bengalensis	Ezhilvendan 2003
Ficus religiosa	Ezhilvendan 2003
Ficus benjamina	Ezhilvendan 2003
Muntingia calabura	Boon and Corlett 1989; Singaravelan and Marimuthu 2006
Madhuca indica	Ezhilvendan 2003
Musa accuminata	Tang et al. 2007
Polyalthia longifolia	Ezhilvendan 2003
Terminalia catappa	Boon and Corlett 1989; Ezhilvendan 2003
Callophyllum inophyllum	Boon and Corlett 1989; Ezhilvendan 2003
Achras sapota	Singaravelan 2002
Psidium guajava	Boon and Corlett 1989; Singaravelan 2002

CONCLUSION

Pteropodid bats play an important role in seed dispersal and pollination in tropical ecosystem, which harbours as much as 70 percent of all known species of organisms in the land. They are the primary pollinators of numerous tropical plants as on now more than 200 genera of trees and shrubs are already known and further more plants are awaiting discovery. Most of the trees pollinated by bats in tropics are commercially important (Fujita and Tuttle 1991). They provide valuable timber for human use, other than that fruits like avocado, durian, wild banana, peaches are bat pollinated. The kapok tree, whose seeds are used for making oil and oil cakes and fibers surrounding the seeds are used for bedding, surgical bandages and the seeds of mahua tree are used for making mahua oil and oil cakes are solely depended on bats for pollination.

In highly fragmented tropical habitats, fruit bats play an important role in maintaining the genetic continuity of plant populations and thus have considerable conservation value (Fleming et al., 2010). The two most revered and worshiped trees by South Indians with lot of medicinal values, *Ficus religiosa* and *Azadirachta indica* depend on bats for the seed dispersal. In orchards bats are generally considered as vermin but closer observation showed that bats mostly consume ripe fruits, which are not suitable for marketing, and the loss due to bat foraging is minimal when compared to other factors (Singaravelan, 2002). In case of *Ceiba pentandra* (Singaravelan and Marimuthu, 2004; Nathan et al. 2005) and *Madhuca latifolia* (Nathan et al. 2009) bats are the only pollinator of these plants.

On conservation point of view, except *Latidens salimalii*, all other fruits bats are generally considered as vermin – as pest in India and categorized in Schedule V. So it is necessary and mandatory to instruct the federal government of India to remove the tag 'vermin' from fruit bats and has to be given priority for conserving them as like other wild animals (Singaravelan et al. 2009).

REFERENCES

Altringham, J.D. 1996. *Bats: Biology and Behaviour*. Oxford University Press, New York.

Balasubramanian, P. 1988. Short-nosed fruit bat *Cynopterus sphinx* (Vahl) feeding on leaves of *Cassia fistula* at Point Calimere Wild Life Sanctuary. *Journal of Bombay Natural History Society*, 85: 183.

Bates, P.J.J. and Harrison, D.L. 1997. *Bats of the Indian Subcontinent*. Harrison Zoological Museum publication, London.

Bhat, H.R.1994. Observations on the food and feeding behaviour of *Cynopterus sphinx* Vahl (Chiroptera, Pteropodidae) at Pune, India. *Mammalia,* 58: 363–370.

Boon, P.P and Corlett, R.T. 1989. Seed dispersal by the lesser short-nosed fruit bat (*Cynopterus brachyotis*, Pteropodidae, Megachiroptera). *Malayan Nature Journal,* 42: 251-256.

Bumrungsri, S., Leelapaibul, W and Racey, P.A. 2007. Resource partitioning in sympatric *Cynopterus* bats in lowland tropical rain forest, Thailand. *Biotropica* 39(2): 241–248.

Chen, Z., Meng, Y., Zhou, F and Li, Y. 2007. Activity rhythms and food habits of Leschenault's rousette (*Rousettus leschenaulti*) in Hainan Island. *Acta Theriologica Sinica*, 27(2): 112–119.

Corlett, R.T. 2004. Flower visitors and pollination in the Oriental (Indomalayan) region. *Biological Reviews* 79: 497-532.

Courts, S.E. 1998. Dietary strategies of Old World fruit bats (Megachiroptera, Pteropodidae): how do they obtain sufficient protein?. *Mammal Review*, 28: 185–194.

Devy, M.S. and Davidar, P. 2003. Pollination systems of trees in Kakachi, a mid-elevation wet evergreen forest in Western Ghats, India. *American Journal of Botany,* 90(4):650–657.

Dumont, E.R. and O'Neal, R. 2004. Food hardness and feeding behavior in Old World fruit bats (Pteropodidae). *Journal of Mammalogy*, 85(1):8–14.

Dumont, E.R.2003. Bats and fruit: An ecomorphological approach. In: Kunz, T.H and Fenton, B. (eds). *Bat Ecology*, University of Chicago Press, Chicago pp 398–429.

Elangovan, V and Marimuthu, G. 2001. Comparative analysis on food, energy and nitrogen intake of a megachiropteran bat, *Cynopterus sphinx*. *Proceedings of National Academy of Science, India*, 71B (II):115–120.

Elangovan, V., Marimuthu, G and Kunz, T.H. 1999. Temporal patterns of individual and group foraging behaviour in the short-nosed fruit bat, *Cynopterus sphinx*, in South India. *Journal of Tropical Ecology*, 15: 681–687.

Elangovan, V., Marimuthu, G and Kunz, T.H. 2001. Temporal patterns of resource use by the short-nosed fruit bat, *Cynopterus sphinx* (Megachiroptera: Pteropodidae). *Journal of Mammalogy*, 82: 161–165.

Elangovan, V., Priya, E.Y.S and Marimuthu G. 2006. Olfactory discrimination ability in the short-nosed fruit bat *Cynopterus sphinx*. *Acta Chiropterologica* 8: 247–253.

Ezhilvendan, S. 2003. Diet and foraging behavior of Indian flying fox, *Pteropus giganteus* in Madurai district. M.Sc Dissertation, Madurai Kamaraj University, Madurai, India.

Fleming, T.H., Geiselman, C and Kress, W.J. 2009. The evolution of bat pollination: a phylogenetic perspective. *Annals of Botany* 104 (6): 1017-1043.

Fujita, M.S. and Tuttle, M.D. 1991. Flying foxes (Chiroptera: Pteropodidae): threatened animals of key ecological importance. *Conservation Biology*, 5: 455–463.

Gentry, A.H. 1974. Flowering phenology and diversity in tropical Bignoniaceae. *Biotropica*, 6: 64–68.

Gould, E. 1978. Foraging behavior of Malaysian nectar-feeding bats. *Biotropica*, 10: 184–193.

von Helversen, O. 1993. Adaptations of flowers to the pollination by glossophagine bats. In: Barthlott, W., Naumann, C.M., Schmidt-Loske, K and Schuchmann, K.L. eds. *Animal–plant interactions in tropical environments*. Bonn: Zoologisches Forschungsinstitut und Museum Alexander Koenig, 41–59.

Howell, D.J and Hodgkin, N. 1976. Feeding adaptations in the hairs and tongues of nectar-feeding bats. *Journal of Morphology*, 148(3): 329–336.

Kunz, T. H. and Ingalls, K.A. 1994. Folivory in bats: an adaptation derived from frugivory. *Functional Ecology*, 8: 665-668.

Marshall, A.G. 1983. Bats, flowers and fruit: evolutionary relationships in the Old World. *Biological Journal of Linnean Society* 20: 115–135.

Marshall, A.G. 1985. Old World phytophagous bats (Megachiroptera) and their food plants: a survey. *Zoological Journal of Linnean Society, London* 83: 351–369.

Molur, S., Marimuthu, G., Srinivasalu, C., Mistry, S., Hutson, A.M., Bates, P.J.J., Walker, S., Padma Priya, K. and Binu Priya, A.R. (eds) 2002. Status of South Asian Chiroptera: Conservation Assessment and Management Plan (C.A.M.P) Workshop Report, 2002. Zoo Outreach Organization, CBSG South Asia and WILD, Coimbatore, India.

Muchala, N. 2006. Nectar bats stows huge tongue in its ribcage, *Nature*, 444: 701–702.

Nathan, P.T., Karuppudurai, T., Raghuram, H and Marimuthu, G. 2009. Bat foraging strategies and pollination of *Madhuca latifolia* (Sapotaceae) in southern India. *Acta Chiropterologica*, 11(2): 435–441.

Nathan, P.T., Raghuram, H., Elangovan, V., Karuppudurai, T and Marimuthu, G. 2005. Bat pollination of kapok tree, *Ceiba pentandra*. *Current Science*, 88: 1679–1681.

Raghuram, H. 2001. Food choice and foraging behavior of the Indian fulvous fruit bat *Rousettus leschenaulti*. M.Sc Dissertation, Madurai Kamaraj University, Madurai, India.

Raghuram, H., Thangadurai, C., Gopukumar, N., Nathar, K and Sripathi, K. 2009. The role of olfaction and vision in the foraging behaviour of an echolocating megachiropteran bat, *Rousettus leschenaulti* (Pteropodidae). *Mammalian Biology*, 74: 9–14.

Rajan, K.E., Nair, N.G and Subbaraj, R. 1999. Seasonal food preference of the Indian short nosed fruit bat *Cynopterus sphinx* (Vahl) (Chiroptera: Pteropodidae). *Journal of Bombay Natural History Society*, 96(1): 24–27.

Shilton, L.A., Altringham, J.D., Compton, S.G and Whittaker, R. J. 1999. Old World fruit bats can be long distance seed dispersers through extended retention of viable seeds in the gut, *Proceedings of the Royal Society London B,* 266: 219-223.

Simmons, N.B. 2005. Order Chiroptera. In: Wilson DE, Reeder DM. eds. *Mammal species of the world. A taxonomic and geographic reference*, 3rd edition. Baltimore: Johns Hopkins University Press, 312–529.

Singaravelan, N. 2002. Foraging behavior of fruit bats in orchards. PhD thesis, Madurai Kamaraj University, Madurai, India.

Singaravelan, N and Marimuthu, G. 2004. Nectar feeding and pollen carrying from *Ceiba pentandra* by pteropodid bats. *Journal of Mammalogy*, 85: 1–7.

Singaravelan, N and Marimuthu, G. 2006. *Muntingia calabura* — an attractive food plant of *Cynopterus sphinx* —deserves planting to lessen orchard damage. *Acta Chiropterologica*, 8: 239–245.

Singaravelan, N and Marimuthu, G. 2008. In situ feeding tactics of short-nosed fruit bats (*Cynopterus sphinx*) on mango fruits: evidence of extractive foraging in a flying mammal. *Journal of Ethology*, 26: 1–7.

Singaravelan, N., Marimuthu, G and Racey, P.A. 2009. Do fruit bats deserve to be listed as vermin in the Indian Wildlife (Protection) and Amended Acts? - A critical review. *Oryx*, 43(4): 608–613.

Subramanya, S. and Radhamani, T.R. 1993. Pollination by birds and bats. *Current Science*, 65: 201–209.

Tan, K.H., Zubaid, A and Kunz, T.H. 1998. Food habits of *Cynopterus brachyotis* (Muller) (Chiroptera: Pteropodidae) in Peninsular Malaysia. *Journal of Tropical Ecology*, 14: 299–307.

Tang, Z., Sheng, L., Cao, M., Liang, B and Zhang, S. 2005. Diet of *Cynopterus sphinx* and *Rousettus leschenaulti* in Xishuangbanna. *Acta Theriologica Sinica*, 25 (4): 367–372.

Tang, Z., Sheng, L., Ma, X., Cao, M., Parsons, S., Ma, J and Zhang, S. 2007. Temporal and spatial patterns of seed dispersal of *Musa acuminata* by *Cynopterus sphinx*. *Acta Chiropterologica,* 9(1): 229–235.

Vaughan, T.A., Ryan, J.M and Czaplewski, N.J. 2010. *Chiroptera*. In: Mammalogy, 5th Edition, Thomson brooks/Cole publishers, pp 255–292.

Wimsatt, W.A ed. (1977) *Biology of bats*, Vol. 3. Academic Press, New York.

In: Bats: Biology, Behavior and Conservation
Editors: J. L. Zupan and S. L. Mlakar, pp. 189-220
ISBN 978-1-61209-749-7

Chapter 5

BATS AND RABIES IN BRAZIL

Karin Corrêa Scheffer**, Rodrigo Fernandes de Barros, Samira M. Achkar, Andréa de Cássia Rodrigues da Silva, Luciana Botelho Chaves, Graciane Maria Medeiros Caporale, Rafael de Novaes Oliveira, Ana Elena Boamorte da Costa, Keila Iamamoto, Willian de Oliveira Fahl and Pedro Carnieli Junior

Pasteur Institute of São Paulo - Brazil

The bats found in Brazil are included in nine families, 64 genera and 168 species (Reis et al., 2007). The most numerous family is *Phyllostomidae*, with 90 species registered, followed by *Molossidae* with 26, *Vespertilionidae* with 24, *Emballonuridae* with 15 species and the families *Noctilionidae*, *Mormoopidae*, *Natalidae*, *Furipteridae* and *Thyropteridae*, totaling another 13 species (Peracchi et al., 2006). With respect to feeding habit, the bats can show very varied feeding habits, and this diversity in diets is not found in any other group of mammals. Among the foods consumed are insects and other arthropods (such as scorpions), small fish, amphibians (frogs and tree frogs), birds, fruits, seeds, leaves, flowers, pollen, nectar, small vertebrates and blood (Gardner, 1977; Hill, Smith, 1988; Fenton, 1992; Peracchi et al., 2006).

The phytophagous bats (nectarivores and frugivores) are found only in the tropical and subtropical regions of the world (Hill, Smith, 1988), where there are plants producing nectar and fruits practicaly the whole year. The ingestion of plant parts is practiced by a large part of species of the Phyllostomidae (Figure 1), so that this family is the group most responsible for the dispersion of seeds and pollination of flowers, especially in arid zones (Humphrey, Bonaccorso, 1979; Peracchi et al., 2006). The frugivorous bats found in Brazil belong to a single family, the *Phyllostomidae* (Figure 2). These bats can consume a large variety of fruits, infrutescences, leaves, seeds and other plant parts (Zortéa, 2007). Due to the great mobility of their intestine (Bernard, Fenton, 2003), ingested seeds are not destroyed (Fleming, Sosa,

* Address all correspondence, proofs, and reprint requests to: Karin Corrêa Scheffer: Instituto Pasteur de São Paulo, Av. Paulista, 393 São Paulo-SP Brazil 01311-000 E-mail: ksferreira@pasteur.saude.sp.gov.br.

1994), where frugivorous bats are important dispersers of seeds (Mello, Kalko, Silva, 2008), which is crucial in the dynamics and regeneration of forests (Whittaker, Jones, 1994; Bredt et al., 1996). Some botanists believe that chiropters are the most important seed dispersers among all the mammals (Huber, 1910; Pijl, 1957).

Figure 1. Frugivorous bat (genus *Artibeus*) feeding on fruits of guava (*Psidium guajava*) in the backyard of a residence in Santa Branca, São Paulo. Photo: JJ Ferrari.

Figure 2. Colony of Frugivorous bats (genus *Artibeus*) grown under natural (palm leaf). Photo: JJ Ferrari.

The pollenivores and nectarivores are bats that have small teeth and obtain carbohydrates from nectar and proteins from pollen of plants, but may also ingest insects. They are recognized by their elongated snout and extremely long tongue. They have facial and body hairs specialized for transporting pollen (Reis et al., 2007).

Insectivory is practiced by species of all bat families found in Brazil. The majority of species is insectivores and forms the largest colonies of bats on the planet, with thousands of individuals (Graham, 1994). There are two types of these bats: aerial insectivorous and foliage gleaner (Wilson, 1973). The aerial insectivorous bats chase insects in full flight and explore free aerial spaces and comprise principally the molossids and vespertilionids, common in rural and urban areas. The foliage gleaners bats fly low and next to vegetation to procure the insects landing on the plants or on the ground. Generally, they possess large ears which allow them to hear the sounds of the movements and sounds produced by the insects (Uieda, 1996).

Despite the term insectivorous, bats with this diet can consume other arthropods, such as arachnids, diplopods and chilopods (Altringham, 1996). Bats can consume up to double their weight in insects in a single night and are therefore the principal natural controllers of arthropods and damages caused by them (Romano, Maidagan, Pire, 1999), where they are extremely important to agriculture and the environment.

In the Neotropical region, there are two piscivorous (or ichthyophagous) species, namely *Noctilio leporinus* (family *Noctilionidae*) and *Myotis vivesi* (family *Vespertilionidae*). These species are skilful in fishing, with large, strong feet in the form of claws. They live close to waterways and fish by means echolocation (Reis et al., 2007). They feed on freshwater or saltwater fish alevin, and aquatic and terrestrial arthropods (Altenbach, 1989; Bordignon, 2006). Due to the plasticity of their feeding habits and to an accelerated metabolism, which leads them to consume large quantities of fish alevin per night (Bordignon, 2006), piscivorous bats are of great environmental importance, because they are major population controllers, be it of fish or arthropods.

Among all the families, only *Phyllostomidae* show species with a carnivorous diet. The feeding behavior of these bats is based on the consumption of small vertebrates such as small rodents, amphibians, small birds, and even other bats (Peracchi et al., 1982; Fenton, 1992; Esbérard, Bergallo, 2004; Oprea et al., 2006). However, a good part of the carnivorous species can also consume arthropods regularly (Peracchi, Albuquerque, 1976) and even sometimes fruits (Uieda et al., 2007), and therefore, they are often considered omnivorous. Thus, the bats that belong to this group are responsible, along with other animals, for controlling populations of small vertebrates and arthropods in forest areas. Of the bats in Brazil, the carnivores are among the larger ones (Reis et al., 2007).

Among the more than one thousand species of bats existing in the world (Simmons, 2005), only three, belonging to the family *Phyllostomidae*, subfamily *Desmondotinae*, possess a hematophagous feeding habit. Of the three hematophagous species, *Desmodus rotundus* is the most common and abundant, with wide distribution in all South America and Central America (Greenhall et al., 1983). This bat feeds on the blood of mammals, especially pigs, horses, and other livestock (Figures 3 and 4), and is responsible for eventual attacks on humans (Aguiar, 2007). *Diphylla ecaudata* possesses the most specialized habits, consuming only the blood of birds of medium and large size (Greenhall et al., 1984). Another species, *Diaemus youngi*, is the rarest species, and has a limited occurrence. It attacks preferentially birds, but can sometimes consume the blood of goats and other livestock (Greenhall, Schutt Jr., 1996; Peracchi et al., 2006).

Figure 3. The vampire bat, *Desmodus rotundus*, is resting on the neck of horse. Photo: JJ Ferrari.

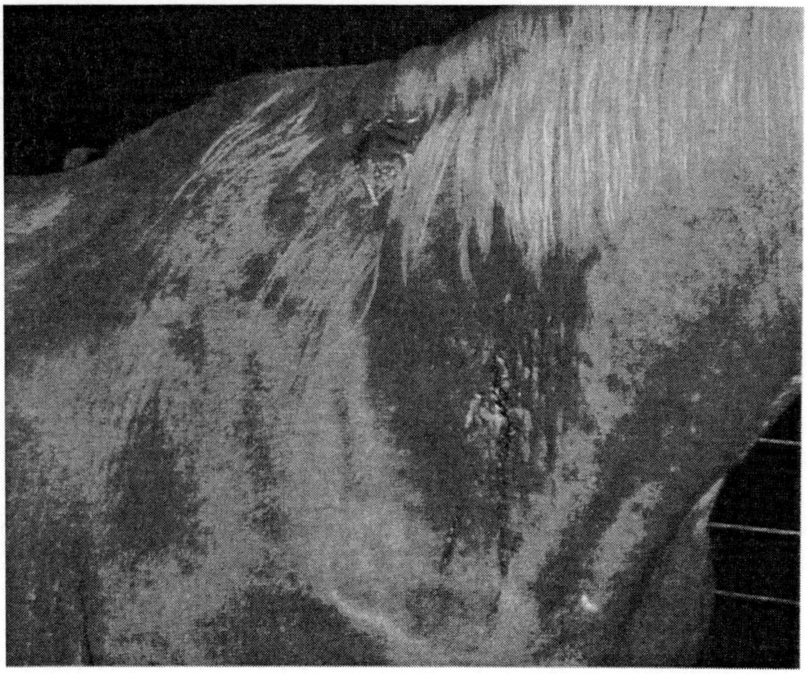

Figure 4. Detail of previous photo. Observe the wound caused by looting in the neck of the horse. Photo by: JJ Ferrari.

The Neotropical region is a biogeographic region that comprises part of southern Mexico and the Baja California peninsula, southern Florida, all the islands of the Caribbean and South America. The Neotropical region is considered the most biodiverse of the planet, also with regard to chiropters, with 288 species (Hutson et al., 2001).

Brazil is the largest country in the Neotropical region and occupies 47% of the South American continent. It has the greatest biodiversity of mammals (Reis et al., 2007), flora, and fauna and shows six principal biomes with specific geographic regions, environmental conditions and vegetation; among them are the Amazon Forest and Atlantic Forest, besides biomes with field formations constituted by typologies of open vegetation, appearing like savannas, which correspond to the Cerrado, predominant in Central Brazil, the steppe/savanna which includes the northeast caatinga, the Mato-Grosso Pantanal and the steppe which corresponds to the Pampas or Campos Sulinos of the plateau of the extreme south of Brazil (Figure 5)(IBGE, 2004).

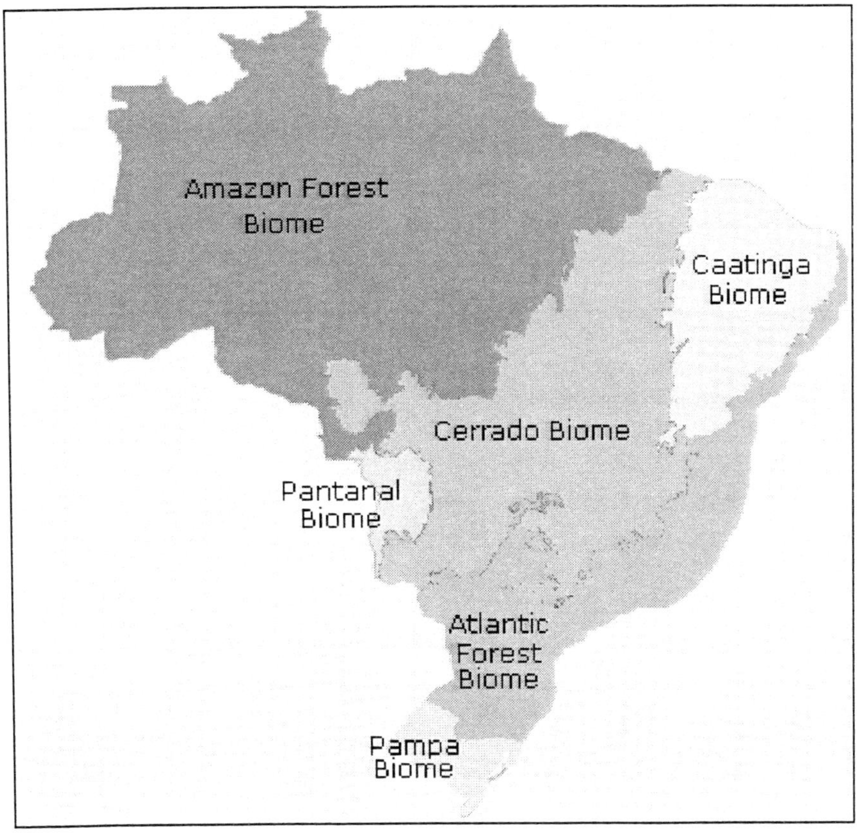

Figure 5. Brazil - Representation of the six continental biomes. Source: IBGE, 2010.

The Amazon Forest is the largest Brazilian biome, occupying almost half of the country's territory (49.3%). It is situated in the northern region of Brazil, encompassing part of five states. It is a tropical forest where dense ombrofilus and open forests predominate, with medium- and large-sized trees and occurrence of vines, bromeliads and orchids (IBGE, 2004). The little entrance of light through the canopies makes low-lying vegetation very scarce. The climate in the Amazon Forest is equatorial, hot and humid, due to the proximity to the equador, with temperatures varying little during the year. Rains are abundant with a rainy period of six months. This biome shelters the largest number of bats with 117 genera and 13 species being endemic (Marinho-Filho, Sazima, 1998). Pedro (1998) in studying the bats endemic to the northern region reported the occurrence of individuals of the families *Emballonuridae*, *Phyllostomidae*, *Thyropteridae* and *Molossidae*.

The Atlantic Forest covers a narrow strip of the Brazilian coast making up 15% of the country, encompassing 17 states. It shows different forms of topography, landscapes and diverse climatic characteristics, such as lowland forests, inland vegetation and high fields, besides ecosystems associated with mangroves, sandbanks vegetation, inland swamps, and coastal and oceanic islands. There is a predominance of seasonal semi-deciduous forests and dense and mixed, ombrofilus forests (with "Araucarias"). In both forest groups, there are, to a lesser degree, seasonal deciduous forests (in which more than 50% of the trees lose their leaves in the dry period) (IBGE, 2004). In this biome, the number of species of endemic bats is small, but with a large diversity of chiropterofauna (Pedro, 1998). Marinho-Filho and Sazima (1998) reported the existence of 96 species with 5 endemic ones in the entire Atlantic Forest in Brazil. Pedro (1998), however, reported the occurrence of two endemic species, *Platyrrhinus recifunus* and *Lasiurus ebenus,* in the southeast region.

In the fragments of Atlantic Forest of the southeast region, there are about 95 species of bats, where seven belong to the family *Emballonuridae*, 53 to the *Phyllostomidae*, two to the *Noctilionidae*, one to the *Furipteridae*, one to the *Natalidae*, one to the *Thyropteridae*, 13 to the *Vespertilionidae* and 17 to the *Molossidae* (Peracchi, Nogueira, 2010).

The Cerrado is the second largest biome in Brazil, with an area of approximately 20% of the country. It has two well-defined seasons: dry winter and rainy summer (IBGE, 2004). It has medium-sized, twisted trees, with rough leaves, tough, rough bark, and roots of up to 15m for obtaining water from deep regions of the soil in dry periods. The bats found in the Cerrado belong to seven families and 80 neotropical species: *Emballonuridae, Noctilionidae, Mormoopidae, Phyllostomidae, Furipteridae, Vespertilionidae and Molossidae*. The family *Phyllostomidae* is the most abundant, with 41 species, followed by *Molossidae* with 16 species and *Vespertilionidae* with 11 species (Marinho-Filho, 1996). Only one species of bats is considered endemic to the Cerrado, namely *Lonchophylla dekeyseri* (Taddei et al., 1983; Marinho-Filho, 1996).

The Caatinga makes up part of the semi-arid region of the Northeast. The vegetation is complex with trees and shrubs with spines like cacti, has open formations with dry, poor soil (IBGE, 2004). Marinho-Filho and Sazima (1998) reported the occurrence of 69 species of bats in the Caatinga without any endemic species.

The Pantanal is the largest continuous floodplain in the world, formed mainly by flooding of the Paraguai River and affluents, where it is flooded for the most part. The mean annual rainfall is approximately 1100 mm. It shows different types of habitats, such as capons, montain ranges, seasonally inundatable fields and temporary or permanent lakes (Damasceno et al., 1999). Studies on the chiropterofauna have reported the occurrence of 61 species of bats in this ecosystem (Marinho-Filho, Sazima, 1998; Camargo, Fischer, 2005). Longo et al. (2007) reported the occurrence of the species *Vampyressa pusilla* (Wagner, 1843), and a specimen of *Vampyrum spectrum* was recently found in Mato Grosso do Sul state (Linnaeus, 1758) (Embrapa, 2010), increasing the list to 63 species.

The Pampas or Southern fields are characterized by the presence of low-lying vegetation (grasses) and small shrubs distant from each other. This vegetation is will meet in several regions of Brazil, but it is restricted to Rio Grande do Sul, occupying 63% of the state's area (IBGE, 2004). The South Region of Brazil shows a total of 70 species of bats distributed in seven families: *Embalonuridae, Noctilionidae, Phyllostomidae, Furipteridae, Thryropteridae, Vespertilionidae* and *Molossidae*. The Paraná state shows the highest species richness of bats with 64 species, followed by Santa Catarina state with 47 species and finally by Rio Grande do Sul state with 40 (Passos et al., 2010).

In Brazil, there are laws that guarantee the protection of bats, but of the 168 species of bats registered in Brazil (Reis et al., 2007), five of two different families are considered vulnerable by IBAMA (2003), these being *Lonchophylla bokermanni* (Sazima et al., 1978), *Lonchophylla dekeyseri* (Taddei, Vizzoto, Sazima, 1983) and *Platyrrhinus recifinus* (Thomas, 1901), belonging to the family *Phyllostomidae,* and *Lasiurus ebenus* (Fazzolari-Corrêa, 1994) and *Myotis ruber* (E. Geoffroy, 1806) of the family *Vespertilionidae* (Reis et al., 2007). However, this number of vulnerable bats may be underestimated, because these animals are exposed to various types of threats, such as deforesting and forest fragmentation, intoxications by insecticides and even capture and killing by humans, where the last is often motivated by myths and superstitions. Therefore, populations of some species have visably decreased in Brazil (Reis et al., 2007).

BATS AND RABIES

Rabies is a viral disease maintained in nature by different species of carnivorous animals, domestic and wild, referred to as "reservoirs", also including bats of different feeding habits (Smith, 1996).

At the beginning of the last century, there was a severe epizootic of rabies in the strip of land between the mountains and the sea, in the state of Santa Catarina, South Brazil, in which about 4000 cattle and 1000 horses and mules died. These animals were meant for working in the field and transporting agricultural products, and thus these deaths led to great economic losses for the local population. The possibility of hematophagous bats playing a role as reservoirs in the propagation of rabies was first raised by Carini. However, this hypothesis was not readily accepted by international scientists, where it was considered a "tropical fantasy" (Carini, 1911).

Haupt and Rehaag (1925) researched the same region where Carini had diagnosed rabies in cattle, and identified the presence of Negri bodies in the central nervous system (CNS) of a hematophagous bat that was feeding on a cow, confirming the hypothesis of Carini. Also ascribed to the two German researchers is the first description, in 1916 in Brazil, of the presence of the rabies virus in a non-hematophagous bat of the species *Phyllostoma superciliatum*, currently classified as *Artibeus lituratus.* However, the finding was not reported until 1935 (Torres, Queiroz Lima, 1935). Contestations arose concerning the identification of the species of the bat involved, if it was in fact the frugivorous bat *Phyllostoma superciliatum* (*A. lituratus*) or the hematophagous *D. rotundus*.

A little more than a decade after the rabies epidemic in the state of Santa Catarina, a mysterious disease attacked the cattle on the island of Trinidad, in the Caribbean, also affecting humans, where it was diagnosed much later as rabies (Carneiro, 1936). The episodes of human rabies on the island of Trinidad were described by Hurst and Pawan (1931). The rabies virus was isolated from the hematophagous bat *D. rotundus* and frugivorous bats *Artibeus planirostris trinitalis, Diclidurus albus* and *Hemiderma* sp. Later, these investigators were able to infect experimentally the bats *D. rotundus* and *A. lituratus* with the rabies virus (Pawan, 1936a).

In the 1930s, occurrences of rabies transmitted by hematophagous bats were registered in Colombia, Bolivia, Mexico, Venezuela, Paraguay and Argentina (Carneiro, 1936) and in other states of Brazil (Carneiro, Freitas Lima, 1927), but, apparently, the report by Pawan (1936a) was the first description about rabies in non-hematophagous bats of Latin America. This finding, however, did not arouse much interest, compared to the great repercussion and preoccupation demonstrated by researchers in relation to rabies transmitted by hematophagous bats (Baer, 1975).

The preoccupation with hematophagous bats in Latin America was understandable due to the enormous economic damages as a result the death of animals with rabies. In addition, there is a decline in productivity by repeated attacks which cause debility of the livestock, mainly due to the loss of blood, secondary to infections of the wounds, myiasis and depreciation of the leather (Acha, 1967; Arellano-Sota, Sureau, Greenhal, 1971).

Hematophagous bats, especially *D. rotundus* as reservoirs of the rabies virus, started to hold a position of distinction in America Latin countries (Torres, Queiroz Lima, 1936), because they are found from northern Chile and Argentina up to northern Mexico and parts of the Caribbean, not existing in other regions of the globe (Koopman, 1988).

There was a small incident in Florida, USA, in 1953, when a child was attacked by an insectivorous bat, *Dasypterus floridanus* (currently known as *Lasiurus intermedius*), which made North American scientists turn their attention to the question of rabies in non-hematophagous bats (Bigler, Hoff, Buff, 1974). The aggressor bat tested positive for rabies, and the child received post-exposure treatment with vaccine and did not develop rabies (Scatterday, 1954; Venters et al., 1954). A short time after this incident, other non-hematophagous bats of the species *Lasiurus cinereus* and *Lasiurus seminola*, both insectivores, tested positive for rabies in the state of Pennsylvania (Witte, 1954).

Since then, public health authorities in the USA became interested in knowing the extent of rabies in non-hematophagous bats, and infection by the rabies virus was confirmed in different species of bats with different feeding habits, including insectivores, frugivores, omnivores, pollenivores and piscivores. The genera of bats most frequently testing positive for rabies include *Lasiurus*, *Tadarida*, *Epitesicus*, *Myotis*, *Plecotus*, *Pipistrellus*, *Mormoops*, and *Euderma*, among others. In the 1960s, a large number of bats of different feeding habits were found testing positive for rabies in the USA, Canada, Mexico, Guatemala, Panáma, Honduras, Colombia and Venezuela in the American continents, and also in Europe (Germany, Turkey and former Yugoslavia), Asia (Thailand) and Africa (South Africa) (Baer, 1975).

In the same period in Brazil, there are records of the isolation of the rabies virus from the bat *Phyllostomus hastatus hastatus*, from Itaguai, Rio of Janeiro (Silva, Rivello, Nilsson, 1961) and various positive cases in different species, such as the following: *Tadarida braziliensis*, captured with symptoms of paralysis, inside a home in São Leopoldo, Rio Grande do Sul (Uieda et al., 1992); the hematophagous bat *Diaemus youngi*, in the state of Alagoas (Silva, Souza, 1968a); an insectivorous bat, *Histiotus velatus*, captured in a furnace, in Alfredo Wagner, Santa Catarina (Amorim, Silva, Silva, 1970); and *Chrotopterus auritus*, captured apparently healthy in the state of Rio de Janeiro (Silva, Alencar, 1968). Other positive diagnoses in insectivorous and frugivorous bats were found in the 1970s, where a bat of the species *Molossus molossus* was captured in broad daylight unable to fly, in a house patio on a farm in Campinas – SP (Rodrigues et al., 1975) and three specimens of *Artibeus lituratus* were captured in a mango tree, in Itajai, Rio de Janeiro (Silva, Silva, 1974).

In the state of São Paulo, between the years of 1984 and 1991, there were seven more cases of rabies in bats, three frugivores and four insectivores (Uieda et al., 1992, 1995). Among these cases was the first record of rabies in *Nyctinomops macrotis* in Brazil (Uieda et al., 1995). Up to to the end of 1998, one more case in a frugivorous bat (Passos et al., 1999) and eight cases in insectivorous bats, including the species *Myotis nigricans*, *Lasiurus borealis*, *Nyctinomops macrotis*, *Molossus ater*, *Histiotus velatus,* and *Tadarida braziliensis,* the first case in the the state of São Paulo, were diagnosed by means of direct immunofluorescence and intracerebral inoculation in mice (Bernardi et al., 1998; Martorelli et al., 1995, 1996; Passos et al., 1998; Silva et al., 1999; Uieda, 1998).

Sodre, Gama and Almeida (2010) reported that in Brazil there are records of 41 species of bats of different feeding habits with positive tests for rabies. Recently, there was the identification and positive diagnosis for rabies in a specimen of *Artibeus obscurus*, from the city of Ribeirão Preto in the northeast region of the state of São Paulo, thereby increasing the list to 42 species positive for rabies[1]).

In the last years, reports of isolating the rabies virus or rabies-like virus from bats of different feeding habits have become frequent in the whole world, where many of these were called emergent lyssaviruses. Even in Japan, a country considered free of rabies, a virus was isolated from the bat *Rhinolophus cornutus*, which showed characteristics very similar to those of a lyssavirus. This virus, according to Iwasaki et al. (2004), caused lethal encephalitis when injected intracerebrally in mice of up to 3 weeks of age, and electron microscopy showed numerous particles in the shape of bullet, with virions showing characteristics of the family *Rhabdoviridae*.

Rabies Virus

Classification

Rabies virus belongs to the genus *Lyssavirus* of the family *Rhabdoviridae*, order *Mononegavirales* (Carstens, 2010).

In the current classification, the genus *Lyssavirus* contains eleven species (Carstens, 2010). Species I – *Rabies virus* (RABV), represented by the classical rabies virus. Species II – *Lagos bat virus* (LBV), virus isolated from frugivorous bats (*Eidolon helvum*, *Micropterus pusillus* and *Epomorphorus wahlbergi*), where the first report was made in 1956, in the region of Lagos, Nigeria (Bingham, 2001; Wunner, 2002). Species III – *Mokola virus* (MOKV), isolated from a musk shrew (*Crocidura* sp) from Nigeria, the only lissavirus to date not isolated from bats. The first isolation was made in 1968 from a pool of organs of musk shrews in Nigeria, followed by isolations in humans also in Nigeria, and felines of Zimbabwe

[1]Data supplied by the Laboratory for the Diagnosis of Rabies of the Instituto Pasteur of São Paulo.

and Ethiopia (Nel, 2001). Species IV – *Duvenhage virus* (DUVV), which in 1970 was isolated from a human in Warmbaths, in the northern region of Pretoria and later isolated from insectivorous bats (*Miniopterus schreibersii* and *Nycteris thebaica*) of South Africa and Zimbabwe (Bingham, 2001).

The viruses isolated in Europe, the "*European bat lyssaviruses*", were also isolated from bats; one of them was referred to as species V – *European bat lyssavirus 1* (EBLV-1), grouping the isolations from bats of the genus *Eptesicus* and the other, species VI – *European bat lyssavirus 2* (EBLV-2), constituted by the grouping of viruses isolated from the genus *Myotis* (King, 2001; Pounder, 2003).

In 1996, a lyssavirus was isolated from the frugivorous bat *Pteropus alecto*, on the eastern coast of Australia, a country considered free of rabies since 1867. This virus was classified as species VII – *Australian bat lyssavirus* (ABLV) (King, 2001; Pounder, 2003). In 2003, the species *Aravan virus* was described, isolated from an insectivorous bat (*Myotis blythi*) from Kyrgyzstan, Central Asia, in 1991 (Arai et al., 2003; Kuzmin et al., 2003). Another species, *Khujand*, was isolated in 2001, in Northeast Tajikistan, Central Asia, also from an insectivorous bat (*Myotis mystacinus*) (Kuzmin, Botvinkin, Khabilov, 2001; Kuzmin et al., 2003). The other two species were isolated in Russia, one obtained in the city of Irkutsk, called *Irkut virus*, isolated from the bat *Murina leucogaster* and the other obtained in the region west of the Caucasus Mountains (about 100 km southeast frm the city of Krasnodar), called *West Caucasian bat virus* (WCBV), isolate from the bat *Miniopterus schreibersi* (Botvinkin et al., 2003). In 2009, a new lyssavirus described as *Shimoni virus* (SHIBV) was isolated from an insectivorous bat (*Hipposideros commersoni*), found in a cavern in the region of Kenya, which has been suggested as a new species (Kuzmin et al., 2010).

To date, among the 11 species of the genus *Lyssavirus*, only the species *Rabies virus* was found on the American continents (Kotait, Carrieri, 2008).

Morphology

The different species of the genus *Lyssavirus* possess negative polarity, linear, non-segmented single-stranded RNA with 11,932 nucleotides and molecular weight of 4.6×10^6 daltons (Da). They are formed by the combination of five genes which encode 5 structural proteins: nucleoprotein (N), phosphoprotein (P), matrix protein (M), glycoprotein (G) and a RNA-dependent RNA polymerase (L) (Tordo, 1996).

In relation to morphology, the rabies virus has the shape of a bullet, with one end flat and the other rounded. The diameter of the virus can vary between 50 and 100 nm and the length from 100 to 430 nm, depending on the species of the virus and the presence of interfering defective particles. In its chemical constitution, the complete virus particle contains 2% to 3% RNA, 67% proteins, 26% lipids and 3% carbohydrates (Kaplan et al., 1986).

The five proteins constitute two principal structures, according to their biological functions: the ribonucleoproteins (RNP) and viral envelope. The RNP or nucleocapsid takes the form of a helicoidal complex composed of single-stranded RNA associated with the proteins N, L and P (Tordo et al., 1986). The viral envelope consists of a lipid bilayer, with which two proteins are associated: protein M and glycoprotein G.

TRANSMISSION

The bats are responsible for the maintenance of the rabies virus in the aerial cycle, and all the species can infect each other and, consequently, transmit the rabies virus, regardless of its classification with respect to feeding habits.

It is likely that the hematophagous bats were already infected with the rabies virus long before the discovery of America by European colonizers, so that humans and animals suffered continuously due rabies being transmitted by bats in tropical America (Constantine, 1988).

The transmission of the virus occurs mainly by direct contact and the main forms of transmission among the species of hematophagous bats are by biting and licking, since these animals posses gregarious habits, characterized mainly by this type of interaction.

The non-hematophagous bats can be infected upon sharing the same shelter with hematophagous bats infected with the rabies virus or even in disputing territory with these bats. Infected non-hematophagous bats, when found alive, dead or prostrate, in urban environments, can accidentally transmit the disease to humans and other animals (Uieda, Harmani, Silva, 1995).

This transmission normally occurs by biting, where it can be a mechanism of defense, or a form of feeding, more common in hematophagous species. Defensive contacts are more frequent among the non-hematophagous species, where minimal wounds of the teeth can be seen in species of bats of small size, which are often not easily seen without a careful examination, possibly leading to the impression of the absence of a bite in the susceptible one.

Bats of the species *Lasionycteris noctivagans* have a major involvement in the transmission of the rabies virus, especially in the USA, and have as a characteristic this supposed absence of bite in the transmission. Experimental studies utilizing isolates from this insectivorous species indicate that an efficient local replication in the dermis occurs, including the infection of fibroblasts and epithelial cells, which can explain the success of an infection in which there is no evidence of aggression by biting. However, after superficial exposure, it is not exactly clear how or which sites the virus invades the peripheral nerves in the skin or subcutaneous tissues (Morimoto et al., 1996).

Still, based on these studies, hypotheses were proposed that isolates from these species have undergone genetic changes that can permit a greater probability of infection after superficial contact. Experiments were carried out utilizing these isolates compared with others of domestic dogs and coyotes, and it was found that the virus isolated from this species of bat showed greater titers in epithelial cells and in muscle tissue, when compared to other species. In addition, changes in molecular structure of the glycoprotein, which can be linked to the increase in infectivity, were also observed (Messenger et al., 2003).

In Brazil, the involvement of hematophagous bats in the transmission rabies to humans, in the beginning of the 1980s, was 2% of the total cases (Schneider, 1990). In the years of 2004 and 2005, there was outbreak human rabies, in the northeast region of the country, where 22 and 42 cases of human rabies transmitted by hematophagous bats were respectively registered. In the last 10 years, 163 cases of rabies in humans no Brazil were reported, where 45% (43 cases) were transmitted by bats (MS, 2010). In the preceding decade 412 cases were registered, of which 12% (49 cases) involved bats as transmitters (Schneider, 2009). On the other hand, the transmission of the rabies virus by non-hematophagous bats to terrestrial wild animals continues being a great question, and that this transmission occurs suggests that control measures should be adopted for this species (Baer, 1975).

Through epidemiological investigations in wild animals, it was demonstrated that the virus can be transmitted specifically to a particular species of host, becoming extremely adapted to this species and less able to infect other species. This host/parasite relation has become known as "compartmentalization" of the rabies virus (Constantine, 1988; Winkler, 1975). It is suggested that this event occurs when the rabies virus of a particular species of bat does not show characteristics similar to those of the virus isolated from other species of bats. These studies became more important with the development of monoclonal antibodies and genetic characterization techniques (Constantine, 1988).

Therefore, the efficient transmission of the disease depends on the quantity of virus present in the saliva of the infected animal and particularly the severity of the bite, mainly if it reaches the muscle tissue, where there is a high density of cell-specific receptors, such as that of acetylcholine (Jackson, 2002).

PATHOGENESIS

The pathogenesis of rabies in bats resembles that observed in other species, although it is more difficult to study in these species, since they are wild animals, which makes it difficult to conduct experimental studies under absolute control conditions.

When an infected animal comes in contact with a susceptible animal, viral replication occurs initially in the myocytes and in the subepithelial tissue as well, depending on the species. On attaining high antigenic levels, extraneural replication occurs, where the nerve endings are reached (Fekadu et al., 1988).

A study in experimental animals with the utilization of established rabies virus strains indicated that the virus is capable of entering peripheral nerves directly, without the necessity of a replicative cycle in extraneural cells and that this could be associated with a shorter incubation period (Shankar, Dietzschold, Koprowski, 1991).

From this moment, the viral glycoprotein is bound to specific receptors, such as the nicotinic acetylcholine receptor, at the level of neuromuscular junctions, and penetrates the sensitive nerve endings. Studies utilizing nerve-muscle cocultures indicate that the neuromuscular junction is the principal site of entrance into neurons (Lewis, Fu, Lentz, 2000).

Two other probable receptors responsible for the interaction of the rabies virus with the host cell has been described in the literature: the *neural cell adhesion molecule* (NCAM)) and the neurotrophic receptor p75 (Tuffereau et al., 1998).

Since in the peripheral nervous system (PNS), the virus is transported to the sensitive ganglia by centripetal path, and reaches the CNS, where there is viral propagation by retrograde axoplasmic flow, by means of synaptic junctions (Kucera et al., 1985; Tsiang, Ceccaldi, Lycke, 1991). In the CNS, there is intense spreading, but it is not homogeneous, which can affect in different ways the various structures of this organ. Later, the virus migrates by centrifugal path toward different organs, particularly involving the parasympathetic nervous system (Jackson, 2002).

In various species of mammals studied, it was found that the organs invaded by the virus during the centrifugal migration include the heart, liver, skin, thymus, kidneys, ovaries, uterus, adrenal glands, lungs, spleen, intestines, smooth and skeletal muscles, hair follicles, epithelium of the tongue, retina and cornea (Brass, 1994; Awasthi et al., 2001; Mattos, Mattos, Rupprecht, 2001).

Scheffer et al. (2007) studied the pathogenesis of rabies in bats, demonstrating the presence of the virus in fragments of the CNS, salivary gland, pectoral muscle, heart, lungs, stomach, kidneys, bladder, genital tract, tongue and interscapular fat, utilizing techniques of viral isolation in laboratory animals and cultured cells.

These studies contribute to reinforcing the theories of other ways of rabies transmission of bats, not only through the CNS and salivary glands, and also point out the importance of maintaining post-exposure prophylactic treatment in humans, when there is contact between humans and bats. With respect to animals that may have come in contact with bats, there is the need to study different treatment schemes.

Evidence of the rabies virus in the lungs reinforces the theory of the transmission of rabies by aerosols among the chiropters, especially in caverns densely populated by infected bats (Constantine, 1962). The evidence of the rabies virus in the bladder, kidneys and stomach shows the importance of the viral distribution in bats and suggests potential routes of transmission of rabies by these animals through urine and feces. The presence of virus in the tongue and salivary glands indicates that the most common transmission route between bats is through biting (Johnson et al., 2006).

Other entrances are rare. However, in 1956, a case of human rabies was registered in an individual who worked in caverns with bats, but without history of a bite or other contact with some suspect animal. After an epidemiological investigation, it was hypothesized that the individual had been infected through inhalation of viral particles in suspension in the cavern (Brass, 1994; Warrell, Warrell, 2004). Infection by airways was also demonstrated under natural conditions, when healthy animals were placed inside caves where large colonies of infected bats lived. Viral antigens were observed in cells of olfactory receptors of bats in Brazil, through infection having occurred in caverns, suggesting that the nasal mucosa is an important entrance of natural infection (Constantine, 1962). However, it is believed that it is necessary to have a very large number of infected bats in an area without ventilation in order for transmission of the rabies virus via airways (Jackson, 2002).

The incubation period of rabies in hematophagous bats infected experimentally is variable, from 7 to 171 days. This variation can be related to the quantity of virus inoculated, the site of entrance, the immune state of the individual, and severity of the lesion (Luna, Chahuayo, Marreros, 1985; Acha, Szyfres, 1986; Brass, 1994).

The clinical manifestations of rabies in hematophagous bats can vary: the classical form, with predominant stage of fury, followed by paralysis and death; classical paralytic form, with predominant stage of silence and absence of fury, and paralysis followed by death; complete absence of symptoms and sudden death; furious form where there is no paralysis and fury is followed by death or even when there is the furious form but with recuperation; and asymptomatic form, where the animal does not show signs of anormality, but is capable of transmitting the disease (Pawan, 1936b).

Constantine (1967), however, reported that the clinical signs in naturally infected bats are predominantly those of paralysis instead of fury and include irritability or depression, debility, anorexia, hypothermia and paralysis. Initial paralysis can involve the legs, wings, neck, eyelids, head or jaw. Neural damage can extend to the bladder, resulting in urinary incontinence and also possibly fecal incontinence. Paralysis develops throughout all the body in the terminal phase death occurs in a few days (Brass, 1994).

There are few reported experiments on the inoculation of non-hematophagous bats with the rabies virus. The insectivores *Tadarida braziliensis mexicana* and *Antrozous pallidus* were susceptible to experimental infection with virus isolated from a bat of the genus *Tadarida.* When inoculated by the intracerebral, subcutaneous and intramuscular routes, the bats showed signs of rabies and died within 6 to 23 days post-inoculation. The virus was isolated from a *pool* of brain and salivary gland tissue of bats found dead. However, these studies were performed with a small number of bats and attempts to isolate the virus were made only from the brain, salivary gland or saliva, and in bats, the most important route of elimination is saliva. (Sulkin, Allen, 1975).

In relation to rabies of hematophagous bats, Silva and Souza (1968b,c) studied the presence of the rabies virus in different non-nervous organs, like Nilsson and Nagata (1975), isolating the virus from brain, salivary glands, interscapular fat, heart, lung, testicles and fetuses.

Knowledge of aspects linked to transmission and pathogenesis of rabies in different species of bats constitutes an important instrument for the control of the disease in these animals, as well as in herbivores, pets and humans.

Diagnosis

Laboratory diagnosis of rabies is essential for the confirmation of the suspect case, as well as for differential diagnosis with other types of encephalitis, should be rapid and precise, since the laboratory results influence not only the medical decision of whether to institute prophylactic treatment in humans, but also the elaboration of measures of control of a possible epizootic in a community (Meslin, Kaplan, 1996).

Intracytoplasmic inclusions called Negri bodies, in CNS infected by the rabies virus, are the pathognomonic sign of the disease (Perl, Good 1991; Lépine, Atanasiu, 1996). Since 1927, Sellers' staining was utilized in the diagnosis of rabies to observe these inclusion bodies (Rupprecht, Hanlone, Hemachuda, 2002). This technique is no longer routinely used, since the World Health Organization (WHO) recommends the direct immunofluorescence (DIF) and virus isolation (VI) methods (Tierkel, Atanasiu, 1996).

DIF is the international reference test, where it is considered the gold standard by WHO as well as the World Organisation for Animal Health (OIE) (Trimarchi, Smith, 2002; OIE, 2008) and is utilized as the preferred method on a global scale (Rudd et al., 2005).

The confirmation of the results obtained by DIF can be obtained by intracerebral inoculation in newborn mice, up to three days old or recently weaned 21-day-old mice with a 20% suspension of CNS of animals suspected of infection by the rabies virus. The intracerebral route is chosen for inoculation in mice because the rabies virus is neurotrophic. After inoculation, the mice are observed daily for 30 days and in positive cases, the animals become ill and generally die around 10 to 20 days after inoculation; the CNS of all mice is submitted to DIF or isolation of the virus in cell culture (Koprowski, 1996).

The use of cell lines for the culture of the rabies virus was first described in 1913 by Levaditi. The cell culture system showed a greater sensitivity in relation to intracerebral inoculation in mice. Since then, other studies developed with various cell lines, including chicken embryo cells (CER), McKoy and murine neuroblastoma (N2A). Rudd and Trimarchi (1987) compared the sensitivity of BHK-21 cells and murine neuroblastoma, and observed a greater diagnostic sensitivity in the results with murine neuroblastoma. According to studies, this greater sensitivity can be explained in part by the fact that murine neuroblastoma cells have receptors similar to those of brain in normal mice.

Various studies were conducted, comparing the techniques of DIF, isolation of the rabies virus in mice (IM) and in cell culture (ICC). The results indicated that the isolation of the wild rabies virus in cell culture is at least as sensitive as inoculation in mice (Webster, 1987). This way, the substitution of IM by ICC has already been done in some laboratories for rabies diagnosis (Webster, Casey, 1996), mainly in developed countries.

At the Pasteur Institute of São Paulo (PI/SP), the technique was standardized for bats and applied in the routine laboratory tests, but for positive specimens and those with doubtful results, there is a need for inoculation in mice (Castilho et al., 2007).

ICC shows some advantages in relation to IM, such as relative ease and rapidity to form the technique, the possibility of being carried out in a small space, lower cost in relation to the mouse inoculation test and producing results in only 4 days (Castilho et al., 2007).

The methods of antigenic typing and genetic analysis allow us to identify variants responsible for individualized episodes and cases, in both humans and animals. These methods can be utilized more like a epidemiological investigation tool for the identification of the source of infection, be it in the urban or in the wild cycle (Kotait, Carrieri, Takaoka, 2009).

Antigenic Characterization of the Rabies Virus

The concept of antigenic variants of rabies virus and the study of their differences and specific reservoirs were consolidated with the development of techniques for the production of monoclonal antibodies (MAb) at the end of the 1970s. These techniques were utilized by Wiktor and Koprowski (1978) to produce the first secretory hybridomas of MAb against protein G and N of rabies virus. Since then MAb were widely utilized for identifying and classifying rabies virus and other lyssaviruses in groups corresponding to antigenic determinants (Wiktor, Koprowski, 1978; Flamand, Wiktor, Koprowski, 1980).

Different panels of MAb were established for differentiation of RABIES VIRUS isolates of terrestrial species and bats in the United States and Western Europe, and to a lesser degree in Africa, Asia, Eastern Europe and Latin America (Lafon, Lafage, 1987; Dietzschold et al., 1988; Vincent, Bussereau, Sureau, 1988; Smith, 1988; Schneider, Barnard, Schneider, 1985).

Besides the utilization of MAb principally in epidemiological investigations, they can be very useful for the diagnosis of rabies in certain circumstances, such as imported cases of human rabies and cases of rabies associated with uncertain exposure (Lumio et al., 1986; Smith et al., 1991) and also routinely in countries with programs of oral vaccination of foxes on a broad scale, in order to establish that no infection is caused by the vaccine virus (Schneider, Barnard, Schneider, 1985; WHO, 1984).

For antigenic characterization by MAb, various panels are available from collaborating centers of WHO and private institutions, each with a different discriminatory power, and therefore it was necessary to chose one that would be useful for the region of the Americas (OPS, 1998).

The centers of disease control of various countries and the "Centro Pan Americano de Zoonoses" (CEPANZO) of the Pan American Health Organization (PAHO) carried out studies during the period of 1987 to 1992, on different isolations of various countries in the Americas (Diaz, Rodriguez, Smith, 1994). Based on these data, they selected a panel of eight MAbs that allow the detection of the most common variants found in Latin America. This panel is available from the collaborating center of WHO, the Centers for Diseases Control and Prevention (CDC), Atlanta GA, USA, for laboratories participating in the consortium of the PAHO reference laboratories in rabies of the Americas (OPS, 1998).

The panel of the CDC for antigenic characterization, by means of reactivity with eight MAb by the indirect immunofluorescence (IIF) technique, defines 12 antigenic profiles, which allows the establishment of the 11 antigenic variants found in the isolates of Latin Americ and differentiates them from laboratory and vaccine strains: *Challenge Virus Standard* (CVS), *Pasteur Virus* (PV), *Street–Alabama–Dufferin* (SAD) and *Evelyn-Rokitnicki-Abelseth* (ERA) (Diaz, Rodriguez, Smith, 1994; Delpietro et al., 1997; De Mattos CC, De Mattos CA, 1998).

ANTIGENIC VARIANTS OF BRAZIL AND LATIN AMERICA

In Bolivia, four antigenic variants (AgV) were identified by antigenic typing of the isolates in the region: AgV1 and 2, typica of dog, and AgV3 and 5, typical of the hematophagous bat *D. rotundus*. Genetic identification was concordant with antigenic typing, demonstrating the presence of three genetic grouping of rabies virus in the country (Favi et al., 2003).

In Venezuela, the variants AgV1, 3 and 5 were identified in the rabies virus isolates in the country, which was confirmed by concordant genetic characterization (De Mattos et al., 1996).

In Chile, various studies characterized the insectivorous bat *Tadarida braziliensis* as the principal reservoir of rabies virus in the country (De Mattos et al., 2000; Favi et al., 2002). Yung, Favi, Fernandez (2002) antigenically and genetically typed isolates of rabies virus and identified the variants AgV3 (*D. rotundus*), AgV4 (*T. braziliensis*) and AgV6, typical of the insectivorous bat *Lasiurus cinereus*. In a retrospective study with isolates of rabies virus performed between 1989 and 2005, antigenic variants AgV1, AgV3, AgV4 and AgV6 were identified in the majority of the isolates, confirming the importance of *T. braziliensis* in the epidemiology of rabies in the country. Furthermore, some isolates did not have reactivity profiles compatible with those of the MAb panel of the CDC (Favi et al., 2008).

In Argentina, antigenic characterization of the isolates of rabies virus in the region identified the antigenic variants (AgV) 2, 3, 4 and 6, and these results were in agreement with the genetic characterization. Besides these variants, isolates of insectivorous bats also showed other profiles not compatible with those established (Delpietro et al., 1997; Cisterna et al., 2005).

In Latin America and Caribbean, the reference laboratories of Argentina, Brazil, Chile, Colombia, Mexico and Venezuela characterized antigenically more than 800 isolated specimens of rabies virus, utilizing practically in all the countries the panel donated by the CDC, with exception of 11 specimens isolated in Argentina, in which a panel provided by the Wistar Institute was used. The accumulated experience in the region revealed that the use of this technique allowed the determination of the geographic distribution of the different antigenic variants of rabies virus, describing new variants and identifying variants recognized in new hosts, information very useful for the epidemiological monitoring of the disease in the region (OPS, 2000).

Since 1996, the laboratory for the diagnosis of rabies of the PI/SP has carried out antigenic characterization of virus isolates of different animal species of various regions of Brazil, utilizing the panel of eight MAb against the viral nucleoprotein, produced by the CDC of Atlanta, USA and defined by WHO. A study of 330 isolates from different species between 1996 and 2000 demonstrated 5 antigenic variants compatible with profiles observed in the pre-established panel, two in dogs, AgV1 and AgV2, and three in bats, AgV3 of *D. rotundus*, AgV4 of *T. braziliensis* and AgV6 of *Lasiurus* spp. Also six other profiles not compatible with the panel were identified. The greatest variability was observed among the specimens isolated from insectivorous bats, and the most common variant isolated among the species was variant 3 of *D. rotundus* (Favoretto et al., 2002).

The observation of incompatibility of certain isolates required the complementation of antigenic studies with genetic analyses, which demonstrated the diversity of rabies virus isolates in Brazil. These isolates, confirmed as antigenic variants, all belong to species 1 of the genus *Lyssavirus*, like all the other isolates on the American continents and Caribbean (Ito et al., 2001; Kobayashi et al., 2005).

In the period of 2000 to 2006, the diagnostic laboratory of PI/SP analyzed 4057 samples of bats found in urban areas of the city of Ribeirão Preto, in the state of São Paulo, of which 64 were positive for rabies and were frugivorous or insectivorous species. The antigenic study of these isolates by the CDC panel identified the majority of the isolates as variants AgV3 and AgV6, and two not compatible with the panel (Carrieri et al., 2006).

Albas et al. (2009) characterized antigenically 18 isolates from non-hematophagous bats from the western part of São Paulo State, and the antigenic variants AgV3 of *D. rotundus* and AgV4 of *T. braziliensis* were detected.

Of the 168 species of bats existing in Brazil, rabies virus has been isolated from 42 species[2] (Sodré, Gama, Almeida, 2010). Of these, approximately 78% have an insectivorous feeding habit.

PI/SP receives annually about 4000 samples from bats for the diagnosis of rabies. In 2008, 42 bats tested positive for rabies. All isolates from bats were submitted to antigenic characterization by IIF utilizing the CDC panel and at least 24 were incompatible with the reactivity profiles determined by the panel. In 2009, 60 bats were diagnosed with rabies and of these approximately 19 showed profiles not compatible with those pre-established by the CDC panel[2].

In the last years, a large number of bats of species with insectivorous feeding habits, diagnosed with rabies in PI/SP when characterized antigenically, about 30 to 40% of these did not show profiles compatible with the panel utilized. Therefore, it is necessary to broaden this panel of MAb for the characterization of new antigenic variants, thereby making it possible to

[2] Data supplied by the Laboratory for the Diagnosis of Rabies of the Instituto Pasteur of São Paulo.

have a better knowledge of the circulating viruses in these species, contributing to a more effective epidemiological monitoring of the disease in Brazil.

The exclusive use of monoclonal antibodies, however, shows certain limitations, for example, the diversity of variants present in insectivorous bats, which is not totally explained with the available monoclonal antibodies. Genomic analysis is necessary, because it provides more detailed information on the evolutionary relation of the isolates, the spatial and temporal changes that can occur and the similarity among the isolates (Kotait, Carrieri, Takaoka, 2009).

Genetic Characterization of Rabies Virus

Standardization of the RT-PCR technique for rabies virus diagnosis in Brazil improved the laboratory and epidemiological monitoring of this disease. Once this technique shows great diagnostic sensitivity and specificity, facilitating the diagnosis in decomposed specimens, the standardization of the ante-mortem rabies diagnosis in humans and the implementation of the DNA sequencing technique (Macedo et al., 2006; Oliveira et al., 2006). Reverse transcription of the viral nucleic acid followed by amplification of the complementary DNA by RT-PCR, and later analysis of the nucleotide sequences in the viral genome constitute the methodological elements for the identification and differentiation of the rabies virus variants (OPS, 1998).

The data provided by the analysis of the nucleotide (nt) sequence along with the epidemiological description of human and animal rabies helps in the epidemiological monitoring of the disease, since the nt substitutions allow the identification of the rabies virus strains that are found associated with the diverse outbreaks and with the preferred hosts of the typed genetic strains. However, the phylogeny of the strains has little value without adequate information on the circumstances in which the epidemiological events occur, particularly in relation to the reservoir and the factors that contribute to the maintenance of the disease (OPS, 1998).

The genetic characterization of rabies virus in Brazil and in all other countries, is done mainly by sequencing the N gene, but a substantial number of works have analyzed the G gene and secondarily the intergenic region G-L. As the size of the generated sequences is variable, some authors utilize the whole genes, others small regions of the amino-terminal as well as the carboxy-terminal regions of the proteins.

The first study related to genetic typing of rabies virus in Brazil was published by Ito et al. (2001), where the authors sequenced 203 nt of the N gene and determined the two principal virus strains in the country, dogs and *D. rotundus*. The authors also noted that the genetic identity of the isolates from dogs utilized in their work was greater than 99%, while in *D. rotundus* is was greater than 96.6%. From this pioneer work, the genetic characterization of rabies virus showed the true diversity of the virus in the country, and others carried on, but a clear dichotomy was established, genetic studies of rabies virus isolated in canids and in bats.

Genetic typing of rabies virus isolated from bats initiated by Ito et al. (2001) diversified to other species of chiropters, besides *D. rotundus*. Kobayashi et al. (2005), in analyzing specimens of rabies virus isolated from frugivorous and insectivorous species and *D.*

rotundus, reported strains associated with the bats *Artibeus spp* (frugivorous), *D. rotundus* and insectivorous bats, suggesting the existence of species-specific strains. Kobayashi et al. (2007) continued the previous work and obtained a strain associated with the insectivorous bats of the genus *Lasiurus*.

Genetic studies with the rabies virus isolates of insectivorous bats species that showed antigenic profiles not compatible with that expected demonstrated that at least one of these profiles does not discriminate some new antigenic variant; the four other profiles were not demonstrated as new antigenic variants (Oliveira et al., 2010).

Oliveira et al. (2010), in a study with different species of bats, reported different strains of rabies virus associated with different species of insectivorous bats, suggesting the existence of species-specific and genus-specific strains, namely the strains *Tadarida braziliensis, Molossus, Nyctinomops, Eptesicus, Histiotus, Lasiurus* and *Myotis*. In the USA, where this type of study is more advanced, by the fact that insectivorous bats are the main transmitters of rabies to humans, 18 species-specific strains are known to be circulating in these animals (Streicker et al., 2010). The strains *T. braziliensis* and *Lasiurus* were also found in other countries of South America and in Mexico (Velasco-Villa et al., 2006; Nadin-Davis, Loza-Rubio, 2006; De Mattos et al., 1996; De Mattos et al., 2000).

During the last decade, the utilization DNA sequencing of the rabies virus isolated in insectivorous bats of Brazil and its utilization in phylogenetic studies brought great advances in the characterization of the virus circulating in insectivorous bats in the Americas, demonstrating the existence of various genus- and species-specific strains (Velasco-Villa et al., 2006; Sato et al., 2004; Nadin-Davis, Loza-Rubio, 2006; Shoji et al., 2004; Kobayashi et al., 2005; De Mattos et al., 1996; De Mattos et al., 2000.).

Currently, it is known that these strains are maintained through independent epidemiological cycles among their specific reservoirs, where the true risk of this finding for public health in our country has not been well studied (Oliveira et al., 2010.). In Brazil, studies of vaccine protection in mice utilizing commercial vaccines for rabies tested vaccine protection for the strains *Molossus* and *Nyctinomops*, demonstrating that the mice were satisfactorily immunized against these strains (Cunha et al., 2010).

The study of rabies virus strains that infect animals of economic interest are for the most part transmitted by *D. rotundus*. Despite that a large number of animals of economic interest infected by rabies virus are diagnosed, only a small number of *D. rotundus* are infected. For this reason, the study of genetic strains of rabies virus that circulated among these bats are studied indirectly, that is, by the use of isolates of the virus from cattle and horses.

Romijn et al. (2003), in partially analyzing the N gene of rabies virus isolates from cattle in the state of Rio de Janeiro, obtained regional groupings of the virus. Bordignon et al. (2005), in the state of Santa Catarina and also partially analyzing the N gene of rabies virus, obtained a grouping of strains typical of the state of Santa Catarina. Sato et al. (2006), in analyzing 599 nt of the G gene of isolates of rabies virus from various species in the states of Maranhão, Pará and Tocantins, identified virus strains associated with *D.rotundus*. Kobayashi et al. (2006), in partially analyzing the N gene of rabies virus isolates from cattle, from various states of the Southeast and Central-West regions, suggest the existence of regional strains formed by the geographic isolation determined by mountains and rivers. Kobayashi et al. (2008), in continuing earlier work, now with 593 isolates of rabies virus from a greater number states and sequencing 202 nt of the N gene, described 24 regional virus strains, originating by geographic isolation.

Carnieli et al. (2009) studied an epizootic of rabies in cattle and horses and characterized 162 specimens of the rabies virus from these animals and concluded that the N and G genes of the virus are conserved over time and in the region of the Atlantic coast of South America. In constrast, Castilho et al. (2010) studied specimens isolated from humans in the Amazon region and belonging to the antigenic variant typical of *D. rotundus* and found 5 genetic strains of the virus. In another study comparing genetically isolates of *D. rotundus* and *Diphylla ecaudata,* it was suggested that the two species share the same genetic strains of rabies virus and that there is possible cross transmission between these two species (Castilho et al., 2010). Fahl et al. (2010) in studying rabies virus isolated from *Artibeus lituratus* concluded that despite the great genetic identity between the virus strains circulating among these species and *D. rotundus*, *A. lituratus* is the specific reservoir of a rabies virus strain.

Monitoring and Control

Although human rabies transmitted by non-hematophagous bats has shown a growing increase in the world due to lyssavirus (Lumio et al., 1986), there is currently no method of control or practices to be employed in focus areas, as there are for other reservoir species. However, when there is mention of control, besides the vaccination of susceptible individuals, there is great emphasis on the preservation of bats, such as dislodging them from their shelter, closing off access to or pruning of tree branches or gathering up flowers and fruits that serve as food, or even the substitution of fruiting plants with other non-fruit-bearing ones (Harmani, Silva, Hayashi, 1996).

With the knowledge of the epidemiology of rabies spread by hematophagous bats, *D. rotundus*, to domestic herbivorous animals and humans, there was a need to establish strategies for the prevention of the disease, where one of these was the development of methods of controlling the populations of these bats. These methods were initially drastic and aimed at eliminating all bats from the shelter, with the use of physical or chemical agents, such as, toxic gases, fire or smoke, dynamite, firearms and distribution of poisons on the walls of shelters. Knowledge of the ecology, biology and ethology of *D. rotundus* allowed the development methods more perfected for the control of their populations, which can be classified as restrictive and selective (Bredt et al., 1998).

Restrictive methods aim at preventing attacks by the use of physical means that act as protection barriers between the animals and the hematophagous bats. These methods do not kill the bats, but only restrict their access to food sources and/or shelters. One example is the use fine-mesh screens, as mechanical barrier, making it impossible for bats to have access to the animals or humans. Although these methods are ecologically correct, they do not show risk to human health and do not require special training, but their application can have disadvantages, by being restricted to small herds and having high costs (Bredt et al., 1998; MAPA, 2009).

The selective methods can be indirect or direct and are carried out with the use of a chemical anticoagulant substance, which upon being ingested by the bat causes internal hemorrhaging leading to death. Both should be executed in a way that only the hematophagous bats, *D. rotundus*, are affected, avoiding damages or disturbances to other species, which play an important role in maintaining ecological balance in nature. By the only indirect method currently available in Brazil, the anticoagulant substance is applied around recent bites of hematophagous bats on herbivorous animals, eliminating only the aggressor bats, considering that they tend to return on consecutive to the same wound for feeding (Bredt et al., 1998; MAPA, 2009).

The direct selective method, currently the most utilized in Brazil, consists of the capture of the hematophagous bat *D. rotundus*, topical application of the anticoagulant substance on its back and its release to return to the shelter, where it will through the habits of social interaction and hygiene, distribute the substance to other bats not captured. This thereby reduces the number of individuals in the colony. The capture should be done, preferentially, next to the food source of the hematophagous bats or directly in their shelters, when these are artificial and in the proximity of natural shelters, where mist nets can be utilized for nighttime capture and hand nets for daytime capture (Bredt et al., 1998; MAPA, 2009).

In Brazil, there is a program for rabies control in herbivores, whose strategy is based on the adoption of procedures of animal health aimed at protecting public health and the development of future policies for the control of this disease. Having notification of the suspect of rabies occurrence in herbivores, or attacks on a herd by hematophagous bats and positive laboratory diagnosis of a sample collected in the focus, measures for control of the focus are initiated, involving epidemiological investigation, collection of samples, control of transmitters, vaccination, and guidance of producers (MAPA, 2009; Kotait et al., 2010).

The prophylactic measures of control and management of bats in urban areas of the focus of rabies in Brazil involve: follow-up of the bat suspected of infection for laboratory diagnosis and identification of the species; prophylaxis of humans who have come in contact with or attacked by a bat, by means of serum vaccination and of animals contacted, depending on the immunological status, by restricting contact with other animals or humans, vaccination and revaccination or euthanasia; geographic and environmental characterization of the region to identify risk factors; and educational actions to inform the public about the positive case, influencing the public about the risk of disease and about the necessary actions of epidemiological monitoring (IP, 2006).

The proximity between humans and bats, in urban and rural spaces, should be understood as an integral part of a broader context of organization of the environment in transformation, considering the dynamics of the factors and productive processes on which this acts. The diseases then appear as the result of new rules of interaction between the living species, established from disrupting the ecological balance. The appropriate utilization of knowlege available today about the biology, ethology and ecology of bats may lead to a reduction in the negative aspects and the exploration of positive aspects that their proximity can offer. Adaptation and transmission of this knowledge to society, as a result of an interactive process, will allow us to attain a harmonic coexistence between humans and bats (MAPA, 2009).

References

Acha, PN. Epidemiology of paralytic rabies and bat rabies. *Boletín de la Oficina International de Epizootias*, 1967; 67(3-4), p.343-382.

Acha, PN, Szyfres, B. *Zoonosis y enfermedades transmisibles comunes al hombre y a los animales.* 2.ed. Washington, D.C.: Organización Panamericana de la Salud; 1986.

Albas, A; Souza, EA; Lourenço, RA; Favoretto, SR; Sodré MM. Antigen profile of rabies virus isolated from different species of non-hematophagous bats in the region of Presidente Prudente, State of São Paulo. *Rev Soc Bras Med Trop*. 2009; 42(1), p.15-17.

Altenbach, JS. Prey capture by fishing bats *Noctilio leporinus* and *Myotis vivesi*. *J Mammal,* 1989; 70, p.421-424.

Altringham, JD. *Bats: Biology and Behaviour.* New York: Oxford University Press. 1996; 262p.

Amorim, AF; Silva, RA; Silva, NM. Isolamento do vírus rábico de morcego insetívoro, *Histiotus velatus*, capturado no Estado de Santa Catarina. *Pesquisa Agropecuária Brasileira*, 1970; 5, p.433-435.

Arai, YT; Kuzmin, IV; Kameoka, Y; Botvinkin, AD. New Lyssavirus genotype from the lesser mouse-eared bat (*Myotis blythi*), Kyrghystan. *Emerg Infect Dis*, 2003; 9(3), p.333-337.

Arellano-Sota, C; Sureau, P; Greenhal, AM. Preferência de la predación del vampiro a la edad y la raza del ganado y la época del año. *Técnica Pecuária*, 1971; 17, p.23-29.

Awasthi, M; Parmar, H; Patankar, T; Castillo, M. Imaging findings in rabies encephalitis. *AJNR Am J Neuroradiol*., 2001; 22(4), p.677-80.

Baer, GM. Rabies in nonhematophagous bats. In: Baer, GM. *The natural history of rabies.* New York: Academic Press, 1975; 2, p.79-97.

Bernard, E; Fenton, MB. Bat mobility and roosts in a fragmented landscape in central Amazonia, Brazil. *Biotropica.* 2003; 35(2), p.262-277.

Bernardi, F; Gomes, AAB; Ito, FH; Sakai, T. Raiva em morcego hematófago. *Ars Veterinaria,* 1998; 14(2), p.186-192.

Bigler, WJ; Hoff, GL; Buff, EE. Chiropteran rabies in Florida: A twenty-year, 1954 to 1973. *Am. J. Trop. Med. Hyg,* 1974; 24, p.347-352.

Bingham, J. The bat lyssaviruses of Africa. *Sixth SEARG meeting,* Lilongwe/Malawi, 2001, p.74-79.

Bordignon, MO. Padrão da atividade e comportamento de forrageamento do morcego-pescador Noctilio leporinus (Linnaeus) (Chiroptera, Noctilionidae) na Baía de Guaratuba, Paraná, Brasil. *Rev Bras Zool,* 2006; 23(1), p.50-57.

Bordignon, J; Brasil-Dos-Anjos, G; Bueno, CR; Salvatiera-Oporto, J; Dávila, AM; Grisard, EC; Zanetti, CR. Detection and characterization of rabies virus in Southern Brazil by PCR amplification and sequencing of the nucleoprotein gene. *Arch Virol*, 2005; 150(4), p.695-708.

Botvinkin, AD; Poleschuk, EM; Kuzmin, IV; Borisova, TI; Gazaryan, SV; Yager, P; Rupprecht, CE. Novel lyssaviruses isolated from bats in Russia. *Emerg Infect Dis,* 2003; 9(12), p.1623-1625.

Brass, DA. *Rabies in bats. Natural history and public health implications.* Connecticut: Livia Press; 1994.

Bredt, AI; Araújo, FAA; Caetano-Júnior, J; Rodrigues, MGR; Yoshiawa, M; Silva, MMS; Harmani, NMS; Massunaga, PNT; Bürer, SP; Potro, VAR; Uieda, W. *Morcegos em áreas urbanas e rurais: manual de manejo e controle.* Brasília: Fundação Nacional de Saúde, Ministério da Saúde, 1998; 117p.

Camargo, G, Fischer, E. Primeiro registro do morcego *Mimon crenulatum* (Phyllostomidae) no Pantanal, sudoeste do Brasil. *Biota Neotropica*, 2005; 5(1), p.4.

Carini, A. Sur une grande épizootie de rage. *Annales de l'Institut Pasteur*, 1911; 5, p.843-846.

Carneiro, V. As epizootias de raiva na América e o papel dos morcegos hematófagos. *Arquivos do Instituto Biológico*, 1936; 7, p.273-321.

Carneiro, V; Freitas Lima, C. Estudos sobre a raiva no Paraná. *Rev Zoot Vet*, 1927; 3(13), p.137-156.

Carnieli, P Jr; Castilho, JG; Fahl, Wde O; Véras, NM; Timenetsky, Mdo C. Genetic characterization of Rabies virus isolated from cattle between 1997 and 2002 in an epizootic area in the state of São Paulo, Brazil. *Virus Res*. 2009; 144(1-2), p.215-224.

Carrieri, ML; Oliveira, RN; Canello, FM; Castilho, JG; Carnielli Jr, P; Achkar, SM et al. Bats rabies in the period between 2000-2006, in Ribeirão Preto County, São Paulo State, Brazil. In: *Reunião Internacional de Raiva nas Américas (RITA) XVII; Brasilia- Brasil*. Anais... Ministério da Saúde; 2006; p.115.

Carstens EB. Ratification vote on taxonomic proposals to the International Committee on Taxonomy of Viruses (2009). *Arch Virol.*, 2010; 155, p.133-146.

Castilho, JG; Carnieli, P Jr; Durymanova, EA; Fahl, Wde O; Oliveira, Rde N; Macedo, CI; da Rosa, ES; Mantilla, A; Carrieri, ML; Kotait, I. Human rabies transmitted by vampire bats: antigenic and genetic characterization of rabies virus isolates from the Amazon region (Brazil and Ecuador). *Virus Res*. 2010;153(1), p.100-105.

Castilho, JG; Iamamoto, K; Lima, JYO; Scheffer, KC; Junior, PC; Oliveira, RN; Macedo, CI; Achkar, SM; Carrieri, ML; Kotait, I. Padronização e aplicação da técnica de isolamento do vírus da raiva em células de neuroblastoma de camundongo (N2A). *Boletim Epidemiológico Paulista* - BEPA, 2007; 4(47), p.12-18.

Cisterna, D; Bonaventura, R; Caillou, S; Pozo, O; Andreau, ML; Fontana, LD; Echegoyen, C; De Mattos, C; De Mattos, C; Russo, S; Novaro, L; Elberger, D; Freire, MC. Antigenic and molecular characterization of rabies virus in Argentina. *Virus Res*. 2005; 109(2), p.139-147.

Constantine, DG. Bat rabies in the southwestern United States. *Public Health Rep.*, 1967; 82, p.867-888.

Constantine, DG. Health precautions for bat researchers. In: Kunz, TH. *Ecological and behavioral methods for the study of bats*. Washington: Smithsonian Inst. Press; 1988; p.491-528.

Constantine, DG. Rabies transmission by non bite route. *Publ Health Rep.*, 1962; 77, p.287-89.

Cunha, EM; Nassar, AF; Lara, Mdo C; Villalobos, EC; Sato, G; Kobayashi, Y; Shoji, Y; Itou, T; Sakai, T; Ito, FH. Pathogenicity of different rabies virus isolates and protection test in vaccinated mice. *Rev Inst Med Trop Sao Paulo*. 2010; 52(5), p.231-236.

Damasceno, GA, Bezerra, MAO, Bortolotto, I, Pott, A. Aspectos florísticos e fitofisionômicos dos capões do Pantanal do Abobral. In: *Anais do II Simpósio sobre Recursos Naturais e Sócioeconômicos do Pantanal – Manejo e Conservação*. CPAP Embrapa e UFMS, Corumbá, 1999; p.203-214.

De Mattos, CA; De Mattos, CC; Smith, JS; Miller, ET; Papo, S; Utrera, A; Osburn, BI. Genetic characterization of rabies field isolates from Venezuela. *J. Clin. Microbiol.* 1996; 34, p.1553-1558.

De Mattos, CA; Favi, M; Yung, V; Pavletic, C; De Mattos, CC. Bat rabies in urban centers in Chile. *J. Wildlife Dis.* 2000; 36, p.231-240.

De Mattos, CC; De Mattos, CA. Uso de anticuerpos monoclonales para La tipificación antigénica da aislamientos de virus rábico. In: *Organización Panamericana de La Salud.*

Consorcio de La OPS de laboratorios de referencia en rabia de las Américas. Washington, DC Bases para la operación. HCP/HCV/R2/015. 1998; p.2-11.

Delpietro, HA; Gury-Dhomen, F; Larghi, OP; Mena-Segura, C; Abramo, L. Monoclonal antibody characterization of rabies virus strains isolated in the River Plate Basin. *Zentralbl Veterinarmed B.* 1997; 44(8), p.477-483.

Diaz, AM; Rodriguez, A; Smith, JS. Antigenic analysis of rabies-virus isolates from Latin America and the Caribbean. *Zentralbl Veterinarmed B.* 1994; 41(3), p.153-160.

Dietzschold, B ; Rupprecht, CE ; Tollis, M ; Lafon, M ; Mattei, J ; Wiktor, TJ; Koprowski, H. Antigenic diversity of the glycoprotein and nucleocapsid proteins of rabies and rabies-related viruses: implications for epidemiology and control of rabies. *Rev. Infect. Dis.* 1988; 10, p.785-798.

Embrapa. Maior morcego das Américas é coletado pela primeira vez no Pantanal Sul. [online]. 2010. Avaiable from: URL:*http://www.cpap.embrapa.br/destaques/2010materia13.html*

Esbérard, CEL; Bergallo, HG. Aspectos sobre biologia de *Tonatia bidens* (Spix) no estado do Rio de Janeiro, sudeste do Brasil (Mammalia, Chiroptera, Phyllostomidae). *Rev Bras Zool,* 2004, 21(2), p.253-259.

Fahl, WO; Carnieli Jr., P, Castilho, JG; Carrieri, ML; Kotait, I; Brandão, PE. Phylogeny of rabies virus strains from *Artibeus* sp and *Desmodus rotundus* bats based on the nucleoprotein and glycoprotein genes In: XX National Meeting of Virology, Brasília. *Journal of the Brazilian Society for Virology- SBV.* , 2009; 14, p.120

Favi, CM; Rodríguez, AL; Espinosa, MC; Yung, PV. Rabies in Chile: 1989-2005. *Rev Chilena Infectol.* 2008; 25(2), p.8-13.

Favi, M; De Mattos, CA; Yung, V; Chala, E; López, LR; De Mattos, CC. First case of human rabies in Chile caused by an insectivorous bat virus variant. *Emerg Infect Dis.* 2002; 8(1), p.79-81.

Favi, M; Nina, A; Yung, V; Fernández, J. Characterization of rabies virus isolates in Bolivia. *Virus Res.* 2003; 97(2), p.135-140.

Favoretto, SR; Carrieri, ML; Cunha, ES; Aguiar, EAC; Silva LHQ; Sodré, MM; Souza, MCAM; Kotait, I. Antigenic typing of brazilian rabies virus samples isolated from animals and humans, 1989-2000. *Rev. Inst. Med. trop. S. Paulo.* 2002; 44 (2), p.91-95.

Fekadu, M; Greer, PW; Chandler, FW; Sanderlin, DW. Use of the avidin-biotin peroxidase system to detect rabies antigen in formalin-fixed paraffinembedded tissues. *J Virol Methods.*, 1988; 19(2), p.91-96.

Fenton, MB. *Bats.* New York: Facts on File. 1992; 207p.

Flamand, A; Wiktor, TJ; Koprowski, H. Use of hybridoma monoclonal antibodies in the detection of antigenic differences between rabies and rabies-related viral proteins. I. The nucleocapsid protein. *J Gen Virol.* 1980; 48, p.97-109.

Fleming, TH; Sosa, VJ. Effects of nectarivorous and frugivorous mammals on reproductive success of plants. *J Mammal,* 1994; 75, p.845-851.

Gardner, AL. Feeding Habits. In: Baker, RJ, Jones-Jr, JK, Carter, DC (eds). *Biology of bats of the New World, Family Phyllostomidae. Part II.* Special Publication of the Museum, Texas Tech University, 1977; 13, p.239-350.

Graham, GL. *Bats of the world.* A golden guide from St. Martin's Press. 1994; 160p.

Greenhall, AM; Schutt Jr, WA. *Diaemus youngi. Mammalian Species.* 1996; 533, p.1-7.

Greenhall, AM; Joermann, G; Schmidt, U; Seider, MR. *Desmodus rotundus. Mammalian Species.* 1983; 202, p.1-6.

Greenhall, AM; Schmidt, U; Joermann, G. *Diphylla ecaudata. Mammalian Species.* 1984; 227, p.1-3.

Harmani, NMS; Silva, MMS; Hayashi, MM. Controle de morcegos em áreas urbanas. *Boletim do Instituto Pasteur,* 1996; 1(2), p.37-43.

Haupt, H; Rehaag, H. Raiva epizoótica nos rebanhos de Santa Catarina, transmitida por morcegos. *Boletim da Sociedade Brasileira de Medicina Veterinária,* 1925; 2, p.17-47.

Hill, JE; Smith, SE. Bats: a community perspective. Cambridge University Press. 1988; 167p.

Huber, J. Matas e madeiras amazônicas. *Boletim do Museu Paraense Emílio Goeldi, série Zoologia,* Belém, 1910, 6, p.91-225.

Humphrey, SR, Bonaccorso, FJ. Population and community ecology. In: Baker, RJ, Jones Jr, JK, Carter, DC (eds). *Biology of bats of the New World family Phyllostomidae Part III.* Special Publications Museum Texas Tech University. 1979; 16, p.409-441.

Hurst, EW; Pawan, JL. An outbreak of rabies in Trinidad. *Lancet,* 1931; 2, p.622-628.

Hutson, AM; Mickleburgh, SP; Racey, PA. Microchiropteran bats: global status survey and conservation action plan. *IUCN/SSC Chiroptera Specialist Group,* 2001, 258p.

IBAMA - Instituto Brasileiro do Meio Ambiente e dos Recursos Naturais Renováveis. Lista nacional das espécies da fauna brasileira ameaçadas de extinção. 2003. [online]. Available from: URL:*http://www.mma.gov.br/port/sbf/fauna/index.cfm*

IBGE - Instituto Brasileiro de Geografia e Estatística. 2010. [online]. Available from: URL:*http://www.ibge.gov.br/home/presidencia/noticias/noticia_visualiza.php?id_noticia=169*

IBGE - Instituto Brasileiro de Geografia e Estatística. *Mapa de biomas do Brasil e o mapa de vegetação do Brasil.* 2004. [online]. Available from: *URL:http://www.ibge.gov.br/home/presidencia/noticias/noticia_visualiza.php?id_noticia=169*

IP - Instituto Pasteur/ Coordenadoria de Controle de Doenças/Secretaria de Estado da Saúde de São Paulo – IP/CCD/SES-SP. Programa de Prevenção e Controle da Raiva Transmitida por Morcegos em áreas Urbanas. *BEPA - Boletim Epidemiológico Paulista,* dezembro de 2006, p.9-13.

Ito, M; Arai, YT; Itou, T; Sakai, T; Ito, FH; Takasaki, T; Kurane, I. Genetic characterization and geographic distribution of rabies virus isolates in Brazil: identification of two reservoirs, dogs and vampire bats. *Virology.* 2001; 284(2), 214-222.

Iwasaki, T; Inoue, S; Tanaka, K; Sato, Y; Morikawa, S; Hayasaka, D; Moriyama, M; Ono, T; Kanai, S; Yamada, A; Kurata, T. Characterization of Oita virus 296/1972 of Rhabdoviridae isolated from a horseshoe bat bearing characteristics of both lyssavirus and vesiculovirus. *Arch Virol.,* 2004; 149(6), p.1139-1154.

Jackson, AC. Pathogenesis. In: Jackson AC, Wunner WH. *Rabies.* San Diego: Academic Press; 2002.

Johnson, N; Wakeley, PR; Brookes, SM; Fooks, AR. European bat lyssavirus type 2 RNA in *Myotis daubentonii. Emerg Infect Dis.* 2006; 12(7), p.1142-1144.

Kaplan, MM. Safety precautions in handling rabies virus. In: Meslin, FX, Kaplan, MM, Koprowski, H (eds.). *Laboratory techniques in rabies.* 4th ed. Geneva: World Health Organization; 1996; p.3-8.

King, A. European bat rabies. *Sixth SEARG meeting,* Lilongwe/Malawi, 2001, p.96-100.

Kobayashi, Y; Sato, G; Mochizuki, N; Hirano, S; Itou, T; Carvalho, AA; Albas, A; Santos, HP; Ito, FH; Sakai T. Molecular and geographic analyses of vampire bat-transmitted cattle rabies in central Brazil. *BMC Vet Res*, 2008; 4, p.44.

Kobayashi, Y; Sato, G; Kato, M; Itou, T; Cunha, EMS; Silva, MV; Mota, CS; Ito, FH; Sakai, T. Genetic diversity of bat rabies viruses in Brazil. *Arch. Virol.* 2007; 153(11), p.1995-2004.

Kobayashi, Y; Ogawa, A; Sato, G; Sato, T; Itou, T; Samara, SI; Carvalho, AA; Nociti, DP; Ito, FH; Sakai T. Geographical distribution of vampire bat-related cattle rabies in Brazil. *J Vet Med Sci*, 2006; 68(10), p.1097-1100.

Kobayashi, Y; Sato, G; Shoji, Y; Sato, T; Itou, T; Cunha, EMS; Samara, SI; Carvalho, AAB; Nociti, DP; Ito, FH; Sakai, T. Molecular Epidemiological Analysis of Bat Rabies Viruses in Brazil. *J. Vet. Med. Sci.* 2005; 67(7), p.647-652.

Koopman, KF. Systematics and distribution. In: Greenhall, AM, Schmidt, U. *Natural history of vampire bats*. Florida: CRC Press, Inc. Boca Raton, 1988, p.7 - 17.

Koprowski, H. The mouse inoculation test. In: Meslin, FX, Kaplan, MM, Koprowski, H. *Laboratory techniques in rabies*. 4.ed. Geneva: World Health Organization, 1996; p.80-87.

Kotait, I; Nogueira Filho, VS; Souza, MCAM; Carrieri, ML; Gomes MN; Peres, NF. *Manual de controle da raiva dos herbívoros*. 9. São Paulo/SP: Instituto Pasteur; 2010.

Kotait, I; Carrieri, ML; Takaoka, N. Raiva – Aspectos gerais e clínica. *Manual Técnico do Instituto Pasteur*. 2009; 8, p.28-29.

Kotait, I; Carrieri, ML. Raiva. In: Trabulsi, LR. *Microbiologia*. 5 ed. São Paulo: Atheneu, 2008; p.689-698.

Kucera, P; Dolivo, M; Coulon, P; Flamand, A. Pathways of the early propagation of virulent and avirulent rabies strains from the eye to the brain. *J Virol*. 1985; 55(1), p.158-162.

Kuzmin, IV; Mayer, AE; Niezgoda, M; Markotter, W; Agwanda, B; Breiman, RF; Rupprecht, CE. Shimoni bat virus, a new representative of the Lyssavirus genus. *Virus Res.*, 2010; 149(2), p.197-210.

Kuzmin, IV; Orciari, LA; Arai, YT; Smith, JS; Hanlon, CA. Bat lyssaviruses (Aravan and Khujand) from Central Asia: phylogenetic relationships according to N, P and G gene sequences. *Virus Res*, 2003; 97(2), p.65-79.

Kuzmin, IV; Botvinkin, AD; Khabilov, TK. The lyssavirus was isolated from a whiskered bat in northern Tajikistan. *Plecotus*, 2001; 4, p.75-81.

Lafon, M., Lafage, M. Antiviral activity of monoclonal antibodies specific for the internal proteins N and NS of rabies virus. *J Gen Virology*. 1987; 68, p.3113-3123.

Lépine, P, Atanasiu, P. Histopathological diagnosis. In: Meslin, FX, Kaplan, MM, Koprowski, H. *Laboratory techniques in rabies*. 4th ed. Geneva: World Health Organization; 1996; p.66-79.

Lewis, P; Fu, Y; Lentz, TL. Rabies virus entry at the neuromuscular junction in nerve–muscle cocultures. *Muscle and Nerve*. 2000; 23(5), p.720–730.

Longo, JM; Fischer, E; Camargo, G; Santos, CF. Ocorrência de *Vampyressa pusilla* (Chiroptera, Phyllostomidae) no Pantanal sul. *Biota Neotrop.*, 2007; 7(3), p.369-372.

Lumio, J; Hillbom, M; Roine, R; Keonen, L; Haltia, M; Valle, M; Neuvonen, E; Lahdevirta, J. Human rabies of bat origin in Europe. *Lancet*. 1986; 15, p.378-382.

Luna, GL; Chahuayo, LM; Marreros, AT. Diez casos de rabia humana em Pucallpa, Enero-Agosto 1984. *Diagnostico*. 1985; 15, p.156-160.

Macedo, CI; Carnieli Jr, P; Brandão, PE; Travassos da Rosa, ES; Oliveira, Rde N; Castilho, JG; Medeiros, R; Machado, RR; Oliveira, RC; Carrieri, ML; Kotait, I. Diagnosis of human rabies cases by polymerase chain reaction of neck-skin samples. *Braz J Infect Dis*. 2006; 10(5), p.341-345.

MAPA - Ministério da Agricultura, Pecuária e Abastecimento/ Secretaria de Defesa Agropecuária. *Controle da Raiva dos Herbívoros*. 2nd ed. Brasília: Ministério da Agricultura, Pecuária e Abastecimento; 2009.

Marinho-Filho, J, Sazima, I. Brazilian bats and conservation biology. A first survey. In: Kunz, TH, Racey, PA. (eds.) *Bat Biology and Conservation*. Washington D.C.: Smithsonian Institution; 1998; p.282-294.

Marinho-Filho, J. The Brazilian cerrado bat fauna and its conservation. *Chiroptera Neotropical*, 1996; 2(1), p.37-39.

Mattos, CA; Mattos, CC; Rupprecht, CE. Rhabdoviruses. In: Knipe, DM; Howley, PM. *FIELD Virology*. 4ed. Philadelphia: Lippincott Williams e Wilkins. 2001; p.1245-1278.

Mello, MAR; Kalko, EKV; Silva, WR. Movements of the bat Sturnira lilium and its role as a seed disperser of Solanaceae in the Brazilian Atlantic Forest. *J Trop Ecol,* 2008; 24, p.225-228.

Meslin, FX; Kaplan, MM. An overview of laboratory techniques in the diagnosis and prevention of rabies and in rabies research. In: Meslin, FX; Kaplan, MM; Koprowski, H. *Laboratory techniques in rabies*. Geneva: World Health Organization, 1996; p.9-27.

Messenger, SL; Smith, JS; Orciari, LA; Yager, PA; Rupprecht, CE. Emerging Pattern of Rabies Deaths and Increased Viral Infectivity. *Emerg Infect Dis*. 2003; 9(2), p.151–154.

Morimoto, K; Patel, M; Corisdeo, S; Hooper, DC; Fu, ZF; Rupprecht, CE; Koprowski, H; Dietzschold, B. Characterization of a unique variant of bat rabies virus responsible for newly emerging human cases in North America. *Proc Natl Acad Sci USA 93*. 1996; p.5653–5658.

MS – Ministério da Saúde. 2010. [online]. Available from: URL:*http://portal.saude.gov.br/portal/saude/default.cfm*

Nadin-Davis, AS; Loza-Rubio, E. The molecular epidemiology of rabies associated with chiropteran hosts in Mexico. Virus Res. 2006; 117, p.215-226.

Nel, LH. Mokola Virus: A brief review of the status quo. *Sixth SEARG meeting,* Lilongwe/Malawi, 2001, p.80-86.

Nilsson, MR; Nagata, CA. Isolamento do vírus rábico do cérebro, glândulas salivares e interescapular, coração, pulmões e testículos de morcegos *Desmodus rotundus* (Geoffroy, 1910), no Estado de São Paulo. *Arq Inst Biol (São Paulo)*.1975; 42(23), p.183-188.

OIE – Organização Mundial da Saúde Animal. Manual of diagnostic test and vaccines for animals. 5th ed. [online]. 2008. Avaiable from: *URL:http://www.oie.int/eng/normes/mmanual/A_00044.html*

Oliveira, Rde N; de Souza, SP; Lobo, RS; Castilho, JG; Macedo, CI; Carnieli, P Jr; Fahl, WO; Achkar, SM; Scheffer, KC; Kotait, I; Carrieri, ML; Brandão, PE. Rabies virus in insectivorous bats: implications of the diversity of the nucleoprotein and glycoprotein genes for molecular epidemiology. *Virology*. 2010; 405(2), p.352-360.

Oliveira, RN; Takaoka, N; Brandão, P; Carnieli, P Jr; Macedo, C; Castilho, J; Carrieri, ML; Kotait, I. Postmortem confirmation of human rabies source. *Emerg Infect Dis.* 2006; 12(5), p.867-869.

OPS - Organización Panamericana de la Salud. OPS. Consorcio de la OPS de laboratórios de referencia en rabia de las Américas. Bases para la operación. HCP/HCV/R2/015; 1998; 43p.

OPS - Organización Panamericana de la Salud. OPS. Los anticuerpos monoclonales en la caracterización y vigilancia de los virus de la rabia en América Latina y el Caribe. *Rev. Panam Salud Publica/ Pan Am J Public Health.* 2000; 8(3), p.214-217.

Passos, FC; Miranda, JMD; Bernardi, IP; Kaku-Oliveira, NY; Munster, LC. Morcegos da região sul do Brasil: análise comparativa da riqueza de espécies, novos registros e atualizações nomenclaturais (Mammalia, Chiroptera). *Iheringia, Sér. Zool.* 2010; 100(1), p.25-34.

Passos, EC; Carrieri, ML; Silva, MMS; Pereira, RG; Melo, JATS; Maule, LJ. Vírus rábico isolado de morcego frugívoro (*Artibeus lituratus*), capturado em 1997 no município d Rio Claro, SP. *Braz. J. Vet. Res. Anim. Sci.* 1999; 36(1).

Passos, EC; Carrieri, ML; Dainovskas, E; Camara, M; Silva, MMS. Isolamento do vírus rábico em morcego insetívoro, *Nyctinomops macrotis*, no município de Diadema, SP (Brasil). *Rev da Saúde Pública,* 1998; 32(1), p.74-76.

Pawan, J. L. The transmission of paralytic rabies in Trinidad by the vampire bat (*Desmodus rotundus murinus* Wagner, 1840). *Annals of Tropical Medicine and Parasitology*, 1936a; 30(4), p.101-130.

Pawan, JL. Rabies in the vampire bat of Trinidad, with special reference to the clinical course and the latency of infection. *Annals of Tropical Medicine and Parasitology*, 1936b; 30(4), p.401-422.

Pedro, WA. *Diversidade de morcegos em habitats florestais fragmentados do Brasil (Chiroptera, Mammalia).* São Carlos, SP: Tese (Doutorado) - Universidade Federal de São Carlos. 1998; 128f.

Peracchi, AL; Nogueira, MR. Anotada Lista dos Morcegos do Estado do Rio de Janeiro. *Chiroptera Neotropical*, 2010; 16(1), p.508-519.

Peracchi, AL, Lima, IP, Reis, NR, Nogueira, MR, Filho, HO. Ordem Chiroptera In: Reis, NR, Peracchi, AL, Pedro, WA, Lima, IP (eds.) *Mamíferos do Brasil.* Londrina, 2006, p.162-164.

Peracchi, AL, Albuquerque, ST, Raimundo, SDL. Contribuição ao conhecimento dos hábitos alimentares de *Trachops cirrhosus* (spix, 1823) (Mammalia: Chiroptera: Phyllostomidae). *Arquivos da Universidade Federal Rural do Rio de Janeiro.* 1982; 5(1), p.1-5.

Peracchi, AL; Albuquerque, ST. Sobre os hábitos alimentares de *Chrotopterus auritus australis* Thomas, 1905 (Mammalia, Chiroptera, Phyllostomidae). *Rev Bras Biol,* 1976; 36(1), p.179-184.

Perl, DP; Good, PF. The pathology of rabies in the central nervous system In: Baer GM. *The Natural History of Rabies.* 2nd ed. Boca Ratón: CRC Press. 1991; p.164-188.

Pijl, LVD. The dispersal of plants by bats (Chiropterocory). *Acta Amazônica.* 1957; 6, p.291-315.

Pounder, DJ. Rabies, lyssaviruses and bats. *Scott. Med. J.,* 2003; 48(4), p. 99-101.

Reis, NR, Shibatta, AO, Peracchi, AL, Pedro, WA, Lima, IP. Sobre os morcegos brasileiros. In: Reis, NR, Peracchi, AL, Pedro, WA, Lima, I.P. (eds.) *Morcegos do Brasil*. Londrina; 2007; p.17-24.

Rodrigues, FM; Nagata, CA; Peixoto, ZMP; Nilsson, MR. Isolamento do vírus da raiva de morcego insetívoro *Molossus obscurus* (Geoffroy, 1805), no Estado de São Paulo. *Arq. Inst. Biol.*, 1975; 42, p.193-196.

Romano, MC; Maidagan, JI; Pire, EF. Behaviour and demography in urban colony of *Tadarida brasiliensis* (Chiroptera, Molossidae) in Rosario, Argentine. *Rev Biol Trop,* 1999; 47(4), p.1121-1127.

Romijn, PC; van der Heide, R; Cattaneo, CA; Silva, Rde C; van der Poel, WH. Study of lyssaviruses of bat origin as a source of rabies for other animal species in the State of Rio De Janeiro, Brazil. *Am J Trop Med Hyg,* 2003; 69(1): p.81-86.

Rudd, RJ; Smith, JS; Yager, PA; Orciari, LA; Trimarchi, CV. A need for standardized rabies-virus diagnostic procedures: Effect of cover-glass mountant on the reliability of antigen detection by the fluorescent antibody test. *Virus Res*. 2005; 111, p.83-88.

Rudd, RJ; Trimarchi, CV. Tissue culture technique for routine isolation of street strain rabies virus. *J Clin Microbiol*, 1987; 25, p.1456-1458.

Rupprecht, CE; Hanlon, CA; Hemachudha, T. Rabies re-examined. *Lancet Infect. Dis.*, 2002; 2(6), p.327-343.

Sato, G; Kobayashi, Y; Shoji, Y; Sato, T; Itou, T; Ito, FH; Santos, HP; Brito, CJ; Sakai, T. Molecular epidemiology of rabies from Maranhão and surrounding states in the northeastern region of Brazil. *Arch Virol*., 2006; 151(11), p.2243-2251.

Sato, G; Itou, T; Shoji, Y; Miura, Y; Mikami, T; Ito, M; Kurane, I; Samara, SM; Carvalho, AAB; Nociti, DP; Ito, FH; Sakai, T. Genetic and Phylogenetic Analysis of Glicoprotein of Rabies Virus Isolated from Several Species in Brazil. *J. Vet. Med. Sci.* 2004; 66(7), p.747-753.

Scatterday, JE. Bat rabies in Florida. *J Am Vet Med Assoc*, 1954; 124(923), p.125.

Scheffer, KC; Carrieri, ML; Albas, A; Santos, HCP; Kotait I; Ito, FH. Rabies virus in naturally infected bats in the state of São Paulo, Southeastern Brazil. *Rev Saúde Pública*. 2007; 41(3), p.389-395.

Schneider, MC; Romijn, PC; Uieda, W; Tamayo, H; Silva, DF; Belotto, A; Silva, JB; Leanes, LF. Rabies transmitted by vampire bats to humans: An emerging zoonotic disease in Latin America? *Rev Panam Salud Publica*. 2009; 25(3), p.260-269.

Schneider, MC. Estudo *de avaliação sobre área de risco para a raiva no Brasil*. 1990. Tese (Mestrado em Saúde Pública) - Escola Nacional de Saúde Pública - Fundação Oswaldo Cruz, Rio de Janeiro. 229f.

Schneider, LG; Barnard, BJH; Schneider, HP. Application of monoclonal antibodies for epidemiological investigations and oral vaccination studies. I. African viruses. II. Artic viruses. III. Oral rabies vaccine. In: Kuwert E, Merieux C, Koprowski H, Bogel K. (eds.) *Rabies in the tropics*. Berlin: Springer-Verlag. 1985; p.47-59.

Shankar, V; Dietzschold, B; Koprowski, H. Direct entry of rabies virus into the central nervous system without prior local replication. *J Virol*. 1991; 65(5), p.2736-2738.

Shoji, Y; Kobayashi, Y; Sato, G; Itou, T; Miura, Y; Mikami, T; Cunha, EM; Samara, SI; Carvalho, AA; Nocitti, DP; Ito, FH; Kurane, I; Sakai, T. Genetic characterization of rabies viruses isolated from frugivorous bat (Artibeus spp.) in Brazil. *J Vet Med Sci.*, 2004; 66(10), p.1271-1273.

Silva, LHQda; Cunha, EMS; Pedro, WA; Cardoso, TC; Souza, MCCde; Ferrari, CIL. Isolamento do vírus rábico em *Molossus ater* (Chiroptera: Mollosidae) no Estado de São Paulo. *Rev. Saúde Pública,* 1999; 33(6), p.626-628.

Silva, RA; Alencar, AO. Isolamento de vírus rábico das glândulas salivares de morcegos carnívoros da espécie *Chrotopterus auritus australis* (Thomas). *Veterinária.* 1968; 21, p.7-10.

Silva, NM; Silva, RA. Ocorrência do vírus da raiva em morcegos frugívoros *Artibeus lituratus lituratus* (Olfers, 1818). In: *Congresso Brasileiro de Medicina Veterinária,* 14, São Paulo, 1974; Anais. p.102.

Silva, RA; Souza, AM. A ocorrência de vírus rábico em morcegos hematófagos a espécie *Diaemus youngi* (Jentink) no Brasil. *Veterinária.* 1968a; 21, p.53-55.

Silva, RA, Souza, AM. Isolamento do vírus rábico de pulmão, coração, rins, bexiga e outros diferentes tecidos de morcegos hematófagos da espécie *Desmodus rotundus*. *Pesquisa Agropecuária Brasileira* 1968b; 3, p.291-301.

Silva, RA; Souza, AM. A pesquisa do vírus da raiva na glândula interescapular de morcegos do Brasil, em condições naturais de infecção. *Pesquisa Agropecuária Brasileira,* 1968c; 3, p.313-315.

Silva, RA; Rivello, GV; Nilsson, MR. Isolamento de vírus rábico de morcego não hematófago da espécie *Phyllostomus hastatus hastatus* (Pallas). *Arq. Inst. Biol. Animal,* 1961; 4, p.115-120.

Simmons, NB. Order Chiroptera. In: Wilson, DE, Reeder, DM (eds). *Mammals Species of the World: a taxonomic and geographic reference.* 3ed, Jonhs Hopkins Universit Press. 2005; 1, p.312-529.

Smith, JS. New aspects of rabies with emphasis on epidemiology, diagnosis, and prevention of the disease in the United States. *Clin Microbiol Review*, 1996; 9(2), p.166-176.

Smith, JS; Fishbein, DB; Rupprecht, CE; Clark K. Unexplained rabies in three immigrants in the United States. A virologic investigation. *New Eng J of Med.* 1991; 324, p.205-211.

Smith, JS. Monoclonal antibody studies of rabies in insectivorous bats of the United States. *Rev Infect Dis.* 1988; 10(4), p.637-643.

Sodré, MM; Gama, AR; Almeida, MF. Updated list of bat species positive for rabies in Brazil. *Rev. Inst. Med. Trop*. São Paulo, 2010; 52(2), p.75-81.

Streicker, DG; Turmelle, AS; Vonhof, MJ; Kuzmin, IV; McCracken, GF; Rupprecht, CE. Host phylogeny constrains cross-species emergence and establishment of rabies virus in bats. *Science.* 2010; 329(5992), p.676-679.

Sulkin, SE, Allen, R. Experimental rabies virus infection in bats. In: Baer, GM. *The natural history of rabies.* New York: Academic Press, 1975; 2, 99-114.

Taddei, VA. Morcegos. *Algumas considerações* sistemáticas e biológicas. *Bol. Téc. CATI*, Campinas; 1983; 172, p.1-31.

Tierkel, ES, Atanasiu, P. Rapid microscopic examination for Negri bodies and preparation of specimens for biological tests. In: Meslin, FX; Kaplan, MM, Koprowski H. *Laboratory techniques in rabies.* 4th ed. World Health Organization: Geneva, 1996; p.55-65.

Tordo, N. Characteristics and molecular biology of the rabies vírus. In: Meslin, FX, Kaplan, MM, Koprowiski, H. *Laboratory techiniques in rabies* Geneva: World Health Organization, 1996; p.28-51.

Tordo, N; Poch, O; Ermine, A; Keith, G. Primary structure of leader RNA and nucleoprotein genes of the rabies genome; segmented homology with VSV. *Nucleic Acids Res.*, 1986; 14, p.2671-2683.

Torres, S; Queiroz Lima, E. A raiva e os morcegos hematófagos. Morcegos que resistem à infecção tornam-se portadores e eliminadores de vírus? *Revista do Departamento Nacional de Produção Animal*, 1936; 3(1-6), p.165-174.

Torres, S; Queiroz Lima, E. A raiva e sua transmissão por morcegos hematophagos e infectados naturalmente. *Revista do Departamento Nacional de Produção Animal*, 1935; 2, p.1-55.

Trimarchi, CV; Smith, JS. Diagnostic Evaluation. In: Jackson, AC; Wunner, HW. *Rabies*. Ellsiever. 2002; p.307–343.

Tsiang, H; Ceccaldi, PE; Lycke, E. Rabies virus infection and transport in human sensory dorsal root ganglia neurons. *J Gen Virol*. 1991; 72, p.1191-1194.

Tuffereau, C; Bénéjean, J; Blondel, D; Kieffer, B; Flamand, A. Low-affinity nervegrowth factor receptor (P75NTR) can serve as a receptor for rabies virus. *EMBO J*. 1998; 17(24), p.7250-7259.

Uieda, W; Sato, TM; Carvalho, MC; Bonato, V. Fruit as unsual food items of the carnivorous bat *Chrotopterus auritus* (Mammalia, Phyllostomidae) from southeastern Brazil. *Rev Bras Zool*, 2007; 24(3), p.844-847.

Uieda, W. Rabies in the insectivorous bat Tadarida brasiliensis in Southeastern in Brazil. *Rev Saúde Pública*. 1998; 32(5), p.484-485.

Uieda, W. Biologia e dinâmica populacional de morcegos hematófagos. *Anais do II Curso de Atualização em raiva dos herbívoros*. Curitiba/Paraná, 1996; 2, p.63-87.

Uieda, W; Harmani, NMS; Silva, MMS. Raiva nos morcegos insetívoros (Molossidae) do Sudeste do Brasil. *Rev. Saúde Pública*. 1995; 29, p.393-397.

Uieda, W; Harmani, NMS; Silva, MMS; Brandão, MM; Aguiar, EAC. Morcegos hematófagos: 2. Um indivíduo macho adulto no centro da cidade de São Paulo. In: *SEMINÁRIO NACIONAL DE RAIVA*. 1992; 1, São Paulo, p. 14. Abstract

Velasco-Villa, A; Orciari, LA; Juárez-Islas, V; Gómez-Sierra, M; Padilla-Medina, I; Flisser, A; Souza, V; Castillo, A; Franka, R; Escalante-Mañe, M; Sauri-González, I; Rupprecht, C.E. Molecular Diversity of Rabies Viruses Associated with Bats in Mexico and Others Countries of the Americas. *J. Clin. Microbiol.* 2006; 44(5), p.1697-1710.

Venters, HD; Hoffert, WR; Scartterday, JE; Hardy, AV. Rabies in bats in Florida. *Am. Jour. Publ. Hlth.*, 1954; 44, p.182-185.

Vincent, J; Bussereau, F; Sureau P. Immunological relationships between rabies virus and rabies-related viruses studied with monoclonal antibodies to Mokola virus. *Ann. Inst.Pasteur/Virol.* 1988; 139, p.157-173.

Warrell, MJ, Warrell, DA. Rabies and other lyssavirus diseases. *The Lancet*. 2004; 363(9413), p.959-969.

Webster, WA, Casey, GA. Virus isolation in neuroblastoma cell culture. In: Meslin, FX, Kaplan, MM, Koprowski, H. *Laboratory techniques in rabies*. Geneva: World Health Organization, 1996; p.96-101.

Webster, WA. A tissue-culture infection test in routine rabies diagnosis. *Canadian J Vet Res.*, 1987; 87, p.113-116.

WHO - World Health Organization. WHO Expert committee on Rabies, Seventh report. Geneva, 1984; (WHO Technical Report Series, n° 709), p.58-60.

Wiktor, TJ, Koprowski, H. Monoclonal antibodies against rabies virus produced by somatic cell hybridization: detection of antigen variants. Proc Nat Acad Sci Wash. 1978; 75, p.3938-3942.

Wilson, DE. Bat Faunas: A trophic comparison. *Systematic Zoology*. 1973; 22(1), p.14-29.

Winkler, WG. Airbone rabies. In: Baer GM. *The natural history of rabies*. New York: Academic Press. 1975; 2, p.115-121.

Whittaker, RJ; Jones, SH. The role of frugivorous bats and birds in the rebuilding of a tropical forest ecosystem, Krakatau, Indonesia. *Journal of Biogeography*. 1994; 21, p.245-258.

Witte, EJ. Bat's rabies in Pennsylvania. *Am J Public Health*, 1954; 44, p.186-187.

Wunner, WH. Rabies Virus. In: Jackson, AC, Wunner, WH (eds.) *Rabies*. San Diego: Academic Press, 2002, p.23-77.

Yung, V; Favi, M; Fernández, J. Genetic and antigenic typing of rabies virus in Chile. Brief report. *Arch Virol*. 2002; 147(11), p.2197-2205.

Zortéa, M. Subfamília Stenodermatinae. In: Reis, NR, Peracchi, AL, Pedro, WA, Lima, IP (eds). *Morcegos do Brasil*. Londrina; 2007; p.107-128.

In: Bats: Biology, Behavior and Conservation
Editors: J. L. Zupan and S. L. Mlakar, pp. 221-251
ISBN 978-1-61209-749-7

Chapter 6

POSTNATAL DEVELOPMENT, WING MORPHOLOGY AND FLIGHT PERFORMANCE OF THE SHORT-NOSED FRUIT BAT, *CYNOPTERUS SPHINX*

***Vadamalai Elangovan*[1], *Yuvana Satya Priya*[1] and *Ganapathy Marimuthu*[2]**

[1]Department of Applied Animal Sciences, Babasaheb Bhimrao Ambedkar University, Vidya Vihar, Rai Bareily Road, Lucknow, Uttar Pradesh, India

[2]Department of Animal Behaviour and Physiology, School of Biological Science, Madurai Kamaraj University, Madurai, Tamil Nadu, India

ABSTRACT

The early life of bats has received very little attention, due to their secretive habits. Members of the Order Chiroptera are altricial and dependent upon the mothers for some weeks or even months following birth. Mother-infant interactions in mammalian species are generally characterized by a marked degree of synchrony in their behaviour. The type of habitat a flying animal chooses to live-in as well as its way of exploiting the habitat is closely related to its body size, wing shape, flight style, flight speed and flight energetics. This chapter deals about postnatal development and age estimation, wing morphology and flight development, and mother-young interactions in the short-nosed fruit bat, *Cynopterus sphinx*.

Adult individuals of *C. sphinx* breed twice in a year, i.e. during September – October and February – March. The young of *C. sphinx* were altricial at birth, their eyes had opened on fifth day from their birth, and the pinnae become unfolded between sixth and ninth day of age. At the age of three days, the mean forearm length and body mass were equivalent to 42.2% and 18.2%, respectively of the values of postpartum females. The length of forearm increased linearly until 36 days and attained 94.1% of mean forearm length of postpartum females at the age of 219 days. Body mass of pups increased linearly until 60 days and attained 72.7% of mean mass of postpartum females at the age of 219 days. Length of total epiphyseal gap of fourth metacarpal-phalangeal joint increased until 15 days of age and subsequently decreased linearly and closed at about 60 days of age. The age predicting equation based on the length of forearm is valid when its dimensions are between 29.4 mm and 52.4 mm (i.e. 3–36 days of age). Whereas the equation for predicting age based on the length of total epiphyseal gap is valid when its dimensions ranged from 47.0 μm to 6.0 μm (i.e. 15–60 days of age). Growth patterns of forearm length and body mass were best fitted with the logistic and Gompertz nonlinear growth models, respectively. There was no significant difference in the growth patterns of body mass and length of total epiphyseal gap with reference to lengths of forearm of captive and wild-grown pups.

Pups of *C. sphinx* were flightless until 25 days of age and they started fluttering at the mean age of 40 days. The wingspan and wing area increased linearly until 45 days of age by which the young bats exhibited clumsy flight with gentle turn. The mean wing area and wingspan of 45-day-old pups were equivalent 55.5% and 67.6%, respectively to the postpartum females. At birth, *C. sphinx* had less developed handwing compared to armwing, however the handwing developed faster than armwing during their postnatal period. Young bats achieved sustained flight at 55 days of age. Wing loading decreased linearly until 35 days of age and thereafter increased to a maximum of 12.82 Nm^{-2} at 125 days of age. The aspect ratio increased linearly until 55 days of age and thereafter stabilized. The logistic growth equation best fitted to the postnatal changes in wingspan and wing area. The predicted minimum power speed (V_{mp}) and maximum range speed (V_{mr}) decreased until onset of flight and thereafter the V_{mp} and V_{mr} increased linearly and approached 96.2% and 96.4%, respectively to the speed of postpartum females at 125 days of age. The requirement of minimum flight power (P_{mp}) and maximum range power (P_{mr}) increased until 85 days of age and thereafter stabilized. The minimum theoretical radius of banked turn (r_{min}) decreased until 35 days of age and thereafter increased linearly and attained 86.5% of r_{min} of postpartum females at the age of 125 days.

Under captivity the females of *C. sphinx* gave birth in the head-down roosting position. Soon after the birth, mothers grasped their infants, covered them with patagium and engaged in cleaning the infants by continuous licking. The newborn babies were found clinging to their mothers tightly using their thumbs and hind feet, and much of the time they firmly attached to a nipple. Mothers carried their newborns with them while foraging. Neonates produced chirping sounds almost immediately after birth. Mothers never approached an alien baby throughout the mother-young reunion experiment. During early postnatal period, the success of mother-young reunion was mainly by the mothers' effort. Mothers always smelt their babies before accepting them and soon after retrieving they licked them for a short period. The time taken for mother-young reunion was shorter during early lactation than mid and late lactations. During the early stage of development, the lactating females showed more care on their young, but the care-giving behaviour declined as the young grew. Such direct contact declined to 6% between 58 and 60 days of lactation. Pups started roosting separately adjacent to their mothers at the age of 30 days and at this stage they scampered frequently in the roost. Pups flew clumsily when they were about 45 days old. After the attainment of clumsy flight, the pups made foraging attempts by biting and licking fruit pieces. Between 45 and 55 days of age, the young bats engaged in both suckling and feeding upon fruits. The young bats exhibited independent foraging from 55 days of their age and weaned completely from their mothers after 65 days of age. After achieving independent foraging, juveniles typically departed from their roost and began to feed upon fruits earlier than adults. The young bats made more number of foraging bouts than their mothers in a night.

INTRODUCTION

Postnatal Development and Age Estimation

Bats are represented by a taxonomically large and ecologically diverse order of mammals, and thus provide excellent opportunities for testing assumptions and predictions from life history theory. Bats exhibit both precocial (Kurta and Kunz, 1987) and altricial (Powers *et al.*, 1991; Hughes *et al.*, 1995) characteristics at birth. At the time of birth many species of bats are altricial and thus unable to fly and forage independently until they are weaned. The important features of postnatal growth and development in bats are the attainment of flight and independence from parental care. Studies on postnatal development and growth rate are particularly important in deriving equations to predict age (Kunz and Hood, 2000). Morphometric variation has been used as an indicator of a diverse of taxonomic and evolutionary studies including ecology, physiology, population biology, geographic variation and sexual dimorphism (Bookstein, 1982; Findley and Wilson, 1982; Willig and Moulton, 1989; Gannon *et al.,* 1992; Emmanuvel Rajan and Marimuthu, 1999; Sterbing, 2002). In ecological studies it is often necessary to determine the exact age of an animal. Without knowing an animal's age it is impossible to establish certain factors, such as growth rates, sexual maturity, development of various behavioural repertoires, and periodicity of reproduction or longevity of an animal. In vertebrates, postnatal growth rates may provide a valuable index of maternal investment, and milk energy output of females during lactation (Kunz and Stern, 1995).

Growth and development of bats have been studied during prenatal and postnatal periods (Orr, 1970; Tuttle and Stevenson, 1982; Kunz and Hood, 2000) under both natural (e.g. Kunz, 1973; O'Farrell and Studier, 1973; Buchler, 1980; Kunz and Robson, 1995; Hoying and Kunz, 1998; Stern and Kunz, 1998; Baptista *et al.*, 2000) and captive (e.g. Jones, 1967; Kleiman, 1969; Taft and Handley, 1991; Hughes *et al.*, 1995; Emmanuvel Rajan and Marimuthu, 1999; Swift, 2001) conditions. Data on growth trajectories of bats also are valuable for estimating or predicting their ages. Postnatal growth data have been used in studies of energy and mineral accretion (Studier and Kunz, 1995; Papadimitriou *et al.*, 1996), milk composition (Kunz *et al.*, 1995), ontogeny of flight (Powers *et al.*, 1991; Kunz and Anthony, 1996), ontogeny of echolocation sounds (Habersetzer and Marimuthu, 1986; Moss *et al.*, 1997) and postnatal growth (Emmanuvel Rajan and Marimuthu, 1999). Baptista *et al.* (2000) studied the postnatal growth in free-ranging *M. lucifugus* using cross-sectional and longitudinal methods. They emphasized that the mark-recapture method (longitudinal sampling) is the most appropriate for deriving growth rates and to estimate age during the postnatal growth period.

Patterns of growth and development vary among species and families of bats (Tuttle and Stevenson, 1982; Kunz and Hood, 2000). A positive relationship between body masses of young and mothers occurs in the greater spear-nosed bat *Phyllostomus hastatus* (Stern and Kunz, 1998). Body mass, length of forearm and length of total epiphyseal gap have been proved as important variables for assessing postnatal growth rates in bats (Kunz and Anthony, 1982; Kunz, 1987; Kunz and Stern, 1995). The growth rates and patterns of bats have been evaluated using different growth models on very few species of bats such as *Tadarida brasiliensis* (Kunz and Robson, 1995), *Plecotus auritus* (McLean and Speakman, 2000) and *M. nattereri* (Swift, 2001). Nonlinear growth equations describe best-fit growth models to the postnatal growth data (Zullinger *et al*., 1984). Most of the above studies have been conducted on microchiropteran bats. This study was conducted to derive age predictive equations from the postnatal growth trajectories and to quantify growth rates in the short-nosed fruit bat *Cynopterus sphinx* (Megachiroptera: Pteropodidae). In addition, we have compared the growth of body mass and length of total epiphyseal gap with respect to the growth of forearm length of captive and wild populations.

Wing Morphology and Flight Development

Studies on wing morphology and development of flight in bats are valuable in understanding foraging strategies. The structure and shape of wings of flying animals influence flight performance and energetics and limit foraging distances (Norberg, 1981). Flight morphology and foraging ecology of bats have a direct relationship (McKenzie *et al.,* 1995; Richmond *et al*., 1998; Rhodes, 2002). Wing morphology of flying animals and habitat use are interrelated (Aldridge and Rautenbach, 1987; Norberg and Rayner, 1987; Block *et al.,* 1991; Landmann and Winding, 1993; Bogdanowicz *et al*., 1999; Rhodes, 2002). The association between wing morphology of bats and their habitat use inferred using aerodynamic theory (Norberg, 1985; Fullard *et al.,* 1991), and habitat use of bats correlated with wing morphology without examining flight performance (McKenzie and Rolfe, 1986; Crome and Richards, 1988; Ševcík, 2003). In microchiropteran bats, the ontogeny of flight is associated with development of their echolocation (Buchler, 1980; Brown *et al.,* 1983; Rother and Schmidt, 1985; Rübsamen, 1987). The development of vocal sound and mother-infant relation is also related with the development of flight (Barclay *et al*., 1979; Brown *et al.,* 1983; Gelf and McCracken, 1986; Habersetzer and Marimuthu, 1986; Jones *et al.,* 1991).

Development of wing morphology and flight behaviour during the postnatal growth have been studied in several species of microchiropteran bats such as *M. lucifugus* (Buchler, 1980), *A. jamaicensis* (Taft and Handley, 1991), *P. pipistrellus* (Hughes *et al.,* 1995), *Phoniscus papuensis* (Rhodes, 1995), *P. hastatus* (Stern *et al.,* 1997) and *P. mimus* (Isaac and Marimuthu, 1997). A few more investigations have simultaneously been conducted on flight behaviour, wing morphology, muscle biochemistry and flight physiology of bats (Carpenter, 1985; Powers *et al.,* 1991). Wing loading is highly correlated with body mass and wing dimensions are used to predict foraging habits (Norberg and Rayner, 1987; Norberg, 1990) and habitats (Kalcounis and Brigham, 1995). Changes in body mass may dramatically affect the flight capability of bats. Increase in body mass and wing loading results with decrease in flight manoeuvrability (Aldridge, 1987; Norberg and Rayner, 1987; Aldridge and Brigham, 1988) and an increase in energy cost (Hughes and Rayner, 1991). Several studies have been carried out on the relations among aerodynamics, wing shape and flight mode (e.g. Pennycuick, 1975, 1989; Norberg and Rayner, 1987; Rayner, 1988; Norberg, 1990).

Growth patterns of bats have been evaluated using different growth models on few species of bats such as *T. brasiliensis* (Kunz and Robson, 1995), *P. auritus* (McLean and Speakman, 2000) and *M. nattereri* (Swift, 2001). The non-linear growth equations describe best-fit growth models to the postnatal growth data. Most of the studies mentioned above, have been conducted on microchiropteran bats and a detailed report on wing morphology, flight development and growth patterns of megachiropteran bats, particularly on Indian megachiropterans are apparently scarce (Elangovan *et al*., 2004). This study documents the postnatal changes in wing morphology and development of flight in captive free-flying *Cynopterus sphinx*. In addition, aerodynamic flight models were used to predict the effects of wing morphology on theoretical flight performance, and the growth patterns of wingspan and wing area were compared using three non-linear equations.

METHODS

Captivity of Bats for Postnatal Development and Age Estimation

Study was conducted at the Department of Animal Behaviour and Physiology, School of Biological Sciences, Madurai Kamaraj University, India (09°58' N, 78°10' E). Advanced stage pregnant females of *C. sphinx* were captured from their foraging area, in the botanical garden, at the university campus between 1830 h and 2300 h. Bats were released into a free flight room (3.5 m long x 2.4 m wide x 3.5 m height) and maintained under 12:12 h light-dark cycles. During dark periods, pieces of fruits such as custard apple (*Annona squamosa)*, guava (*Psidium guajava)*, banana (*Musa paradisiaca)*, papaya (*Carica papaya)*, sapota (*Achras sapota)* and grape (*Vitis vinifera)* were placed on trays about 1.5 m above the floor level. Discarded fruits, faeces and bolus were removed at 0800 h of the following mornings.

Morphological Measurements and Statistical Analysis

Newborn young were soft and fragile and therefore the young bats were gently removed with great care from their mothers and morphological measurements such as body mass, length of forearm and length of total epiphyseal gap of the fourth metacarpal-phalangeal joint were made for *C. sphinx* between 3 and 219 days at an interval of 3 – 5 days. Mothers and pups were tagged with different colour beads for individual recognition. Each tag was prepared with a thin, flexible aluminum wire loaded with plastic beads (3 mm). In order to resist pressure the wires were inserted into soft, flexible, transparent plastic tubes with the beads well exposed. The tags were then placed around the necks of mothers and pups and the two ends of the wire were sufficiently tightened. The bats did not show any adverse reaction on such tagging. Length of forearm was measured with vernier calipers to the nearest 0.1 mm and body mass was measured to the nearest 0.1 g using a spring balance (Avinet, Dryden, New York). Length of the total epiphyseal gap in the fourth metacarpal-phalangeal joint was measured to the nearest 1.0 μm using a binocular microscope equipped with an ocular micrometer and substage illumination to view the transilluminated wing (Kunz and Anthony, 1982). Observations were made using a night vision scope (FJW Optical Systems, Inc) and dim red light under captive condition. On completion of the study the tags were removed and all bats were released during dark hours at the site of capture.

In addition to captive study, wild grown pups of *C. sphinx* were collected with their mothers by erecting mist nets adjacent to their day roosts (palm trees) in the botanical garden at the time of emergence. A total of 69 mother-pup pairs were captured at five-day intervals during the study period. Upon capture, their morphometric measures such as length of forearm, body mass and length of total epiphyseal gap were taken and then released after attaching them with their mothers. Since the ages of various sized free-ranging pups were unknown, we considered the length of forearm of captive and wild pups as an index to correlate the growth patterns of their body mass and lengths of total epiphyseal gap by comparing the slopes of the linear regression (Emmanuvel Rajan and Marimuthu, 1999). Postnatal growth curves were constructed based on the growth of forearm, body mass and concurrent changes in the total epiphyseal gap of the fourth metacarpal-phalangeal joint. The linear increase in length of forearm and linear decrease in length of total gap were used to derive regressions and to derive age predictive equations. Ninety-five percent confidence intervals and prediction bands were plotted for the regression equations for lengths of forearm and total gap. Prediction intervals give the range of values within which the true age of a single *C. sphinx* will fall, whereas confidence intervals indicate the range within which the mean true age of bats of a given size will fall (Kunz and Anthony, 1982). In order to determine the difference in the growth pattern of young that grew under captive and natural conditions, the values of their body mass and epiphyseal gap relative to length of forearm were compared by performing a *t*-test.

The best-fit postnatal growth models of the length of forearm and body mass were fitted with three standard nonlinear growth equations (logistic, Gompertz and von Bertalanffy) using Pisces (1998) software 'Simply Growth 1.7' (Pisces Conservation Ltd, Lymington, United Kingdom). The growth models were fitted using mean values across the individuals for length of forearm and body mass, and the growth patterns were compared among the three models. Goodness of fit was taken inversely related to the sum of squares of the models after checking systematic deviations (Zullinger *et al.*, 1984; Boyd and Myhill, 1987; Hughes *et al.*, 1995; Kunz and Robson, 1995; McLean and Speakman, 2000; Swift, 2001). Linear regression analyses were performed using SigmaStat for Windows (Version 2.03, SPSS Inc., 1995), and in age predicting equations 95% prediction and confidence limits were plotted using SigmaPlot (Version 2.0, Jandel Corporation, 1994). Means are expressed with ± SD throughout the text.

Captivity, Parturition and Wing Measurements for Flight Performance Study

A group of seven pregnant *C. sphinx* was collected from their foraging area at Madurai Kamaraj University campus, India. Bats were released into a free flight room (3.5 m long x 2.4 m wide x 3.5 m height) and maintained under 12:12 h light and dark cycles. Bats were fed with locally available fruits during the dark period and the discarded materials, faeces and bolus were removed at 0800 h of the following morning. For individual recognition, bats were marked with thin aluminium neck collars (laboratory made) containing different-coloured reflective tapes. Five days after the onset of parturition, the pups were gently removed from their mothers and morphological parameters including body mass (g), wingspan (cm), and wing area (cm^2) were measured at 5-day intervals from 5 to 125 days of age. Beginning on the day of parturition, bats were observed thrice a day to study their behaviour. Wing area (cm^2) was measured by placing the young bat on its ventral side over a black sheet; the right wing was extended with leading edge perpendicular to the body axis and traced. The traced area was carefully detached and used to calculate wing area using a leaf area meter (Area Meter AM100, Analytical Development Company Ltd, Herts, United Kingdom). Wingspan was measured as two times the distance from the body axis to the wing tip. Tip length ratio was calculated dividing handwing length (distance from the wrist to the wing tip) by the armwing length (distance from the shoulder joint to the wrist). Tip area ratio was calculated dividing the handwing area (area of membrane spanned by the second to fifth digit) by armwing area (area of wing between the fifth digit, the body and the legs). Aspect ratio was calculated as wingspan squared divided by wing area. Wing loading (Nm^{-2}) was calculated multiplying the mass of individual by 9.8 (acceleration due to gravity) and divided by the wing area. Body mass was measured to the nearest 0.1 g using a spring balance (Avinet, Dryden, New York).

Flight Tests and Statistical Analysis

Flight tests were made once in five days from 20 to 75 days of age to test the stages of flight development. A foam pad (3 cm thick, 1.85 m width, 2 m length) was placed on the floor of the flight chamber (1.85 m width, 14 m length, 3.5 m height) to cushion the fall of young. Young bats were brought to the flight chamber and suspended by their hind feet on the flight-launching apparatus (1.5 m height). The distance travelled by bat in the flight chamber was measured. Each bat was tested at least three times and the best performance of each individual was recorded. The flight ability of pups was rated by following the method of Powers *et al.* (1991) as follows: Group I (Flop): bat dropped to the floor exhibiting no flapping behaviour and no horizontal displacement. Group II (Flutter): bat achieved no horizontal displacement but exhibited wing-flapping behaviour. Group III (Flap): bat exhibited wing-flapping behaviour but landed somewhere in the flight chamber. Group IV (Fly): bat flew in circles around the flight chamber, hit a wall and landed somewhere within the flight chamber. After finishing the experiments the tags were removed and the bats were released after sunset at the site of capture.

Postnatal growth patterns of wingspan and wing area were fitted with three standard non-linear growth equations (logistic, Gompertz and von Bertalanffy) using the Pisces (1998) software 'Simply Growth 1.7' (Pisces Conservation Ltd, Lymington, United Kingdom). The growth models were fitted using mean values across the individuals' wingspan and wing area and the growth patterns were compared among the three models. Goodness of fit was taken inversely related to the sum of squares of the models after checking systematic deviations (Zullinger *et al.*, 1984).

Aerodynamics of Flight

Theoretical flight performance was predicted using aerodynamic models (Norberg and Rayner, 1987; Pennycuick, 1989). Mechanical power (P_{mp} and P_{mr}) output of flight was estimated as follows:

$$P_{mp} = 24.01M^{1.555}\,B^{-1.794}\,S^{0.314}\text{ watts}$$

$$P_{mr} = 27.26M^{1.531}\,B^{-1.762}\,S^{0.324}\text{ watts}$$

where M is body mass, B is wingspan, and S is wing area. Since theoretical mechanical flight power varied with speed, we examined variations in power at two characteristic speeds: the minimum power speed (V_{mp}) and the maximum range speed (V_{mr}), calculated as follows:

$$V_{mp} = 6.58M^{0.422}\ B^{-0.479}\ S^{-0.148}\ \text{m/s}$$

$$V_{mr} = 8.71M^{0.423}\ B^{-0.498}\ S^{-0.144}\ \text{m/s}$$

where V_{mp} represents speed at which power for flight was minimized and V_{mr} indicates distance traveled for a given amount of energy maximized (Pennycuick, 1975). Minimum theoretical radius of a banked turn (r_{min}) was calculated using the model of Norberg (1990) as follows:

$$r_{min} = (Mg/S)(2/gpC_L)$$

where (Mg/S) is wing loading, p is air density (kg m^{-3}), and C_L is a non-dimensional lift coefficient. Values are expressed as mean ± SD throughout the text.

RESULTS

Postnatal Development and Age Estimation

Parturition and Morphological Features of Young

Each female gave birth to a single pup. A total of 12 young were born in captivity, one died at birth and two died at the age of three days. The remaining nine young (three males and six females) survived till the end of the study. Newborn pups were naked, pink with closed eyes and folded pinnae, and thus they were altricial. The pups positioned themselves firmly on the ventral side of their mothers. Eyes had opened and the pups began to move at the age of 5.0 ± 0.87 days. The pinnae were unfolded at the age of 7.3 ± 1.22 days. The short, fine, soft hairs of pups were distinguishable at about ninth day and thereafter, developed grey fur, which appeared similar to that of sub-adults.

Forearm, Body Mass and Total Epiphyseal Gap

The forearm length of pups at 3-day-old ranged from 27.0 to 31.0 mm ($\bar{X}$ = 29.4 ± 1.63 mm), body mass ranged from 7.7 to 8.7 g ($\bar{X}$ = 8.2 ± 0.45 g) and dimension of total epiphyseal gap varied from 35 to 42 μm ($\bar{X}$ = 38.0 ± 3.01 μm). The mean values of the length of forearm and body mass of 3-day-old pups were 42.2% and 18.2% of postpartum females, respectively. Even though body mass increased throughout the study period, growth was rapid and linear until 60 days of age and thereafter it became nonlinear (Figure 1).

A considerable amount of deviation occurred in body mass during later period of growth. Pups obtained 40% of mean mass of postpartum females at seventh week of age. At the age of 219 days the young bats attained 32.9 ± 1.80 g of body mass which was 72.7% of mean mass of postpartum females. The length of forearm increased linearly until 36 days (up to 52.4 ± 1.13 mm) and thereafter it became nonlinear (Figure 2).

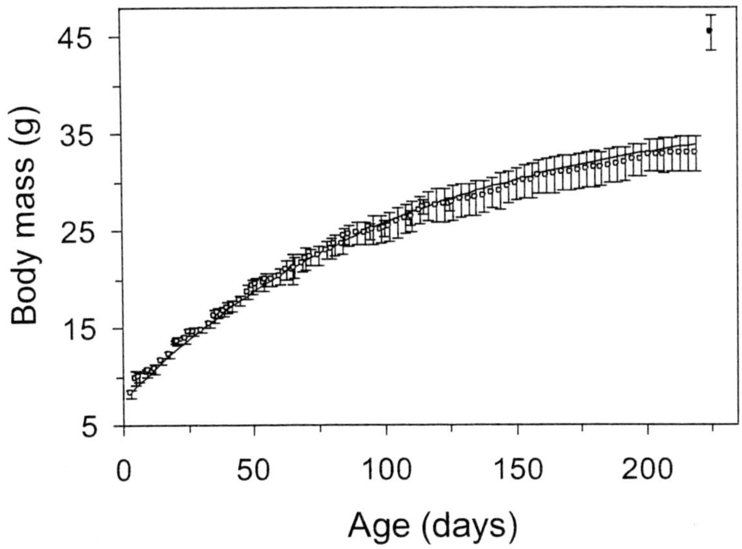

Figure 1. Empirical growth curve from 3 to 219 days of age for body mass (n = 90) of captive young *C. sphinx*. The open circles represent the mean observed values of nine individuals. The uninterrupted lines represent the nonlinear growth of Gompertz model. Postpartum females' mean values ± SD (n = 9) are shown as solid circles with vertical bars.

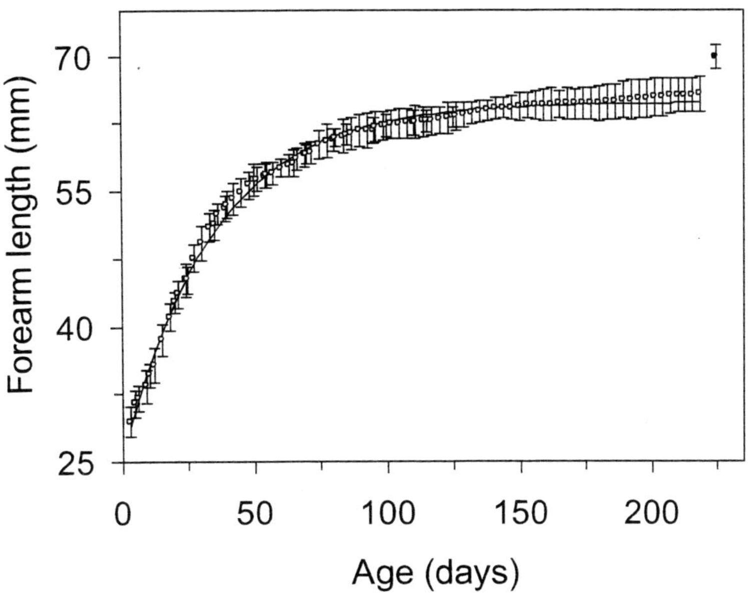

Figure 2. Empirical growth curve from 3 to 219 days of age for length of forearm (n = 90) of captive young *C. sphinx*. The open circles represent the mean observed values of nine individuals. The uninterrupted lines represent the nonlinear growth of logistic model. Postpartum females' mean values ± SD (n = 9) are shown as solid circles with vertical bars.

By the time of first flight (ca. 45 days) the pups attained mean forearm length of 54.9 ± 1.58 mm, which was 79% of forearm length of postpartum females. At the age of 219 days the pups attained 94.1% of forearm length of postpartum females. The length of total epiphyseal gap of fourth metacarpal-phalangeal joint showed a linear increase for the first 15 days and then decreased with increasing age until 60 days, indicating the formation of secondary centre of ossification (Figure 3).

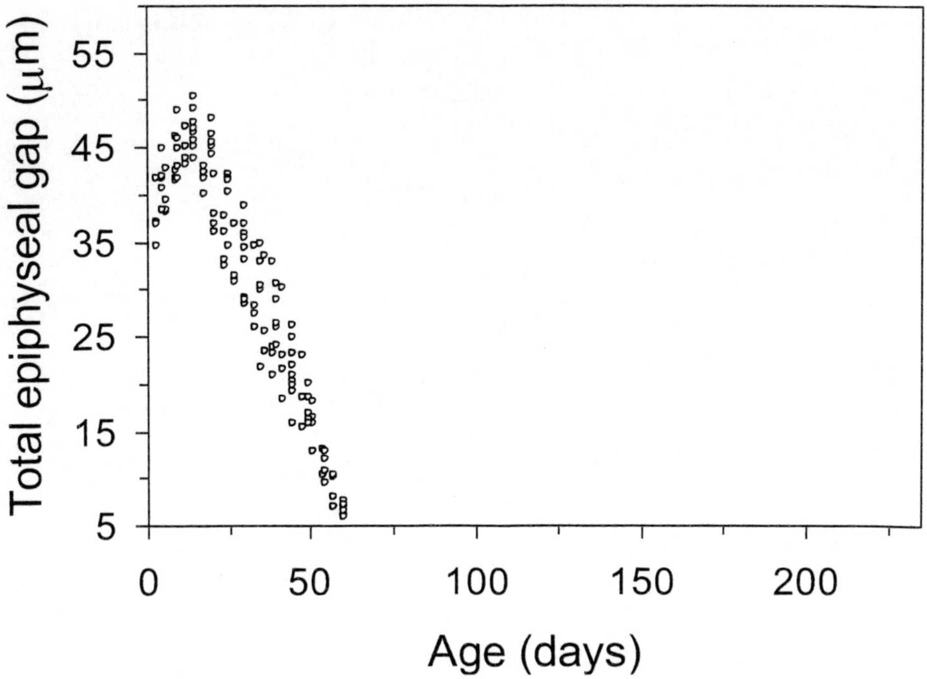

Figure 3. Postnatal changes in the length of total epiphyseal gap of the fourth metacarpal-phalangeal joint from 3 to 60 days of age (n = 138) of captive young *C. sphinx*.

A linear regression equation predicts age of young *C. sphinx* on the basis of length of forearm from 29.4 ± 1.63 mm to 52.4 ± 1.13 mm (3–36 days) with 95% confidence and prediction limits (Figure 4). Another equation was derived to estimate the age from the length of total epiphyseal gap between 15 and 60 days of age when the dimensions decreased linearly from 47.0 ± 2.1 μm to 6.0 ± 1.7 μm (Figure 5). At a mean length of forearm (41.5 mm), the extent of deviation from the estimated value at 95% confidence limit was ± 1.0 day. At the extreme lengths of forearm (29.4 mm and 52.4 mm) the deviations were ± 1.55 days and ± 1.67 days, respectively.

At the mean length of total epiphyseal gap (26.0 ± 13.0 μm), the estimation of age with 95% confidence limit was ± 1.43 days. At its extremes (6 μm and 50 μm) the extent of deviations was ± 2.53 days and ± 2.55 days, respectively. Analysis of coefficients of determination for the relationships between length of forearm and age, and between length of total epiphyseal gap and age revealed that the length of forearm can be used reliably to estimate the age of young *C. sphinx* up to 36 days old, thus until the length of forearm ≤ 52.4 mm (Figure 4). Whereas, the age predicting equation based on length of total epiphyseal gap is valid for young ranging from 15 to 60 days, when the forearm length exceeds 52.4 mm.

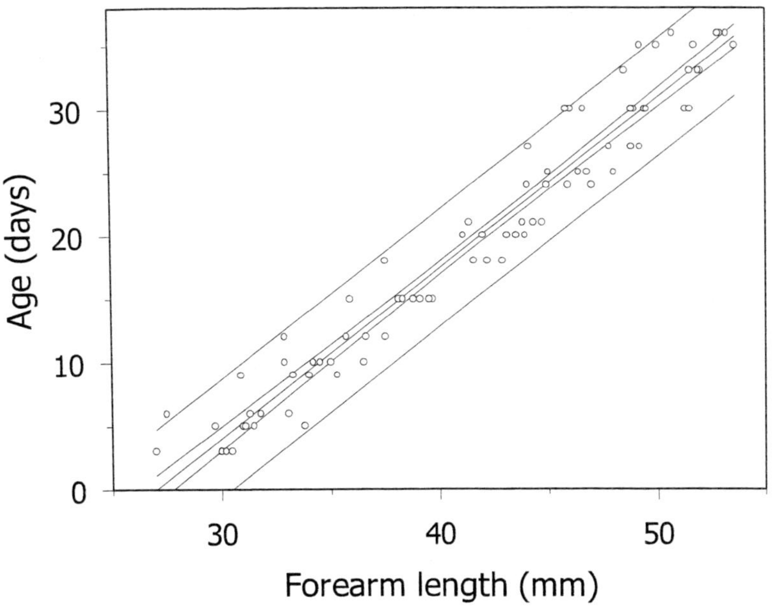

Figure 4. Regression line estimating the age of *C. sphinx* from the values of length of forearm. Based on the length of forearm age predictive equation is valid for the dimensions ranging from 29.4 ± 1.63 mm to 52.4 ± 1.13 mm; age = –36.3 + (1.34 x length of forearm); $r^2 = 0.95$, $df = 82$, $P < 0.001$. Narrow and wide bands indicate 95% confidence and prediction limits, respectively.

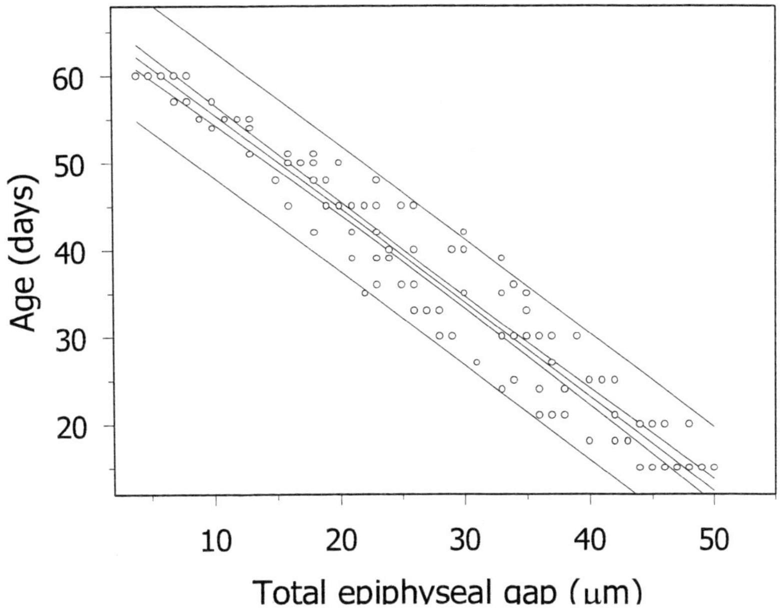

Figure 5. Regression line estimating the age of *C. sphinx* from the values of length of total epiphyseal gap. The age predictive equation based on the length of total epiphyseal gap is valid when the dimensions range from 6 ± 1.7 μm to 47 ± 2.1 μm; age = 66.5 – (1.1 x total gap); $r^2 = 0.93$, $df = 112$, $P < 0.001$. Narrow and wide bands indicate 95% confidence and prediction limits, respectively.

Table 2.1. Growth parameters of *C. spinx*, derived from the logistic, Gompertz and von Bertalanffy nonlinear growth modes

Growth model	Parameters	Length of forearm versus age				Body mass versus age			
		Estimate	SE	Coefficient of variation (%)	Model sum of squares	Estimate	SE	Coefficient of variation (%)	Model sum of squares
	A	64.32	0.220	0.342		31.48	0.389	1.235	
Logistic	K	0.040	0.001	3.56	7331.40	0.025	0.0007	3.000	6744.19
	I	4.28	0.848	19.82		34.63	1.236	3.568	
	A	64.39	0.223	0.346		31.31	0.334	1.067	
Gompertz	K	0.036	0.063	175.28	7516.26	0.019	0.013	66.24	3745.44
	I	- 4.12	34.83	845.04		13.62	17.26	126.67	
	A	64.87	0.256	0.394		36.335	0.867	2.38	
von Bertalanffy	K	0.028	0.0011	3.70	7473.61	0.0096	0.0006	5.72	4225.77
	T_0	- 17.91	...	...		- 26.324	...	...	

A=asymptotic size of length of forearm (mm) or body mass (g), K= growth rate constant, I= inflection pint T_0 = time when length of forearm or body mass is 0.

Curves derived from the three models for body mass were similar in shape, since the correlation coefficient of the predicted values for each model was ≥0.99 (Table 1). Nevertheless, the Gompertz equation appeared to be the most appropriate model on the basis of statistical criteria describing postnatal changes in body mass of *C. sphinx* (Figure 1).

$$\text{Body mass}_{(t)} = 31.31e^{-e^{-0.0196(t-13.62)}}$$

where 'e' is 2.718 and 't' is time in days.

The predicted values for the length of forearm of three nonlinear growth models had correlation coefficients ≥0.99. Because of these high correlations in the lengths of forearm, it was difficult to graphically distinguish among the three models. However, after deriving an equation based on each model the logistic equation best-fitted to express the postnatal changes in the length of forearm of *C. sphinx* (Figure 2).

$$\text{Length of forearm}_{(t)} = 64.32[e^{-0.04(t-4.28)} + 1]^{-1}$$

where 'e' is 2.718 and 't' is time in days.

The negative value of an inflection point in the equation describing the length of the forearm argues against the suitability of Gompertz model (inflection point: -4.12) for postnatal growth rates in *C. sphinx*. Coefficients of variation for the estimates of growth parameters were consistently less when derived from the logistic growth model (0.342), compared to the von Bertalanffy (0.394) model. The values of body mass (Figure 6a) and total epiphyseal gap (Figure 6b) were compared with the respective lengths of forearm of the young that were grown under captive and natural conditions. There was no significant difference on the growth of these two parameters under both conditions (for body mass t = 1.97, df = 167, P > 0.05; for epiphyseal gap t = 1.97, df = 127, P > 0.05).

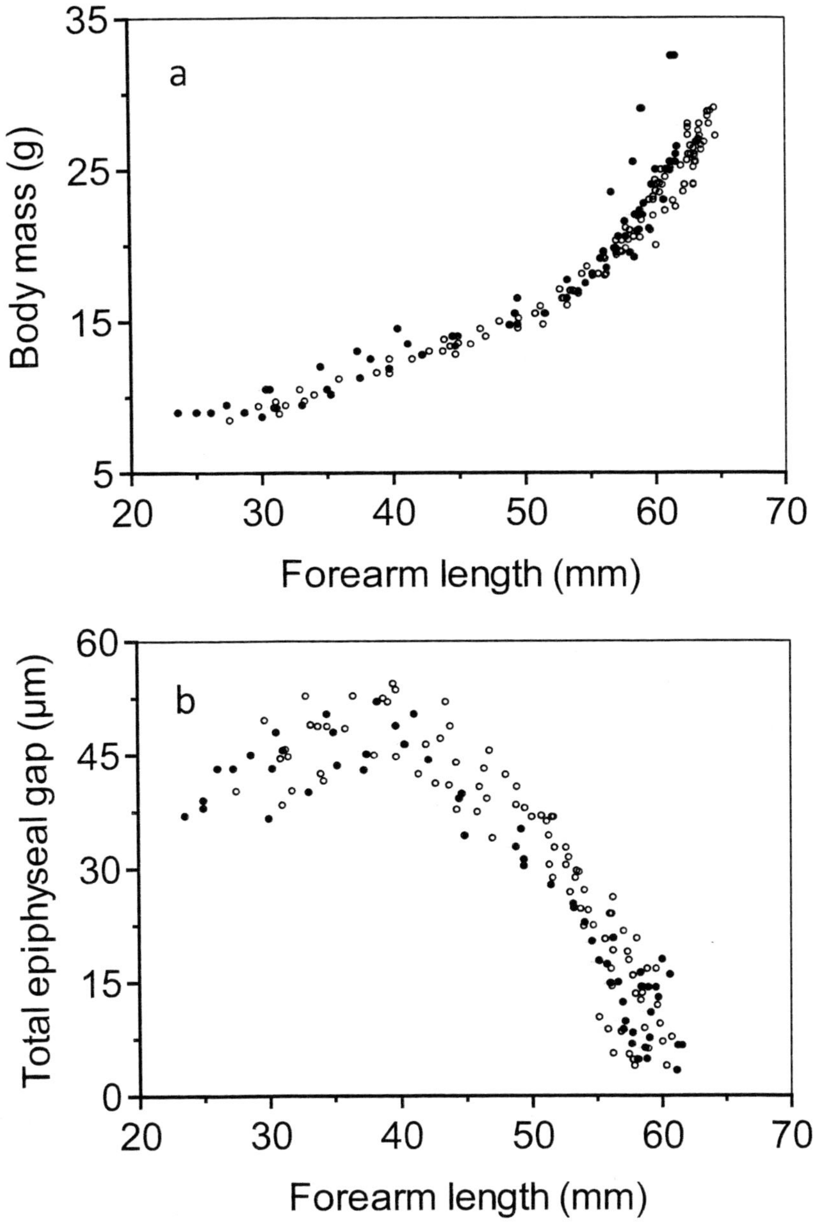

Figure 6. Comparisons of body mass (a) and the length of total epiphyseal gap of the fourth metacarpal-phalangeal joint (b) of young *C. sphinx* under captive (open circles) and wild (solid circles) conditions with their respective forearm lengths; for body mass n = 225 (captive) and 69 (wild); for the length of total epiphyseal gap n = 126 (captive) and 69 (wild).

Wing Morphology and Theoretical Flight Performance

Parturition started seven days after the bats were released into the free flight room and the parturition period extended over 22 days. Each female gave birth to a single pup. A total of seven young were born in captivity, two died at birth. The remaining five individuals (two males and three females) survived until end of the study. The mothers enshrouded their young with their plagiopatagium for the first three weeks of age and thereafter the pups started roosting separately adjacent to their mothers. The body mass of 5-day-old pups ranged from 8.8 to 10.5 g ($\overline{X}$ = 9.3 ± 0.68 g) and the mean value of the body mass of 5-day-old pups was 21.8% of postpartum females. Pups attained 42.9% of mean mass of postpartum females at the age of 45 days when they exhibited their first flight. At the age of 125 days the young bats attained 28.0 ± 1.01 g of body mass which was 65.3% of mean mass of postpartum females.

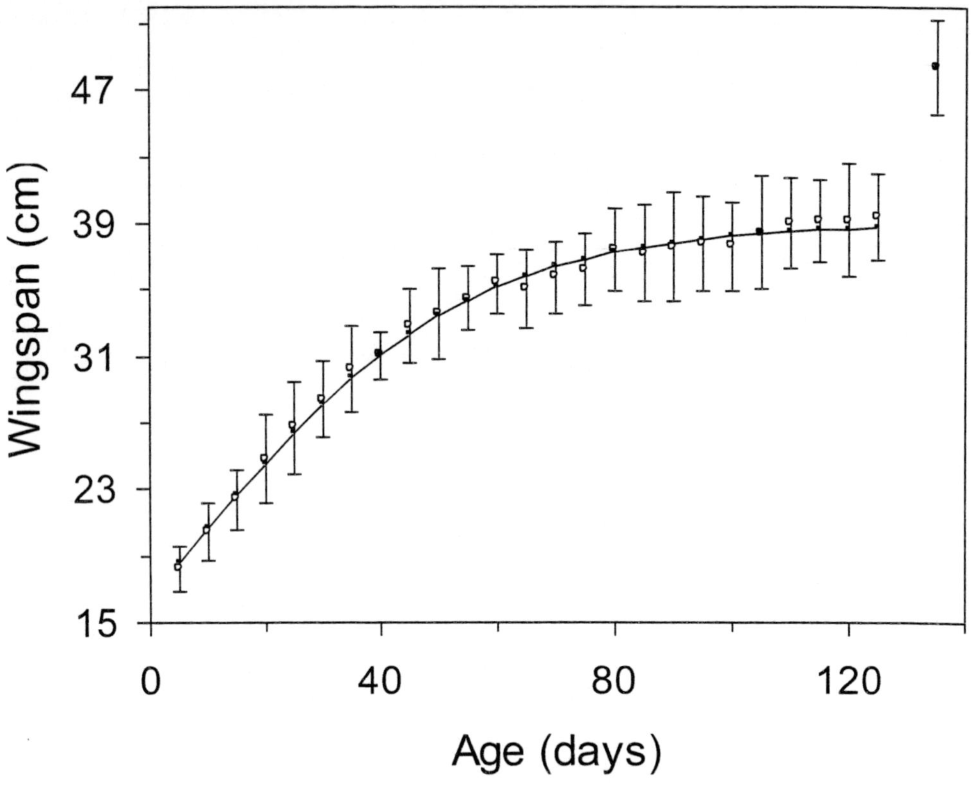

Figure 7. Age related changes in wingspan during postnatal growth of *C. sphinx*. The growth pattern of wingspan was best described by the logistic equation (illustrated by the fitted curves). Mean values across five individuals are given with ± SD. Postpartum females' mean values ± SD (n = 5) are shown as solid circles with vertical bars.

Young did not show any displacement until 25 days of age. However 30 and 35-day-old young showed mean displacement (flop) of 24.0 ± 16.36 cm and 126.0 ± 76.68 cm, respectively. They fluttered a mean distance of 700.0 ± 384.06 cm at 40 days of age. Bats that reached 45-day-old flew reasonably well for a mean distance of 1033.3 ± 478.5 cm. At this stage they maintained their flight with gentle turns. At the age of 50 days they made few circling flights in the flight chamber and often hung at roof. The young bats achieved sustained flight at 55 days of age. It involved virtually continual flight manoeuvres that became indistinguishable from those of adults. Bats started to forage on their own when they achieved sustained flight. The mean wingspan of five-day-old pup was 18.2 ± 1.34 cm and showed a mean increase of 0.365 cm/day until 45 days of age, thereafter the growth rate decreased and attained an apparent stability (Figure 7). The mean wing area of five-day-old pup was 63.7 ± 8.17 cm^2 and showed a linear increase of 2.40 cm^2/day until 45 days, after which the rate of growth decreased and stabilized (Figure 8).

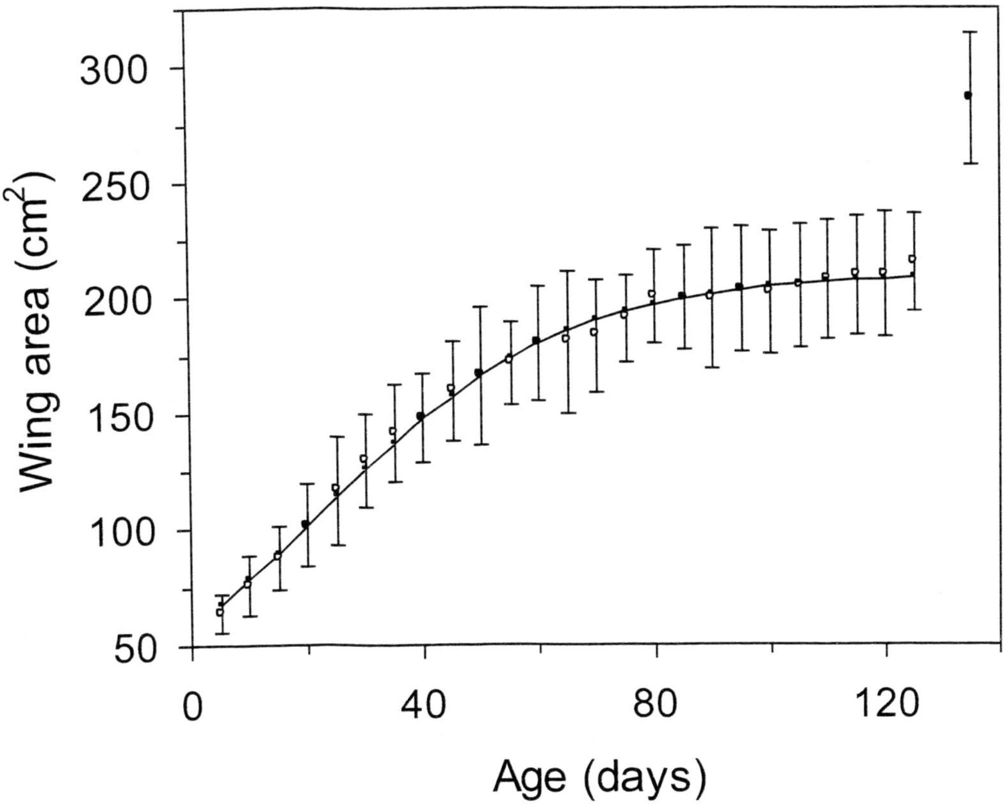

Figure 8. Age related changes in wing area during postnatal growth of *C. sphinx*. The growth pattern of wing area was best described by the logistic equation (illustrated by the fitted curves). Mean values across five individuals are given with ± SD. Postpartum females' mean values ± SD (n = 5) are shown as solid circles with vertical bars.

The five-day-old pups had an average armwing length of 5.2 ± 0.62 cm, which was 38.5% of postpartum females. The armwing length increased linearly until 35 days with mean growth rate of 0.113 cm/day and thereafter the growth rate decreased and attained 80.7% of postpartum females' value at 125 days of age (Figure 9a). The mean handwing length of five-day-old pups was 3.9 ± 0.43 cm, which was 36.6% of postpartum females (Figure 9b).

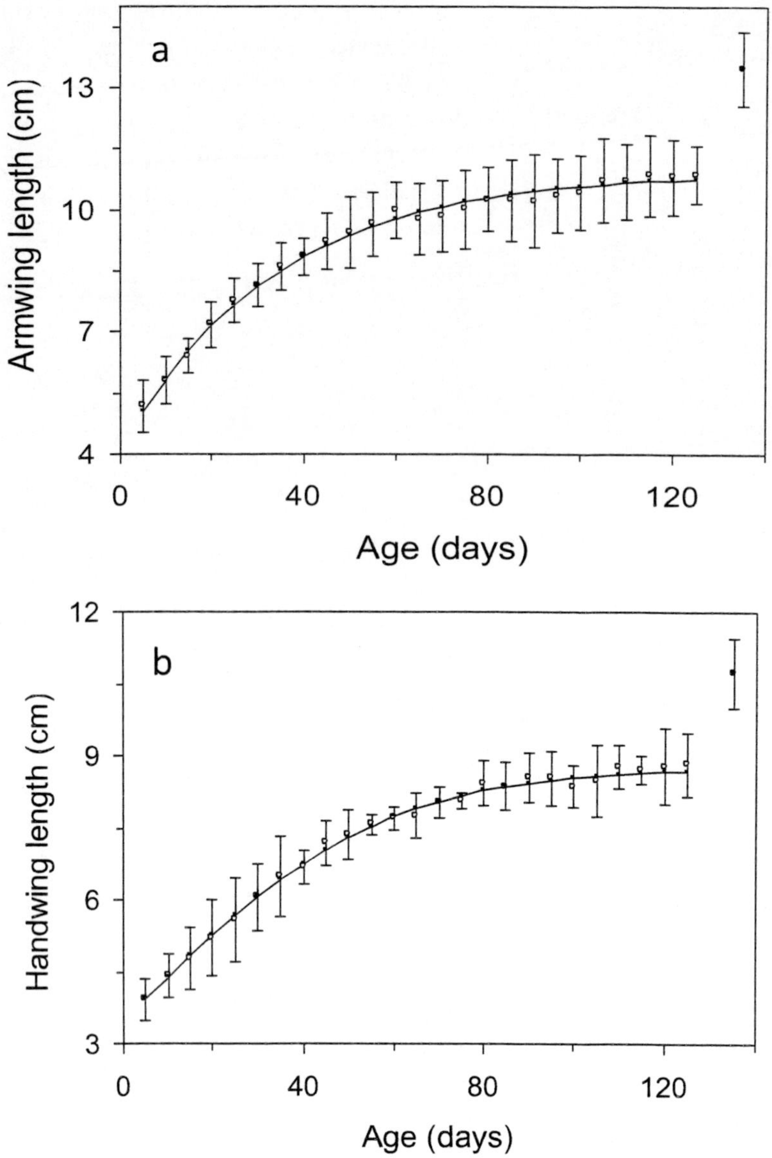

Figure 9. Postnatal changes in armwing length (a) and handwing length (b) in relation to age over 5 to 125 days in *C. sphinx*. Mean values across five individuals are given with ± SD. Postpartum females' mean values ± SD (n = 5) are shown as solid circles with vertical bars.

Unlike the armwing length, the handwing length increased linearly until 50 days of age with a mean growth rate of 0.078 cm/day. The mean armwing and handwing areas of five-day-old pups were 22.1 ± 3.18 cm^2 and 9.8 ± 1.58 cm^2, which were 24.3% and 18.8% of postpartum females, respectively (Figure 10aandb). The armwing area increased linearly until 35 days of age with a mean growth rate of 0.794 cm^2/day, while the handwing area increased linearly until 50 days with a mean growth rate of 0.480 cm^2/day. The linear growth of armwing and handwing areas showed similar pattern but handwing grows slightly faster than armwing until onset of sustained flight. The mean tip length and tip area ratios of five-day-old pups were 0.77 ± 0.152 and 0.45 ± 0.073, respectively. At the time of sustained flight the tip length and tip area ratios reached 0.79 ± 0.049 and 0.58 ± 0.024, respectively.

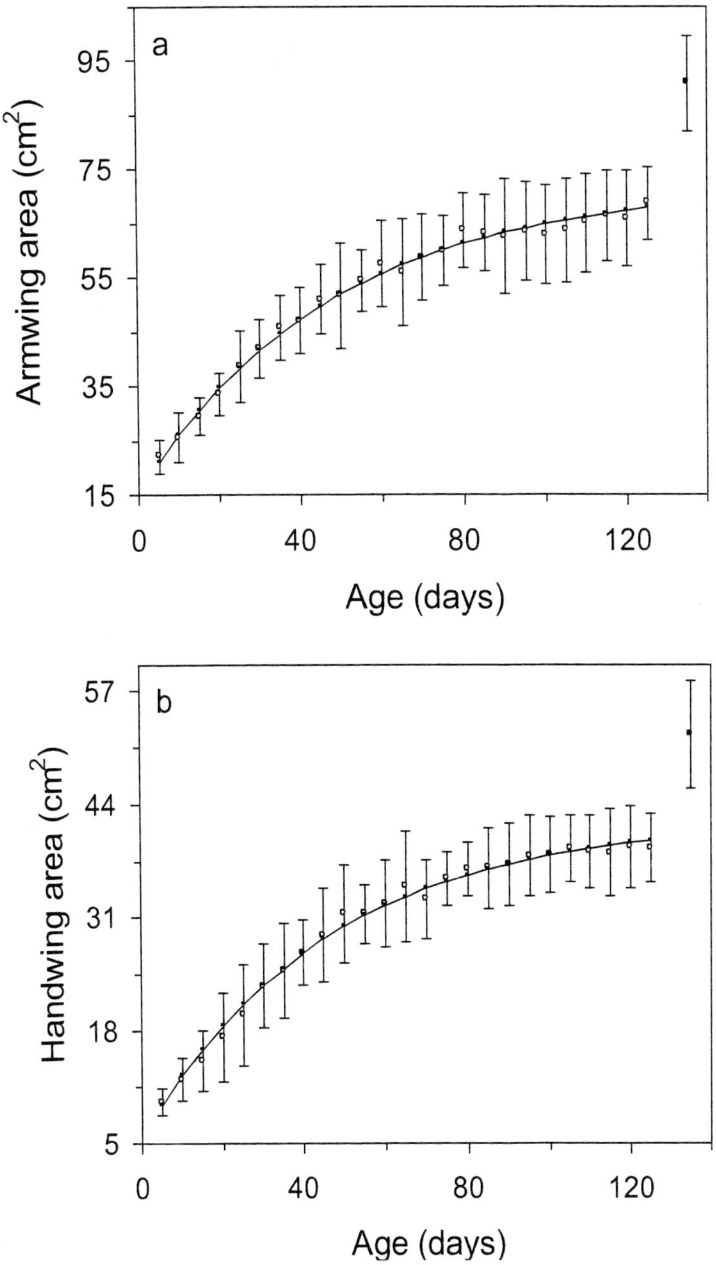

Figure 10. Postnatal changes in armwing area (a) and handwing area (b) in relation to age over 5 to 125 days in *C. sphinx*. Mean values across five individuals are given with ± SD. Postpartum females' mean values ± SD (n = 5) are shown as solid circles with vertical bars.

The five-day-old pups had an average aspect ratio of 5.3 ± 0.53 which was equivalent to 63.8% of postpartum females. The aspect ratio increased linearly until 55 days of age and thereafter maintained its stability (Figure 11a). Five-day-old young had a high wing loading of 14.51 ± 1.61 Nm^{-2}. The wing loading decreased linearly until 35 days of age and thereafter it increased to a maximum of 12.82 ± 0.81 Nm^{-2} at 125 days of age (Figure 11b). Bats achieved sustained flight at the age of 52.0 ± 2.74 days when they attained a mean wing loading of 11.47 ± 1.04 Nm^{-2} which was equal to 77.4% of adults. At this stage bats could negotiate turns. The wing tracings of a single individual (female) at different ages were superimposed to show the growth of wing membrane (Figure 12). In general the growth rate of wing was rapid until 55 days of age, by which the young bats achieved sustained flight, and then the growth stabilized.

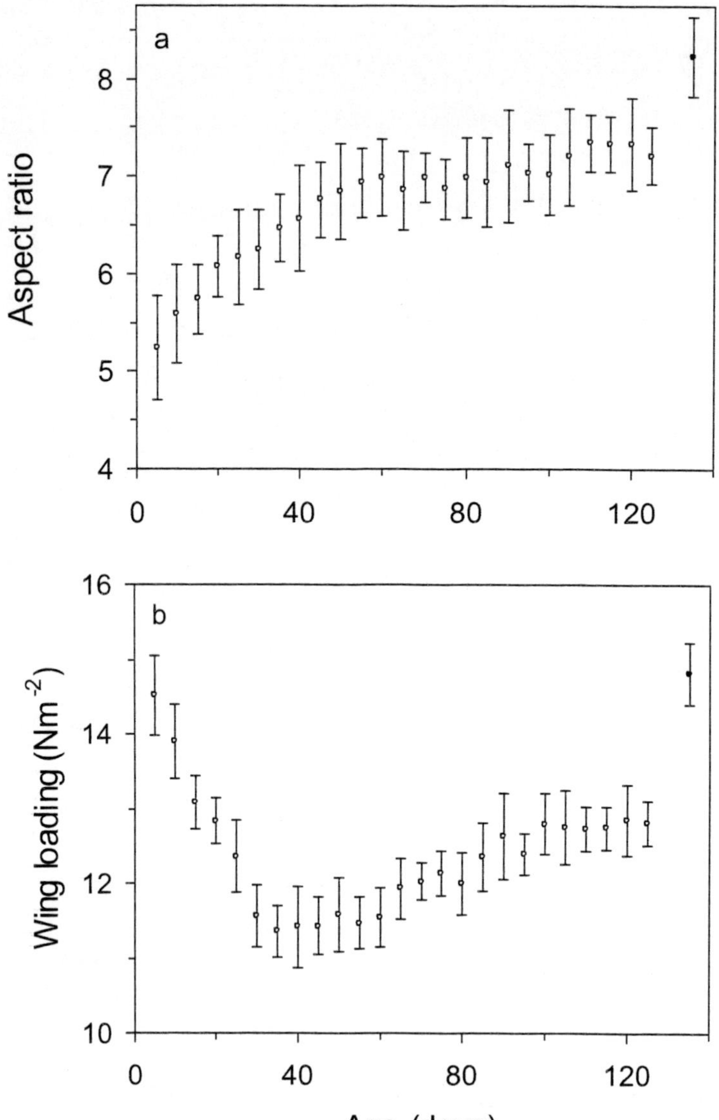

Figure 11. Changes in aspect ratio (a) and wing loading (b) in relation to age over 5 to 125 days in *C. sphinx*. Mean values across five individuals are given with ± SD. Postpartum females' mean values ± SD (n = 5) are shown as solid circles with vertical bars.

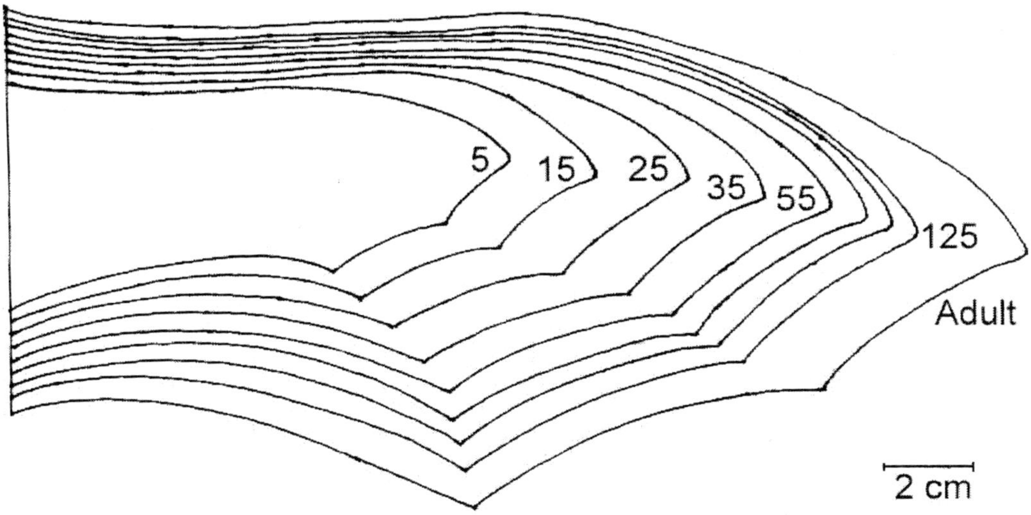

Figure 12. A series of wing tracings of a single female *C. sphinx* at the ages of 5, 15, 25, 35, 55, 80, 105 and 125. The outermost tracing is of an adult female bat.

Table 4.1. Growth of *C. sphinx*, derived from the logistic, Gompertz and von Bertalanffy non-linear growth models

Model	Parameters	Wingspan versus age				Wing area versus age			
		Estimate	SD	Coefficient of variation (%)	Model sum of squares	Estimate	SD	Coefficient of variation (%)	Model sum of squares
Logistic	A	39.04	0.459	1.176		210.13	0.450	0.214	
	K	0.042	0.003	7.774	5499.7	0.046	0.0005	1.059	164643.9
	I	7.25	1.309	18.048		21.36	0.176	0.826	
Gompertz	A	39.43	0.467	1.185		214.56	0.459	0.214	
	K	0.034	0.082	241.178	5505.5	0.0346	0.052	151.061	165172.2
	I	-2.46	35.01	-1422.354		10.99	18.39	167.382	
von Bertalanffy	A	40.18	0.670	1.666		223.660	0.808	0.361	
	K	0.026	0.003	9.718	5505.2	0.02	0.0003	1.439	165074.8
	T_0	-17.57	...	...		-8.08	...	...	

A=asymptotic size of wingspan (cm) or wing area (cm^2). K=growth rate constant, I=inflection point; t_0=time when wingspand or wing are is 0.

A summary of growth parameters derived from three non-linear growth models is given in Table 2. Curves derived from the three models for wingspan were similar in shape, since the correlation coefficient of the predicted values of the models was 0.937. The logistic equation appeared to be the most appropriate model on the basis of statistical criteria describing postnatal changes in wingspan of *C. sphinx* (Figure 7). Based on the sum of square the logistic (5499.7) growth equation fitted well than von Bertalanffy (5505.2) and Gompertz

(5505.5) models. The coefficient of variations for the estimate of growth parameters was consistently less when derived from the logistic (1.17) than Gompertz (1.18) and von Bertalanffy (1.66) models.

$$\text{Wingspan}_{(t)} = 39.039[e^{-0.0416(t-7.251)} + 1]^{-1}$$

where 'e' is 2.71 and 't' is time in days.

The predicted values for the wing area of three non-linear growth models had correlation coefficients of 0.924. However, based on the statistical criteria the logistic equation was a best fitted growth model to express the postnatal changes in the wing area of *C. sphinx* (Figure 8).

$$\text{Wing area}_{(t)} = 210.131[e^{-0.0463(t-21.360)} + 1]^{-1}$$

where 'e' is 2.71 and 't' is time in days.

The total sum of square was least to the logistic (164643.9) than von Bertalanffy (165074.8) and Gompertz (165172.2) equations. Coefficients of variation for the estimates of growth parameters were consistently less when derived from the logistic growth model (0.214), which describes well than Gompertz (0.214) and von Bertalanffy (0.361) models.

The predicted mechanical power (P_{mp} and P_{mr}) increased slowly for the first 35 days of age and thereafter increased linearly until 85 days of age (Figure 13aandb). The mean minimum predicted flight power (P_{mp}) and maximum range power (P_{mr}) of 35-day-old pups were 0.38 ± 0.022 W and 0.46 ± 0.025 W, respectively. After 35 days of age, both of these predicted power requirements (P_{mp} and P_{mr}) increased linearly and approached 68.3% to the power requirements of postpartum females at the age of 125 days. The predicted minimum power speed (V_{mp}) and maximum range speed (V_{mr}) decreased until 45 days of age. The mean predicted flight speeds V_{mp} (Figure 14a) and V_{mr} (Figure 14b) of 45-day-old pups were 1.94 ± 0.086 ms^{-1} and 2.62 ± 0.118 ms^{-1}, respectively. The predicted flight speeds (V_{mp} and V_{mr}) increased slowly and approached 2.03 ± 0.063 ms^{-1} and 2.74 ± 0.088 ms^{-1} at the age of 125 days. The mean V_{mp} and V_{mr} of 125 days old bats were equivalent to 96.2% and 96.4%, respectively to the speed of postpartum females. The minimum theoretical radius of a banked turn (r_{min}) decreased until 35 days of age to a minimum of 0.46 ± 0.055 m and thereafter increased linearly and attained 0.51 ± 0.033 m at the age of 125 days (Figure 15), which was equivalent to 86.5% of postpartum females.

Most of the studies carried out so far on postnatal development and age estimation have been restricted mainly to microchiropterans; for example on Vespertilionidae (Krátky, 1981; De Fanis and Jones, 1995; Hughes *et al.*, 1995; Isaac and Marimuthu, 1996; Hoying and Kunz, 1998; Swift, 2001), on Molossidae (Kunz *et al.*, 1995), on Phyllostomidae (Stern and Kunz, 1998), on Megadermatidae (Emmanuvel Rajan and Marimuthu, 1999) and on Hipposideridae (Cheng and Lee, 2002). Studies on growth aspects in megachiropterans are much limited (Kunz and Stern, 1995; Kunz and Hood, 2000). This is the first detailed study concerning the growth rate of a megachiropteran.

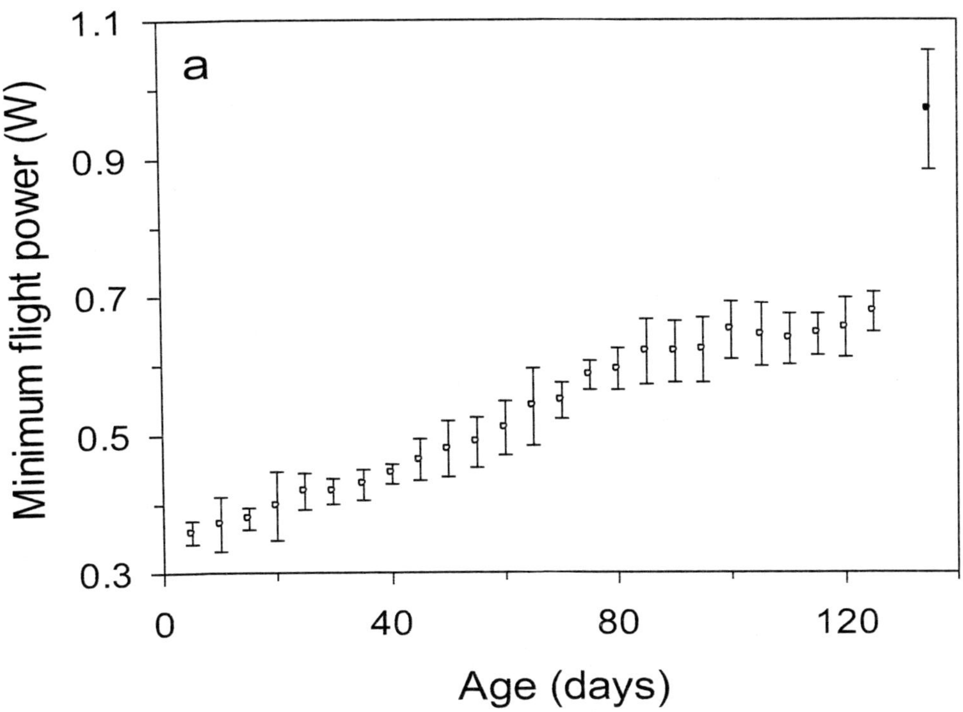

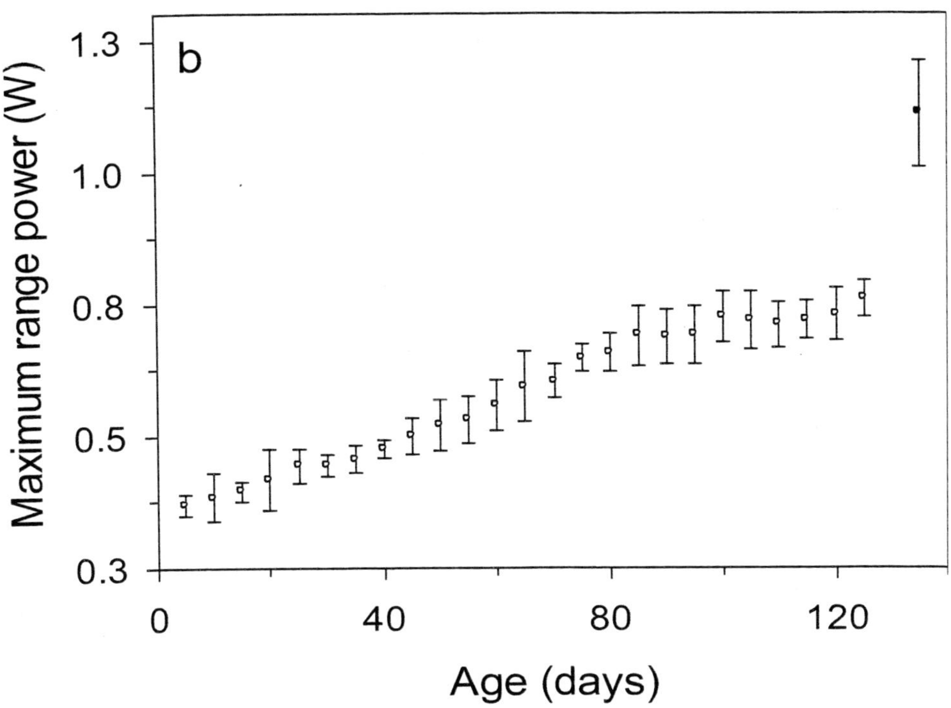

Figure 13. Changes in mechanical power (W) requirement for flight and theoretical flight speed (ms^{-1}) in relation to the age of *C. sphinx* from 5 to 125 days: (a) minimum flight power (P_{mp}) and (b) maximum range power (P_{mr}). Mean values across five individuals are given with ± SD. Postpartum females' mean values ± SD (n = 5) are shown as solid circles with vertical bars.

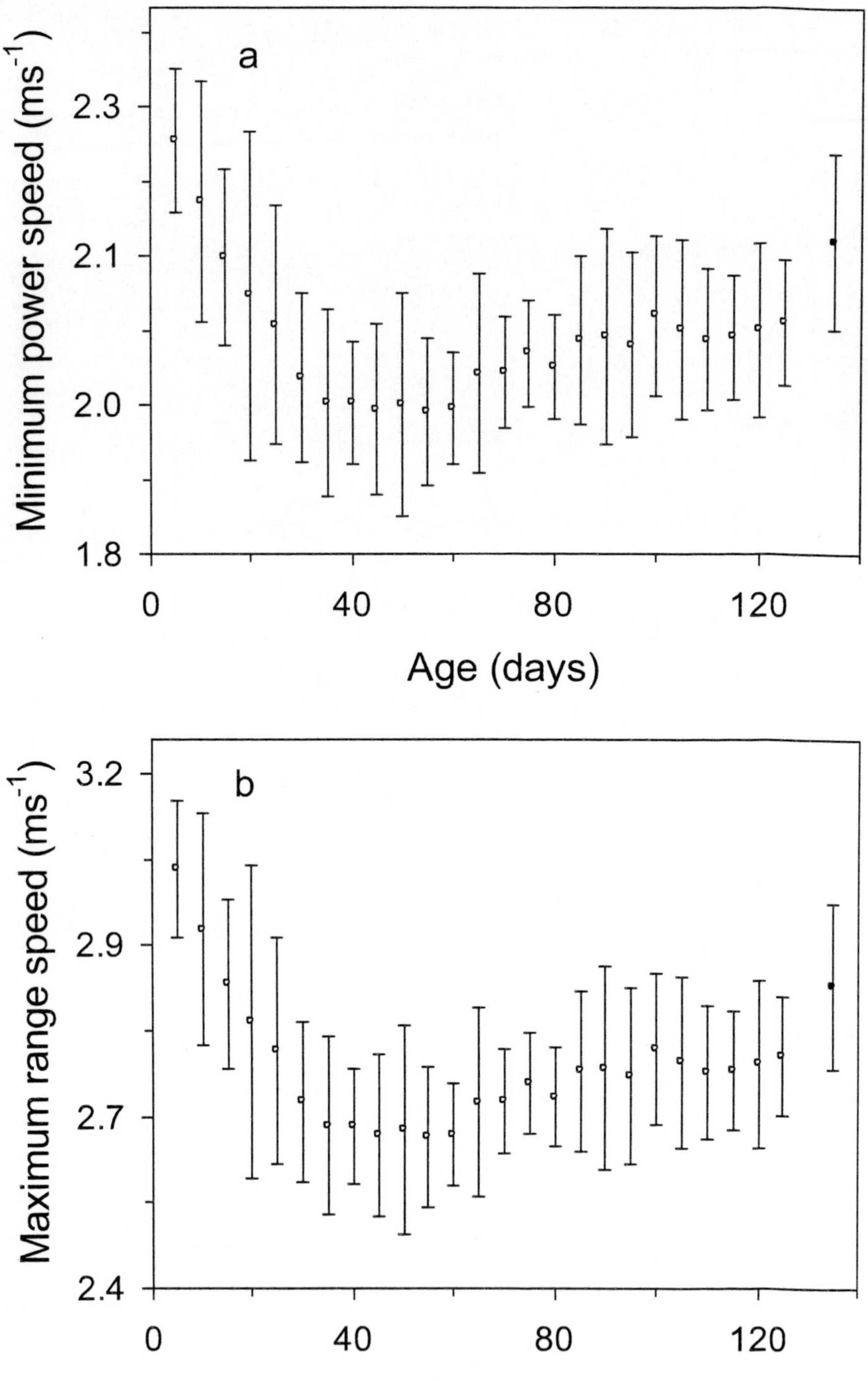

Figure 14. Changes in theoretical flight speed (ms^{-1}) in relation to the age of *C. sphinx* from 5 to 125 days: (a) minimum power speed (V_{mp}) and (b) maximum range speed (V_{mr}). Mean values across five individuals are given with ± SD. Postpartum females' mean values ± SD (n = 5) are shown as solid circles with vertical bars.

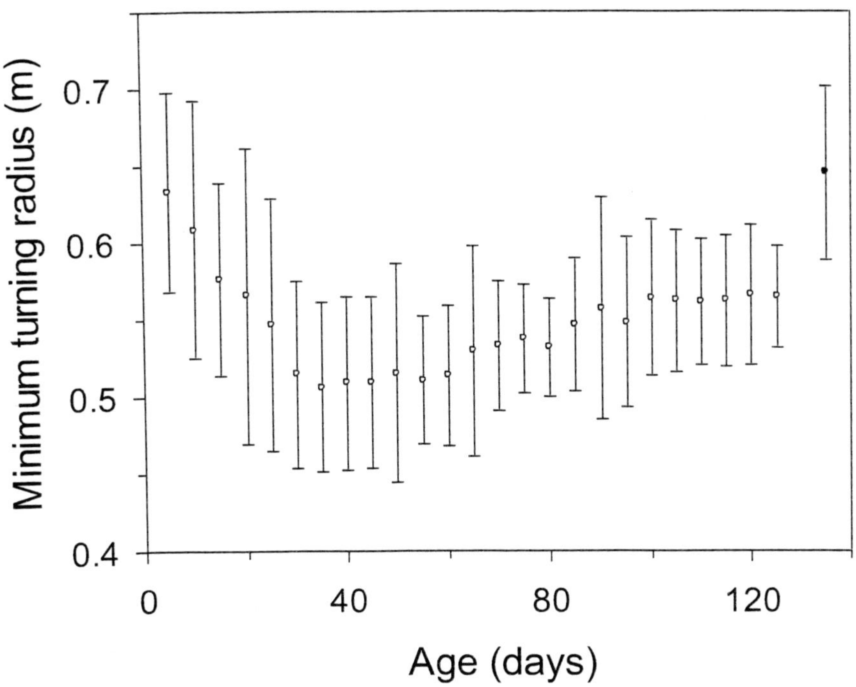

Figure 15. Minimum theoretical radius of a banked turn (r_{min}) in relation to the age of *C. sphinx* from 5 to 125 days. Mean values across five individuals are given with ± SD. Postpartum females' mean values ± SD (n = 5) are shown as solid circles with vertical bars.

The pattern of postnatal growth and development in *C. sphinx* showed the basic trend of linear growth of forearm and body mass during the preflight period. Although at the age of three days, *C. sphinx* weighs 18.2% of mean mass of postpartum females which is slightly lesser than most of microchiropterans, the forearm length was 42.2% of postpartum females as in microchiropterans (Kurta and Kunz, 1987). Juveniles of several species of microchiropteran bats typically began to fly when they have attained 70% of adult body mass and over 95% of adult skeletal size and wing dimension (Maeda, 1972; Barclay, 1995; Kunz and Stern, 1995). Swift (2001) has argued that no juvenile bats of any species yet been shown to exhibit flight before they attain at least 90% of adult skeletal size. However, this study provides the first record that young *C. sphinx* began to fly when they achieved about 40% of adult body mass and nearly 79% of adult skeletal size. Such occurrence of early flight suggests that megachiropterans presumably do not need to manoeuvre so accurately during their foraging sallies as microchiropterans do. This observation substantiates the report of Orr (1970), who stated that megachiropterans are more advanced compared to microchiropterans during the course of development. Though one can argue that such advanced growth in captive condition might be due to the overabundance of food, the absence of significant difference in the growth pattern between captive and free-ranging bats nullifies such an argument.

In theory, asymptotic body mass is achieved by bats when the postnatal growth rate becomes zero. Asymptotic mass of young bats is usually less than adult mass, because it does not include accretionary growth after the first year (Kunz and Stern, 1995). Kunz and Stern (1995) found that growth rates in body mass were negatively correlated with asymptotic body size, thus the smaller species growing faster than larger ones. The values of the lengths of forearm and epiphyseal gap during their linear phases of growth increased during the preflight period and decreased during the postflight period, and reliable for deriving equations to estimate the age of young bats (e.g. Kunz and Anthony, 1982; De Paz, 1986; Kunz and Robson, 1995; Isaac and Marimuthu, 1996; Hoying and Kunz, 1998; Emmanuvel Rajan and Marimuthu, 1999; Baptista *et al*., 2000; Cheng and Lee, 2002). Even though there were no significant differences in the postnatal growth pattern in *C. sphinx* between captive and free-ranging bats, a few reports state that there is a marked difference between these two conditions. For example, Buchler (1980) observed that *M. lucifugus* began to fly 10 days later in captivity than in the wild. Habersetzer and Marimuthu (1986) noted a slower growth rate in *H. speoris* maintained in an outdoor enclosure compared with individuals of same species living in the natural caves. Orr (1954) observed that the young *Antrozous pallidus* began to fly a week earlier under natural conditions compared to the bats kept in captivity. However, the growth trajectories of body mass and total gap length in the fourth metacarpal-phalangeal joint in relation to the growth in the length of forearm of the Indian false vampire bat *M. lyra* (Emmanuvel Rajan and Marimuthu, 1999) are apparently similar in both captive and natural conditions. Since it is difficult to mimic the diet and foraging environment of insectivorous bats in captivity, such species usually show slower growth in captivity. On the other hand, it is relatively easy to provide and replenish regularly an acceptable diet to carnivorous and frugivorous bats. When Kunz and Stern (1995) removed the effect of body mass they found no significant difference in growth conditions among 33 species of free-ranging and captive bats.

Baptista *et al*. (2000) compared two sampling methods, longitudinal and cross-sectional, to predict the age of free-ranging bats. Their study suggests that the longitudinal method (based on mark-recapture sampling) is more reliable for age estimation compared to cross-sectional (grab sampling) method. Even though in the current study the bats were maintained in captivity, the method employed was equivalent to longitudinal sampling. Since there is no significant difference in the growth pattern between captive and free-ranging bats, the equations to predict the age of young *C. sphinx* may reliably be used in behavioural, physiological and ecological studies.

In the present study, the analysis of postnatal growth based on three models showed that the logistic model best described the growth patterns of length of forearm, while Gompertz model best suited for the body mass of *C. sphinx*. The logistic growth model best fitted to the growth rates of forearm length as well as body mass of *T. brasiliensis* (Kunz and Robson, 1995), *P. pipistrellus* (Boyd and Myhill, 1987) and *Plecotus auritus* (De Fanis and Jones, 1995). Similarly, McLean and Speakman (2000) reported that growth of forearm in *P. auritus* best described by the logistic equation, and this model reflects a rapid growth and attainment of asymptote than von Bertalanffy equation. However, Kunz and Robson (1995) pointed out that the logistic model might not be appropriate in all growth studies. Hughes *et al*. (1995) found that growth patterns of body mass and forearm length of *P. pipistrellus sensu lato* were best fitted by the logistic and Gompertz functions, respectively. Even though these nonlinear models are useful to compare between different species, such studies should be based on the same models (Zullinger *et al*., 1984).

Similar to most microchiropteran species (Kurta and Kunz, 1987), *C. sphinx* is altricial at birth (Elangovan *et al*., 2002). Bat wings are underdeveloped at early stages of life and therefore they require a few weeks to attain flight. The developmental pattern of wing morphology of *C. sphinx* is in agreement with those reviewed for pteropodids (Norberg and Rayner, 1987; Elangovan *et al.*, 2004). The young *C. sphinx* became volant at the age of 6 – 7 weeks, which suggests that their wing membranes at this stage have developed adequately to evince flight. The handwing of *C. sphinx* is not completely formed compared to the relatively well-developed armwing at birth similar to other bat species (Powers *et al.*, 1991; Taft and Handley, 1991; Isaac and Marimuthu, 1997; Elangovan *et al.,* 2004). Even though the handwing grew relatively faster, it was still shorter than the armwing at the completion of postnatal development. Since the handwing is necessary to produce thrust during the down stroke, its rapid growth at the time of initiation of flight is expected (Norberg, 1976; Norberg and Rayner, 1987; Powers *et al.,* 1991). However, wings of bats permit little flexure or folding of the handwing without serious loss or lift. According to Norberg and Rayner (1987) "flexure can be greatest (retardation least) with relatively long armwings or short handwings". Hence in fast flying bats like *R. leschenaulti* the tip length ratio is <1 (Norberg and Rayner, 1987).

The higher quantitative increase of growth of wing area compared to body mass explains the linear decrease of wing loading until 50 days. At later stages when quantitative growth of body mass increased, growth of the wing area stabilized, which caused an increase in wing loading and subsequently converged to the adult value after 5 months. In addition, flight muscles are less developed at birth. The wet mass and cross-sectional fiber area of flight muscles of *M. lucifugus* reach the adult level when the young begin to fly (Powers *et al.,* 1991). Stern *et al.* (1997) and McLean and Speakman (2000) suggested that such differential development of wings, body mass, and flight muscles may facilitate learning the art of flight. Several microchiropterans exhibit similar patterns in the changes of wing loading (Jones, 1967; Davis, 1969; Hughes *et al.,* 1989a; Powers *et al.,* 1991; De Fanis and Jones, 1995; Stern *et al.*, 1997; McLean and Speakman, 2000).

Aspect ratio generally is considered an indication of efficiency of a wing (Norberg, 1990; Norberg and Rayner, 1987). Aspect ratio is not constrained before the young bat flies, allowing considerable variability. When flying age is reached, young bats that stayed back in the day roost practiced their flights at night hours (Bradbury, 1977). The young of a few species of bat accompany their mothers at least during their initial foraging flights (Bateman and Vaughan, 1974; Vaughan and Vaughan, 1987; Brigham and Brigham, 1989). Such early exercise may lead to further changes in the wing parameters resulting in the conformation of the aspect ratio to that of adults. Similar changes in aspect ratio occur in several species of bats (O'Farrell and Studier, 1973; Yokoyama *et al.,* 1975; Hughes *et al.,* 1989a; Powers *et al.,* 1991; Isaac and Marimuthu, 1997).

The logistic model best fitted to the growth patterns of both wingspan and wing area of *C. sphinx.* Suitability of logistic model reflects a rapid growth and attainment of asymptote than other growth equations (McLean and Speakman, 2000). It supports the growth pattern of wingspan and wing area of *C. sphinx*, which shows rapid growth until onset of flight and early attainment of asymptote. Usually the logistic model best describes growth of forearm length of different species of bats such as *P. pipistrellus* (Boyd and Myhill, 1987; Hughes *et al.*, 1995), *Plecotus auritus* (De Fanis and Jones, 1995; McLean and Speakman, 2000), *T. brasiliensis* (Kunz and Robson, 1995), and *C. sphinx* (Elangovan *et al.,* 2003). In contrast, growth of body mass varies among the three models not only between different species but also within the same species of bats. For example, Kunz and Stern (1995) preferred the logistic model to analyze postnatal growth in body mass of 33 species. The logistic (Boyd and

Myhill, 1987) as well as Gompertz models (Hughes *et al.*, 1995) best fit with growth of body mass of *P. pipistrellus*. In *P. auritus* both logistic (De Fanis and Jones, 1995) and von Bertalanffy (McLean and Speakman, 2000) models work best to describe growth of body mass. Even though these models are useful to compare between different species, such studies would be more comparable if based on the same model (Zullinger *et al.*, 1984).

The wing loading decreased linearly until 35 – 45 days of age due to faster development of wings than body mass in *C. sphinx*. It influenced the predicted flight performance including the minimum power speed and maximum range speed, and minimum turning radius to reach minima between 35 and 45 days of age. Such drops in flight performances correspond to wing loading observed in *P. auritus* (McLean and Speakman, 2000). However, the minimum flight power and maximum range power did not correspond to the pattern of wing loading, instead they increased linearly until 85 days of age and maintained their stability afterwards in *C. sphinx*. The age at which the flight parameters including wing loading, predictive flight speed and turning radius reached their lowest level corresponded to the onset of clumsy flight of *C. sphinx*. Juveniles of several species of microchiropteran bats typically start to fly when they attain 70% of adult body mass and over 95% of adult skeletal size (Barclay, 1995). However, the young *C. sphinx* in the present study began to fly when they have achieved about 42.9% of adult body mass and nearly 76.9% of adult skeletal size. This study substantiates the earlier report that megachiropterans are more advanced than microchiropterans in the pace of development (Orr, 1970).

REFERENCES

Aldridge, H. D. J. N. 1987. Turning flight of bats. *Journal of Experimental Biology* 128: 419 – 425.

Aldridge, H. D. J. N and Brigham, R. M. 1988. Load carrying and maneuverability in an insectivorous bat: a test of the 5% 'rule' of radio-telemetry. *Journal of Mammalogy* 69: 379 – 382.

Aldridge, H. D. J. N and Rautenbach, I. L. 1987. Morphology, echolocation and resource partitioning in insectivorus bats. *Journal of Animal Ecology* 56: 763 – 778.

Baptista, T. L., Richardson, C. S and Kunz, T. H. 2000. Postnatal growth and age estimation in free-ranging bats: a comparison of longitudinal and cross-sectional sampling methods. *Journal of Mammalogy* 81: 709 – 718.

Barclay, R. M. R., Fenton, M. B and Thomas, D. W. 1979. Social behavior of the little brown bat, *Myotis lucifugus*. 2. Vocal communication. *Behavioral Ecology and Sociobiology* 6: 137 – 146.

Bateman, G. C and Vaughan T. A. 1974. Nightly activity of mormoopid bats. *Journal of Mammalogy* 55: 45 – 65.

Block, W. B., Brennan, L. A and Gutierrez, R. J. 1991. Ecomorphological relationships of a guild of ground-foraging birds in northern California, U.S.A. Oecologia 87: 449 – 458.

Bogdanowicz, W., Fenton, M. B and Daleszczyk, K. 1999. The relationships between echolocation calls, morphology and diet in insectivorous bats. *Journal of Zoology* (London) 247: 381 – 393.

Bookstein, F. L. 1982. Foundations of morphometrics. *Annual Review of Ecology and Systematics* 13: 124 – 130.

Boyd, J. L and Myhill, D. G. 1987. Variation in the post-natal growth of pipistrelle bats (*Pipistrellus pipistrellus*). *Journal of Zoology* (London) 213: 750 – 755.

Bradbury, J. W and Vehrencamp, S. L. 1977. Social organization and foraging in emballonurid bats. IV. Parental investment patterns. *Behavioral Ecology and Sociobiology* 2: 19 – 29.

Brigham, R. M and Brigham, A. C. 1989. Evidence for association between a mother bat and its young during and after foraging. *The American Midland Naturalist* 121: 205 – 207.

Brown, P. E., Brown, T. W and Grinnell, A. D. 1983. Echolocation, development, and vocal communication in the lesser bulldog bat, *Noctilio albiventris*. *Behavioral Ecology and Sociobiology* 13: 287 – 298.

Buchler, E. R. 1980. The development of flight, foraging and echolocation in the little brown bat (*Myotis lucifugus*). *Behavioral Ecology and Sociobiology* 6: 211 – 218.

Carpenter, R. E. 1985. Flight physiology of flying foxes, *Pteropus poliocephalus.Journal of Experimental Biology* 114: 619 – 647.

Cheng, H. C and Lee, L. L. 2002. Postnatal growth, age estimation, and sexual maturity in the formosan leaf-nosed bat (*Hipposideros terasensis*). *Journal of Mammalogy* 83: 785 – 793.

Crome, F. H. J and Richards, G. C. 1988. Bats and gaps: microchiropteran community structure in Queensland rainforest. *Ecology* 69: 1960 – 1969.

Davis, R. 1969. Growth and development of young pallid bats, *Antrozous pallidus*. *Journal of Mammalogy* 50: 729 – 736.

De Fanis, E and Jones, G. 1995. Post-natal growth, mother-infant interactions and development of vocalizations in the vespertilionid bat *Plecotus auritus*. *Journal of Zoology* (London) 235: 85 – 97.

Elangovan, V., Raghuram, H., Yuvana Satya Priya, E and Marimuthu, G. 2002. Postnatal growth, age estimation and development of foraging behaviour in the fulvous fruit bat *Rousettus leschenaulti*. *Journal of Biosciences* 27: 695 – 702.

Elangovan, V., Yuvana Satya Priya, E., Raghuram, H and Marimuthu, G. 2003. Postnatal development in the Indian short-nosed fruit bat *Cynopterus sphinx*: growth rate and age estimation. *Acta Chiropterologica* 5: 107 – 116.

Elangovan, V., Raghuram, H., Yuvana Satya Priya, E and Marimuthu, G. 2004. Wing morphology and flight performance in *Rousettus leschenaulti*. *Journal of Mammalogy* 85: 806 – 812.

Emmanuvel Rajan, K and Marimuthu, G. 1999. Postnatal growth and age estimation in the Indian false vampire bat (*Megaderma lyra*). *Journal of Zoology* (London) 248: 529 – 534.

Findley, J. S and Wilson, D. E. 1982. Ecological significance of chiropteran morphology. Pp. 243 – 260 *in*: Ecology of bats. (T. H. Kunz, ed.). Plenum Press, New York.

Fullard, J. H., Koehler, C., Surlykke, A and McKenzie, N. L. 1991. Echolocation ecology and flight morphology of insectivorous bats (Chiroptera) in south-western Australia. *Australian Journal of Zoology* 39: 427 – 438.

Gannon, M. R., Willig, M. R and Jones, Jr. K. J. 1992. Morphometric variation, measurement error and fluctuating asymmetry in the red fig eating bat (*Stenoderma rufum*). *Texas Journal of Science* 44: 389 – 404.

Gelfand, D. L and McCracken, G. F. 1986. Individual variation in the isolation calls of Mexican free-tailed bat pups (*Tadarida brasiliensis mexicana*). *Animal Behaviour* 34: 1078 – 1086.

Habersetzer, J and Marimuthu, G. 1986. Ontogeny of sounds in the echolocating bat *Hipposideros speoris*. *Journal of Comparative Physiology* A 158: 247 – 257.

Hoying, K. M and Kunz, T. H. 1998. Variation in size at birth and post-natal growth in the insectivorous bat *Pipistrellus subflavus* (Chiroptera: Vespertilionidae). *Journal of Zoology* (London) 245: 15 – 27.

Hughes, P. M., Ransome, R. D and Jones, G. 1989. Aerodynamic constraints on flight ontogeny in free-living greater horseshoe bats, *Rhinolophus ferrumequinum*. Pp. 255 – 262 *in:* European bat research 1987. (V. Hanak, I. Horacek and J. Gaisler, eds.). Charles University Press, Praha.

Hughes, P. M and Rayner, J. M. V. 1991. Addition of artificial loads to long-eared bats *Plecotus auritus*: handicapping flight performance. *Journal of Experimental Biology* 161: 285 – 298.

Hughes, P. M., Rayner, J. M. V and Jones, G. 1995. Ontogeny of 'true' flight and other aspects of growth in the bat *Pipistrellus pipistrellus*. *Journal of Zoology* (London) 245: 291 – 318.

Isaac, S. S and Marimuthu, G. 1996. Postnatal growth and age estimation in the Indian pygmy bat *Pipistrellus mimus*. *Journal of Mammalogy* 77: 199 – 204.

Isaac, S. S and Marimuthu, G. 1997. Development of wing morphology in the Indian pygmy bat *Pipistrellus mimus*. *Journal of Biosciences* 22: 193 – 202.

Jones, C. 1967. Growth, development and wing loading in the evening bat, *Nycticeius humeralis* (Rafinesque). *Journal of Mammalogy* 48: 1 – 19.

Jones, G., Hughes, P. M and Rayner, J. M. V. 1991. The development of vocalizations in *Pipistrellus pipistrellus* (Chiroptera: Vespertilionidae) during post-natal growth and the maintenance of individual vocal signatures. *Journal of Zoology* (London) 225: 71 – 84.

Kalcounis, M. C and Brigham, R. M. 1995. Intraspecific variation in wing loading affects habitat use by little brown bats (*Myotis lucifugus*). *Canadian Journal of Zoology* 73: 89 – 95.

Kleiman, D. G. 1969. Maternal care, growth rate and development in the noctule (*Nyctalus noctula*), pipistrelle (*Pipistrellus pipistrellus*), and serotine (*Eptesicus serotinus*) bats. *Journal of Zoology* (London) 157: 187 – 211.

Krátky, J. 1981. Postnatale Entwicklung der Wasserfledermaus, *Myotis daubentoni* Kuhl, 1819 und bisherige Kenntnis dieser Problematik im Rahmen der Unterordnung Microchiroptera (Mammalia: Chiroptera). *Folia Musei Rerum Naturalium Bohemiae Occidentalis* 16: 3 – 34.

Kunz, T. H. 1973. Population studies of cave bat (*Myotis velifer*) reproduction, growth and development. *Occasional Papers of the Museum of Natural History*, University of Kansas, 15: 1 – 43.

Kunz, T. H. 1987. Post-natal growth and energetics of suckling bats. In: *Recent advances in the study of bats*, Fenton, M. B., Racey, P. A. and Rayner, J. M. V. (Eds.). 395 - 420. Cambridge: Cambridge University Press.

Kunz, T. H and Hood, W. R. 2000. Parental care and postnatal growth in the Chiroptera. Pp. 415 – 468 *in: Reproductive biology of bats*. (E. G. Crichton and P. H. Krutzsch, eds.). Academic Press, London.

Kunz, T. H and Anthony, E. L. P. 1982. Age estimation and post-natal growth in the bat *Myotis lucifugus*. *Journal of Mammalogy* 63: 23 – 32.

Kunz, T. H and Anthony, E. L. P. 1996. Variation in nightly emergence behavior in the little brown bat, *Myotis lucifugus* (Chiroptera: Vespertilionidae). Pp. 225 – 236 *in: J. Knox Jones, Jr Memorial* (H. H. Genoways and R. J. Baker, eds.). Texas Tech University Press, Lubbock.

Kunz, T. H and Robson, S. K. 1995. Postnatal growth and development in the Mexican free-tailed bat *Tadarida brasiliensis mexicana*: birth size, growth rate and age estimation. *Journal of Mammalogy* 76: 769 – 783.

Kunz, T. H and Stern, A. A. 1995. Maternal investment and post-natal growth in bats. *Symposium of the Zoological Society of London* 67: 123 – 138.

Kunz, T. H., Oftedal, O. T., Robson, S. K., Kretzmann, M. B and Kirk, C. 1995. Changes in the milk composition in three species of insectivorous bats. *Journal of Comparative Physiology B* 164: 543 – 551.

Kurta, A and Kunz, T. H. 1987. Size of bats at birth and maternal investment during pregnancy. *Symposia of the Zoological Society of London* 57: 79–106.

Landmann, A and Winding, N. 1993. Niche segregation in high altitude Himalayan chats (Aves, Turdidae): does morphology match ecology? *Oecologia* 95: 506 – 519.

Maeda, K. 1972. Growth and development of the large noctule, *Nyctalus lasiopterus* Schreber. *Mammalia* 36: 269 – 278.

McKenzie, N. L and Rolfe, J. K. 1986. Structure of bat guilds in the Kimberly mangroves, Australia. *Journal of Animal Ecology* 55: 401 – 420.

McKenzie, N. L., Gunnell, A. C., Yani, M and Williams, M. R. 1995. Correspondence between flight morphology and foraging ecology in some Palaeotropical bats. *Australian Journal of Zoology* 43: 241 – 257.

McLean, J. A and Speakman, J. R. 2000. Morphological changes during postnatal growth and reproduction in the brown long-eared bat *Plecotus auritus*: implications for wing loading and predicted flight performance. *Journal of Natural History* 34: 773 – 791.

Moss, C. F., Redish, D., Gounden, C. T and Kunz, T. H. 1997. Ontogeny of vocal signals in the little brown bat, *Myotis lucifugus*. *Animal Behaviour* 54: 131 – 141.

Norberg, U. M. 1976. Aerodynamics, kinematics and energetics of horizontal flapping flight in the long-eared bat *Plecotus auritus*. *Journal of Experimental Biology* 65: 179 – 212.

Norberg, U. M. 1981. Flight, morphology and the ecological niche in some birds and bats. *Symposium of the Zoological Society of London* 48: 173 – 197.

Norberg, U. M. 1985. Gliding, flying and soaring. Pp. 129 – 158 *in*: *Functional vertebrate morphology*. (M. Hildebrand, D. M. Bramble, K. F. Liem and D. B. Wake, eds.). Harvard University Press, Cambridge.

Norberg, U. M. 1990. Vertebrate flight. Springer-Verlag, Berlin.Norberg, U. M and Rayner, J. M. V. 1987. Ecological morphology and flight in bats (Mammalia: Chiroptera): wing adaptations, flight performance, foraging strategy and echolocation. *Philosophical Transactions of the Royal Society* B 316: 335 – 427.

O'Farrell, M. J and Studier, E. H. 1973. Reproduction, growth and development in *Myotis thysanodes* and *Myotis lucifugus* (Chiroptera: Vespertilionidae). Ecology 54: 131 – 141.

Orr, R. T. 1970. Development: prenatal and postnatal. Pp. 217 – 231 *in: Biology of bats*. (W. A. Wimsatt, ed.). Vol. 1. Academic Press, New York.

Papadimitriou, H., Swartz, S. M and Kunz, T. H. 1996. Ontogenetic and anatomic variation in mineralization of the wing skeleton of the Mexican free-tailed bat, *Tadarida brasiliensis*. *Journal of Zoology* (London) 240: 411 – 426.

Pennycuick, C. J. 1975. Mechanics of flight. Pp. 1 – 75 *in: Avian biology*. (D. S. Farner and J. R. King, eds.). Academic Press, New York.

Pennycuick, C. J. 1989. *Bird flight performance: a practical calculation manual*. Oxford University Press, Oxford.

Pisces, 1998. *Simply growth 1.7. Pisces* Conservation Ltd. Lymington, United Kingdom.

Powers, L. V., Kandarian, S. C and Kunz, T. H. 1991. Ontogeny of flight in the little brown bat, *Myotis lucifugus*, behavior, morphology and muscle histochemistry. *Journal of Comparative Physiology A* 168: 675 – 685.

Rayner, J. M. V. 1988. Form and function in avian flight. Pp. 1 – 66 *in*: *Current ornithology*. (R. F. Johnston, ed.). Plenum Press, New York.

Rhodes, M. P. 2002.Assessment of sources of variance and patterns of overlap in microchiropteran wing morphology in southeast Queensland, Australia. *Canadian Journal of Zoology* 80: 450 – 460.

Rhodes, M. P. 1995. Wing morphology and flight behaviour of the golden-tipped bat, *Phoniscus papuensis* (Dobson) (Chiroptera: Vespertilionidae). *Australian Journal of Zoology* 43: 657 – 663.

Richmond, J. Q., Banack, S. A and Grant, G. S. 1998. Comparative analysis of wing morphology, flight behaviour and habitat use in flying foxes (Genus: *Pteropus*). *Australian Journal of Zoology* 46: 283 – 289.

Rother, G and Schmidt, U. 1985. Die ontogenetische Entwicklung der Vokalisation bei *Phyllostomus discolor* (Chiroptera*). Zeitschrift für Säugetierkunde* 50: 17 – 26.

Rübsamen, R. 1987. Ontogenesis of the echolocation system in the rufous horseshoe bat, *Rhinolophus rouxi* (Audition and vocalization in early postnatal development). *Journal of Comparative Physiology A* 161: 899 – 913.

Ševcík, M. 2003. Does wing morphology reflect different foraging strategies in sibling bat species *Plecotus auritus* and *P. austriacus*? *Folia Zoologica* 52: 121 – 126.

Sterbing, S. J. 2002. Postnatal development of vocalizations and hearing in the phyllostomid bat, *Carollia perspicillata*. *Journal of Mammalogy* 83: 516 – 525.

Stern, A. A and Kunz, T. H. 1998. Intraspecific variation in postnatal growth in the greater spear-nosed bat. *Journal of Mammalogy* 79: 755 – 763.

Stern, A. A., Kunz, T. H., Studier, E. H and Oftedal, O. T. 1997. Milk composition and lactational output in the greater spear-nosed bat, *Phyllostomus hastatus*. *Journal of Comparative Physiology A* 167: 389 – 398.

Studier, E. H and Kunz, T. H. 1995. Nitrogen and mineral accretion in suckling bats, *Tadarida brasiliensis* and *Myotis velifer*. *Journal of Mammalogy* 76: 32 – 42.

Swift, S. M. 2001. Growth rate and development in infant Natterer's bats (*Myotis nattereri*) reared in a flight room. *Acta Chiropterologica* 3: 217 – 223.

Taft, L. K and Handley, Jr. C. O. 1991. Reproduction in a captive colony. Pp. 19 – 41 *in: Demography and natural history of the common fruit bat, Artibeus jamaicensis* on Barro Colorado Island. (C. O. Handley Jr., D. E. Wilson and A. L. Gardner, eds.). Smithsonian Institution Press, Washington.

Tuttle, M. D and Stevenson, D. 1982. *Growth and survival of bats*. Pp. 105 – 150 *in: Ecology of bats* (T. H. Kunz, ed.). Plenum Press, New York.

Vaughan, T. A and Vaughan, R. P. 1987. Parental behaviour in the African Yellow-winged bat (*Lavia frons*). *Journal of Mammalogy* 68: 217 – 223.

Willig, M. R and Moulton, M. P. 1989. The role of stochastic and deterministic processes in structuring neotropical bat communities. *Journal of Mammalogy* 70: 323 – 329.

Yokoyama, K., Uchida, T. A and Shiraishi, S. 1975. Functional morphology of wings from the standpoint of adaptation for flight in Chiroptera. I. Relative growth and ossification in the forelimb, wing loading, and aspect ratio. *Zoological Magazine* 84: 233 – 247.

Zullinger, E. M., Ricklefs, R. E., Redford, K. H and Mace, G. M. 1984. Fitting sigmoidal equations to mammalian growth curves. *Journal of Mammalogy* 65: 607 – 639.

In: Bats: Biology, Behavior and Conservation
Editors: J. L. Zupan and S. L. Mlakar, pp. 253-273
ISBN 978-1-61209-749-7

Chapter 7

HEMATOPHAGOUS BATS

Karin Corrêa Scheffer**, Rodrigo Fernandes de Barros, Alexandre Mendes Batista, Carla Isabel Macedo, Samira M. Achkar, Keila Iamamoto, Juliana Galera Castilho, Willian de Oliveira Fahl, Rafael de Novaes Oliveira, Ana Elena Boamorte da Costa, and Pedro Carnieli Junior
Pasteur Institute of São Paulo - Brazil

There is great interest in mammals due to their diversity, beauty, utility and also possible problems and diseases that they can cause. One of the orders most characteristic of mammals, because they have specialized structures for true flight, is the order *Chiroptera*.

The order *Chiroptera* is divided into two suborders; *Megachiroptera* and *Microchiroptera*. Among the microchiropters, there are 17 families, but only 9 families of bats occur in Brazil. Hematophagous bats are inserted in the family *Phyllostomidae*, subfamily *Desmodontinae* (Reis, 2007). The subfamily *Desmodontinae* comprises three species that possess a hematophagous feeding habit (Gardner, 1977), and they are *Desmodus rotundus* (E. Geofroy, 1810), *Diphylla ecaudata* (Spix, 1823) and *Diaemus youngi* (Jentink, 1893).

The *Desmodontinae* have characteristics that distinguish them from other families of Neotropical bats: rudimentary nasal appendix, of discoid structure in the shape of a horseshoe or as a protuberance (Altenbach, 1979; Greenhall et al. 1983; Greenhall, 1988). They do not have a tail, and the uropatagium is reduced. The legs, forearms and thumbs are long, where the last are thickened and used as feet for walking, jumping or climbing like a quadruped. Pads on the thumbs may or may not be present (Altenbach, 1979; Greenhall et al. 1983).

These animals are very specialized. They are obligatory blood feeders of homeothermal animals and show adaptations related to hematophagy, where the most notable are a reduction in the number and size of molars and pre-molars and the great development of the inner upper incisors, which have a triangular shape, sharp apex and cutting edges, extended postero-

* Address all correspondence, proofs, and reprint requests to: Karin Corrêa Scheffer: Instituto Pasteur de São Paulo, Av. Paulista, 393 São Paulo-SP Brazil 01311-000 E-mail: ksferreira@pasteur.saude.sp.gov.br.

laterally (Taddei, 1983; Fernandez et al., 1998). Due to this shape, at the moment of the bite, small pieces of skin and tissue are removed from its prey, in the shape of a "U," to obtain its food. The puncture made measures less than 5 mm in diameter and from 1 to 5 mm in depth, capable of breaking small vessels. To avoid the problem of coagulation, while feeding, hematophagous bats have in their saliva three proteins that retard the process of coagulation (Sodré, 2010). The first impedes the formation of the clot, the second prevents the blood cells from aggregating, and the third delays the constriction of the vessels at the wound site. The bat licks the wound constantly to ingest the blood that flows, and the contact of its saliva keeps the blood flowing. The meal lasts 20 to 30 min (Sodré, 2010).

The inferior lip is sulcated and devoid of papillae, with the tongue sulcated which allows the blood to flow inside the mouth by capillary action; stomach and kidneys are specialized in the absorption and processing of blood plasma. The stomach of a hematophagous bat rapidly processes the absorption of the liquid portion of the blood (plasma), which then passes to the kidneys and then to the bladder, and thus about two minutes after initiating the meal, the bat begins to urinate, avoiding a possible overload. Digestion occurs with the help of enterobacteria capable of decomposing the different parts of the blood (Greenhall et al., 1983; Bernard, 2005; Sodré 2010).

The hematophagous bat cannot survive for more than two days without feeding, and because it is an animal specialized in a single type of food, some characteristics are adapted for its consumption. It has a heat sensor, which consists of small sensors thermically isolated, situated in the snout; these help it to detect warmer points, where the blood vessels are closest to the skin surface, certainly for a better choice of the site for the bite (Greenhall et al., 1983; Bernard, 2005; Sodré, 2010).

Hematophagous bats have a visual acuity comparable to some nocturnal rodents, where it is likely that they use vision for long-distance orientation (Suthers, 1966; Manske, Schmidt, 1976). As with other bats, hematophagous ones also emit echolocalization signals, emitted by the mouth, for spatial orientation. The audition of these animals, however, is better adapted for low frequencies, between 100 Hz and 10 Hz (Schmidt et al., 1991). Their olfactory capacity is comparable to that of other small mammals, but is more sensitive than that of an insectivorous bat (Sodré, 2010).

The most important shelters for hematophagous bats are the diurnal ones, where these animals spend more than half of their daily cycle, from 05:00 am until 06:00 pm (Uieda, 1996). For this reason, during the day, hematophagous bats lodge in dark and protected locations, preferring caves, burrows, furnaces, abandoned mines and wells, mainly humid places, in which there is vegetation and water below. They can also be found in cracks in rocks, hollows of trees, in storm drains, under bridges and even old tanks or abandoned houses, where the last places are less frequent (Trajano, Sobrinho, 1980). These types of shelters have suitable environmental conditions of temperature (approximately 20 ºC), relative humidity (70 to 80%), light (complete absence and little or no perturbation), which is of extreme importance for survival of the colonies (Uieda, 1996).

The shelters occupied by hematophagous bats can be identified by the presence of pasty feces, yellowish brown, which later turn black similar to grease or pitch, and with a strong characteristic odor, mixed with the smell of ammonia. This odor can be detected from a distance, facilitating the localization of the shelter (Trajano, Sobrinho, 1980).

All types of shelters and refuges can be occupied by other species of bats, but the colonies do not mix and remain in different locations (Trajano, Sobrinho, 1980). The coexistence of three species hematophagous bats is probably facilitated by the division of the food resources and/or occupation of different diurnal shelters (Sodré, 2010).

The hematophagous bat goes out at night to feed, after complete darkness, avoiding hours with moonlight. Generally, they prefer the first hours of the night, but when there is a moon they go out later and until the first hours of daybreak (Trajano, Sobrinho, 1980). However, under favorable environmental conditions, feeding activity of these animals can occur throughout the night, beginning about two hours after sundown until one hour before sunrise (Uieda, 1992).

During the night, the bats do not spend all the available time feeding. On darker nights, they can use part of this time exploring their habitat, to get better awareness of the potential of alternative food sources (other types of prey) and of shelter, because in a situation where their main food source is lost, the bats feed on alternative sources, thereby allowing them to survive. Besides, part of the available time can also be used in social and reproductive interactions (Uieda, 1996).

The chiropters have been recognized as hosts of microorganisms that can potentially cause diseases in humans. Rabies is the most common zoonosis associated with bats, especially the hematophagous ones (Calisher et al., 2006). Another important public health problem is histoplasmosis, which is an infection caused by the fungus *Histoplasma capsulatum* found in soil containing feces of bats and birds. There are records of microepidemics of histoplasmosis in persons who were infected after visiting caves or who stayed at houses whose roof liner were greatly contaminated with feces of bats or birds (Dias, 2009).

Other agents isolated or detected in various species of chiropters are: *Salmonella* spp., *Escherichia coli*, *Campylobacter* spp. (Adesiyun et al., 2009), *Leptospira* spp. (Bessa, 2008; Matthias, 2005), *Trypanosoma* (Barros, 2008) and approximately 60 viruses (Calisher et al., 2006). Also detected in hematophagous bats besides rabies virus are coronavirus (Brandão et al., 2008) and Venezuelan equine encephalitis virus (Correa-Giron et al., 1970).

Desmodus Rotundus

Desmodus rotundus (Figure 1), popularly known as *common vampire bat*, is a species with greater occurrence, where it is abundantly studied because of its association with the transmission of rabies virus to herbivores in Latin America and because it feeds preferentially on the blood of mammals (Taddei, 1983).

In contrast with other mammals, vampiro has a tubular stomach and gastrointestinal tract in the shape of a "T," which has a large distension capacity. Besides, since it is a strictly hematophagous species, its feeding habit requires a large integrative capacity of the brain, where its neocortex and cerebellum are more developed than in other species of bats (Bernard, 2005).

In general, *D. rotundus* has a body covered by dense, short hairs, very soft, with a rust-brown coloration on the dorsal part of the body and light gray on the ventral part, but depending on the region or on the shelter, it can have other colors, varying in shade from reddish, golden to orangish (Bernard, 2005). There are even records of various albino individuals in the northeast, southeast and south of Brazil (Uieda, Chaves, Santos, 2002). The total length varies between 69 and 90 mm; forearm of 52 to 63 mm and on average, 35 to 40 cm in wingspan (Greenhall, 1972). It weighs between 25 and 40 g, where the females are larger and heavier than the males (Greenhall et al., 1983). It can be considered of medium size, compared with the approximately 160 other species of this group of mammals that exist in Brazil (Sodré, 2010).

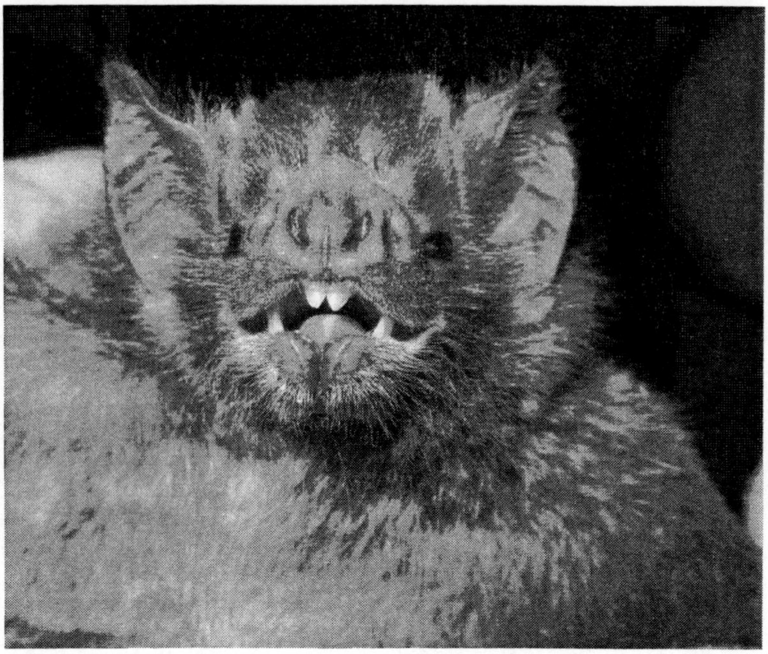

Source: MAPA e PANAFTOSA, 2007.

Figure 1. Desmodus rotundus.

The ears are relatively small, separated and pointed (Greenhall et. al., 1983). The eyes are large, but smaller than those of other species. The lower lip has a median sulcus in the shape of a "V" and this format helps the bat during feeding (Greenhall et. al., 1983; Sodré, 2010).

It has long thumb, the longest of the species, with three well-developed pads on the ventral surface (Vizotto, Taddei, 1973). This thumb is used to support it with the wings closed, allowing it to walk in a quadruped position when on the ground (Greenhall, 1972).

The interfemural membrane is narrow, poorly developed, showing short, spaced hairs its dorsal surface. The calcaneous, which is located on the side of its feet is reduced and resembles a small wart (Sodré, 2010).

It possesses a dental formula of incisors 1/2, canines 1/1, pre-molars 1/2 and molars 1/1, totaling 20 teeth. The upper incisors are cutting and wider than the canines and projected forward, with sharp triangular points and concave posterior edge. The lower incisors do not have more than two cusps. The upper and lower canines are wide, pointed and sharp on their posterior edge (Wilkinson, 1986).

Studies have revealed the existence of sexual dimorphism and occurrence of a greater number of females (Alencar et. al, 1994; Nuñez, Vianna, 1997; Gomes, Uieda, 2004). The species *D. rotundus* is considered polyestrus, without a defined period of reproduction (Taddei et. al., 1991; Alencar et. al., 1994), where it can reproduce during the whole year, although the birth of the majority of the pups appears to occur in the warmer and rainy months (Gomes, Uieda, 2004). Gestation is longer compared to other bats, which can be as long as seven months, followed by the birth of one pup at a time, occasionally with twins. The pup is born with hair, milk teeth (appearing with hooks), eyes open (Greenhall et. al., 1983), and it grows fast in the first two months of life, completing its development by five months (Turner, 1975; Lord, 1992; Sodré, 2010). In the first month, the pup generally stays stuck to the nipple of its mother, and between 50 and 60 days it begins to learn to fly. Sometimes, the female leaves its pup in the shelter, in "daycare", where it is common for other females to take care of the pups that are not theirs. The change in diet from milk to blood already occurs in the second month when the mother begins to regurgitate blood directly into its mouth. As of the fourth month, it begins to accompany its monther on night outings in search of prey (Sodré, 2010).

As with all hematophagous bats, its saliva contains an anticoagulant substance called draculin (Ciprandi et al., 2003), which prevents the coagulation of blood of the animal, allowing the blood to flow from the wound for a longer time. They have two longitudinal sulci on the underside of their tongue and, upon drinking blood, the sides of the tongue fold downward such that the surface becomes convex, forming a tube which they use to suck the blood (Mayen, 2003). With slight movements of the tongue in and out the mouth, a partial vacuum is formed in the buccal cavity, and the blood flows through the longitudinal sulci of the underside of the tongue, passing the upper side and to the back of the mouth. At this point, the blood is swallowed (Wilkinson, 1986).

D. rotundus is exclusive to the American continents, occurring from Sonora, Nuevo Leon and Tamaulipas in Mexico, Ilha Margarita in Venezuela, Trinidad, Bolivia, northern Chile, Brazil, Paraguay, Uruguay down to northern Argentina (Peracchi, 2006). This species does not tolerate cold climates and does not occur in locations where the mean temperature is lower than 10 ºC in the coldest month of the year (Greenhall et al., 1983). They do not hibernate, because food consumption and muscle activity increase when these animals are exposed to temperatures between 3 and 5 ºC (Wimsatt, 1962). Hematophagous bats have an unforeseeable thermal regulatory capacity, and when the environmental temperature declines, body temperature is normally maintained between 33 and 37 ºC by the increase in motor activity. If body temperature reaches 20 ºC, the *D. rotundus* are not able to warm themselves up again (Wimsatt, 1962). They are also very sensitive to high temperatures, where exposure to 37 to 38 ºC can be lethal (Greenhall et al., 1983).

Hematophagous bats are nocturnal, they normally shelter in dark and humid refuges (Taddei, 1983), live in small colonies (Greenhall et. al., 1983) with approximately 10 to 50 individuals; however, groupings with 100 or more bats can occur mainly in mountain regions with many caverns, where the control of their populations is not done regularly (Uieda et. al., 1996). In some areas, it is possible to find larger ones of 200, 300 and even 400 individuals (Uieda, Chaves, Santos, 2002).

They take refuge in locations difficult to access. They utilize many types of shelters and can share this space with about 45 other species of bats (Arellano-Sota, 1988; Greenhall, et. al., 1983). Modifications in ecological conditions, due to intense human activity, such as deforestation and various constructions, which are factors that indirectly or gradually act on the populations of hematophagous bats, determining an increase in the number of individuals, and alteration of their habits, among other consequences (Taddei, 1983). Therefore, of the three species of hematophagous bats, *D. rotundus* has been shown to be the most versatile at exploring a variety of types of shelter. One interesting aspect is the predominance of the use of artificial shelters, that is, those built by humans. In the state of São Paulo and in various other regions in Brazil, more than 50% of the groupings of this species have been found in constructions such as storm drains, galleries, abandoned mines, coal furnaces and abandoned houses (Uieda, 1996).

The diurnal shelters are very important for these animals, once they spend a greater part of their life cycle at this location. At these sites, they rest, have social interactions, reproduce and are protected from predators and harsh environmental conditions, such as rains and winds (Uieda, 1996). *D. rotundus* bats are also known to use nocturnal shelters. Which are generally used temporarily (they can change periodically) and are frequently situated close to food sources (Sazima, 1978), such as barns, stables, storehouses, cellars and nearby vegetation. They serve as places for rest, adaptation, and elimination of eventual excess weight in the form of urine and feces, where they can be utilized before and/or after feeding (Sazima, 1978; Uieda, 1992). These shelters can be identified during the day by the presence of fresh feces on the floor and walls. Through the feces, nocturnal shelters are also known as places for digestion. Their use by *D. rotundus* varies from one region to the next and can depend on the existence of suitable places close to sources of food (Sazima, 1978).

As usually found among gregarious mammals, *D. rotundus* shows a social structure characterized by hierarchy of dominance, based on the formation of a harem, where one dominant male protects a group of females (about 12) and their pups (Wilkinson, 1988). In general, colonies with more than 50 individuals can contain several groups of 10 to 20 females with pups. The colonies of the maternity type shelters are composed of females, their pups and dominant males (Sodré, 2010).

The dominant male is at the top of the shelter, surrounded by females, and the other males stay in peripheral locations in the same colony or are expelled from the shelters when they reach an age of 12 to 18 months and find a new shelter always at least 3 km from where they were born (Wilkinson, 1988). They form small groupings of young males that have still not reached sexual maturity, where these groupings are located close to the harem. Thus, the young males remain waiting for an opportunity of disputing the place of the dominant male or forming their own harem (Sodré, 2010).

Dominant males, besides having greater access to females, also feed in places close to the shelter, while the peripheral males fly large distances to feed, thereby being able to overlap its forraging area with those of other colonies (Wilkinson, 1988).

After leaving the diurnal shelter, the bats fly to find food. The flights of *D. rotundus* are generally made at a height of 0.5 to 1.5 m (Greenhall, Schmidt, Lopez-Forment, 1969), since their prey are mammals that rest on the ground, such cattle and horses (Sazima, 1978).

Hematophagous bats normally forrage in an area of 5 to 8 km around the diurnal shelters (Crespo et. al., 1961), and in certain regions this distance can increase to 15 to 20 km (Málaga-Alba, 1954). Under favorable environmental conditions, feeding activity of the bats *D. rotundus* can begin at about two hours after sundown and end around one hour before sunrise (Uieda, 1992), but it was also found that in the summer bats leave the shelters after 9 pm and in the winter after 10 pm (Villa-R, 1966), confirming that the bats only go out to forrage when darkness is complete, where they can make a preliminary flight to check the moonlight (Crespo et. al., 1961).

Studies of the activity of hematophagous bats in enclosures indicate that the period when bats forrage is related to the absence of moonlight (Crespo et. al., 1972). On very moonlit nights, hematophagous bats can forego feeding for one night, but they cannot stand more than two nights without eating, especially in the dry season (Uieda, 1996).

The olfactory sense appears to be of importance in the localization and selecting prey (Schmidt, 1978). Therefore, *D. rotundus* can locate livestock resting areas of herds (stables and pastures) by the odor of manure, and they then find the prey by sight. This species would be capable of detecting visually a cow at least 130 m away (Turner, 1975).

Before feeding, *D. rotundus* makes a "recognizance flight" around the livestock in open locations (Greenhall et. al., 1971). It is believed that these flights are an adaptation behavior, in which the bats examine and choose their prey (Sazima, 1978).

One important factor in the selection of prey is its accessibility. *D. rotundus* chooses and attacks the most accessible prey in a herd, which are generally the animals of docile temperament and which are sleeping on the periphery of the herd (Turner, 1975; Sazima, 1978).

The approach of *D. rotundus* to its prey can be made in two ways: resting directly on the body of the prey or on the ground close to the animal (Uieda, 1996). The mode of approach can indicate the sites chosen on the body of the victim to bleed it (Uieda, 1996). Reaction of the animals to the approach of the bats generally occurs when the bats rest on their body. The cattle or horses herd react with movements of the head, tail and musculature of the skin (Crespo et. al., 1971). During the approach to the prey, *D. rotundus* remains cautious to any reaction of the victim. Any sign of danger, it leaves the site until the danger passes or abandons this animal and flies off to find another more accessible prey (Uieda, 1996). *D. rotundus* can spend about 40 min to choose a site on the body of the prey to bite (Greenhall, 1972).

On feeding, this bat prefers the extremities such as ears, neck, anal region, vulva, teats, snout, and tail, among others (Greenhall et. al., 1983). The prey is perforated with the sharpened incisors leaving a characteristic wound.

A bat can ingest between 15 and 25 mL in one prey and an animal can be visited by various bats on the same night (Constantine, 1979). The time necessary for feeding by hematophagous bats depends on the reactions of the victim. *D. rotundus* generally spends about 30 min, but this time can be as long as an hour (Uieda, 1982).

While the bat feeds, they eliminate urine, which can be a way of emptying the stomach more rapidly, making it possible to consume more food or making flying easier during forraging if necessary (Uieda, 1996).

The hematophagous bat can reopen wounds made on previous nights, because that can be done in a few minutes, which decreases the time to possible risks during forraging such as kicks or bites (Greenhall et. al., 1983).

In the shelter, collective physical contact is common, when the individuals lick themselves for body hygiene. This behavior occurs mainly among the females, stimulating the regurgitation of food (Sodré, 2010). Females with pups and young animals belonging to the same group that are unable to feed on one night, receive blood from other well-fed females through regurgitation, thereby helping these individuals to survive. This behavior is important in social interaction, because it involves the recognition and reciprocity of the individuals that share blood, and is a form of parental care among females and their pups (Mills, 1980; Schmidt, Schmidt, Manske, 1980; Wilkinson, 1988; Schmidt, 1988; Altringham, 1996). The females that do not cooperate in the sharing of food are expelled from the group (Sodré, 2010).

By the fact of having an exclusively hematophagous feeding habit, *D. rotundus* despite not constituting a danger to large-sized animals in which loss of blood is not critical, causes harm to these animals through eventual infections of the wounds and transmission of diseases. The most important diseases are some trypanosomiases such as "murrina" transmitted by *Trypanosoma hippicum* and "mal de caderas" transmitted by *Trypanosoma evansi* and rabies (Greenhall et. al., 1983). The direct and indirect economic damages caused by rabies in herbivores transmitted by *D. rotundus* are substantial (Mayen, 2003).

These animals are not found on the list of threatened species (red book) for Brazil, according to Environment Ministry (MMA, 2003), or on the list of IUCN (Barquez et. al., 2008; IUCN, 2010).

DIPHYLLA ECAUDATA

Diphylla ecaudata (Figure 2) is considered the second species in importance, both in number of individuals and in distribution. It is relatively rare and does not cause great economic damages. It is known as the *hairy-legged vampire bat,* due to the rudimentary interfemural membrane with long hairs (Vizotto, Taddei, 1973).

Source: MAPA e PANAFTOSA, 2007.

Figure 2. Diphylla ecaudata.

The fur is dense covering the forearms, legs and interfemural membrane, where it is long and soft. The coloration of the fur varies from light brown in the dorsal region and is always lighter in the ventral region. The length of the head to body varies from 75 to 93 mm with males and females being similar in size. The length of the forearm is from 50 to 56 mm and of the wingspan about 30 cm. The body weight is 24 to 43 g (Villa-R, 1967; Greenhall, Schmidt, Joermann, 1984).

D. ecaudata resembles *D. rotundus* externally, but can be distinguished by having a small size, besides short, rounded ears, large, brillant eyes, and short thumbs without pads and short calcaneous. However, the most outstanding characteristic is the shape of the mandible and the dental formula - incisors 2/2, canines 1/1, pre-molars 1/2 and molars 2/2, totaling 26 teeth, as well as the structure of the lower incisors when compared to those of *D. rotundus*. The external upper incisors are cutting and pointed, with sharpened edges, but less than that in *D. rotundus*, and internal upper incisors are diminutive, that is, not very noticeable. However, the lower incisors are larger than those in *D. rotundus*, forming a continuous and convex line, separated from the canines by distinct spaces with the internal lower incisors with four cusps and the external lower incisor with four to seven cusps (Greenhall, Schmidt, Joermann, 1984).

Due to its diet, these bats have a series of morphological modifications in the digestive tract, with an elongated stomach, with an absorption surface much greater than that of other species of bats (Gomes, 2008).

The distribution of this species is more restricted than that of other hematophagous bats (Aguiar, 2006) *D. ecaudata* occurs from the southern Texas through eastern Mexico, part of Central America to South America, at least part of Peru and Brazil (citing occurrence in 13 of the 26 states), with abundance in the Amazon region (Villa-R, 1968; Peracchi et. al., 2006). Like *D. rotundus,* it does not tolerate very cold climates, inhabiting places with mean temperatures between 10 and 35 °C. McNab (1969; 1973), in studying the capacity of thermoregulation of *D. ecaudata* after feeding, showed that oxygen intake by these animals is increased by about 50%, and that mean body temperature (32.4 °C) increases by approximately 2 °C.

D. ecaudata basically explores often a type of shelter, which includes caverns, where they can be rarely found in hollows of trees, abandoned mines, coal furnaces and abandoned houses. This demonstrates a strong selection of shelters and it is possible that this accounts for its relative scarcity in nature (Uieda, 1996). The same shelter is often inhabited by *D. rotundus* and a variety of other phyllostomids (Felten, 1956).

They live in small colonies of approximately 3 to 12 individuals, and occasionally, colonies of 50 to 70 individuals have been recorded (Figure 3). They are often found cohabitating with *D. rotundus*, almost exclusively in caves and mines, rarely in hollows of trees (Bredt et. al., 1996). They are timid bats, with rapid movements, and they fly off rapidly when disturbed. In contrast to *D. rotundus*, when perturbed in their habitat, these bats move to another place and do not hide in cracks (Greenhall et. al., 1984). Even in colonies with many individuals, they maintain a solitary habit, where they do not aggregate into large groups (Aguiar, 2006).

Of the three species of hematophagous bats, *D. ecaudata* is the one that shows more specialized feeding habits, preferentially feeding on fresh blood of birds (Uieda, 1993; Greenhall, Schutt, 1996). Ruschi (1951) observed these animals in captivity and proposed the hypothesis that under poor conditions or with the lack of their prey, they can rarely complete their diet with the blood of mammals such as pigs, cattle and horses.

Like *D. rotundus*, the choice of prey depends directly on their acessibility to the predator. They can also be guided by their olfactory sense following the odor emitted by the feces of their prey and localize their prey in trees (Uieda, 1993). Generally, they attack domestic birds and wild birds that perch in trees, preferentially, on branches free of foliage, because they are more accessible (Uieda, 1993), such as on perches; the birds are bitten in the region of the cloaca or on lower part of the legs (Ruschi, 1951; Dalquest, 1953). The places most often bled vary from one type of prey to the other; in chickens they are the edges of the cloaca, toes and tarsi (Uieda, 1996). The approach to their prey can also be done in two ways, with a light and manoeverable flight, landing on the perch or directly on the body of the perched bird (Uieda, 1982). After landing, the bat moves rapidly to the position of the head and down to the region of the cloaca firmly grabbing the tail feathers with its thumbs. These bats remain about 10 to 40 min attached feeding.

Source: MAPA e PANAFTOSA, 2007.

Figure 3. *Dyphylla ecaudata* colony found in cave of Guararema's city/São Paulo/Br.

The reactions of the chickens can be a little clucking by the presence of the bats, where they can stay standing on the perch, while they turn round themselves (Uieda, 1982, 1993). Occasionally, the chickens strike with pecking in the direction of the aggressor (Uieda, 1993). A greater mortality is due to attack by *D. ecaudata* and *D. rotundus* on the same night; this is because of the size of the bite and to the repetitive attacks on the same bird by these animals (Uieda 1993).

Like the majority of bats, *D. ecaudata* restricts its feeding activity to the darkest period of the night, and reduces this activity in relation to environmental factors, such as moonlight, torrential rains and strong winds (Uieda, 1982). Arellano and coworkers (2007), kept these animals in captivity and observed the altruism of these animals with regurgitatioin of blood between individuals that did not feed, as observed in *D. rotundus*.

Practically nothing is known about the reproduction of *D. ecaudata*, but pregnant females have been found during the middle of the year; the time of gestation is unknown. It produces only one pup at a time and perhaps two pups per year (Uieda, Chaves, Santos, 2002).

D. ecaudata is not found on the list of threatened species (red book) for Brazil, according to data from the Environment Ministry (MMA, 2003), or on the list of the IUCN (Barquez et. al., 2008; IUCN, 2010).

DIAEMUS YOUNGI

Diaemus youngi (Figure 4) or popularly known as *white-winged vampire bat* because it has white spots on the points of the wings as well as on the membrane between the second and third fingers (Greenhall, Schutt, 1996).

Source: Uieda, 2010.

Figure 4. *Diaemus youngi*.

Its fur is short and silky and of brilliant light brown coloration on the dorsum and a little lighter on the venter (Uieda, Chaves, Santos, 2002), where it can also vary from light to dark brown (Greenhall, Schutt, 1996).

It is a medium-sized bat with a robust build, a length of about 83 mm from head to the body and wingspan of 35 cm; its weight varies from 30 to 50 g and has a forearm of 51 to 53 mm (Uieda, Chaves, Santos, 2002).

It bears a strong resemblance to *D. rotundus* and *D. ecaudata*, but can be distinguished by other characteristics which include the length and shape of the ears, where these are longer and more pointed, the external edge that does not possess folds as in *D. rotundus*, and the presence of a tragus with many hairs on the anterior side. The size of the thumb is relatively short when compared to *D. rotundus*, possessing a single pad, as well as the absence of a calcaneous. It has a narrow interfemural membrane with few hairs (Greenhall, Schutt, 1996).

Both sexes have glands located bilaterally inside the mouth (in the cheeks), which are only seen when the bat is disturbed (Greenhall, Schutt, 1996), then releasing a volatile and nauseating substance with a musky and disaggreable odor. Works have demonstrated the presence of amino acids in these glands, such as aspartate, glutamate, norvaline, alanine, and a type of lysine, which are believed to be involved in a mechanism against predators, since these bats, when handled, release an offensive odor and concomitantly emit cries of agony (Goodwin, Greenhall, 1961; Greenhall, 1988). Meanwhile, Schutt (1995) report that these glands may be involved with marking territory and of individual recognition in the group.

The dental formula of *D. youngi* is incisors 1/2, canines 1/1, pre-molars 1/2, molars 2/1, totaling 22 teeth. However, in many individuals, the upper posterior molar is vestigial, and in older individuals, the molars are often lost. The lower incisors tend to be very irregular and variable in both types and cusps. The upper incisors and canines are pointed and with cutting edges utilized to scrape off hairs and feathers of prey, perforate the skin and feed itself. This bat has a prominent medial cusp on the external upper molar, which is absent in *D. rotundus* as well as *D. ecaudata* (de La Torre, 1956; Goodwin, Greenhall, 1961; Husson, 1962; Koopman, 1988). Like *D. rotundus*, it has a pronounced split in the lower lip. The stomach of *D. youngi* is elongated, but is less tubular and more conical shape than the stomach and intestine of *D. rotundus* (Forman et al, 1979).

The distribution of this species is wide with occurrences from northeast Mexico, passing through Central America and reaching South America, from the Amazon basin down to northern Argentina. In contrast to *D. rotundus*, which is a very abundant and common species, *D. youngi* is locally rare regardless of its wide distribution, and there is a lack of population, biological and ecological data (Greenhall, Schutt, 1996; Aguiar et. al., 2006). In Brazil, there are records of the occurrence of *D. youngi* in 13 of the 26 states (Aguiar et. al., 2006), occurring in all the Brazilian biomes (Marinho-Filho, Sazima, 1998).

Its colonies contain about 8 to 12 individuals, where colonies with up to 30 individuals can be found (Greenhall, Schutt, 1996). They have a dominance-hierarchy behavior with displays and behavior patterns not related to other species of bats (Schutt et. al., 1999).

It appears to be the rarest of the three hematophagous species. This rarity can be related to its specialization on blood of birds and to the use of hollows of trees as the diurnal shelter most utilized (Uieda, Chaves, Santos, 2002). Caverns have also been cited as shelters of this species, but their use is not frequent (Uieda, 1996; Uieda, Chaves, Santos, 2002).

Trajano (1984) found in southeast Brazil *D. youngi* bats in caverns, in association with eight other species of bats, including *Peropteryx macrotis* (Wagner 1843), *Chrotopterus auritus* (Peters, 1856), *Anoura caudifer* (E Geoffroy, 1818), *Carollia perspicillata* (Linnaeus, 1758), *Sturnira lilium* (E. Geoffroy, 1810), *Artibeus lituratus* (Olfers,1818), *Platyrrhinus lineatus* (Geoffroy, 1810) and *Desmodus rotundus*.

Source: Gitti, 2007.

Figure 5. Specie *Diaemus youngi* sucking chicken blood.

D. youngi preferentially attacks domestic and wild birds that show a behavor of perching at heights higher than 3 m, but can also attack birds close to the ground (Figure 5) (Costa et.

al., 2008). Although it feeds on fresh blood and has a preference for blood of birds, in captivity, feeding on cattle blood was observed (Uieda, 1993; Greenhall, Schutt, 1996).

In studies conducted in Trindad and Tobago, *D. youngi* was seen preying on birds and mammals, as well as goats and occasionally the livestock (Goodwin, Greenhall, 1961). In other studies of 23 bats captured and observed in captivity, 13 fed on blood of mammals (cattle and pigs, and some samples not identifiable), eight fed on blood of birds in combination with blood of mammals, and only two fed exclusively on blood of birds (Greenhall, 1970; 1988).

In Tamaulipas, in Mexico, a vampire bat with white wings was captured while feeding on a chicken (Villa-R., 1967). Besides, its behavior feeding on birds free in nature (domestic pigeons, collared pigeons and half-grown chickens) have been observed in Brazil (Sazima, Uieda, 1980; Uieda, 1992, 1993, 1994) and Trinidad and Tobago (Greenhall, 1988). It is not known if the saliva of the vampire bat with white wings contains anesthetic or enzymatic components to aid in the bite or licking or if it softens the keratinized reticulated scales that cover the surface of the toes of the birds (Greenhall, Schutt, 1996).

This bat generally flies around its prey at high heights, sometimes drawing near and then flying from the perched birds. The olfactory sense is also of great importance in the localization of the prey. *D. youngi* makes its approach landing directly on the perch or branches close to the prey (Uieda, 1993), and its landing directly on the prey has not been observed, likely because it is more robust and heavier (Uieda, 1996).

After the landing, the bat rapidly grabs the perch, moving slightly toward the inner side of the perch in the direction of the bird, chooses a site to bite, bites and begins its meal (Uieda,1992). Like in *D. ecaudata*, the sites most often bled depend on the type of prey. In caipira chickens bred in backyards, they prefer to bleed the toes, tarsi, cristae, barbels and base of the neck. However, in Angola chickens, the sites chosen are toes and tarsi, and in turkeys, the toes, tarsi and the naked chest (Uieda, 1996). Ducks and geese, because they sleep on the ground, are not attacked (Uieda, 1992; Uieda, Chaves, Santos, 2002).

Occasionally, feeding is interrupted when the prey is momentarily perturbed. In this case, *D. youngi* generally withdraws to the lower part of the branch or perch and aligns itself with this. This behavior is believed to be defensive, with the branch or perch protecting the bat from pecking by the birds. In other cases, it was reported that when the bird was disturbed, the vampire bat with white wings hid under the wing and in the belly of the bird (Uieda, 1992).

The time necessary for its feeding depends on the reactions of the prey during the meal, requiring 15 to 30 min (Uieda, 1996). However, it can remain with the prey for more than an hour (Uieda, 1982), and the mean consumption of blood of birds is 33 mL (Uieda, 1996). There are no reports of death of the birds due to the predatory action of *D. youngi*. The reports also involved the other two species (Uieda, Chaves, Santos, 2002).

The conformation of the bones of lower limbs of *D. youngi* also appears to be adapted for quadruped locomotion. However, the fact that the species *D. youngi and D. ecaudata* possess a shorter thumb, can be a factor that impedes the exhibition of agile acrobatics, for example, jumping from the ground to initiate flight, as with *D. rotundus* (Schutt et. al., 1993).

Despite that reproduction in this species is not well known (Greenhall, Schutt, 1996), reproductive males were found in the rainy season in the Cerrado of Central Brazil, and there are no reports of captures in the dry period (Aguiar et. al., 2006; Costa et. al., 2008). According to Bredt et al. (1996), these bats show a gestation of approximately 7 months, with a reproduction peak per year, with partum in the warmer months, producing only a single pup at each birth.

D. youngi is not found on the list of threatened species (red book) for Brazil, according to the Environment Ministry (MMA, 2003), and is also not on the list of the IUCN (Barquez et. al., 2008; IUCN, 2010). However, it is considered a threatened species in two states of Brazil, Parana (Margarido, Braga, 2004) and Rio de Janeiro (Bergallo et. al., 2000).

Gonçalves et al. (2002) and Aguiar et al. (2006) suggest that due to its similarity to *D. rotundus,* the species *D. youngi* is not being negatively affected by current control measures of rabies in herbivores which involve population control of *D. rotundus*. Due to limitation in technical abilities for correct identification of species, techniques involved in this activity can be applying vampiricidal paste in individuals of the species *D. youngi*.

The rabies virus has already been reported in all three species of hematophagous bats, but the only reports of human rabies transmitted by *D. youngi* were in Trinidad and Tobago (Constantine, 1988; Goodwin, Greenhall, 1961).

How to Identify Species of Hematophagous Bats

Due to the three species being very similar in appearance, the main external morphologic characteristics were combined here for the identification of the subfamily *Desmodontinae* and also those that differentiate the species of this subfamily.

Subfamily *Desmodontinae* – nasal appendix rudimentary, of discoid structure in the shape of a horseshoe; upper incisors strictly contiguous, very large and falciform; sulcated lower lip.

Species *Desmodus rotundus* – Uropatagium more developed inthe median region (Figure 6); dactylopatagium without light spots (Figure 7); ears pointed at the apices; tibia and dorsum of the feet with few hairs; calcaneous almost imperceivable (Figure 8); thumb well developed and with 3 pads (Figure 9); 2nd phalange of the third finger smaller (16.0 – 20.0); incisors 1/2; lower incisors well separated, 2 to 2, and with 2 cusps on the surface (Figure 10).

Species *Diaemus youngi* – Uropatagium more developed in the median region (Figure 6); dactylopatagium long and wide, with whitened spots on the corners (Figure 7); ears pointed at the apices, tibia and dorsum of the feet with few hairs; calcaneous almost imperceivable (Figure 8); thumb shorter that the foot with a single pad (Figure 9); 2nd phalange of the third finger higher (25.0 – 29.0); incisors 1/2; lower incisors less separated, 2 to 2, without cusps on the surface (Figure 10).

Species *Diphylla ecaudata* – Uropatagium rudimentary, almost imperceivable in its median part (Figure 6); dactylopatagium without light spots (Figure 7); ears short, rounded at apices and very hairy; tibia and dorsum of the feet covered by dense hairs; calcaneous short (Figure 8); thumb developed and without pad (Figure 9); incisors 2/2; lower incisors together and similar, with many cusps on the surface (4, 6 or more) (Figure 10).

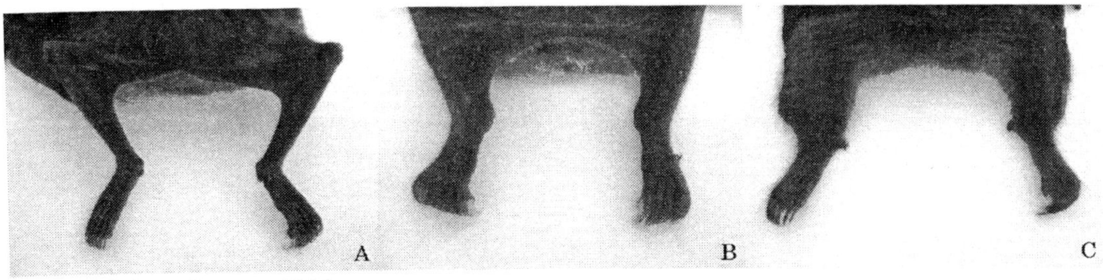

Source: Instituto Pasteur, 2010

Figure 6. Uropatagium from *Desmodus rotundus* (A); *Diaemus youngi* (B) and *Diphylla ecaudata* (C) species.

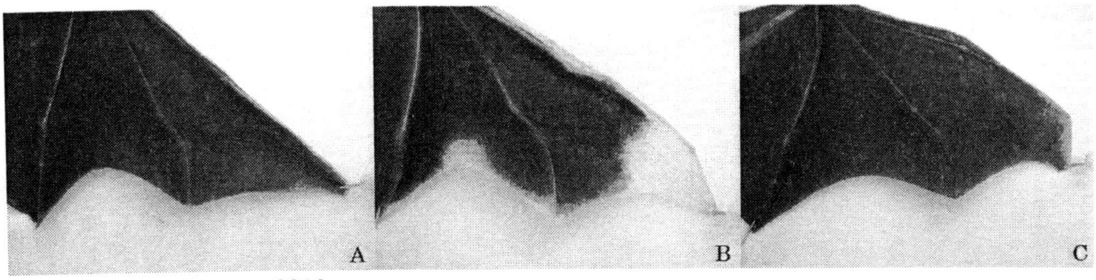

Source: Instituto Pasteur, 2010.

Figure 7. Dactylopatagium from *Desmodus rotundus* (A); *Diaemus youngi* (B) and *Diphylla ecaudata* (C) species.

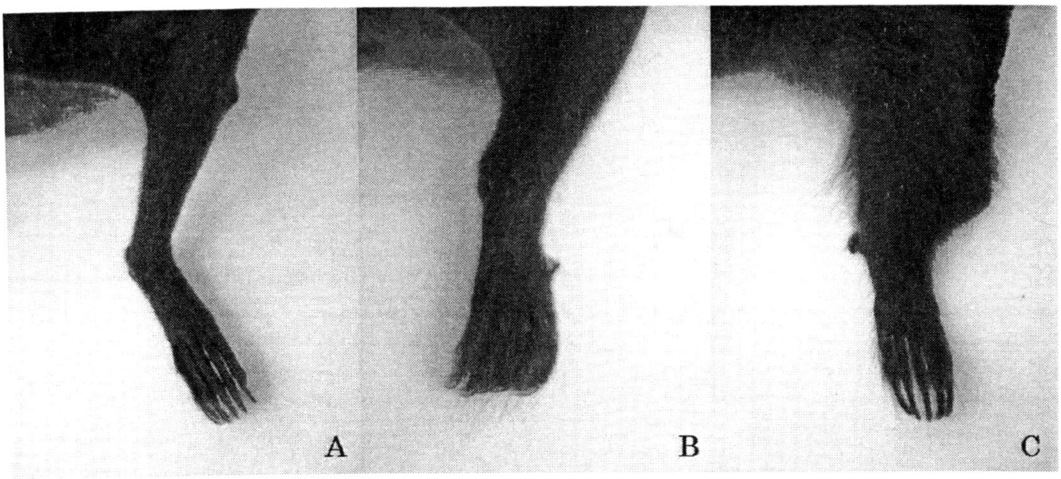

Source: Instituto Pasteur, 2010.

Figure 8. Tibia, dorsum of the feet and calcar from *Desmodus rotundus* (A); *Diaemus youngi* (B) and *Diphylla ecaudata* (C) species.

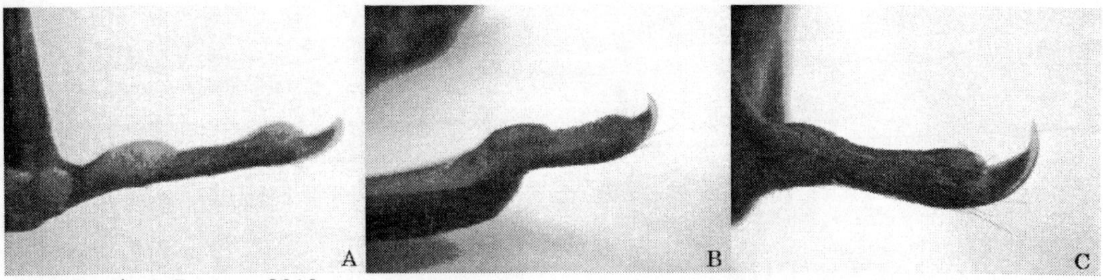

Source: Instituto Pasteur, 2010.

Figure 9. Thumb (first finger) from *Desmodus rotundus* (A); *Diaemus youngi* (B) and *Diphylla ecaudata* (C) species.

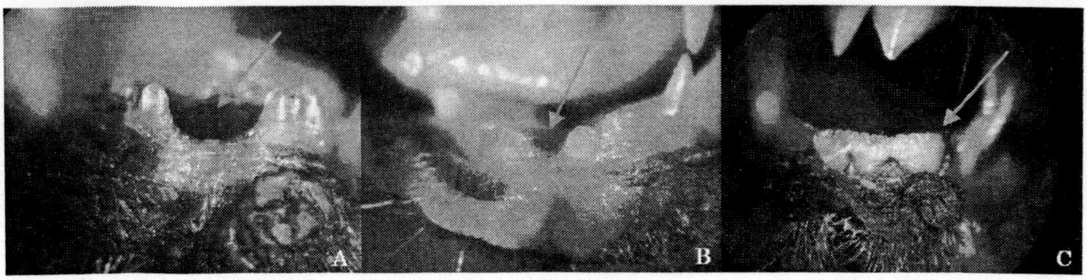

Source: Instituto Pasteur, 2010.

Figure 10. Lower incisors teeth from *Desmodus rotundus* (A); *Diaemus youngi* (B) and *Diphylla ecaudata* (C) species; arrows showing the central gap and cusps teeth.

References

Adesiyun, A.A.; Stewart-Johnson, A. and Thompson, N.N. (2009). Isolation of enteric pathogens from bats in Trinidad. *J Wildl Dis*, 45(4), 952-961.

Aguiar, L.M.S.; Camargo, W.R. and Portella A.S. (2006). Occurrence of white-winged vampire bat, Diaemus youngi (Mammalia, Chiroptera), in the Cerrado of Distrito Federal, Brazil. *Rev Bras Zoo,* 23(3), 893–896.

Alencar, A.O.; Silva, G.A.P.; Da Arruda, M.M.; Soares, A.J. and Guerra, D.Q. (1994). Aspectos biológicos e ecológicos de *Desmodus rotundus* (Chiroptera) no nordeste do Brasil. *Pesq Vet Bras*, 14(4), 95-103.

Altenbach, J.S. (1979). Locomotor morphology of the vampire bat *Desmodus rotundus*. *Special Publications American Society of Mammalogists*. 6, 1-137.

Altringham, J.D. (1996). Bats, biology and behavior. Oxford, *Oxford Univ Press*, 262.

Arellano, C.E.; Vidal, J.C.L.; Cabrales, J.A.; Medellín, R.A. and Laundré, J. W. (2007). Food sharing behavior in the hairy-legged vampire bat Diphylla ecaudata. *Acta Chiropterologica*, 9 (1), 314-319.

Arellano-Sota, C. (1988). Biology, ecology, and control of the vampire bat. *Rev Infect Dis*, 10(4), 615-619.

Barquez, R. ; Perez, S. ; Miller, B. and Diaz, M. (2008). *Desmodus rotundus*. In: IUCN 2010. IUCN Red List of Threatened Species. Version 2010.4. <www.iucnredlist.org>. Downloaded on 01 November 2010.

Barros, J.H.S.B.; Romijn, P.C.; Baptista, C.; Pinto, A.G.S.P. and Madeira, M.F. (2008). Relato de infecção natural de morcegos por flagelados tripanosomatideos em diferentes municípios do Estado do Rio de Janeiro. *Rev Soc Bras Med Trop*, 41(6), 683-685.

Bergallo, H.G.; Geise, L.; Bonvicino, C.R.; Cerqueira, R.; D'Andrea, O.S.; Esberárd. C.E.L.; Fernandez, F.A.Z.; Grelle, C.E.; Peracchi, A.; Siciliano, S. and Vaz, S.M. (2000). Mamíferos (125-136). In: Bergallo, H.G.; Rocha, C.F.D; Alves, M.A.S. and Sluys, M.V. *A fauna ameaçada de extinção do Estado do Rio de Janeiro*. Rio de Janeiro: *Ed. UERJ*.

Bernard, E. (2005). *Morcegos vampiros: sangue, raiva e preconceito*. Ciência Hoje, Rio de Janeiro, 36(214), 44-49.

Bessa, T.A.F.; Chapola, E.G.B.; Spichler, A.; Husch, A.C.; Silva, M.M.S. and Almeida, M.F. Leptospirose e morcegos em área urbana no município de São Paulo – Brasil. In: 35° CONBRAVET- 19 a 22 de outubro de 2008 – Gramado-RS.

Brandão, P.E.; Scheffer, K.; Villarreal, L.Y.; Achkar, S.; Oliveira, R.N.; Fahl, W.O.; Castilho, J.G.; Kotait, I. and Richtzenhain, L.J. (2008). A Coronavirus Detected in the Vampire Bat *Desmodus rotundus. Braz J Infect Dis*, 12(6), 466-468.

Bredt, A.I.; Araújo, F.A.A.; Caetano-Júnio, J.; Rodrigues, M.G.R.; Yoshiawa, M.; Silva, M.M.S.; Harmani, N.M.S.; Massunaga, P.N.T.; Bürer, S.P.; Potro, V.A.R. and Uieda, W. (1996). Morcegos em áreas urbanas e rurais: manual de manejo e controle. Brasília, *Fundação Nacional de Saúde, Ministério da Saúde*, 117.

Calisher, H.; Childs, J.E.; Field, H.E.; Holmes, K.V. and Schountz, T. (2006). Bats: Important reservoir hosts of emerging viruses. *Clin Microbiol Rev*, 19(3), 531-545.

Ciprandi, A.; Horn, F. and Termignoni, C. (2003). Saliva de animais hematófagos: fonte de novos anticoagulantes. *Rev Bras Hematol Hemoter*, 4, 250-262.

Constantine, D.G. (1979). Bat rabies and bat management. Bull. *Soc Vector Ecol*, 4, 1-9.

Constantine, D.G. (1988). Transmission of pathogenic microorganisms by vampire bats. In: A.M. Greenhall and U. Schmidt (Eds.), *Natural History of Vampire Bats* (167-189). Boca Raton: CRC Press.

Correa-Giron, P.; Calisher, C. H. and Baer, G. M. (1972). Epidemic strain of Venezuelan equine encephalomyelitis virus from a vampire bat captured in Oaxaca, Mexico, 1970. *Science,* 175,546-547.

Costa, L. M.; Oliveira, D. M.; Dias e Fernandes, A. F. P. and Esberard, C. E. L. (2008). Occurrence of Diaemus youngi (Jentink 1893), Chiroptera, in the State of Rio de Janeiro. *Biota Neotrop*, 8(1).

Crespo, J.A.; Vanella, J.M.; Blood, B.J. and Carlo, J.M. (1961). Observaciónes ecologicas del vampire *Desmodus r. rotundus* (Geoffroy) en el norte de Cordoba. *Rev Mus Argentino Cien Nat. "Bernardino Rivadavia", Inst Nac Inves Cien Nat*, 6, 131-160.

Crespo, R.F.; Burns, R.J. and Linhart, S.B. (1972). Comportamiento del vampire (Desmodus rotundus) durante su alimentacion em ganado bovino em cautiverio. *Tec Pecu*, 18, 40-44.

Crespo, R.F.; Linhart, S.B.; Burns, R.J. and Mitchell, G.C. (1972). Foraging behavior of the common vampire bat related to moonlight. *J Mammal*, 53(2), 366-368.

Dalquest, W.W. (1953). Mamíferos do estado mexicano de San Luis Potosi. *Louisiana State University Estudos Biológicos da Série 1:1*, 229.

De La Torre, L. (1956). The dental formula of the bats of the genus *Diaemus*. In: *Proceedings of the Biological Society of Washington*, 69, 191-192

Dias, M.A.G. (2009). Aspectos epidemiológicos de *Histoplasma capsulatum* em morcegos em áreas urbanas do estado de São Paulo. Dissertação de Mestrado. Pós graduação Interunidades em Biotecnologia USP/Instituto Butantan/IPT.

Falcão, F.C. (2007). Mammalia, Chiroptera, Phyllostomidae, Diaemus youngi: First record for the state of Bahia, northeastern Brazil. *Check List* 3(4), 330–332.

Felten, H. (1956). Fledermäuse (Mammalia, Chiroptera) aus El Salvador, Teil IV. *Senck Biol*, 37, 341-367.

Fernandez, A.; Tablante, A.; Bartoli, F.; Beguin, S. and Aptiz-Castro, R. (1998). Expression of biological activity of draculin, the anticoagulant factor from vampire bat saliva is stricly dependent on the appropriate glycosylation of the native molecule. *Biochemica et Biophysica Acta*, 2(1425), 291-299.

Forman, G.L.; Phillips C.J.; and Rouk C.S. (1979). Alimentary tract. In: Baker, R.J.; Jones, J.K. and Carter, D.C. *Biology of bats of the New World family Phyllostomatidae. Part III* (16, 1. 441). Special Publications, the Museum, Texas tech University.

Gardner, A.L. (1977). Feeding habits. In: Baker, R.J.; Jones Jr, J.K. and Carter, D.C. (Eds) *Biology of bats of the New World family Phyllostomidae, Part II (13, 1-364)*. Texas: *Spec Publs Mus Tesas Tech Univ*.

Gitti, C.B. (2007). *Apresentação Bioecologia de Morcegos In: MAPA e PANAFTOSA – I Curso sobre o controle da raiva silvestre* (*D. rotundus)* no Brasil, 2007 , Guararema São Paulo – CDRoom.

Gomes, C.I.D. (2008). Metabolismo energético e resposta ao jejum do morcego hematófago *Diphylla ecaudata. Dissertação de Mestrado*, Brasília, 63.

Gomes, M.N. and Uieda, W. (2004). Diurnal roosts, colony composition, sexual sixe dimorphism and reproduction of common vampire bat *Desmodus rotundus* (E. Geoffroy) (Chiroptera, Phyllostomidae) from State of São Paulo, Southeastern Brazil. *Rev Bras Zool*, 21(3) 38-43.

Gonçalves, M.A.; Sa-Neto, R.J. and Brazil, T.K. (2002). Outbreak of agressions and transmission of rabies in human beings by vampire bats in northeastern Brazil. *Rev Soc Bras Med Trop*, 35, 461-464.

Goodwin, R.A. and Des Prez, R.M. (1978). Histoplasmosis, State of the art. American. *Rev Respir Dis*, 117, 929-956.

Goodwin, G. G. and Greenhall, A. M. (1961). A review of the bats of Trinidad and Tobago. *Bull Amer Mus Nat Hist*, 122 (3), 187-302.

Greenhall, A. M. (1972). The biting and feeding habits of the vampire bat, *Desmodus rotundus*. *J Zool*, 168, 451-461.

Greenhall, A.M. and Schutt, W.A. (1996). *Diaemus youngi*. *Mammal Species*, 533, p.1-7.

Greenhall, A.M. (1988). Feeding and Feeding Behavior. In: Greenhall, A.M. and Schmidt, U. (Eds) *The Natural History of Vampire Bats: Desmodus, Diaemus, Diphylla* (6750, 11-131). *Boca Raton*, Florida: *CRC Press*.

Greenhall, A.M.; Schmidit, U. and López-Forment, W. (1971). Attacking behavior of the vampire bat, Desmodus rotundus, under field conditions in Mexico. *Biotropica*, 3, 136-141.

Greenhall, A.M; Joermann, G. and Schmidt, U. (1983). *Desmodus rotundus*. *Mammal Species*, 2002, 1-6.

Greenhall, A.M.; Schmidt, U.; Lopez-Forment, W. (1969). Field observations on the mode of attack of the vampire bat (Desmodus rotundus) in Mexico. *An Inst Biol,* 40, 245-252.

Greenhall, A.M. (1970). The use of precipitan test to determine host preferences of vampire bats, Desmodus rotundus and Diaemus youngi. Bijdragen tot de Dierkunde, 40, 36-39.

Greenhall, A.M.; Joermann, G.; Schmidt, U. and Seider, M.R. (1983). *Desmodus rotundus. Mammal Species*, 202, 1-6.

Greenhall, A.M.; Schmidt, U .and Joermann, G. (1984). *Diphylla ecaudata. Mammal Species*, 227, 1-3.

IUCN 2010. IUCN Red List of Threatened Species. Version 2010.4. <*www.iucnredlist.org*>. Downloaded on 02 November 2010.

Koopman, K.F. (1988). Systematics and distribution (7-17). In: Greenhall, A.M. and Schmidt, U. (Eds.), *Natural History of Vampire Bats*. (246) Florida, Boca Ratón: *CRC Press*.

Lord, R.D. (1992). Seasonal reproduction of vampire bats and its relation to seasonality of bovine rabies. *J Wildlife Dis*, 28(2), 292-294.

Málaga-Alba, A. (1954). Vampire bat as a carrier of rabies. *Amer J Publ Helath,* 44, 909-918.

Manske, U and Schmidt, U. (1976). Visual acuity of the vapire bat, *Desmodus rotundus*, and its dependence upon light intensity. *Z Tierpsychol,* 42, 215-221.

MAPA e PANAFTOSA – I curso sobre o controle da raiva Silvestre (*D. rotundus)* no Brasil, 2007, Guararema São Paulo – CDRoom.

Margarido, T.C. and Braga F.G. (2004). Mamíferos (27-142). In: Mickich S.B.; Bérnils, R.S. (Eds), *Livro vermelho da fauna ameaçada no estado do Paraná* (763). Curitiba, Governo do Estado do Paraná, IAP, SEMA.

Marinho-Filho, J. and Sazima, I. (1998). Brazilian bats and conservation biology: A first survey. In: Kunz, T.H. and Racey, P.A. (Eds). ***Bat Biology and Conservation***. (282-294), Washington: *Smithsonisn Institution Press.*

Mathias, M.A.; Díaz, M.M.; Campos, K.J.; Calderon, M.; Willig, M.R.; Pacheco, V.; Gotuzzo, E.; Gilman, R.H. and Vinetz, J.M. (2005). Diversity of bat-associated Leptospira in the Peruvian Amazon inferred by bayesian phylogenetic analysis of 16S ribosomal DNA sequences. *Am J Trop Med Hyg,* 73(5), 964-974.

Mayen, F. (2003). Haematophagous bats in Brazil, treir role in rabies transmission, impact on public health, livestock industry and alternatives to an indiscriminate reduction of bat population. *J Vet Med B*, 50, 469-472.

McNab, B. K. (1973). Energetics and the distribution of vampires. *J Mammal*, 54(1), 131.

McNab, B. K. (1969). The economics of temperature regulation in neotropical bats. *Camp Biochem Physiol,* 31, 227-268.

Mills, R. (1980). Parturition and social interaction among captive vampire bats, Desmodus rotundus. *J Mammal*, 61, 336-337.

MMA – Ministério do Meio Ambiente – Lista das espécies da fauna brasileira ameaçadas de extinção – Instrução Normativa nº 3, de 27 de maio de 2003. Available: *www.ibama.gov.br/fauna/downloads/listassp.pdf. Acess in October, 2010*

Nowak, R.M. (1991). *Walker's mammals of the world. Baltimore*, Johns Hopkins University Press, 1:1, 642.

Nuñez, H.A. and Vianna, M.L. (1997). Estacionalidad reproductiva en el vampire común *Desmodus rotundus* em el Valle de Lerma (Salta, Argentina). *Rev Biol Trop*, 45(3), 1231-1235.

Paula, A.V. and Aidé, M.A. (1985). As microepidemias de histoplasmose do Estado do Rio Janeiro. *J Bras Med*, 49(1), 18-26, 28.

Peracchi, A.L.; Lima, I.P.; Reis, N.R.; Nogueira, M.R. and Filho, H.O. (2006). Ordem Chiroptera In: Reis, N.R.; Peracchi, A.L.; Pedro, W.A. e Lima, I.P. (Eds) *Mamíferos do Brasil*. Londrina, p.162-164.

Reis, N.R.; Peracchi, A.L.; Pedro, W.A. and Lima, I.P. (2007). *Morcegos do Brasil*. Londrina, 253.

Ruschi, A. (1951). Morcegos do estado do Espírito Santo – Descrição de Diphylla ecaudata Spix e algumas observações a seu respeito. *Boletim do Museu de Biologia Prof. Mello Leitão*, 3, 1-6.

Sazima, I. and Uieda, W. (1980). Feeding behavior of the white-winged vampire bat, Diaemus youngii, on poultry. *J Mammal*, 61(1), 102-104.

Sazima, I. (1978). Aspectos do comportamento alimentar do morcego hematófago, Desmodus rotundus. *Bo Zool Univ São Paulo,* 3, 97-119.

Schmidt, C. (1988). Reproduction. In: Reenhall, A. M. G. and Schmidt (Eds), U. *Natural history of vampire bats* (246), Boca Raton, Florida: *CRC Press*.

Schmidt, C.; Schmidt, U. and Manske, U. (1980). Observations of the behavior of orphaned juveniles in the common vampire bat (*Desmodus rotundus*). In: Wilson, D. E. and Gardner, A. L. (Eds), *Proceedings of the Fifth International Bat Research Conference* (434). Lubbock, Texas: *Texas Tech Press*.

Schmidt, U. (1978). *Vampirfledermäuse: familie Desmodontinae* (Chiroptera). Wittenberg Lutherstadt, A Zimsen Verlag, 99.

Schmidt, U.; Schlegel, P.; Schweizer, H. and Neuweiler, G. (1991). Audition in vampire bats, *Desmodus rotundus*. *Journal of Comparative Physiology A: Neuroethology, Sensory, Neural, and Behavioral Physiology* 168(1), 45-51.

Schutt, W. A. (1995). *The Chiropteran hindlimb: functional, behavioral, and evolutionary correlates of morphology*. Ithaca, New York: Cornell University.

Schutt, W.; Muradali, F.; Mondol, N.; Joseph, K. and Brockmann, K. (1999). Behavior and Maintenance of Captive White-Winged Vampire Bats, Diaemus youngi. *J Mammal*, 80(1), 71-81.

Schutt, W.A.; Hermanson, J.W.; Bertram, J.E.A.; Cullinane, D.; Chang, Y.H.; Muradali, F. and Altenbach, J.S. (1993). Aspects of locomotor morphology, performance and behavior in two vampire bat, *Desmodus rotundus* and *Diaemus youngi*. *Bat Research News*, 34, 127-128A.

Sodré, M.M. (2010). Biologia do morcego hematófago *Desmodus rotundus* (E. Geoffroy 18810). In: Kotait, I.; Filho, V.S.N.; Souza, M.C.A.M.; Carrieri, M.L.; Gomes, M.N. and Peres, N.F. *Manual de controle da raiva dos herbívoros – Manual Técnico do Instituto Pasteur*. 9, 25-39.

Suthers, R.A.(1966). Optomotor responses by echolocating bats. *Science*, 152, 1102-1104.

Taddei, V.A. (1983). Morcegos: algumas considerações sistemáticas e biológicas. *Bol Tec Cati*, 172, 1-31.

Taddei, V.A.; Gonçalves, C.A.; Pedro, W.A.; Tadei, W.J.; Kotait, I. and Arieta, C. (1991). Distribuição do morcegos vampiro Desmodus rotundus no Estado de São Paulo e a raiva dos animais domésticos. Campinas: Impresso Especial da CATI, 107.

Trajano, E. and Sobrinho, R.B. (1980). O Morcego Hematófago. *Boletim Técnico CATI*, 142, 25.

Trajano, E. (1984). Ecologia de populações de morcegos cavernícolas em uma região cárstica do Sudeste do Brasil. *Rev Brasil Zool,* 2, 255-320.

Turner, DC. (1975). The vampire bat. A Field study in behavior and ecology. Baltimore, *The Johns Hopkins University Press*, 145.

Uieda, W.; Chaves, M.E. and Santos, C.F. (2002). Guia das principais espécies de morcegos brasileiros. Manuscrito preliminar para o 1° Curso de Treinamento "Manejo ecológico de morcegos em áreas urbanas e rurais". *Botucatu*, 27.

Uieda, W.; Hayashi, M.M.; Gomes, L.H. and Silva, M.M.S. (1996). Espécies de quirópteros diagnosticadas com raiva no Brasil. *Boletim do Instituto Pasteur*, 2(1), 17-36.

Uieda, W. Photo *Diaemus youngi.* [20 oct 2010] [online]. Available from:URL *http://www.morcegolivre.vet.br/diaemusyoungi.html*

Uieda, W. (1992). Período de atividade alimentar e tipos de presa dos morcegos hematófagos (Phyllostomidae) no Sudeste do Brasil. *Rev Brasil Biol*, 52(4), 563-573.

Uieda, W. (1993). Comportamento alimentar do morcego hematófago Diaemus youngi, em aves domésticas. *Rev Brasil Biol*, 53(4), 529-538.

Uieda, W. (1994). Comportamento alimentar de morcegos hematófagos ao atacar aves, caprinos e suínos, em condições de cativeiro. *Instituto Biológico Universidade Estadual de Campinas*, 1978.

Uieda, W. (1996). Biologia e dinâmica populacional de morcegos hematófagos. *Anais do II Curso de Atualização em raiva dos herbívoros*, 2, 63-87.

Uieda, W. (1982). *Aspectos do comportamento alimentar das três espécies de morcegos hematófagos* (Chiroptera, Phyllostomidae). (Dissertação de Mestrado), Instituto de Biologia Universidade Estadual de Campinas.

Uieda, W.; Buck, S. and Sazima, I. (1992). Feeding behavior of the vampire bats *Diaemus youngi* and *Diphylla ecaudata* on smaller birds in captivity. Ciência e Cultura. *Journal of Brazilian Association for the Advancement of Science* 44:410–412

Villa-R, B. (1966). *Los murciélagos de México. Univ Nac Autónoma México*, 492.

Villa-R, B. (1967). *Los murciélagos de Mexico*. Instituto de Biología, xvi, 491.

Villa-R, B. (1968). Ethology and ecology of vampire bats. *Int Un Conser Nature Publs.* n.serie Morges, 13, 104-10.

Vizotto, L.D. and Taddei, V.A. (1973). Chave para a determinação de quirópteros brasileiros. São José do Rio Preto, Francal, 72.

Wilkinson, G.S. (1986). Social grooming in the common vampire bat, *Desmodus rotundus. Animal Behaviour*, 34(6), 1880-1889.

Wilkinson, G.S. (1988). Social organization and behavior. In: Greenhall, A.M. and Schimidt, U. (Eds.) *Natural history of vampire bats* (85-97). Florida: *CRC Press*.

Wimsatt, W.A. and Guerriere, A. (1962). Observations on the feeding capacities and excretory functions of captive vampire bats. *J Mammal,* 43, 17.

In: Bats: Biology, Behavior and Conservation
Editors: J. L. Zupan and S. L. Mlakar, 275-297
ISBN 978-1-61209-749-7

Chapter 8

REGULATORS IN EPIDYDIMAL MATURATION AND PROLONGED STORAGE SPERM PROCESSES IN THE BAT *CORYNORHINUS MEXICANUS*

E. Arenas-Ríos,[1] M. A. León-Galván,[2] R. López-Wilchis,[2] and A. Rosado[1]
[1]Departamento de Biología de la Reproducción
[2]Departamento de Biología; División de Ciencias Biológicas y de la Salud, Universidad Autónoma Metropolitana-Iztapalapa

ABSTRACT

Prolonged sperm storage is a reproductive particularity in *vespertilionid* and *rhinolophid* bats. Several hypotheses have been proposed to explain this important reproductive phenomenon. The initial proposal that prolonged storage of spermatozoa is a natural consequence of hibernating hypothermia has now been refuted because many species of bats remain unusually active during hibernation period. In *Corynorhinus mexicanus* during winter only a daily torpor is present, but no a true hibernation period and sperm storage, this occurs both in the male (epididymides) and in the female (uterus). It is known that the presence of reactive oxygen species (ROS), affects the spermatozoa in complex ways, but it is also beneficial, even necessary, for normal sperm physiology. This becomes more interesting if we remember that some of the most important physiological properties of mammalian sperm cells develop gradually as the cells progress through the epididymis. In the *Corynorhinus mexicanus* the three enzymes related to ROS modulation are present in the cephalic and caudal region of the epididymides during the progress of the epididymis sperms and their maturation, but interesting to say the activity of the superoxide dismutase (SOD) is not present or perhaps it is inhibited in both epididymal segments in the post testicular phase of the sperm storage function. The SOD is present during the period of maturation and during the transport of the sperms, but it is totally or almost totally inhibited or not present in both epididymal segments during the post testicular phase of storage, and an absence in the sperm cells SOD activity (We try to avoid the possibility of producing peroxidative damage in the sperm cells storage during this last phase). It is interesting to observe that the glutathione peroxidase (GPX) presents a complete pattern contrary to that of the

SOD; its activity is low during the period of production and maturation of the epididymal sperms which depend perhaps on the testicular activity, but present an important activity in both segments of the epididymis during the storage phase. The activity of catalase (CAT) is preserved relatively high during the whole reproductive cycle indicating its importance in the protection of the spermatic cells against the effect of H_2O_2. However it is important to mention that its activity is significantly higher in the tail of the epididymis during the maturation and progress process (January-July) and particularly during the storage process (August-September, and perhaps the first half of October). In sperm cells from the cephalic region, the CAT activity reaches its highest levels at the beginning of September and the lowest at the end of the same month, date in which the CAT activity in sperm from the caudal region reaches its highest level. In the epididiymal fluid the CAT activity shows a similar pattern in the cephalic region from that found for SOD and GPX in both regions. We observed high levels in September, and they remained low around the middle of September, rising gradually until they arrived at their highest level at the end of October. The redox balance associated with the microenvironment, through which the sperms pass, must be specific and differentially controlled to assure its adequate function.

Introduction

To perpetuate the species, reproduction is indispensable in the superior organisms, being obligatory, the union of the masculine and feminine gametes, this process is known as sexual reproduction. This reproduction type consists, in the sum of adaptations that will depend on the species and of the environment in which they are developed.

In the peculiar case of the bats, they are adapted to a wide variety of environments, from tropical, subtropical environments, even temperate and cold environment; and also they can be found in jungles, forests or deserts including not regular environments. Each one of the environmental adaptations will be accompanied by a reproductive characteristic pattern that will assure its success as a species.

In this way, the females as well as the males develop different general patterns of reproduction; this is recognized by the number of reproductive cycles and by the impermanence with which these appear throughout the yearly period. One of these general patterns is characteristic of the species whose reproductive activity in the males (spermatogenesis) happens in a continuous way (being correlated with more than two estrous cycles a year). In another general pattern where most of the well-known species of bats are to be found, the reproduction is defined as seasonal; inside this modality two particular categories exist; one of which is characteristic of the bats in that the testicular recrudescence and its involution are carried out in an outline bimodal (in connection with the polyestric bimodal of the females). In the second category are included the species that present a single period of testicular activity during one year (coinciding with the monoestric pattern of the females). This last category is observed mainly in the mammals that inhabit cold regions [24].

In cold areas, the decrease of the environmental temperature during the winter season, provokes a reduction in the size of insect populations which are the basic food for most of the chiroptera species inhabiting those places. The decrease in food and the low temperature environment are factors that on the whole exercise a direct effect on the bats, and for this reasons some species are forced to migrate to areas with more favorable climate conditions and only certain species are adapted to respond to these changes executing a behavior of physiologic lethargy, this varies in agreement with the intensity of the unfavorable conditions and with the species [56].

Various species that go through this lethargy, their hibernation period is separated by the reproductive season, such as in the case of the European chiroptera of the gender Miniopterus whose matings finish before entering into hibernation, begins in the following spring [55]. In contrast, in other members of the families Vespertilionidae and Rhinolophidae, the reproduction season extends until the winter torpor period, when the lethargy has a marked influence on the reproductive physiology of the individuals, providing the existence of an unusual masculine reproductive pattern, characterized by a temporary asyncrony among the sexual functions [45, 58], as well as asynchrony between the reproductive functions of male and female. In order to limit the births and breed at a time which is more appropriate when food is available for the mothers and their young.

THE *CORYNORHINUS MEXICANUS* BAT AND THE PROLONGED SPERM STORAGE

The *Corynorhinus mexicanus* bat is an endemic species of Mexico that inhabits high and humid areas of the mountainous main systems, habitually located 2000 msnm.

The female reproductive period is a stational monoestric type; and the males, reveal a single yearly reproductive cycle, but in both sexes a temporary asyncrony shows between the reproductive functions, in that the males have a long period of sperm storage in the epididymis of up to four months [11, 47].

It is important to underline, in general for mammals as well as humans, the time that the epididymal sperm storage lasts, around 11 days [62]. In this way, some species that have longer period of spermatic storage, like *Corynorhinus mexicanus* are; transforming into a peculiar phenomenon, in that the sperms prolong for several months, even when the period of mating has passed.

The phenomenon of prolonged sperm storage doesn't only happen in males, because it has been reported that in certain species of bats there is a prolonged period of receptivity on the part of the females [24], and in those that are recently inseminated at the end of the autumn don't have an ovulation and maintain a reservoir of live sperms in the reproductive feminine tract during the winter period, while they carry out a behavioral lethargy that means they save energy during the cold season of the year, when the food shortage is important; such is the case of the female *Corynorhinus mexicanus* [50]. This reproductive adaptation is known as deferred fertilization, and if you calculate the time when the sperms are stored firstly in the epididymis and later on in the uterus, this results in an extraordinarily long period of time compared with other mammals showing that during this time the gametes go through different environments while they maintain their fertilizing capacity. In the case of *Corynorhinus mexicanus* the estimated time of the permanence of sperms is approximately 5 months [20, 46].

Initial intents to explain this important reproductive strategy, sustains the idea that the prolonged sperm storage could be a natural consequence of the descent of the corporal temperature in the hibernators. However, it is very well-known that many tropical and sub-topical species exist in high areas that are not hibernating and they also present prolonged storage of sperms in the epididymis [35, 43, 73]. It is important to indicate that this species live at high altitudes, which make them similar to those in warm and cold areas. In addition to this, some species of hibernating bats that store sperms, wake up regularly during this hibernation period in search of food [12] that implies dynamic changes in the metabolism and energy consumption that generate a great quantity of products among those are the reactivate oxygen species (ROS), same that take place inevitably during the physiologic processes that involve the oxygen consumption, normal in the aerobic organisms.

When the generation of ROS exceeds the defense systems, you are correlated with some problems in the spermatic functionality [30]. However when these are produced in appropriate quantities, they have been linked to the aspects of the inter and intra cellular signaling [27]. In this form, one way or another, the balance will depend on the oxidants and antioxidants. Oxidants in which you find the ROS and antioxidants which counter them.

REACTIVE OXYGEN SPECIES AND FREE RADICALS

They include reduced forms of oxygen, as well as the reactive products of reaction of the oxygen with other molecules. Some, but not all the products of these reactions are free radicals.

What are Free Radicals?

All the atoms and molecules are formed by small well-known particles such as subatomic particles; on one hand, the protons and neutrons, and on the other, the electrons that turn around the core through "orbitals". The quantum mechanics ends up correcting the concept of orbital, and you substitute it for that of distribution of probabilities. That is to say, an electron can meet with certain probabilities in any point around the core; the fields of space where the probability of finding it is bigger, corresponds to what constitutes the orbit (figure 1) in the precedent models outside of whose contour the probability of finding an electron is minimum.

There are spherical orbits (type s), like in the case of the only electron of the hydrogen atom, the plainest; orbital in an eight form (type p) which, contrary to orbital s, are multiple and directional in the space according to the three Cartesian, and even more complex orbits (type d; figure 1).

Something that should be taken into account, is that, you can never have more than two electrons in the same orbital, In this way, the global distribution of electrons of an atom or molecule is the sum of many similar orbitals. When an orbital field is full the pair of electrons should turn anti-clockwise (spin), to eliminate its magnetic field, but this doesn't always happen, when in an orbital there is an electron it is said that it doesn’t match up, and in this, when an atom or molecule has one or more unpaired electrons it receives the name of free radical [38].

The free radicals can react with other radicals, a case in which both species cancel their magnetic field with each other causing both radicals to stop, or it can happen that the reaction only involves a free radical that will steal electrons of neighboring molecules forming new free radicals [38].

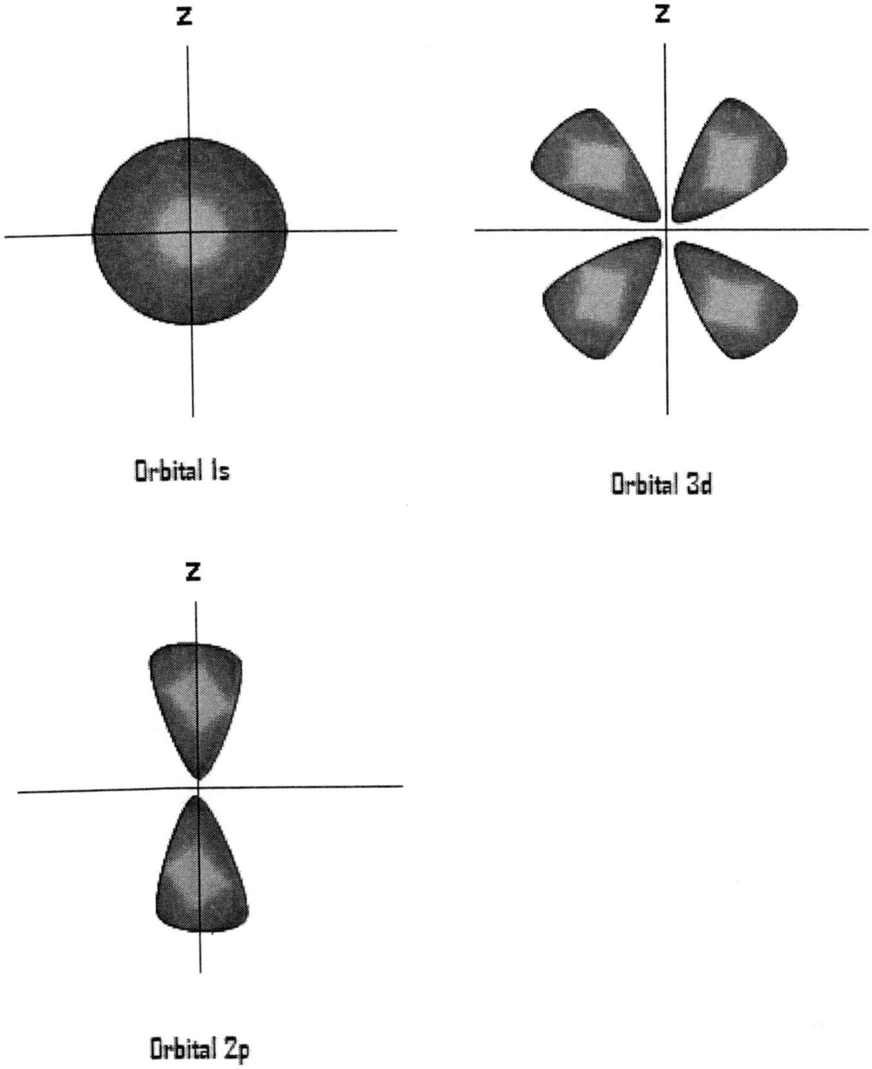

Figure 1. The atomic orbits are mathematical descriptions of the probability of finding an electron of an atom or molecule in a certain place. The orbital ones exposed show the space distribution of electrons s, p and d.

The molecular oxygen (O_2) is a biradical, having two unpaired electrons of parallel spin. The molecules that are not radical possess electron pairs with opposed spines, and in this way, both electrons can be accepted. However, in the particular case of oxygen this is possible but a little rare, because the reactions which involve oxygen are regularly univalent, accepting a sole electron. The product of this univalent reduction is that called superoxide ($O_2^{\cdot -}$), a molecule that alone generates another ROS, the hydrogen peroxide (H_2O_2), is the last, for on the other hand, it is not a free radical it is considered also part of ROS, because it is produce fundamentally by the divalent reduction in value of O_2 and will give it origin another free radical capturing an electron and a proton, other ROS principals, that named ion hydroxyl ($OH^{\cdot}$; figure 2).

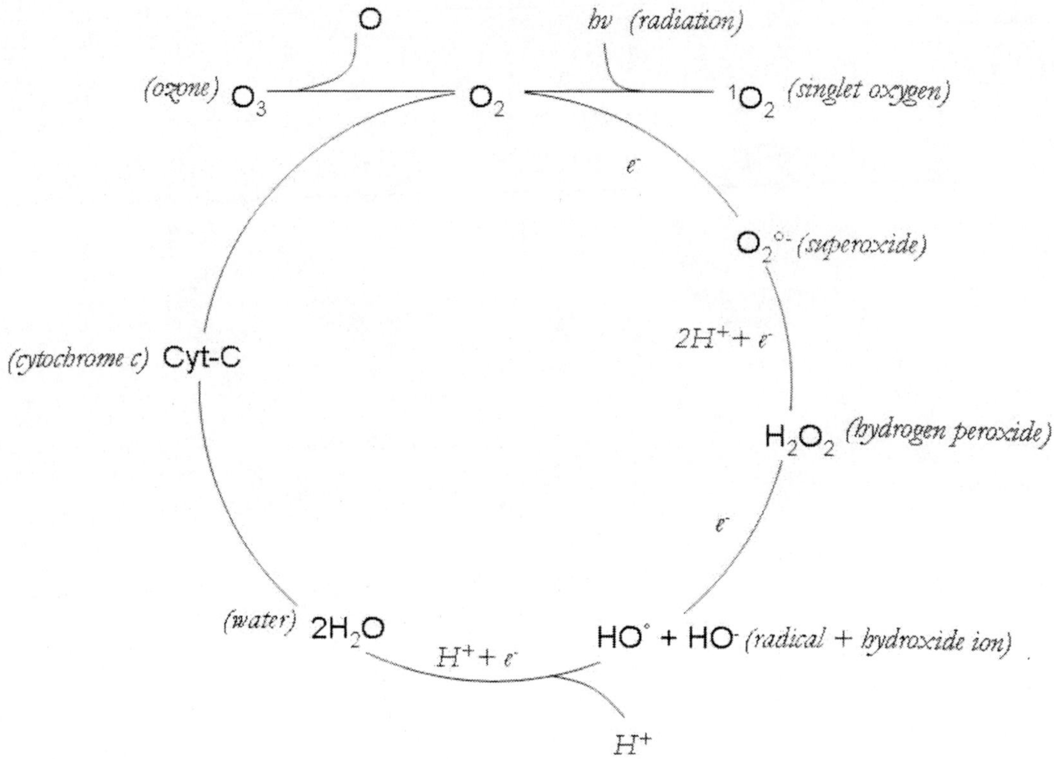

Figure 2. Formation of main ROS starting from molecular oxygen. Modificated from Domínguez, 2007 [25].

The superoxide anion is produced in vivo constantly in small quantities, with the action of some enzymes within which can be found cytocromo P^{450-} oxygenase dependent [22], and for the proteolytical conversion of dehydrogenase xanthine/ xanthine oxidase that provide another enzymatic sources of $O_2^{\cdot-}$, H_2O_2, and therefore $OH^{\cdot}$ [88].

The non enzymatic source of $O_2^{\cdot-}$ happens, when an electron is transferred to oxygen by reducing the coenzymes in value or prosthetic groups, or by previously reducing xenobiotics with certain enzymes. The mitochondrial respiratory chain contains various redox centers which take the oxygen electrons, constituting in this way the principal sources of $O_2^{\cdot-}$ in most of the tissues [80].

The produced ROS in this way are usually related with toxic products and are involved in the cause and sicknesses and ageing [59]. Some cells have other mechanisms for producing ROS with physiological purposes, such as the phagocytic leukocytes which contain a type of membranal enzyme called: NADPH oxidase, the same that $O_2^{\cdot-}$ produces in phagocytic blisters during "the reaction of death" [13,66]. However, among the cells involves in the reproductive aspects and that produce ROS with physiological ends we find spermatozoa.

ROS PRODUCTION FOR SPERMATOZOIDS

From 1950 Tosic and Walton demonstrated that the bull spermatozoa, pig and ram can produce H_2O_2 by means of an oxidase which acts in L-aromatic amino acids. Later on

Holland and its collaborators (1981) proved that rabbit spermatozoids are capable of producing H_2O_2 by dismutation of $O_2^{\cdot-}$ resulting from the activity of the superoxide dismutase enzyme (SOD) [8].

On the other hand we can observed that, the extracellular addition of NADPH can pioneer the sperm capacitation due to H_2O_2 formation that has related with the tyrosine phosphorylation, and that the production of $O_2^{\cdot-}$ can be maintained with the addition of NADPH or NADH, and interrupted by flavoproteins inhibitors, in the same way as what happen with NADPH oxidase in leukocytes, observing more that this production was not affected by mitochondrial inhibitors or by diaphorases (enzymes which can be found mainly in: cytosol and spermatozoa membranes, transferring electrons from NADH or NADPH to an electronic acceptor similar to 2,6-diclorofenol indophenol, the same which play an important part in the ROS formation), and the effect of NADPH was localized in the membranal fraction [6]. It was seen that to add NADPH to elevated concentration (5-10 nmol / l) causes lipid peroxidation and oxidation to DNA [3, 81, 82].

These observations were also corroborated even more by the NADPH membrane oxidase activity that later on is revalidated by the work of Banfi and collaborators (2004) [16] which informs that the gene of NADPH oxidase, member of the NOX5 family, is expressed in primary spermatocytes of humans. However, the appearance of this enzyme has not been confirmed in mature spermatozoa.

Nevertheless, work where Western blot was used to determine the presence of proteins that unite with antibodies anti components of very complex NADPH oxidase of leukocytes (NOX2) including gp91phox, p67 (phox), p47 phox and p40 (phox); propose that this unusual oxidase is regulated by the availability of p40 (phox) and is independent from p22 (phox); as well it sustains that this oxidase reaches its maximum activity in testicular spermatozoa, but diminishes gradually conforming with the advances of the sperm maturity, that is to say one of the indispensable components for the NADPH oxidase activity, apparently, is absent in the epididymal spermatozoa [72]. The confirmation of this activity pattern of the oxidase NADPH in the different phases of sperm maturation, results in great importance, because in agreement with this work, the participation of ROS spermatozoids, product of the oxidase activity, is inversely proportional to its maturity state.

In this way, the production of H_2O_2 by spermatozoa, is not only the result of the NADPH oxidase activity, because it has been confirmed that it depends also in a big way on the mitochondrial activity [27], and inclusive, it has shown that what has been reported about the rat epididymal sperm, where the NADPH oxidase activity occurs as the primordial source of ROS, is in reality the product of the Cytochrome-B5 reductase activity localized in epithelial cells which could have infected the spermatic suspension [14,15]. This finishes with the majority of the evidences concerning the production of ROS given that the NADPH oxidase activity in rodents epididymal sperm, and centers the attention on the mitochondria as the principal sources of ROS.

Damage Caused by Ros to the Spermatozoa

When the interaction of ROS with susceptible molecules to oxidation exceeds the defense systems, this provokes a state that is named oxidative stress (figure 3). This oxidative stress is produced when a cascade of intracellular events generate which can result in the adaptation or in the cell damage. In the first of these cases, it is necessary to find the positive regulation of the elements that form a part of this anti-oxidant defense system, this is, in an attempt to restore the oxidant / anti-oxidant balance of the cell. However in the second of these cases, the oxidative stress can provoke injury or inclusive death.

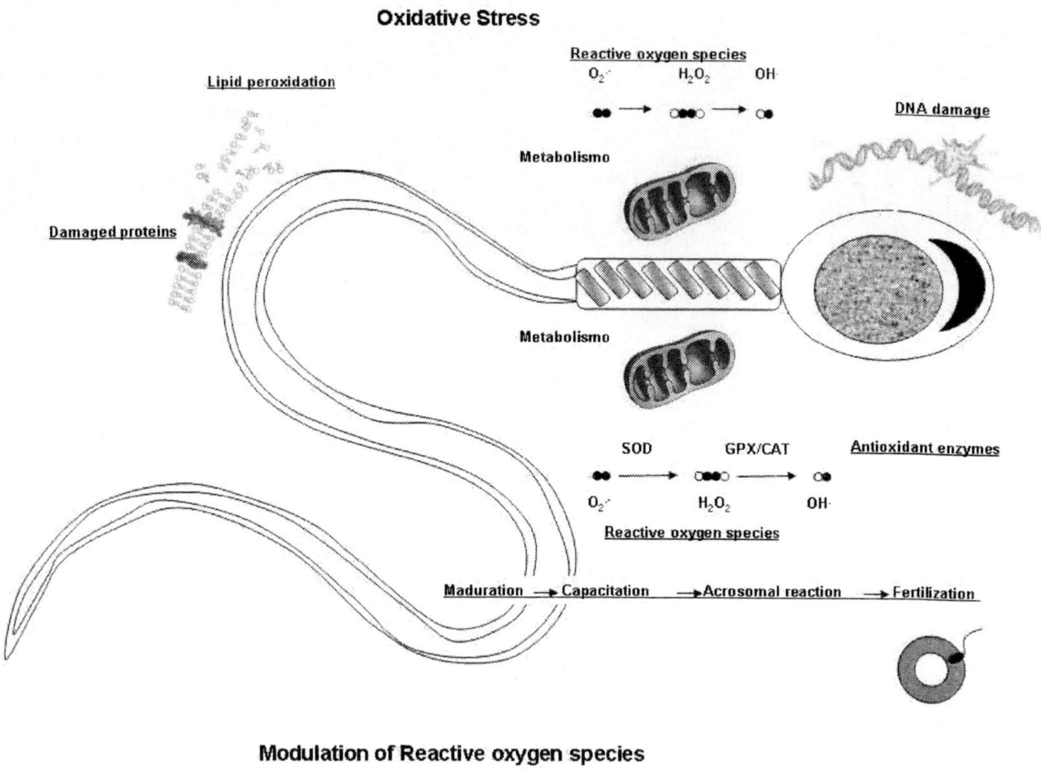

Figure 3. Noxious and beneficial effect of the reactive oxygen species in the spermatozoa, modulation by the antioxidant enzymatic system (SOD: superoxide dismutase; CAT: catalase; GPX: glutathione peroxidase).

ROS can react with a ample range of biological molecules amongst these can be found the unsaturated fatty acids of the membranes, the sulfhydryl protein and the nucleic acids [38]. For this reason, ROS are involved in a large quantity of related illnesses with masculine infertility [28].

From the first reports about the adverse effects of ROS to the spermatozoa [51], up to this date, a large amount of studies have been published to this respect, accepting totally the overproduction of ROS in a sperm suspension, this will be associated with problems in the sperm function and the subfertility, due overalls to the injury membranes (lipid peroxidation) and DNA oxidation [1, 49, 70, 74].

The exposition of human spermatozoa in ROS induces the loss of its mobility, this event can be related directly with the lipid peroxidation [34], due probably to the changes of the permeability and integrity of the membrane and successively to the problems to maintain a flagella movement, added to this, the lipid peroxidation destabilizes other dependent sperm functions of the membranal integrity, among these are: the capacitation, the acrosomal reaction and the fusion of the sperm with the oocyte [4,5].

It has been observed in human and hamster sperm that the effect provoked by ROS to the DNA includes: oxidative damage in its nitrogenous bases (modification and/or the suppression), rupture of the DNA chain and chromosomal re-arrangement. These damages have been observed, in both types of DNA: nuclear and mitochondrial [37].

Using the polymerase chain reaction technique to calculate the amount of damage to the DNA, it have been observed that the mitochondrial genome has resulted in being much more susceptible to damage caused by ROS than the nuclear genome [64], in that this study, the nuclear genome resulted in been particularly resistant to the oxidative damage.

This resistance radicates in the hyper compactation in the DNA of the spermatozoa obtained during the differentiation of the sperms, where the nuclear histones is progressively replaced by other known proteins such as protamines [29]. The protamines possess a great amount of cysteine residues which can be oxidized during there movement through the epididymis, formed a series of disulfide bridges inter and intra molecular which serve to stabilize the DNA nuclear, making it in that way, more resistant to oxidative damage [18].

This is reason which the DNA sperm obtained from the cephalic region of the epididymis results in being more susceptible to oxidative damage than those obtained from the caudal region [21, 44]. For this reason, the DNA damage in ejaculated sperms can come from three principal sources: a bad compactation and an inconclusive apoptosis during the spermatogenesis, as well as, a disbalance between the protection system and ROS [63].

Antioxidants Enzymes in the Maturation Sperm

The sperms obtained from the testicle are not capable of exhibiting progressive mobility neither to capacitate, being they acquire this ability while they pass trough the epididymis, this process is known as sperm maturation. Other relative changes in the maturation, include the termination of the nuclear condensation and changes in the expression and distribution of the sperm surface molecules [23, 54], events that occur during the time they remain in this organ.

The time taken for the sperms to cross the epididymis, have been studied in humans, labeling these cells with thymidine, where time was estimated between 1 and 21 days (11 days on the average) [62]. However, in the case of various bat species vespertilionids and rinolofidos these show a temporary asyncrony in the development and the function of the male reproductive organs [45, 47]; where particularly, the spermatogenesis develops in summer mean while the maximum development of the sexual accessory glands, the libido and the mating occur in autumn; this asynchrony results in an unusually long period of sperm storage in the epididymis, this can extend for several months after the testicles have involved totally [58].

During this transit through the epididymis, the sperm are in serious risk, due to the continuing effect from ROS (oxidative stress), deteriorating the fertilizing capacity and the genetic integrity [2]. Nevertheless as been commented, ROS can be produced by the same sperms and by some non phagocytic cells that can be presented in the epididymal tissue [53, 57].

ROS plays an important part in the maturation regulation, being that while the sperms acquire the capacity to move, these suffer an amount of changes obtaining their fertilizing capacity [87]; this is, due to an increment in the AMPc synthesis and to the phosphorylation / dephosphorylation events that can be found principally in the proteins of the flagellum [6, 75, 76]. The tyrosine phosphorylation in the cells, are pioneered by the union of ligands with its receptor; many studios have suggested that the stabilization of the sperm structure during the epididymal maturation, is completed principally by the oxidation of the thioles groups during its transit throughout the epididymis [67, 69]. In the same manner, the capacity of the spermatozoids to suffer tyrosine phosphorylation, also is increased during the transit of the sperms, from the cephalic region until the caudal region of the epididymis [48, 86].

In some cellular types the increase in the tyrosine phosphorylation was suggested as a result of the oxidation and direct activation of the kinase proteins [31, 36]. However, it has been observed that the active center of the phosphatase proteins contains a large amount of residues of cysteine for the phosphatase activity [77, 83], in this way, the inhibitory effects of the oxidants on the phosphatase activity has been considered the most probable mechanism and indirectly responsible for the tyrosine phosphorylation [68, 83]. And from there it is greatly important to maintain a balance between the cellular production of ROS and its destruction.

Of the components in charge of the regulation of ROS, the most important can be found, the anti-oxidant enzymes: SOD, catalase (CAT) and glutathione peroxidase (GPX).

From the GPX a large number of isoforms have been found, of these, those that are related with the epididymis are: GPX1, 3, 4 and 5 [84]; where GPX5 has been localized in the epithelium of the epididymis and in the sperms [32], have being secreted towards the epididymal lumen, where they can be found in a free form or associated with spermatozoa, this last, increasing as it advances from the cephalic region to the caudal region of the epididymis, remaining until the feminine reproductive tract [42, 60, 85].

The activity of GPX1 is related principally with the H_2O_2 regulation or terbutyl hydropexodide in total extracts of the epididymis, and GPX3 also is expressed in epididymis, although being in the cephalic region its expression very low, not like that in of the caudal region (principal place of expression in mouse), where remains in the cytosol of the epithelial cells. This activity being the dependent enzyme of androgens in the body region and tale of the epididymis, and independent from the cephalic region [52, 65].

GPX4 or PHGPX are present principally in the epididymal sperms, in the head and middle piece of many, where it exhibits a thiol oxidase activity of protaminas, being larger in these coming from sperms obtained from the cephalic region that those obtained from the caudal region, thus like this, they contribute to the nuclear condensation during the transit trough the epididymis, being that, the compactation of the cromatin constitutes one of the main ways in the maturation of the epididymal sperms [33, 61]. These data suggest a primordial paper in the sperm maturation during the part of GPX in the metabolism of hydroperoxides and thiol oxidation, to sum up with its anti-oxidant participation.

Experiments carried out in the rat epididymis show that the cellular synthesis of the prostaglandin is partially regulated by hydroperoxides and intracellular levels, for these reason, PHGPX would be participating in this regulation by means of the reduction of the same [71]. Extremely important activity in the regulation of the maturation.

The participation of CAT in the epididymis, is one of the most controversial, in that although it has shown expressions at very low levels in human and rat sperms [9, 41, 79], this, is absent in rabbit sperm [39], mouse [8] and bull [19]; but it is present in the spermatic environment during its transit through the epididymis and in the ejaculation [19, 41, 89), finding that the expression patterns of this enzyme, involves not only the epididymal epithelium, but also other tissues of the masculine reproductive tract with the exception of the seminal blisters [90]. However, the expression of CAT RNAm is so low that its part in this enzyme is undervalued and as an anti-oxidant in these tissues.

The SOD is expressed at high levels in the epididymis, and doesn't vary significantly in its different regions [40]. In the caudal region of the epididymis, can be related with the protection of the sperms against the oxidative stress, and associated with the plasmatic membrane of the epididymal sperms, being able to promoted the production of H_2O_2 that would be participating in the maturation associated with the tyrosine phosphorylation [48].

In this way, the prolonged storage of sperm in bats appears to be more interesting if we take into account that the maturation develops gradually in a specific way, when the spermatozoa advance from the cephalic region until the caudal region of the epididymis; completing its maturation regularly, before reaching the caudal region, after that they are stored until ejaculation.

Enzymes Anti-ROS Like Regulators in the Processes of Maturation snd Prolonged Storage of Spermatozoa in the *Corynorhinus Mexicanus* Bat

As has been mentioned previously, in the Corynorhinus mexicanus bat females present a reproductive activity of the monoestric type; mean whilst the males, reveal only a yearly reproductive cycle and a temporal asyncrony between its primary and secondary reproductive functions, observing that the maximum size of the testicles can be seen in the August month, mean while the maximum development of the epididymis and the group of sexual accessory glands, can be seen 1 to 3 months after; this implies the storage of spermatozoa in the epididymis, of up to four months (Figure 4)[11, 46,47,50].

In the *Corynorhinus mexicanus* bats during its reproductive cycle, show that the spermatic cells are present between August and January, observing a clear reduction in value in the spermatic concentration at the end of October in the cephalic region and a clear tendency to increase in the caudal region (Table 1)[10,20, 46].

It is interesting to notes, the process of sperm prolonged storage of course firstly in the epididymis and after mating in the genital feminine tract, a characteristic that occurs in females due to the existence of a postponement in the ovulation with respect to the moment of mating; in the manner that the time passed from the moment when the sperm has been produced in the testicle and fertilization occurs have been estimated at 5 months [46].

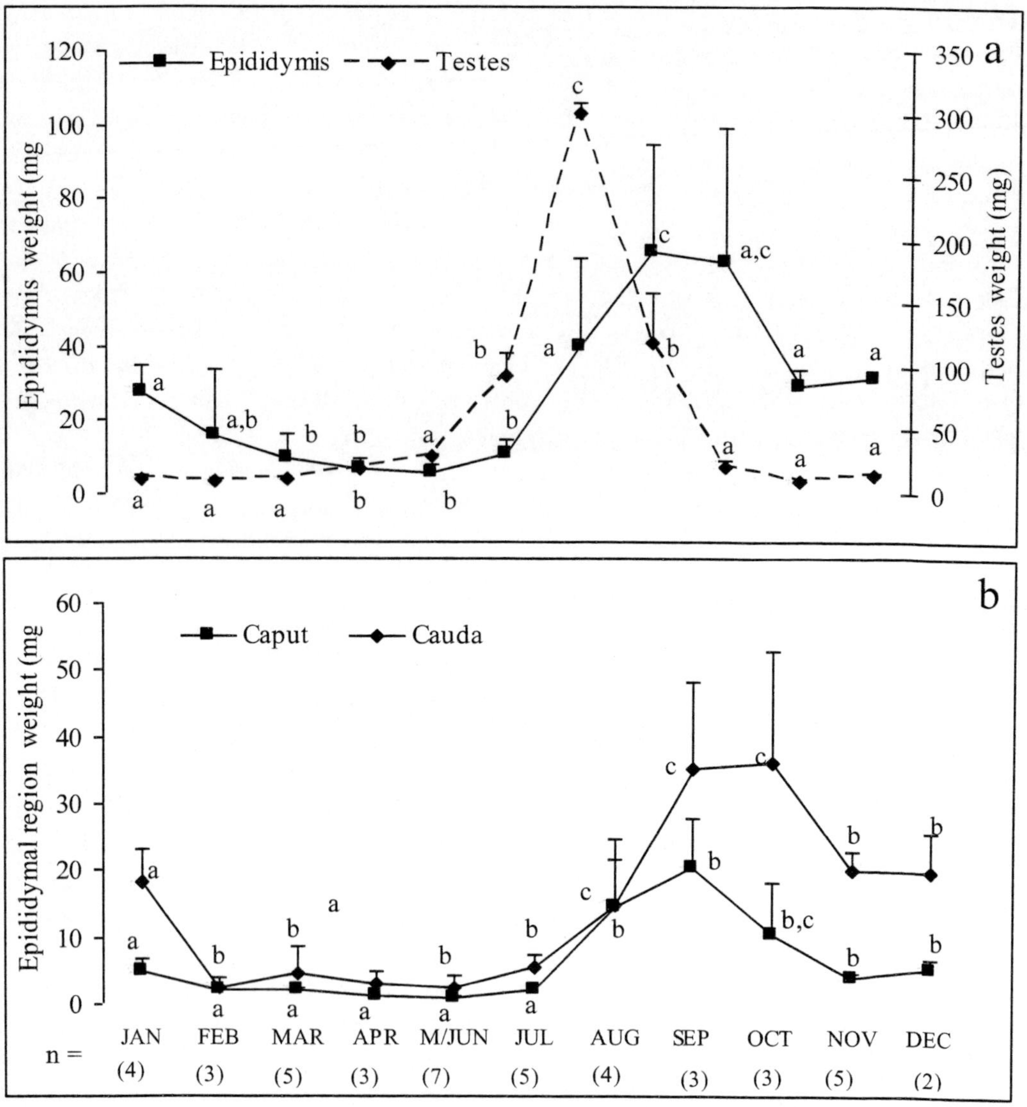

Figure 4. Changes in reproductive organs of adult male *Corynorhinus mexicanus*: (*a*) testicular and epididymal mass and (*b*) epididymal segment (caput and tail) mass. The epididymal segments were obtained after ligation of the epididymides at the deferent-cauda, caudacorpus, corpus-caput, and caput-testis junctions; see text for more details. For simplicity, we report only the mean + SE mass of one reproductive organ or epididymal region per bat, using all bats in the sample, and the values in parentheses indicate the number of bats. Different letters indicate statistically significant differences ($P < 0.05$) between values in the same trace (ANOVA plus Bonferroni post hoc test applied to the testis data). Modificated from Arenas-Ríos, 2005 [11].

Several hypothesis have been planted to try to explain and understand how it is that the epididymal sperm and/or the inseminated in the genital female tract of the bat with prolonged retention of the sperm can maintain themselves viable for such a long time. One of these hypotheses refers to the increase of the osmolarity in the tubule lumen of the caudal region of the epididymis, being able to prolong the life of the sperms inducing the partial dehydration [25]. For another part studies released on the *Corynorhinus mexicanus* bat have shown:

Table 1. Determination of the concentration of spermatozoa (X10^6), recovered of the different areas of the epididymis of male mature *Corynorhinus mexicanus* bats, captured monthly during all its reproductive cycle. The segments of the epididymis were obtained after tying the epididymises like you continue: among conduit deferential-area throughput; area throughput-body; cephalic body-area and area cephalic-union with the testicle

Date	Presencia SPZ	Epididymal region Caput [SPZ] (X10^6)	% droplet cytoplasm	Body [SPZ] (X10^6)	% droplet cytoplasm	Cauda [SPZ] (X10^6)	% droplet cytoplasm
Jul 31							
Aug 14	No						
Aug 28	Yes						
Sep 11	Yes	20.5±5.5*a*	84.6 ± 4.0^{1}	9.8 ±2.8*b*	45.7 ± 7.6^{2}	77.2±28.9*d*	19.7 ± 3.1^{3}
Sep 23	Yes	23.3±4.8*a*	79.6 ± 8.7^{1}	31.6±7.9*a*	37.3 ± 14.1^{2}	184.2±41.3*e*	15.3 ± 5.8^{3}
Oct 16	Yes	6.3 ±1.6*b*	65.3 ± 5.0^{1}	21.1±5.7*a*	14.7 ± 4.3^{3}	117±27.6*d*	6.7 ± 3.0^{4}
Oct 30	Yes	2.5±0.3*c*		6.2±3.0*b*	10.5 ± 2.8^{3}	114.2±20.5*d*	1.7 ± 1.2^{5}
Nov	Yes					90.3±7.6	
Dec	Yes					72.6±3.8	
Jan	Yes					25.3	
Feb	No						

Note: The data of spermatic concentration corresponding to the months between November and February, correspond to the work published by León-Galván et al., 1999; The shaded part corresponds to the work of Cervantes et al., 2008; The register of appearance of spermatozoa corresponds to the work published by Arenas-Ríos, 2009 [10,20,47].

The numbers indicate the average ± DS of 3-4 standalone experiments.

The number of spermatozoa and the percent of cytoplasmatic drop were independently submitted to ANOVA followed by the test of Bonferroni.

Different letters (concentration of spermatozoa) or different numbers (percent of drop cytoplasmatic) indicate significant differences.

Table 2. Malondialdehído production for pig spermatozoa and *Corynorhinus mexicanus* bat spermatozoa, in appearance and absence of genital feminine (FGF) fluid of bat

FGF (μl)	*n*	Pig sperm MDA	*n*	Bat sperm MDA
0	4	1.14±0.23*a*	3	0.64±0.11*
50	4	0.71±0.18*b*	——	—
100	4	0.50±0.08*b*	3	—
200	4	0.28±0.04*c*		0.25±0.07**

Nota: The lipid peroxidation activity is expressed in nanomols of produced malondialdehído (MDA; I average out ± DS) by 10^8 spermatozoa for 22 hours. FGF was obtained by means of laundry of captured females of October to November. For the pig spermatozoa, the securities continued by different letters indicate differences statistically significant ($P < 0.05$; Kruskal-Wallis and Student-Newman-Keuls). Modifided from León-Galván et al. 1999 [47].

*You differ significant ($P < 0.05$) of obtained securities with pig spermatozoa (t proves Student).

** Significant Differences ($P < 0.01$) of obtained securities of bat spermatozoa incubated in absence of FGF (t proves Student).

On side that the secretions produced in the female genital tract, maintain the inseminated spermatozoa viable and with fertilized capacity during a long period of time, this effect is due to the capacity of these mentioned secretions to inhibit the lipid peroxidation, this aspect also had an effect in pig sperm, which are known to be highly sensitive to suffer lipid peroxidation (Table 2) [46]. These said determinations were made measuring the lipid peroxidation that gives place to the degradation of the polyunsaturated fatty acids of the plasmatic membranes in lipidic peroxides and aldehydes like the malondialdehyde (MDA), quantifying in this way the concentration of MDA by a colorimetric method using the technique of the thiobarbituric acid following the technique of Barber and Bernheim (1967)[17] modified by Álvarez and Storey (1982)[7].

On the other hand the anti-oxidant enzymes SOD, CAT and GPX are present in the cephalic and caudal region of the epididymis, being possibly more important in the modulation of ROS in the processes of maturation and storage of spermatozoa (Figure 3)[11], it can be observed that depending on the reproductive cycle phase, two different activity patterns of the anti-oxidant enzymes were observed: CAT is active during the whole yearly cycle, having its activity particularly high during the post testicular phase; the SOD activity shows itself higher during the phase which coincides with the transfer and maturation of spermatozoa, being almost totally absent or inhibited in both epididymal regions in the spermatic storage; and the activity of GPX is low during the testicular phase and the spermatic maturation and high in both epididymal segments, during the spermatic storage period (Figure 3)[11].

As well it determines the production of ROS in epididymal spermatozoa of the C. mexicanus bat (cephalic region (figure 4A) and caudal region (figure 4B)) between the different dates of capture, with the finality of correlating these dates with those obtained from activity of different enzymes. However, different significative statistics were not found in the production of ROS between the different dates (head: $F=1.193$, $P > 0.05$; queue: $F = 1.780$, $P > 0.05$); although a clear tendency to raise this production can be seen around the middle of September, achieving it highest level around the middle of October and maintaining these levels until the end of the same month [10]; with the objective of verifying the participation of the NADPH oxidase enzyme as the sources of production of ROS, this production was stimulated by adding NADPH (cephalic region (figure 4A) and caudal region (figure 4B)), without finding any differences significant statistical differences between those that were not stimulated (cephalic: $F=0.751$, $P > 0.05$; caudal: $F = 1.377$, $P > 0.05$). However, we observed the same standard of intermonthly changes (figure 4)[10].

The production of ROS in the epididymal sperm of the *Corynorhinus mexicanus* bat (figure 4), indicates, in agreement with what has been said before this could be participating in the process of maturation of the spermatozoa in the cephalic region of the epididymis (figure 4A). However, the production of ROS remains in the spermatozoa obtained from the caudal region (figure 4B), even in the close dates from when the mating to place [50]. This could confirm the information reported by Cervantes and collaborators (table 1y3)[20], where they determined the presence of cytoplasmic drop and the sequential induction of capacitation and acrosomal reaction as indicated in maturation in sperms obtained from the different regions of the epididymis through the annual reproductive cycle of the C. mexicanus bat, where it can be found that, a high percentage of spermatozoa persists with cytoplasmic drop when arriving at the caudal region, including a low indication of capacitation and acrosomal reaction in spermatozoids obtained from the body that raises significantly during its stay in the tail (table 1y3); revealing that, the process of sperm maturation in the *Corynorhinus mexicanus* bat, contrary to that reported in then general species of mammals, is completed in the caudal region of the epididymis, this could explain the necessity for the longer period storage of the epididymis.

Upon no finding significant statistical differences between the results from the production of ROS in epididymal sperm of the *Corynorhinus mexicanus* bat, independently, from being stimulated or not with NADPH (figure 4), we can sustain that the production of ROS doesn't depend on the activity of one NADPH oxidase, as had been put forward by diverse investigators [3, 81,82], because NADPH oxidase appears or is inactive or absent, this fact is strengthened by Banfi and colaborators (2004) [16] who found that the NADPH oxidase gene, member of the NOX5 family, is expressed in primary spermatocytes in humans, but, this was not confirmed in mature spermatozoa. As well as the information reported by Shukla and colaborators (2005) [72] who found that this oxidase reaches its maximum activity in testicular spermatozoa, but diminishes gradually conforming with the advance of sperm maturation, because apparently the NADPH oxidase is absent in epididymal sperm, so that in relation to the NADPH oxidase activity, relating before to the production of ROS for epididymal sperm, this could be in reality the activity of the cytochrome-B5 reductase localized in epithelial cells which could have contaminated the sperm suspension [14,15].

For this reason we can sustain that the production of ROS for the epididymal spermatozoa in *Corynorhinus mexicanus* bat could possibly have its origin in the mitochondrial activity [27]; and that the redox balance associated with the micro environment epididymal by means that the spermatozoids transfer, must be specific and differentially controlled to assure its adequate function.

Table 3. Percent of spermatozoa that enter qualification and reaction recovered acrosomal of the different areas of the epididymis of the *Corynorhinus mexicanus* bat. Modificated from Cervantes et. al. 2008 [20]

Date	**Epididymal region**			
	Capacitation		Acrosomal reaction	
	Body	Cauda	Body	Cauda
Sep 11	6.0±3.4*a*	16.0±3.2*b*	7.6±2.8^1	12.0±3.2^1
Sep 23	7.3±3.1*a*	29.3±4.8*c*	22.0±3.9^2	27.3±7.1^2
Oct 16	13.3±3.8*b*	27.3±5.6*c*	18.3±3.7^2	34.4±2.5^3
Oct 30	19.4±4.3*b*	33.2±5.3*c*	22.2±1.8^2	32.3±3.8^3

Note: The numbers indicate the average ± DS of 3-4 standalone experiments.

The percent of enabled spermatozoa and with reaction acrosomal, they were independently submitted to ANOVA followed by the test of Bonferroni.

Different letters (percent of enabled spermatozoa) or different numbers (percent of spermatozoa with reaction acrosomal) indicate significant ($P < 0.05$) differences.

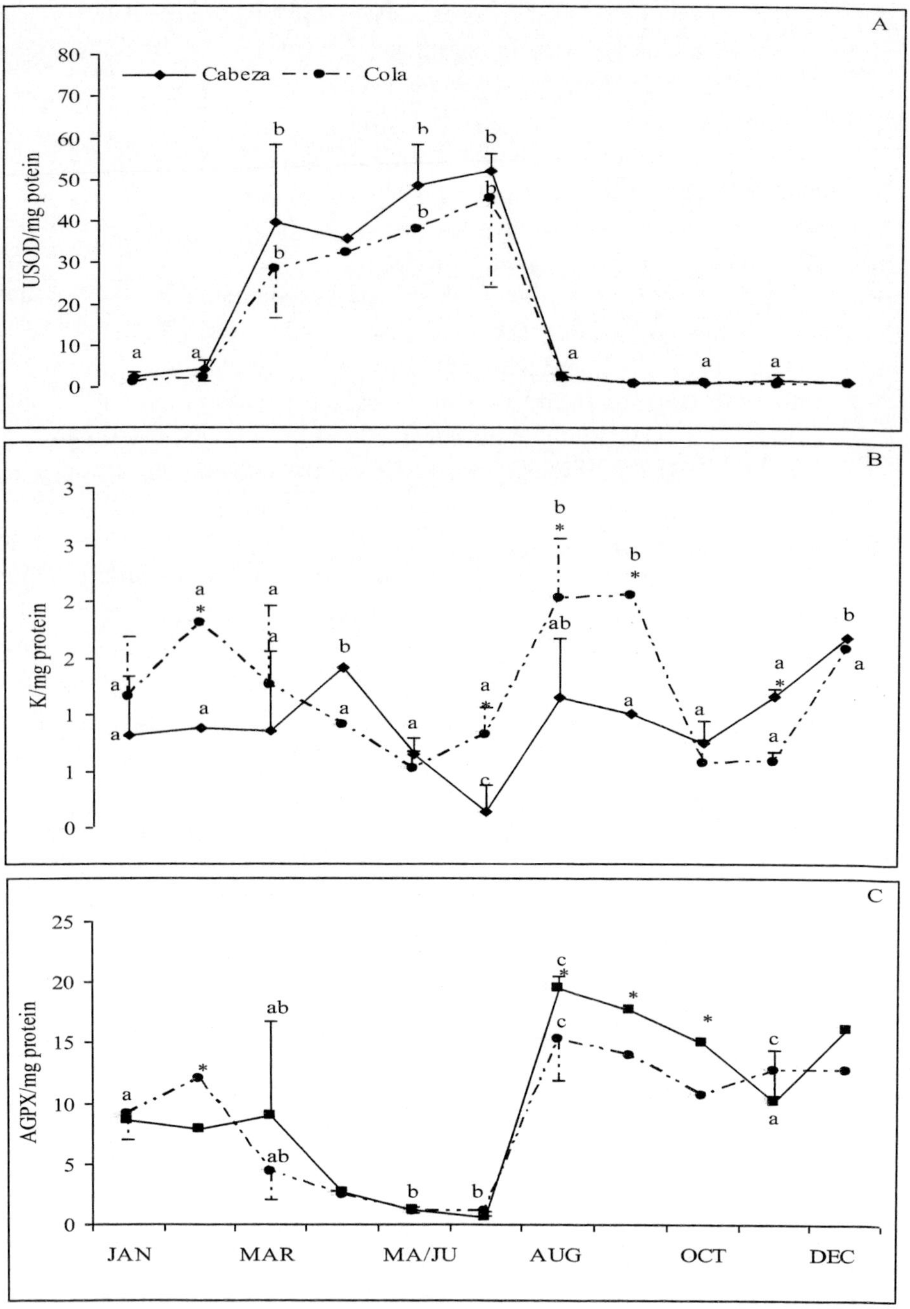

Figure 5. Enzymes anti-ROS in head and queue of epididymis of male mature *Corynorhinus mexicanus* bats captured the reproductive cycle monthly yearly: (To) SOD expressed as drives of SOD, where an activity drive is defined as the quantity of enzyme that waive 50% of the formazán reason; (B) CAT expressed as the changes in k, the constant of the reaction of first command (Cohen et al., 1970); and (C) GPX expressed as the activity of GPX (nmoles of rusty NADPH / minute). Senna reports the average ± you are. Different letters indicate statistic ($P < 0.05$) differences among the securities of the same outline (ANOVA followed by Bonferroni). The asterisks indicate statistic ($P < 0.05$) differences among the obtained securities starting from both areas of the epididymis (Test of t Student). Modificated from Arenas-Ríos et. al. 2005 [11].

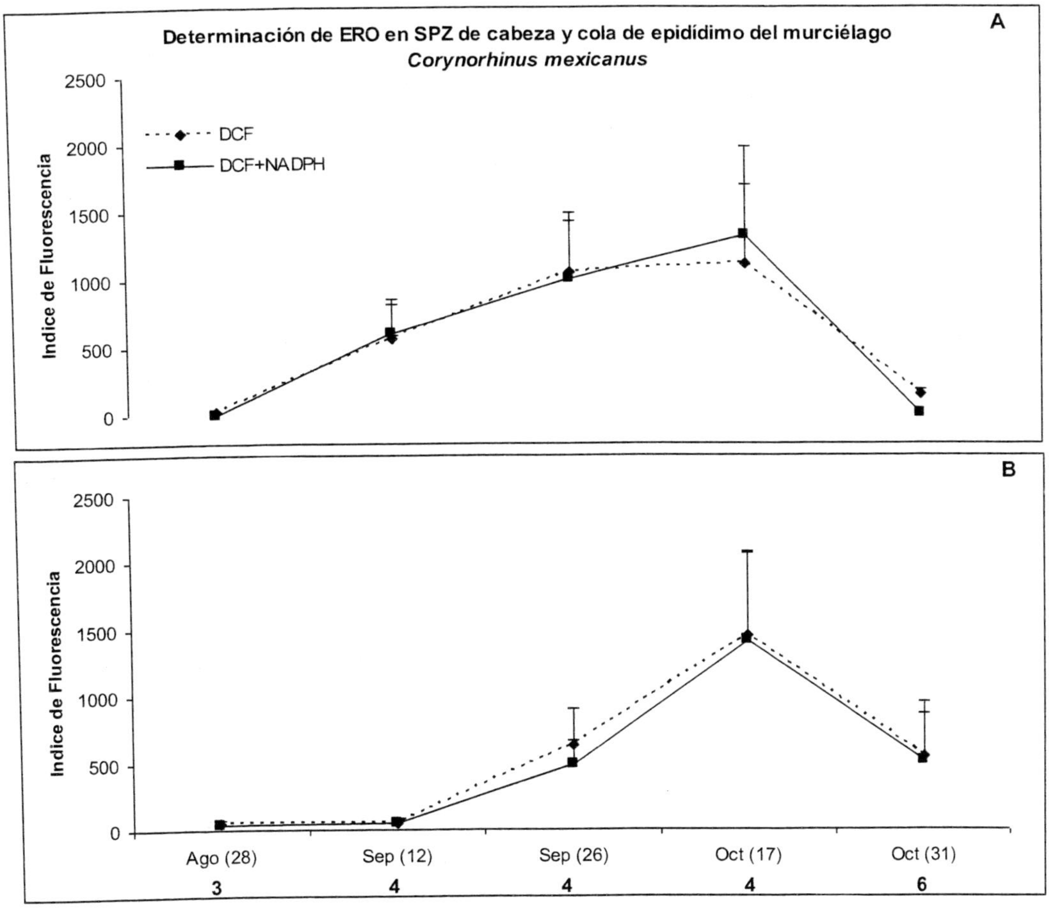

Figure 6. Determination of ROS in epididymal spermatozoa: A) cephalic and B) caudal regions, from C. mexicanus bats among final of August and final of October. The points represent the mean ± ES.

Different significative statistics were not found in the production of ROS between the different dates (ANOVA; head: F=1 .193; P > 0.05; queue: F=1 .780; P > 0.05), neither among produced ROS with DCF+NADPH (ANOVA; head: F=0 .751; P > 0.05; queue: F=1 .377; P > 0.05). The numbers in the inferior part of the graph indicate the number of determinations for date. Modificated from Arenas-Ríos, 2009 [10]

CONCLUSION

In addition to promoting sperm maturation and providing a place for sperm storage, the epididymis plays a role in the transport of spermatozoa along the duct and protects spermatozoa from harmful substances during its transport from the rete testis to the epididymal cauda. Many important tasks related to these processes appear to be under redox control. In *Corynorhinus mexicanus,* redox equilibrium of the micro-environments associated with the milieus by which mammalian spermatozoa must progress during its transit throughout the epididymis seems to be specifically and differentially controlled in relation with the compartmentalization of epididymal functions. Our data on ROS related enzyme activities (GPX, CAT and SOD) stress the existence of a careful differentially regulated equilibrium between the activities of these enzymes in the cauda and in the caput epididymidis that seems to be specifically related to the precise maturation/storage function of the different epididymal regions

REFERENCES

[1] Agarwal, A., Saleh, R.A., Bedaiwy, M.A., 2003. Role of reactive oxygen species in the pathophysiology of human reproduction. *Fertil. Steril.* 79, 829–843.

[2] Aitken, R.J., 1999. The Amoroso lecture. The human spermatozoon- a cell in crisis?. *J. Reprod. Fertil.* 115(1), 1–7.

[3] Aitken, R.J., Gordon, E., Harkiss, D., Twigg, J.P., Milne, P., Jennings, Z., Irvine, D. 1998. Relative impact of oxidative stress on the functional competence and genomic integrity of human spermatozoa. *Biol. Reprod.* 59, 1037–1046.

[4] Aitken, R.J., Harkiss, D., Buckingham, D.W., 1993a. Analysis of lipid peroxidation mechanisms in human spermatozoa. *Mol. Reprod. Dev.* 35, 302–315.

[5] Aitken, R.J., Harkiss, D., Buckingham, D., 1993b. Relationship between iron-catalysed lipid pROSxidation potential and human sperm function. *J. Reprod. Fertil.* 98, 257–265.

[6] Aitken, R.J., Ryan, A.L., Curry, B.J., Baker, M.A., 2003. Multiple forms of redox activity in populations of human spermatozoa. *Mol. Hum. Reprod.* 9, 645–661.

[7] Alvarez, J.G., Storey, B.T., 1982. Spontaneous lipid peroxidation in rabbit epididymal spermatozoa: its effect on sperm motility. *Biol. Reprod.* 27, 1102–1108.

[8] Alvarez, J.G., Storey, B.T., 1984. Lipid peroxidation and the reactions of superoxide and hydrogen peroxide in mouse spermatozoa. *Biol. Reprod.* 30, 833-841

[9] Alvarez, J.G., Touchstone, J.C., Blasco, L., Storey, B.T., 1987. Spontaneous lipid peroxidation and production of hydrogen peroxide and superoxide in human spermatozoa: superoxide dismutase as major enzyme protectant against oxygen toxicity. *J. Androl.* 8, 338-348.

[10] Arenas-Ríos, E., 2009. Enzimas Anti-especies reactivas de oxígeno, como reguladores en los procesos de espermatogénesis, maduración y almacenamiento prolongado de espermatozoides en el murciélago *Corynorhinus mexicanus*. *Tesis Doctoral UAM,* México.

[11] Arenas- Ríos, E., León-Galván, M.A., Mercado, P.E., Rosado, G.A., 2005. Superoxide dismutase, catalase, and glutathione peroxidase during epididymal maturation and prolonged storage of spermatozoa in the mexican big-eared bat (Corynorhinus mexicanus). *Can. J. Zool.* 83, 1556-1565.

[12] Avery, M.I., 1985. Winter activity of pipistrelle bats. *J. Anim. Ecol.* 54(3), 721–738.

[13] Babior, B.M., El Benna, J., Chanock, S.J., Smith, R.M., 1997. The NADPH oxidase of leukocytes: The respirtory burst oxidase. In Scandalios JG (ed) Oxidative stress and the molecular biology of antioxidant defenses. Cold Spring Harbor Laboratory Press, New York, pp 737-784

[14] Baker, M.A., Krutskikh, A., Curry, B.J., Hetherington, L., Aitken, R.J., 2005. Identification of cytochrome-b5 reductase as the enzyme responsible for NADH-

dependent lucigenin chemiluminescence in human spermatozoa. *Biol. Reprod.* 73, 334–342.

[15] Baker, M.A., Krutskikh, A., Curry, B.J., McLaughlin, E.A., Aitken, R.J., 2004. Identification of cytochrome p450-reductase as the enzyme responsible for NADPH-dependent lucigenin and tetrazolium salt reduction in rat epididymal sperm preparations. *Biol. Reprod.* 71,307-318.

[16] Banfi, B., Tirone, F., Durussel, I., Knisz, J., Moskwa, P., Molnar, G.Z., Krause, K.H., Cox, J.A., 2004. Mechanism of Ca^{2+} activation of the NADPH oxidase 5 (NOX5). *J. Biol. Chem.* 279,18583-18591.

[17] Barber, A.A., Bernheim, F., 1967. Lipid peroxidation. Its measurement, occurrence and significance in animal tissues. *Adv. Gerontol. Res.* 2, 355–403.

[18] Bennetts, L.E., Aitken, R.J., 2005. A comparative study of oxidative DNA damage in mammalian spermatozoa. *Mol. Reprod. Dev.* 71, 77–87.

[19] Bilodeau, O.J.F., Chattierjee, S., Marc-Andre, S., Gagnon, C., 2000. Levels of antioxidant defenses are decreased in bovine spermatozoa after a cycle of freezing and thawing. *Mol. Reprod. Develop.* 55, 282–288.

[20] Cervantes, D.L.I., Arenas-Ríos, E., León-Galván, M.A., López-Wilchis, R., Ambríz, G.D., Rosado, G.A., 2008. Spermatozoa epididymal maturation in the mexican big-eared bat (Corynorhinus mexicanus). *Syst. Biol. Reprod. Med.* 54, 196-204.

[21] Chen, H., Cheung, M.P.L., Chow, A.L.M., Liu, W., O.W.S., 2002. Protection of sperm DNA against oxidative stress in vivo by accessory sex gland secretions in male hamsters. *Reprod.* 124, 491-499.

[22] Coon, M.J., Ding, X., Pernecky, S.J., Vaz, A.D.N., 1992. Cytochrome P450: Progress and predictions. *FASEB J.* 6, 669–673.

[23] Cooper, T.G., 1995. Role of the epididymis in mediating changes in the male gamete during maturation. *Adv. Exp. Med. Biol.* 377, 87–101.

[24] Crichton, E.G., 2000. Sperm storage and fertilization. In: *Reproductive Biology of Bats,* E.G. Crichton and H. Krutzsch editors. Academic Press, London, United Kingdom. 295-320.

[25] Crichton, E.G., Hinton, B.T., Pallone, T.L., Hammerstedt, R., 1994. Hyperosmolality and sperm storage in hibernating bats: prolongation of sperm life by dehydration. *Am. J. Physiol.* 267, R1363–R1370.

[26] Domínguez, M.M.A., 2007. *Determinación de la actividad de la enzima gamma glutamil transpeptidasa en plasma seminal del cerdo criollo del istmo de tehuantepec* (Sus scrofa) Universidad Autónoma "Benito Juárez de Oaxaca", Oaxaca, México.

[27] Ford, W.C.L., 2004. Regulation of sperm function by reactive oxygen species. *Hum. Reprod. Up.* 10(5), 387–399.

[28] Foresta, C., Flohe, L., Garolla, A., Roveri, A., Ursini, F., Maiorino, M., 2002. Male fertility is linked to the selenoprotein phospholipids hydroperoxide glutathione peroxidase. *Biol. Reprod.* 67, 967–971.

[29] Fuentes-Mascorro, G., Serrano, F.H., Rosado, G.A., 2000. Sperm cromatin. *Arch. Androl.* 45, 215-225.

[30] Fujii, J., Iuchi, Y., Matsuki, S., Ishii, T., 2003. Cooperative function of antioxidant and redox systems against oxidative stress in male reproductive tissues. *Asian J. Androl.* 5,231-242.

[31] Gamou, S., Shimizu, N., 1995. Hydrogen peroxide preferentially enhances the tyrosine phosphorylation of epidermal growth factor receptor. *FEBS Lett*. 357, 161–164.

[32] Ghyselinck, N.B., Dufaure, J.P., 1990. A mouse cDNA sequence for epididymal androgen-regulated proteins related to glutathione peroxidase. *Nucleic Acids Res*. 18, 7144.

[33] Godeas, C., Tramer, F., Micali, F., Soranzo, M., Sandri, G., Panfili, E., 1997. Distribution and possible novel role of phospholipid hydroperoxide glutathione peroxidase in rat epididymal spermatozoa. *Biol. Reprod.* 57,1502–1508.

[34] Gómez, E., Irvine. D.S., and Aitken, R.J., 1998. Evaluation of a spectrophotometric assay for the measurement of malondialdehyde and 4-hydroxyalkenals in human spermatozoa: relationships with semen quality and sperm function. *Int. J. Androl.* 21, 81–94.

[35] Gopalakrishna, A., Bhatia, D., 1980. Storage of spermatozoa in the epididymis of the bat *Hipposieros speoris* (Schneider). *Curr Sci* (Bangalore). 49, 951–953.

[36] Guyton, K.Z., Liu, Y., Gorospe, M., Xu, Q., Holbrook, N.J., 1996. Activation of mitogen-activated protein kinase by H_2O_2. Role in cell survival following oxidant injury. *J. Biol. Chem*. 271, 4138–4142.

[37] Halliwell, B., Aruoma, O.I., 1991. DNA damage by oxygen-derived species, Its mechanism and measured in mammalian system. *FEBS lett*. 281, 9-19.

[38] Halliwell, B., Gutteridge, J.M.C., 1999. *Free Radicals in Biology and Medicine*. 3rd edn. Oxford University Press, Oxford

[39] Holland, M.K., Alvarez, J.G., Storey, B.T., 1981. Oxygen metabolism of mammalian spermatozoa generation of hydrogen peroxide by rabbit epididymal spermatozoa. *Biochem. J.* 198, 273-280.

[40] Jervis, K.M., Robaire, B., 2001. Dynamic changes in gene expression along the rat epididymis. *Biol. Reprod.* 65, 696–703.

[41] Jeulin, C., Soufir, J.C., Laval-Martin, D., Calvayrac, R., 1989. Catalase activity in human spermatozoa and seminal plasma. *Gamete Res*. 24,185–196.

[42] Jiménez, C., Lefrancois, A.M., Ghyselinck, N.B., Dufaure, J.P., 1992. Characterization and hormonal regulation of 24 kDa protein synthesis by the adult murine epididymis. *J. Endocrinol.* 133,197–203.

[43] Karim, K.B., Banerjee, S., 1985. Storage of spermatozoa in the epididymis of the tropical bat, *Rhinopoma hardwickei hardwickei* (Gray). *Anat. Rec*. 211, 95A. [Abstract: Yamaguchi H, Fukuda R, Kasai J y Tanaka T (1997) Increased oxidative deoxyribonucleic acid damage in the spermatozoa of infertile male patients. *Fertil. Steril*. 68(3), 519–524.

[44] Kodama, H., Yamaguchi, R., Fukuda, J., Kasai, H., Tanaka, 1997. Increased oxidative deoxyribonucleic acid damage in the spermatozoa of infertile male patients. *Fertil. Steril.* 68, 519-524.

[45] Krutzsch, P.H., 2000. Anatomy, physiology and cyclicity of the male reproductive tract. In: Crichton y Krutzsch, editors. *Reproductive biology of bats.* Academic Press, London, United Kingdom. Pp 91–155.

[46] León-Galván, M.A., Fonseca, T., López, W.R., Rosado, A. 1999. Prolonged storage of spermatozoa in the genital tract of female Mexican big-eared bats (Corynorhinus mexicanus): the role of lipid peroxidation. *Can. J. Zool.* 77, 7-12.

[47] León-Galván, M.A., López-Wilchis, M.A., Hernández, O.P., Arenas-Ríos, E., Rosado, A. 2005. Male reproductive cycle of mexican big-eared bats, Corynorhinus mexicanus (chiroptera: vespertilionidae) *The southwest. Natural.* 50(4), 453-460.

[48] Lewis, B., Aitken, R.J., 2001. Impact of epididymal maturation on the tyrosine phosphorylation patterns exhibited by rat spermatozoa. *Biol. Reprod.* 64,1545–1556.

[49] Lopes, S., Jurisicova, A., Sun, J.G., Casper, R.F., 1998. Reactive oxygen species: potential cause for DNA fragmentation in human spermatozoa. *Hum. Reprod.* 13, 896–900.

[50] López-Wilchis R., 1989. Biología de Plecotus mexicanus (Chiroptera Vespertilidae) en el estado de Tlaxcala. México. *Tesis Doctoral Universidad Nacional Autónoma de México.* 227

[51] Macleod, J., 1943. The role of oxygen in the metabolism and motility of human spermatozoa. *Am. J. Physiol.* 138, 512–518.

[52] Maser, R.L., Magenheimer, B.S., Calvet, J.P., 1994. Mouse plasma glutathione peroxidase. cDNA sequence analysis and renal proximal tubular expression and secretion. *J. Biol. Chem.* 269, 27066–27073.

[53] Matsubara, T., Ziff, M., 1986. Increased superoxide release from human endothelial cells in response to cytokines. *J. Immunol.* 137(10), 3295–3298.

[54] Moore, H.D.M., 1996. The influence of the epididymis on human and animal sperm maturation and storage. *Hum. Reprod,* 11,103–110.

[55] Neuweiler, G. 2000. *The Biology of Bats*. New York: Oxford University Press. 320 pp.

[56] Racey, P.A. 1982. Ecology of bat reproduction. In Kunz, T.H., editor. (Ed.). *Ecology of bats.* Plenum Press. New York. 57–104.

[57] Plante, M., De Lamirande, E., Gagnon, C.A., 1994. Reactive oxygen species released by activated neutrophils, but not by deficient spermatozoa, are sufficient to affect normal sperm motility. *Fertil. Steril.* 62, 387–393.

[58] Racey, P.A., and Entwhistle, 2000. Life-history and reproductive strategies of bats. In:Crichton y Krutzsch, editors. *Reproductive biology of bats*. Academic Press, London, United Kingdom. Pp 363–414.

[59] Raha, S., Robinson, B.H., 2000. Mitochondria, oxygen free radicals and ageing. *Trends Biochem. Sci.* 25, 502-508.

[60] Rejraji, H., Vernet, P., Drevet, J.R., 2002. GPX5 is present in the mouse caput and cauda epididymidis lumen at three different locations. *Mol. Reprod. Dev.* 63, 96–103.

[61] Rousseaux, J., Rousseaux-Prevost, R., 1995. Molecular localization of free thiols in human sperm chromatin. *Biol. Reprod.* 52,1066–1072.

[62] Rowley, M.J., Teshima, F., Heller, C.G., 1970. Duration of transit of spermatozoa through the human male ductular system. *Fertil. Steril.* 21, 390–396.

[63] Sakkas, D., Mariethoz, E., Manicardi, G., Bizarro, D., Bianchi, P.G., Bianchi, U., 1999). Origin of DNA damage in ejaculated human spermatozoa. *Rev. Reprod.* 4, 2801-2807.

[64] Sawyer, D.E., Mercer, B.G.,Wiklendt, A.M., Aitken, R.J., 2003. Analysis of gene-specific DNA damage and single-strand DNA breaks induced by pro-oxidant treatment of human spermatozoa in vitro. *Mut. Res.* 529, 21–34.

[65] Schwaab, V., Baud, E., Ghyselinck, N., Mattei, M.G., Dufaure, J.P., Drevet, J.R., 1995. Cloning of the mouse gene encoding plasma glutathione pROSxidase: organization. *Gene* 167,25–31.

[66] Segal, A.W., Abo, A., 1993. The biochemical basis of the NADPH oxidase of phagocytes. *Trends Biochem.* Sci. 18, 43–47.

[67] Seligman, J., Shalgi, R. 1991. Protein thiols in spermatozoa and epididymal fluid of rats. *J. Reprod. Fertil.* 93, 399–408.

[68] Seligman, J., Zipser, Y., Kosower, S.N., 2004. Tyrosine phosphorylation, thiol status, and protein tyrosine phosphatase in rat epididymal spermatozoa *Biol. Reprod.* 71, 1009–1015.

[69] Shalgi, R., Seligman, J., Kosower, N.S., 1989. Dynamics of the thiol status of rat spermatozoa during maturation: analysis with the fluorescent labeling agent monobromobimane. *Biol. Reprod.* 40,1037–1045.

[70] Shen, H.M., Ong, C.N., 2000. Detection of oxidative DNA damage in human sperm and its association with sperm function and male infertility. *Free Rad. Biol. Med.* 28,529–536.

[71] Shitashige, M., Morita, I., Murota, S., 1998. Different substrate utilization between prostaglandin endoperoxide H synthase-1 and -2 in NIH3T3 fibroblasts. *Biochim. Biophys. Acta.* 1389,57–66.

[72] Shukla, S., Jha, R.K., Laloraya, M., Kumar, P.G., 2005. Identification of nonmitochondrial NADPH oxidase and the spatio-temporal organization of its components in mouse spermatozoa. *Biochem. Biophys. Res. Commun.* 331, 476-483.

[73] Singh, K., Krishna, A., 1995, Inhibitory effects of melatonin on testosterone but not on androstenedione production during winter in the vespertilionid bat, *Scotophilus heathi. J. Pineal Res.* 19(3),127–132.

[74] Storey, B.T., 1997. Biochemistry of the induction and prevention of lipoperoxidative damage in human spermatozoa. *Mol. Hum. Reprod.* 3, 203–213.

[75] Tash, J.S., 1989. Protein phosphorylation: the second messenger signal transducer of flagellar motility. *Cell Motil. Cytoskel.* 14, 332–339.

[76] Tash, J.S., Bracho, G.E., 1994. Regulation of sperm motility: emerging evidence for a major role for protein phosphatases. *J. Androl.* 15, 505–509.

[77] Tonks, N.K., 2003. PTP1B: from the sidelines to the front lines! *FEBS Lett.* 546, 140–148

[78] Tosic, J., Walton, A., 1950. Metabolism of spermatozoa the formation and elimination of hydrogen peroxide by spermatozoa and elimination of and effects on motility and survival. *Biochem. J.* 47,199–212.

[79] Tramer, F., Rocco, F., Micali, F., Sandri, G., Panfili, E., 1998. Antioxidant systems in rat epididymal spermatozoa. *Biol. Reprod.* 59, 753-758.

[80] Turrens, J.F., 2003. Mitochondrial formation of reactive oxygen species. *J. Physiol.* 552(2), 335-344.

[81] Twigg, J., Fulton, N., Gomez, E., Irving, D.S., Aitken, R.J., 1998a. Analysis of the impact of intracellular reactive oxygen species generation on thestructural and functional integrity of human spermatozoa: Lipid peroxidation, DNA fragmentation and effectiveness of antioxidants. *Hum. Reprod.* 13,1429–1436.

[82] Twigg, J.P., Irvine, D.S., Aitken, R.J., 1998b. Oxidative damage to DNA in human spermatozoa does not preclude pronucleus formation at intracytoplasmic sperm injection. *Hum. Reprod.* 13,1864–1871.

[83] Van Montfort, R.L., Congreve, M., Tisi, D., Carr, R., Jhoti, H., 2003. Oxidation state of the active-site cysteine in protein tyrosine phosphatase 1B. *Nature* 423, 773–777.

[84] Vernet, P., Aitken, R.J., Drevet, J.R., 2004. Antioxidant strategies in the epididymis. *Mol. Cell. Endocrinol.* 216, 31-39.

[85] Vernet, P., Faure, J., Dufaure, J.P., Drevet, J.R., 1997. Tissue and developmental distribution dependence upon testicular factors and attachment to spermatozoa of GPX5 a murine epididymis-specific glutathione peroxidase. *Mol. Reprod. Dev.* 47, 87–98.

[86] Visconti, P.E., Bailey, J.L., Moore, G.D., Pan, D., Old-Clarke, P., Kopf, G.S., 1995. Capacitation of mouse spermatozoa. I. Correlation between the capacitation state and protein tyrosine phosphorylation. *Develop.* 121,1129–1137.

[87] Yanagimachi, R., 1994. Mammalian fertilization. In: Knobil K, Neill JD (eds.), *The Physiology of Reproduction*. New York: Raven Press; 189–317.

[88] Yokoyama, Y., Beckman, J.S., Beckman, T.K., Wheat, J.K., Cash, T.G., Freeman, B.A., Parks, D.A., 1990. Circulating xanthine oxidase:potential mediator of ischemic injury. *Am. J. Physiol.* 258, G564-570.

[89] Zini, A., de Lamirande, E., Gagnon, C., 1993. Reactive oxygen species in semen of infertile patients: levels of superoxide dismutase- and catalase-like activities in seminal plasma and spermatozoa. *Int. J. Androl.* 16, 183–188.

[90] Zini, A., Schlegel, P.N., 1997. Identification and characterization of antioxidant enzyme mRNAs in the rat epididymis. *Int. J. Androl.* 20, 86–91.

In: Bats: Biology, Behavior and Conservation
Editors: J. L. Zupan and S. L. Mlakar, pp. 299-313
ISBN 978-1-61209-749-7

Chapter 9

METABOLIC ADAPTATIONS IN FRUGIVOROUS BATS

***André Otavio Peres Protzek*[1] *and Eliana de Cássia Pinheiro*[2]**
[1]Department of Anatomy, Cell Biology and Physiology and Biophysics, Institute of Biology, UNICAMP – State University of Campinas, Campinas, SP, Brazil
[2]Department of Physiological Sciences, Institute of Biological Science, University of Brasília, Brasília, DF, Brazil

ABSTRACT

Frugivorous bats like Artibeus lituratus and Artibeus jamaicensis are known to digest carbohydrate-rich meals rapidly and efficiently. These bats feed on variety of fruits with preference for figs, which presents an availability of 14.4 g of carbohydrate per 100 g wet mass. Ingesting the equivalent of their body weight every night, frugivorous bats have to deal with an overload of glucose daily and must have several physiological adaptations that include specialized gastrointestinal tract structures and hormonal response capable of avoiding post-prandial hyperglycemia.

It has been shown that frugivorous bats present high paracellular absorption of an analog of D-glucose (3-O-methyl-D-glucose) in their intestine. This pathway seems to contribute at least for 55% of total glucose uptake in Rousettus aegyptiacus and 70% in Artibeus lituratus. This mechanism must have an important role considering that these bats have shorter intestinal tracts than similarly sized non-flying mammals.

Besides the high intestinal absorptive capacity, some authors also pointed to a specialization of the endocrine pancreas of R. aegyptiacus as an additive mechanism to deal with high glucose influx. It has been demonstrated that the endocrine pancreas of R. aegyptiacus represents 9.1% of the total pancreas volume, far more than in some other mammals. In a similar way, A. lituratus presents an apparent large β-cell distribution with different islets sizes and shapes, similar to R. aegyptiacus. However, the pattern of β-cell distribution in A. lituratus differs from that found in other mammals. Their β cells seem to be located, mainly, at the periphery of the islets, a pattern not observed in R. aegyptiacus, rodents, humans, non-human primates or pigs, where β cells are located throughout the islets, but especially in the core. This pattern could reflect physiological and environmental adjustments to accommodate changing demands for insulin. This species also presents elevated glucose tolerance, high insulin sensitivity and efficient insulin signalization in the pectoral muscle and hepatic tissue. Another important adaptation in A. lituratus is the ability to maintain normoglycemia even after 6 days of starvation, possibly because of its high hepatic glycogen stores and increased hepatic gluconeogenesis.

In conclusion, and from an evolutionary point of view, the rapid and efficient digestion, high sensitivity to glucose and insulin and the presence of an apparent large β-cell distribution represent adequate adaptations that guarantee the proper control of glucose homeostasis in fruit-eating bats, constantly challenged by a high influx of glucose from its carbohydrate-rich diet.

INTRODUCTION

Maintenance of normoglycemia in mammals is essential for the metabolism of tissues that use glucose as their primary source of energy, such as the central nervous system and renal medulla (Nordlie et al. 1999; Sawdon and Yeo, 2007), regardless of the nutritional state (absorptive, post-absorptive or fasting). The maintenance of blood glucose requires a precise interaction between the nervous system, pancreas, liver, muscle and fat tissue. These tissues are essential for delivery and glucose uptake (Tirone and Brunicardi, 2001) through activation of specific metabolic pathways such as those responsible for the storage of nutrients from the diet or mobilization of body reserves.

The type of the diet can modulate the metabolic pathways (Felig, 1979), such that, in general, the metabolic pattern during fed state is different in high carbohydrate or protein diet (Beardsall et al. 2006). Moreover, nutritional status also influences the activation of metabolic pathways (Yeo and Sawdon, 2007), whereas in fed state glycogenesis is the primary pathway involved in glucose homeostasis, and during fasting, gluconeogenesis and glycogenolysis predominate (Beardsall et al., 2006).

Another important modulator of glucose homeostasis is hormonal action, which controls both glucose uptake and production (Beardsall et al. 2006; Klover and Mooney, 2004). The endocrine pancreas, represented by the Islets of Langerhans has a fundamental role in the flow of substrates, both in fed state, when insulin is secreted, as in fasting, when glucagon is more active.

ENDOCRINE PANCREAS

The Islets of Langerhans represent 1-2% of the total mass of human pancreas and are essential for energy metabolism. Each islet is composed predominantly of β cells, which synthesize and secrete insulin, and α cells that synthesize and secrete glucagon (Cabrera et al., 2006).

Insulin, an anabolic polypeptide hormone, has special action in skeletal muscle, hepatic and adipose tissue. Insulin is secreted primarily in response to an increase in plasma nutrients concentrations, mainly glucose. Thus, in a healthy organism, normoglycemia is finely regulated by insulin secretion and actions, since plasma glucose substrate is the main regulator of the secretory process (Deeney et al. 2000; Rutter, 2001). The immediate metabolic effects of insulin include: increase in glucose uptake by muscular and adipose tissue, increase in glucose oxidation, elevation of protein, fatty acids, triacylglycerol and glycogen synthesis, inhibition of lipolysis, proteolysis, gluconeogenesis and glycogenolysis.

Thus, insulin stimulates the use and storage of surplus energy substrates. Medium and long term actions of insulin are: gene expression alteration, cell differentiation and proliferation, increasing in production of nitric oxide by the endothelium, prevention of apoptosis or cell death and promotion of cell survival (Beardsall et al. , 2006; Beardsall et al., 2006, Bernal-Mizrachi et al., 2001, Cabrera et al. 2006; Sawdon and Yeo, 2007; Zecchin et al., 2004).

Mechanism of Insulin Secretion and Action

Insulin secretion is stimulated by the degradation of different energy substrates that can be metabolized by the β cell, glucose being the main secretagogue. The secretory process is initiated with the influx of glucose into the β cell through glucose transmembrane transporters isoform 2 (GLUT2). Within the β cell, glucose is phosphorylated and converted into glucose-6-phosphate (G-6-P) by the enzymes hexokinase I and hexokinase IV (glucokinase). Glucokinase regulates the flow through the glycolytic pathway and hence insulin secretion, acting as the glucose sensor in β cell (Boschero, 1996). G-6-P is completely oxidized into β cell, resulting ultimately ATP, increasing ATP/ADP ratio, which leads to closure of ATP sensitive potassium channels (K_{ATP}). This event promotes the depolarization of plasma membrane, the opening of voltage-dependent calcium channel, an influx of this ion and activation of secretory machinery secretion, with migration of granules containing insulin to plasma membrane and subsequent extrusion of their content (Figure 1).

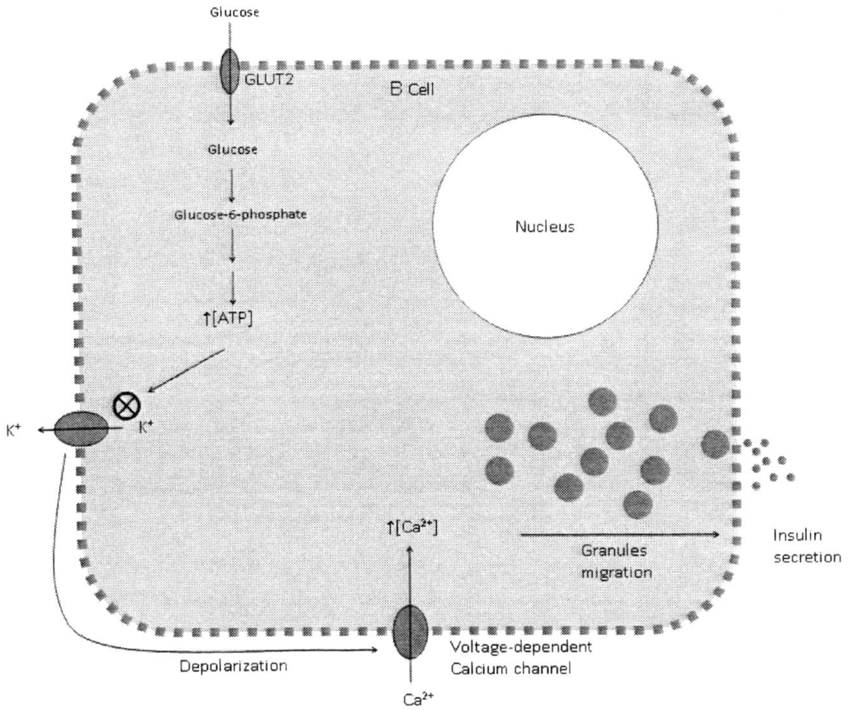

Figure 1. *Insulin secretion*: glucose influx and metabolization in β cell open ATP sensitive potassium channels, depolarizing plasma membrane and allowing calcium influx, resulting in insulin secretion.

Although glucose is the main secretagogue of insulin, other energy substrates such as amino acids and free fatty acids or hormones and neurotransmitters can modulate insulin secretion. Insulin is precisely secreted and finely regulates energy substrates levels in blood, in different conditions, such as fasting, feeding, pregnancy, exercise and growth (Boschero, 1996).

Insulin plays its role after coupling with a membrane receptors present in almost all tissues. The insulin receptor is a heterotetramer protein composed of two α subunits, located in the extracellular portion, and two transmembrane β subunits with tyrosine kinase activity (Kahn, 1985). The α subunits have binding sites for insulin and inhibitory activity on the β subunits. The coupling of insulin with α subunits promotes conformational change and autophosphorylation of the β subunits in several tyrosine residues (1158, 1162 and 1163) increasing their kinase activity (Patti and Kahn, 1998, White et al., 1984) (Figure 2).

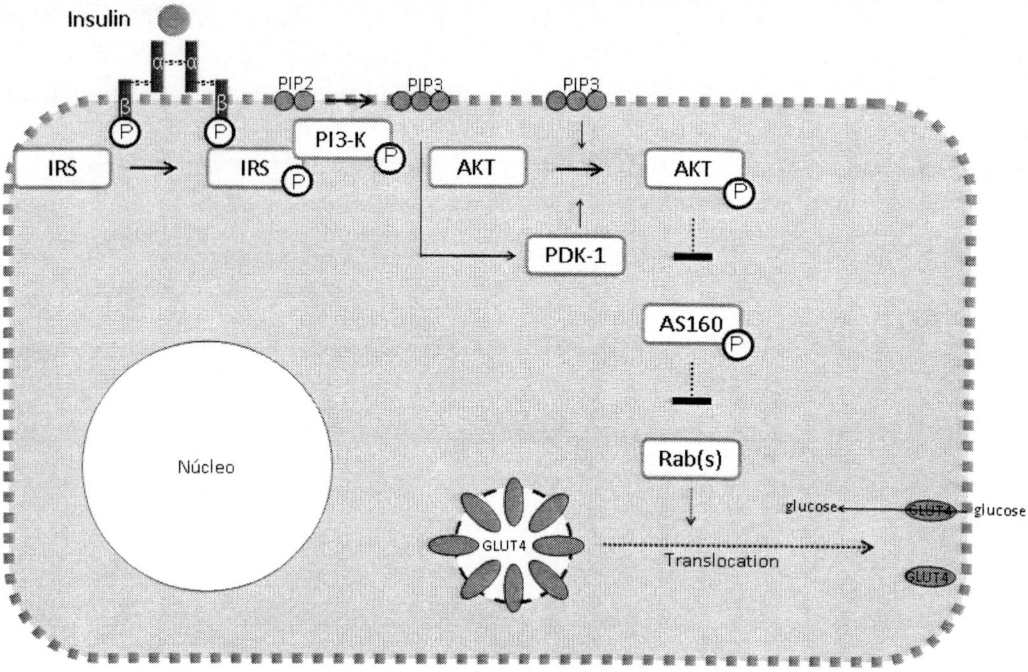

Figure 2. *Insulin signaling pathway*. Insulin binds to its membrane receptor, activating its intrinsic tyrosine kinase activity, which results in autophosphorylation followed by phosphorylation of intracellular substrates such as IRSs (insulin receptor substrate). IRSs, in turn, interact with signaling molecules containing SH2 domains (Src homology 2) and PI3K (phosphatidylinositol 3-kinase). PI3K mediates major metabolic actions of insulin activating downstream effectors as protein Akt (protein kinase B). Akt, when phosphorylated, inhibits via phosphorylation protein AS160, allowing Rab proteins (GTPase activating protein) acting in vesicle trafficking of glucose transporter GLUT4.

The autophosphorylation of insulin receptor β subunits promotes a tyrosine phosphorylation cascade. So far, ten insulin receptors substrates have been identified, and four of which belong to the insulin receptor substrates superfamily, IRS proteins. When phosphorylated, IRS activates recognition sites for molecules containing SH2 domains such as phosphatidylinositol 3-kinase (PI3-K) (Zecchin et al., 2004). The PI3K mediates the major metabolic actions of insulin by downstream effectors as the serine-threonine protein Akt (protein kinase B).

PI3-kinase has a strong influence on the modulation of insulin-stimulated glucose uptake (Zecchin et al., 2004) and is considered the only essential molecule for intracellular glucose uptake (Czech and Corvera, 1999). The interaction of phosphorylated IRS sites with PI3-kinase activates the phosphorilation of phosphoinositide, generating phosphatidylinositol 3-phosphate, phosphatidylinositol 3,4-phosphate and phosphatidylinositol 3,4,5-phosphate (Zecchin et al., 2004). One of the enzymes regulated by phosphatidylinositol 3,4,5-phosphate is the PDK-1 (phosphoinositide-dependent kinase 1), a serine threonine kinase that phosphorylates and enhances the activation of Akt (Zecchin et al., 2004).

In mammals, Akt was observed in three isoforms: Akt1 (PKBa), Akt2 (PKBb) and Akt3 (PKBg), which are distributed differently in tissues (Vanhaesebroeck and Alessi, 2000). When phosphorilated, Akt catalyzes the activation of several metabolic pathways (Whiteman et al., 2002) promoting glucose uptake (via GLUT4). Recently, a protein phosphorylated by Akt, AS160 was associated with the translocation of GLUT4 in muscle and adipose tissue. When activated by Akt, AS160 induces traffic and exocytosis of GLUT4 vesicles to the plasma membrane. On the other hand, when inhibited, Akt promotes partial reduction of GLUT4 transport stimulated by insulin in muscle and fat cells (Hill et al., 1999).

Pancreas	Liver	Muscle	Adipose
↑ β cell mass ↑ Insulin secretion ↓ Apoptosis	↓ Gluconeogenesis ↑ Glycogenesis ↓ Glycogenolysis	↑ Glucose uptake ↑ Glycogenesis ↑ Protein synthesis ↑ Lipogenesis	↑ Glucose uptake ↑ Protein synthesis ↑ Lipogenesis

Figure 3. A summary of the actions of Akt in insulin-sensitive tissues (Cho et al. 2001; Downward, 1998, Whiteman et al. 2002; Zecchin et al., 2004).

INFLUENCE OF THE DIET ON HORMONE SECRETION AND METABOLIC PATHWAYS ACTIVATION

Nutrients from the diet can induce the secretion of hormones that modulate the activity of a metabolic pathway. The consumption of a high-carbohydrate diet promotes a rise in blood glucose, which efficiently stimulates insulin secretion, resulting in glucose uptake, its oxidation and synthesis of glycogen, proteins and lipids (Yeo and Sawdon, 2007).

Moreover, the consumption of a high protein diet promotes an elevation of plasma amino acids concentration that, initially, stimulates glucagon secretion, reinforced by lower plasma insulin and glucose levels. Glucagon, in turn, promotes the activation of hepatic gluconeogenesis, increasing glucose production, with amino acids from the diet as precursors. It results in a modest blood glucose increase, which stimulates insulin secretion and finally glucose uptake. In this case, insulin secretion is lower when compared with that obtained when animals are submitted to high-carbohydrate diets (Kettelhut et al., 1980).

During starvation, high carbohydrate or high protein diets induce glycogenolysis and gluconeogenesis in response to glucagon and other hyperglycemic hormone secretion, resulting in normoglycemia (Beadrsall et al. 2006; Kettelhut et al., 1980). It is noteworthy that most high protein diet animals are more resistant to starvation, probably because they present high gluconeogenesis activity in the absorptive period and still maintained during fasting periods (Kettelhut et al., 1980). Anyway, regardless of the diet type, the mobilization of amino acids from body proteins is a key point to the maintenance of blood glucose by gluconeogenesis during fasting (Gazola et al., 2007).

Besides having a rich carbohydrate diet the fruit-eating bat *Artibeus lituratus* is more resistant to starvation (Pinheiro et al., 2006) than the vampire bat *Desmodus rotundus* (Freitas et al. 2003), whose diet is rich in protein. This fact is, possibly, due to large liver glycogen (16% of hepatic tissue) reserves, possibly involved in the maintenance of plasma glucose levels when *Artibeus lituratus* is submitted to short periods of fasting (24-48 h) (Pinheiro et al., 2006). During more prolonged fasting (48-96 h) liver gluconeogenesis seems to be responsible for glucose homeostasis in this frugivorous bat (Pinheiro et al., 2006). Also, large free fatty acid mobilization especially from the carcass possibly has important contribution to energy homeostasis of this species, besides saving glucose. These results resemble the metabolic behavior of most vertebrates feeding on high carbohydrate diet, including man. Thus, although *A. lituratus* submitted to fasting for up to 6 days presents a shortage of about 40-50% in blood glucose concentration, its glycemia remains within the limits compatible with mammals life (approximately 2.8 mMol/L) (Pike et al. 2006).

This capability to maintain normoglycemia even after six days of fasting must be enable *Artibeus lituratus* wide range of distribution, from Mexico to northern Argentina, Bolivia, Leeward Islands, Three Marias Islands and all regions of Brazil (Simmons et al., 2005).

A. lituratus is one of the best known bat species in Brazil due to its abundance, including in urban environments (Reis et al., 2007). A study conducted in Southern Brazil showed that 80-88% of its diet is composed by fruits, although this species also feeds on insects, leaves, flowers, pollen and nectar (Passos and Graciolli, 2004). The preference for fruits, associated with the behavior of removing them from the parent plant and leading them to a shelter to consume it, makes this bat an excellent dispersal of seeds of different plant species, demonstrating an important role in forest recovery (Reis et al., 2007).

An important feature of fruit-eating bats is their high intake of fruits (carbohydrate) every night, consuming the equivalent of its own body weight in only one night of foraging (Morrison, 1978; Van Der Westhuizen, 1976). This dietary pattern would eventually result in weight increase that could unbalance their performance during the flight. Another impact of this feeding behavior in their physiology could be an overload in plasma glucose, which could negatively affect insulin independent tissues. To deal with the daily overload of carbohydrate, frugivorous bats must have several physiological adaptations associated with both gastrointestinal and endocrine pancreas.

It has been seen that frugivorous bats as *Artibeus lituratus* are able to digest a carbohydrate rich meal in a rapid and efficient manner (Keegan, 1977; Craick and Markovich, 2000) and that this species presents a specialized gastrointestinal tract (Keegan, 1997; Van Der Westhuizen, 1976). In 1997 a study conducted by Karasov and Hume showed that, similar to most vertebrates, *A. lituratus* possesses intestinal sugar transporters such as the Na^+-coupled glucose transporter (SGLT1) in the apical membrane and GLUT2 in the basolateral membrane. More recently, a study using another fruit-eating bat (*Rousettus aegyptiacus*) showed high paracellular absorption of an analog of D-glucose (3-O-methyl-D-glucose) in its intestine that contributes, at least, 55% of total glucose uptake (Tracy et al., 2007). A similar mechanism was also found in *A. lituratus* and accounts for approximately 70% of total glucose transport, suggesting that high paracellular absorption of carbohydrates may be a general pattern among frugivorous bats (Caviedes-Vidal et al., 2007, 2008). This mechanism must have an important role considering that these bats have shorter intestinal tracts than similarly sized non-flying mammals (Keegan and Modinger, 1979), which could reduce the body weight during flight.

The rapid intestinal absorption of glucose should be reflected in large fluctuations in blood glucose in these bats. However, we did not observe major blood glucose variation in A. lituratus comparing fed and fasted states (Pinheiro et al, 2006) suggesting that this species copes well with the large influx of glucose into the plasma after each meal.

A study conducted by Michelmore et al., (1998) showed that endocrine components of the pancreas (Islets of Langerhans) of *R. aegyptiacus* represents 9.1% of the total pancreas volume, which is much higher than the observed in other mammalian species (1-2%) (Berne and Levy, 2009), suggesting that this species also has higher amount of α and β cells. If this hypothesis is correct, we could expect increased capacity for insulin and glucagon secretion, as part of physiological adaptation for glucose homeostasis control in fruit-eating bats.

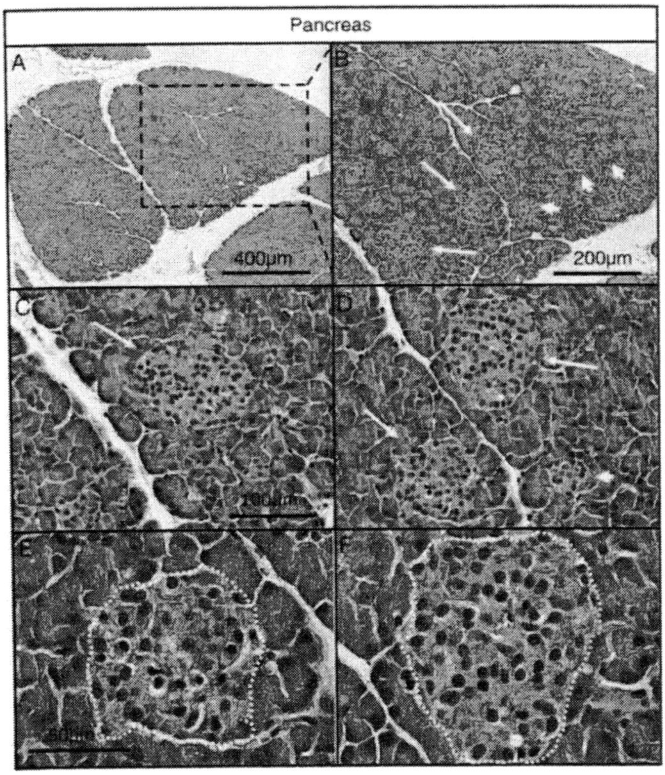

Figure 4. Light microscopy from pancreas of *A. lituratus*. Panoramic view of a histological section stained for hematoxiylin/eosin (A). Pancreas with big and small islets (arrows and arrowheads, respectively) (B–D). Islets (endocrine tissue) surrounded by exocrine tissue (E, F). Magnification ×100 in A, ×200 in B, ×400 in C, D and ×1000 in E, F.

More recently, interesting findings on morpho-physiological aspects related to glucose homeostasis in *A. lituratus* (Protzek et al., 2010), particularly those associated to insulin regulation shed some light on this issue.

The authors showed that similar to *R. aegyptiacus* (Michelmore et al., 1998), *A. lituratus* presents different islets sizes (Figure 4B–D). On the other hand, the pattern of β-cell distribution located mainly at the periphery of the islets (Figure 5D–G) differs from that found in other mammals including *R. aegyptiacus* (Michelmore et al., 1998), rodents (Bani Sacchi and Bani, 1985; Cabrera et al., 2006), humans, non-human primates or pigs (Cabrera et al., 2006) where β cells are located especially in the core of the islets. The variety of islet cells topography among species might reflect evolutionary adaptations to different dietary habits, other environmental constraints (Cabrera et al., 2006) or may reflect physiological and evolutionary adjustments to accommodate changing demands for insulin (Kim et al., 2009).

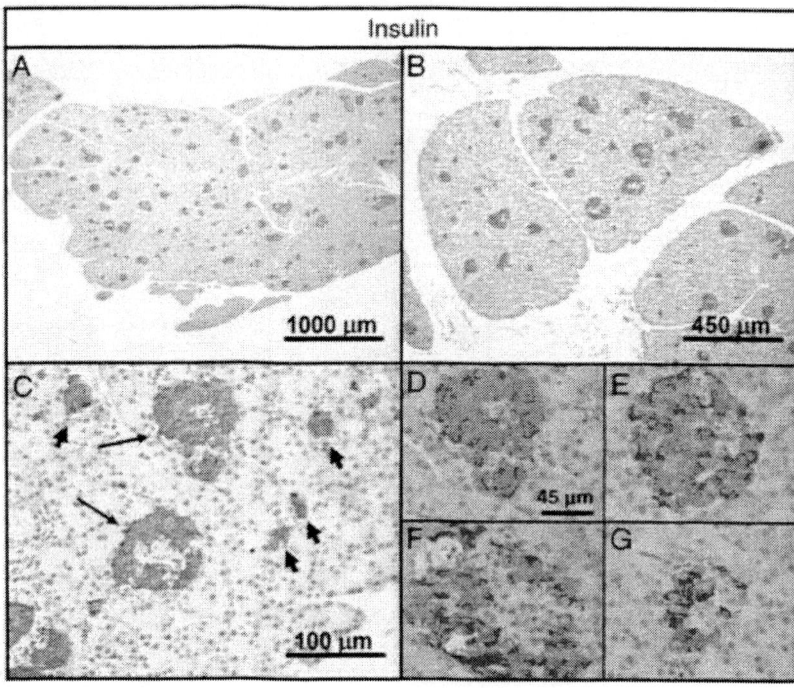

Figure 5. β cells in *A. lituratus* pancreas. Panoramic histological section of the pancreas immunostained for insulin (A, B). Morphological aspect of pancreatic islets (C). Note that there are different size of islet and that the pancreatic β cells localize preferably at the periphery of the islet (C, arrows). Small islets or group of β cells were commonly observed along pancreas section (C – arrowheads). Observe the peripheral localization of the β cells in islets (D–G). Magnification ×40 in A, ×100 in B, ×400 in C and ×1000 in D–G.

Analyzing the ability of glucose in stimulating insulin secretion in *A. lituratus* fragments of pancreas (Protzek et al., 2010) it was observed that, at basal glucose concentrations (2.8 mmol/L), the amount of insulin released to the incubation medium was close to 16% of the total insulin content. These values are significantly higher than those found in humans (Karam et al., 1974), rats (Rafacho et al., 2009), and mice islets (Ullrich et al., 2005), whose secrete around 1–2% of the total insulin islet content. The same pattern was observed when pancreas fragments were incubated with 5.6 and 8.3 mmol/L glucose (29 and 34% of total insulin content, respectively) (Protzek et al., 2010). These findings could indicate that *A. lituratus* β cells are more responsive to glucose than other mammals.

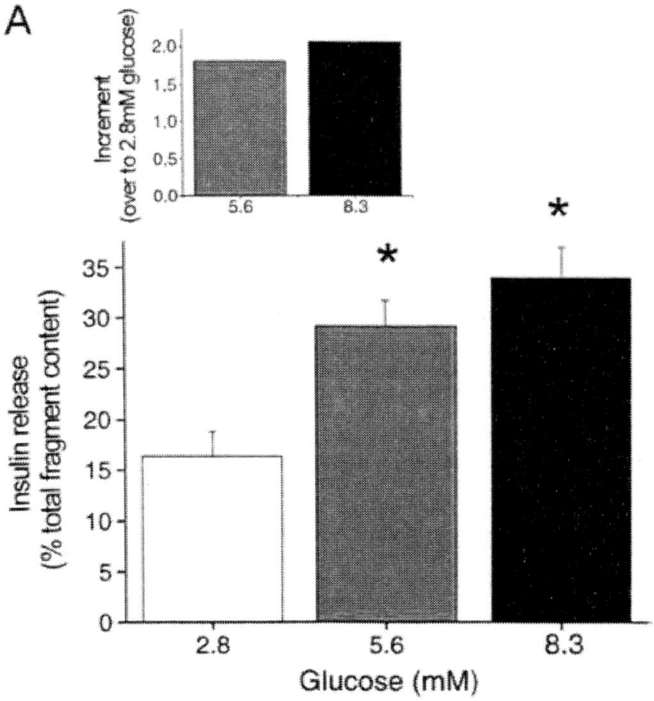

Figure 6. Insulin secretion from pancreas fragments of *A. lituratus*. Insulin response to low glucose (2.8 mmol/L), 5.6 mmol/L glucose and stimulatory 8.3mmol/L glucose levels. Increment on insulin secretion from 2.8 to 5.6 or 8.3 mmol/L glucose (upper figure). Values represent mean±SEM. n=6 fragments from two independent experiments. *Significantly different vs. 2.8 mmol/L. $p<0.05$.

The serine/threonine kinase Akt is a key protein of the insulin signalization implicated in glucose homeostasis, cell growth, differentiation, and survival (Cho et al., 2001; Downward, 1998; Schenck et al., 2008). Phosphorylated Akt participates in GLUT4 translocation to plasma membrane (Watson and Pessin, 2007), increasing the glucose influx in muscle and adipose tissues (Kido et al., 2001; Saltiel and Kahn, 2001).

Immunoblotting experiments analyzing Akt phosphorylation after insulin administration indicated that insulin signaling in skeletal muscle is more effective than in hepatic tissue (Protzek et al., 2010). These findings also show that *A. lituratus* pectoral muscles not only respond to insulin, but its signaling pathway is similar to that observed in many vertebrates, such as humans (Holness et al., 2000), rats (Rafacho et al., 2007), mice (Hult et al., 2009) and fish (Moon, 2001). Albeit less striking than in muscle, we observed an increase in hepatic Akt phosphorylation level after insulin administration, in accordance with the response observed for other vertebrates (Ropelle et al., 2009) corroborating the high ability of *A. lituratus* in storing liver glycogen (Pinheiro et al., 2006).

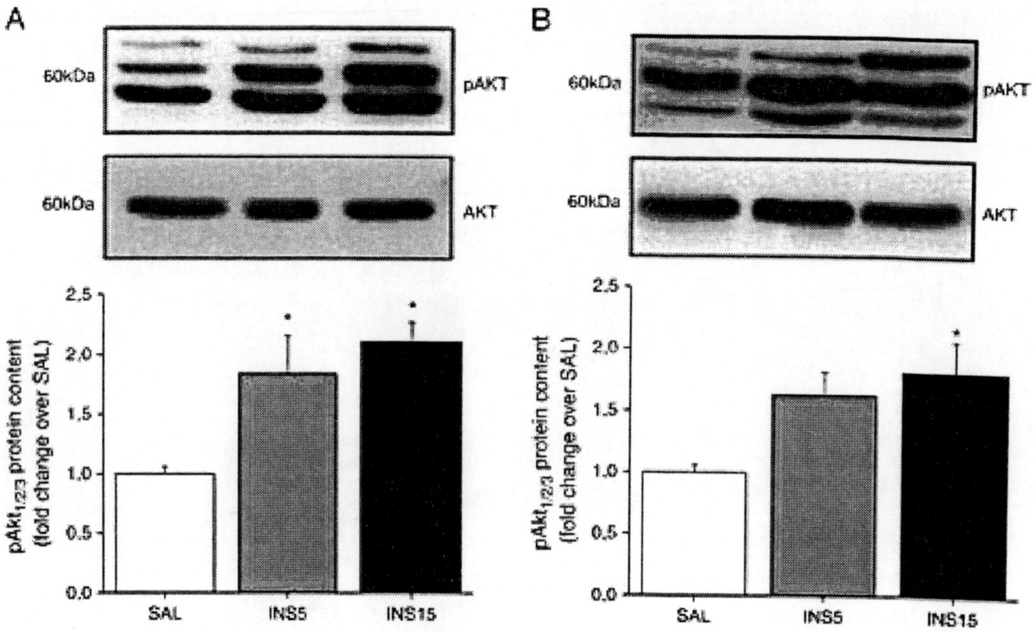

Figure 7. Protein content of total Akt and phosphorylated Akt in muscle and hepatic tissues. Total and phosphorylated levels of Akt in pectoral muscle (A) and hepatic tissue (B) from fed *A. lituratus* after 5 (INS5, n=3) or 15 min (INS15, n=3) of insulin administration (0.56 U/kg). Values represent mean±SEM of 4 independent bats. *significantly different vs. saline (SAL, n=3). $p<0.05$. (AOU) =optical arbitrary units.

When 12h-fasted *A. lituratus* was submitted to an intraperitoneal glucose tolerance test (*ip*GTT, 2 g/kg) there was no increase in blood glucose after 30 min of glucose injection (Protzek et al., 2010). Only 3 g of glucose/kg b.w. was able to elicit a significant increase in blood glucose after 30 min. of the injection (Figure 8A). The concentration of 3 g of glucose/kg b.w. is higher than that commonly used for rats (2 g/kg) (Rafacho et al., 2007) and mice (Fujiwara et al., 2007) during an *ip*GTT. These results suggest an efficient uptake of glucose by insulin-responsive tissues (mainly muscle, live and adipose tissue) and could reflect a well coupled between circulating insulin level and its action. The rapid restoration of blood glucose to basal levels strengthens the high glucose tolerance in *A. lituratus*, reflectiong on its ability to maintain glycemic control even after a carbohydrate rich meal and is in line with the high hepatic glycogen stores found in this species (Pinheiro et al., 2006).

The ability of *A. Lituratus* to handle a large influx of carbohydrates is not observed in the other fruit-eating bat species, *R. aegyptiacus*. When *R. aegyptiacus* was submitted to a glucose overload its blood glucose levels reached values 5-fold higher than those observed in rats and mice (Keegan, 1977). This finding seems to indicate that *R. aegyptiacus* is not as tolerant to glucose as *A. lituratus*. Normal adult rats exhibit an increase in blood glucose after 30 min of 2 g/kg of intraperitoneal glucose load (Rafacho et al., 2007). Wild-type, insulin hypersensitive (Fujiwara et al., 2007) and taurine supplemented mice (Ribeiro et al., 2009) also show a significant elevation in blood glucose after 30 min of 1.5 or 2 g/kg of intraperitoneal glucose load. It seems that *A. Lituratus* is more tolerating to glucose than most other mammals studied until now.

A. lituratus seems also to present high peripheral insulin sensitivity as judged by intraperitoneal insulin test (*ip*ITT - 0.56U/kg of insulin), which was able to induce a sharp and fast decrease in blood glucose levels (Protzek et al., 2010). Mice (Fujiwara et al., 2007) or rats (Rafacho et al., 2007) challenged with higher insulin doses (2U/kg) exhibited a slower glucose disappearance than observed in *A. lituratus* (Protzek *et al.*, 2010). Thus *A. lituratus* seems to be more tolerant to glucose and more sensitive to insulin than rodents as mice and rats.

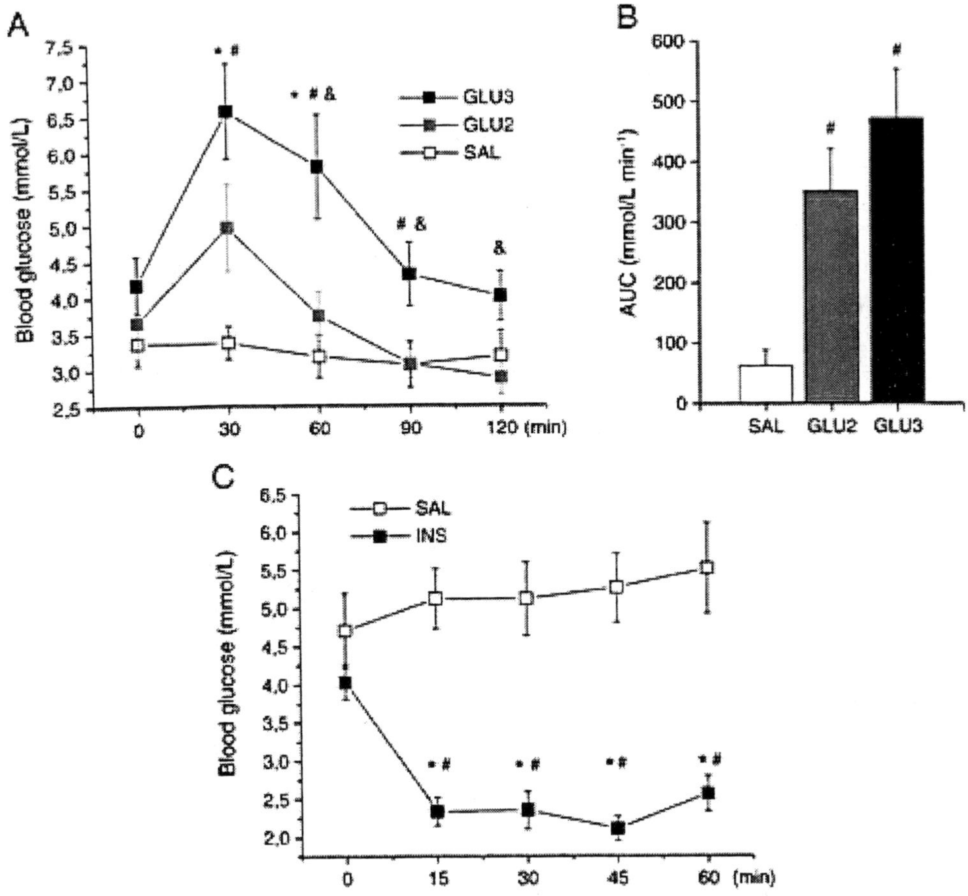

Figure 8. Peripheral sensitivity to glucose and insulin. Glycemic values under intraperitoneal glucose tolerance test (*ip*GTT) in 12 h-fasted (F12) *A. lituratus* after 2 (GLU2, n=14) or 3 g/kg b.w. (GLU3, n=15), or saline (SAL, n=14) administration (A). Area-under-the-curve (AUC) for *ip*GTT (B). Blood glucose values during intraperitoneal insulin tolerance test (*ip*ITT) in fed bats after insulin (0.56 U/kg, b.w.) (INS, n=17) or SAL (n=17) administration (C). Values represent mean±SEM. n=14 for SAL and GLU2 in A and B, 15 for GLU3 in A and B and 17 for SAL and INS in C. #significantly different vs. SAL, and vs. GLU2 and *vs. T0. $p<0.05$.

Another interesting finding in *A. lituratus* was that levels of circulating insulin, were similar between fed and fasting bats (Protzek et al., 2010). This unexpected result could mean that a regular meal did not induce an increase in blood glucose capable to stimulate insulin secretion sufficient to increase its circulating levels, and/or that the amount of circulating insulin is quite enough to deal with the glucose absorbed after a regular meal. Data obtained from *ip*GTT and measurements of plasma insulin in fed and fasting states seem to corroborate the hypotheses above.

It is noteworthy that literature shows that hyperinsulinemia always characterizes insulin resistance (Rafacho et al., 2008b), but this does not seem to be the case of *A. lituratus*, since many results cited above appear to indicate that this species exhibits high glucose tolerance and high insulin sensitivity.

However, we cannot rule out the possibility that an increase in insulin secretion occurred immediately after meal ingestion and absorption and we were unable to detect it due the delay between ingestion and blood collection, place of collection and/or a high inactivation of insulin during its first passage through the liver.

CONCLUSION

All results described here showing that *Artibeus lituratus* presents a rapid and efficient digestion of an overload of carbohydrate from the diet, a high sensitivity to glucose and to insulin and a large number of β-cells scattered throughout the pancreas seems to reveal that this species of fruit-eating bat, constantly submitted to high influx of glucose from its carbohydrate-rich diet, presents physiological adaptations that guarantee a proper control of glucose homeostasis.

REFERENCES

Bani Sacchi T, Bani D. 1985. New views on the identification of the various cell types in the pancreatic islets of the rat. An ultrastructural and morphometrical study. *Acta Anat* (Basel) 122: 1-17.

Beardsall K, K.Yuen, Willians R, Dunger D. 2006. Applied physiology of glucose control. *Curr. Paediatr* 16: 434-438.

Bernal-Mizrachi E, Wen W, Stahlhut S, Welling CM, Permutt MA. 2001. Islet beta cell expression of constitutively active Akt1/PKB alpha induces striking hypertrophy, hyperplasia, and hyperinsulinemia. *J Clin Invest* 108: 1631-1638.

Berne RM, Levy MN. 2009. Physiology (sixth edition). Rio de Janeiro: Guanabara koogan.

Boschero AC. 1996. Acoplamento excitação-secreção nas células beta pancreáticas (revisão). *Arquivos Brasileiros de Endocrinologia e Metabologia* 40: 149-155.

Cabrera O, Berman DM, Kenyon NS, Ricordi C, Berggren PO, Caicedo A. 2006. The unique cytoarchitecture of human pancreatic islets has implications for islet cell function. *Proc Natl Acad Sci U S A* 103: 2334-2339.

Caviedes-Vidal E, McWhorter TJ, Lavin SR, Chediack JG, Tracy CR, Karasov WH. 2007. The digestive adaptation of flying vertebrates: high intestinal paracellular absorption compensates for smaller guts. *Proc Natl Acad Sci U S A* 104: 19132-19137.

Caviedes-Vidal E, Karasov WH, Chediack JG, Fasulo V, Cruz-Neto AP, Otani L. 2008. Paracellular absorption: a bat breaks the mammal paradigm. *PLoS One 3:* e1425.

Cho H, Thorvaldsen JL, Chu Q, Feng F, Birnbaum MJ. 2001. Akt1/PKBalpha is required for normal growth but dispensable for maintenance of glucose homeostasis in mice. *J Biol Chem* 276: 38349-38352.

Craik JD, Markovich D. 2000. Rapid GLUT-1 mediated glucose transport in erythrocytes from the grey-headed fruit bat (Pteropus poliocephalus). *Comp Biochem Physiol A Mol Integr Physiol* 126: 45-55.

Czech MP, Corvera S. 1999. Signaling mechanisms that regulate glucose transport. *J. Biol. Chem.* 274.

Deeney JT, Prentki M, Corkey BE. 2000. Metabolic control of beta-cell function. *Semin Cell Dev Biol* 11: 267-275.

Downward J. 1998. Mechanisms and consequences of activation of protein kinase B/A. *Curr Opin Cell Biol* 10: 262-267.

Felig P. 1979. *Starvation Endocrinol.* New York: Grune and Stratton.

Freitas MB, Welker AF, Millan SF, Pinheiro EC. 2003. Metabolic responses induced by fasting in the common vampire bat Desmodus rotundus. *J Comp Physiol B* 173: 703-707.

Fujiwara Y, Hiroyama M, Sanbe A, Aoyagi T, Birumachi J, Yamauchi J, Tsujimoto G, Tanoue A. 2007. Insulin hypersensitivity in mice lacking the V1b vasopressin receptor. *J Physiol* 584: 235-244.

Gazola VA, Garcia RF, Hartmann EM, Barrena HC, Albuquerque GG, Souza HM, Bazotte RB. 2007. Glycemia recovery with oral amino acid administration during experimental short-term insulin-induced hypoglycemia. *J Diabetes Complications* 21: 320-325.

Hill MM, Clark SF, Tucker DF, Birnbaum MJ, James DE, Macaulay SL. 1999. A role for protein kinase Bbeta/Akt2 in insulin-stimulated GLUT4 translocation in adipocytes. *Mol Cell Biol* 19: 7771-7781.

Holness MJ, Langdown ML, Sugden MC. 2000. Early-life programming of susceptibility to dysregulation of glucose metabolism and the development of Type 2 diabetes mellitus. *Biochem J* 349 Pt 3: 657-665.

Hult M, Ortsater H, Schuster G, Graedler F, Beckers J, Adamski J, Ploner A, Jornvall H, Bergsten P, Oppermann U. 2009. Short-term glucocorticoid treatment increases insulin secretion in islets derived from lean mice through multiple pathways and mechanisms. *Mol Cell Endocrinol* 301: 109-116.

Kahn CR. 1985. The molecular mechanism of insulin action. *Annu Rev Med* 36: 429-451.

Karam JH, Grodsky GM, Ching KN, Schmid F, Burrill K, Forsham PH. 1974. "Staircase" glucose stimulation of insulin secretion in obesity. Measure of beta-cell sensitivity and capacity. *Diabetes* 23: 763-770.

Keegan DJ. 1977. Aspects of the assimilation of sugar by Rousettus aegyptiacus. *Comp. Biochem. Physiol.* 58a: 349-352.

Keegan DJ, Modinger R. 1979. Microvilli of the intestinal mucosal cells of *Rousettus aegyptiacus*. *S. Afr. J. Zool* 14: 223-220.

Kettelhut IC, Foss MC, Migliorini RH. 1980. Glucose homeostasis in a carnivorous animal (cat) and in rats fed a high-protein diet. *Am J Physiol* 239: R437-444.

Kido Y, Nakae J, Accili D. 2001. Clinical review 125: The insulin receptor and its cellular targets. *J Clin Endocrinol Metab* 86: 972-979.

Kim A, Miller K, Jo J, Kilimnik G, Wojcik P, Hara M. 2009. Islet architecture: A comparative study. *Islets* 1: 129-136.

Klover PJ, Mooney RA. 2004. Hepatocytes: critical for glucose homeostasis. *Int J Biochem Cell Biol* 36: 753-758.

Michelmore AJ, Keegan DJ, Kramer B. 1998. Immunocytochemical identification of endocrine cells in the pancreas of the fruit bat, Rousettus aegyptiacus. *Gen Comp Endocrinol* 110: 319-325.

Moon TW. 2001. Glucose intolerance in teleost fish: fact or fiction? *Comp Biochem Physiol B Biochem Mol Biol* 129: 243-249.

Nordlie RC, Foster JD, Lange AJ. 1999. Regulation of glucose production by the liver. *Annu Rev Nutr* 19: 379-406.

Patti ME, Kahn CR. 1998. The insulin receptor--a critical link in glucose homeostasis and insulin action. *J Basic Clin Physiol Pharmacol* 9: 89-109.

Pinheiro EC, Taddei VA, Migliorini RH, Kettelhut IC. 2006. Effect of fasting on carbohydrate metabolism in frugivorous bats (Artibeus lituratus and Artibeus jamaicensis). *Comp Biochem Physiol B Biochem Mol Biol* 143: 279-284.

Protzek AO, Rafacho A, Viscelli BA, Bosqueiro JR, Cappelli AP, Paula FM, Boschero AC, Pinheiro EC. 2010a. Insulin and glucose sensitivity, insulin secretion and beta-cell distribution in endocrine pancreas of the fruit bat Artibeus lituratus. *Comp Biochem Physiol A Mol Integr Physiol* 157: 142-148.

Rafacho A, Giozzet VA, Boschero AC, Bosqueiro JR. 2008a. Functional alterations in endocrine pancreas of rats with different degrees of dexamethasone-induced insulin resistance. *Pancreas* 36: 284-293.

Rafacho A, Roma LP, Taboga SR, Boschero AC, Bosqueiro JR. 2007. Dexamethasone-induced insulin resistance is associated with increased connexin 36 mRNA and protein expression in pancreatic rat islets. *Can J Physiol Pharmacol* 85: 536-545.

Rafacho A, Ribeiro DL, Boschero AC, Taboga SR, Bosqueiro JR. 2008b. Increased pancreatic islet mass is accompanied by activation of the insulin receptor substrate-2/serine-threonine kinase pathway and augmented cyclin D2 protein levels in insulin-resistant rats. *Int J Exp Pathol* 89: 264-275.

Rafacho A, Cestari TM, Taboga SR, Boschero AC, Bosqueiro JR. 2009. High doses of dexamethasone induce increased beta-cell proliferation in pancreatic rat islets. *Am J Physiol Endocrinol Metab* 296: E681-689.

Reis NR, Peracchi AL, Pedro WA, Lima IP. 2007. *Morcegos do Brasil*. Londrina.

Ribeiro RA, Bonfleur ML, Amaral AG, Vanzela EC, Rocco SA, Boschero AC, Carneiro EM. 2009. Taurine supplementation enhances nutrient-induced insulin secretion in pancreatic mice islets. *Diabetes Metab Res Rev* 25: 370-379.

Ropelle ER, Pauli JR, Cintra DE, Frederico MJ, de Pinho RA, Velloso LA, De Souza CT. 2009. Acute exercise modulates the Foxo1/PGC-1alpha pathway in the liver of diet-induced obesity rats. *J Physiol* 587: 2069-2076.

Rutter GA. 2001. Nutrient-secretion coupling in the pancreatic islet beta-cell: recent advances. *Mol Aspects Med* 22: 247-284.

Saltiel AR, Kahn CR. 2001. Insulin signalling and the regulation of glucose and lipid metabolism. *Nature* 414: 799-806.

Schenck A, Goto-Silva L, Collinet C, Rhinn M, Giner A, Habermann B, Brand M, Zerial M. 2008. The endosomal protein Appl1 mediates Akt substrate specificity and cell survival in vertebrate development. *Cell* 133: 486-497.

Tirone TA, Brunicardi FC. 2001. Overview of glucose regulation. World J Surg 25: 461-467.

Tracy CR, McWhorter TJ, Korine C, Wojciechowski MS, Pinshow B, Karasov WH. 2007. Absorption of sugars in the Egyptian fruit bat (Rousettus aegyptiacus): a paradox explained. *J Exp Biol* 210: 1726-1734.

Ullrich S, et al. 2005. Serum- and glucocorticoid-inducible kinase 1 (SGK1) mediates glucocorticoid-induced inhibition of insulin secretion. *Diabetes* 54: 1090-1099.

Van Der Westhuizen J. 1976. The feeding pattern of the fruit bat Rousettus aegyptiacus. *S. Afr. J. Med. Sci.* 41: 271-278.

In: Bats: Biology, Behavior and Conservation
Editors: J. L. Zupan and S. L. Mlakar, pp. 315-326
ISBN 978-1-61209-749-7

Chapter 10

THE CONTRIBUTION OF FOSSILS TO THE RECONSTRUCTION OF BAT POPULATION DYNAMICS

***Juan Manuel López-García*[*1], *Ivan Horacek*[≠2] and *Paloma Sevilla*[‡3]**
[1]Institut de Paleoecologia Humana i Evolució Social (IPHES),
Universitat Rovira i Virgili (URV). Campus Catalunya, Tarragona, Spain
[2]Department of Zoology, Charles University, Czech Republic
[3]Departamento de Paleontología, Facultad de Ciencias Geológicas (UCM) e Instituto de Geología Económica (CSIC), Madrid, Spain

ABSTRACT

Modern research in conservation biology of bats permits to establish the status of these mammals on the basis of their demography and population dynamics. Most concern goes to those species that are endangered or are more vulnerable due to evidence of the decline in their populations or the degradation of their natural habitats. Taking Europe as an example, nearly a third of its bat species have been decreasing in numbers during the last decade according to the IUCN Red List of Endangered Species. Researchers control population numbers year after year and try to identify the reason for the changes they observe. However, short term fluctuations in population numbers are common in all species, and the importance of these may be difficult to interpret. Tracking population numbers during longer periods of time might be the best means to establish real trends and to identify the possible causes of these changes. The data obtained from the bat remains found in fossil and sub-fossil localities provide the means for this long-term tracking of populations, some good examples available in published papers. In this chapter we present an overview of how bat fossils can be useful in conservation biology

* Institut de Paleoecologia Humana i Evolució Social (IPHES), Universitat Rovira i Virgili (URV). Campus Catalunya, Avinguda Catalunya 35, E-43002 Tarragona, Spain. jmlopez@iphes.cat.
≠ Department of Zoology, Charles University, Vinicna 7, CZ 128 44 Praha 2, Czech Republic. ivan.horacek@seznam.cz.
‡ Departamento de Paleontología, Facultad de Ciencias Geológicas (UCM) e Instituto de Geología Económica (CSIC), C/ Jose Antonio Novais 2, 28040 Madrid, Spain. psevilla@geo.ucm.es.

with some reference examples, and the limits of the use of these mammals for this purpose will be discussed.

Keywords: fossil Chiroptera, Quaternary population dynamics, past human-bat interactions

INTRODUCTION

Bats, with their more than 50 million years of evolution have been subject to fluctuations in their diversity and abundance finally placing them as the second most diverse group of mammals. Their striking similarity in morphology is surprising considering the high number of known species, some still being described. The long duration of bat taxa compared to that of other small mammals may lead to the idea that they are comparatively resistant to environmental changes and less endangered than other groups. However, this diversity is the consequence of a high specialisation to environmental factors that have had less influence in their general morphology, thus their similarity. But as highly specialised organisms, they are sensible to environmental change, even if it does not seem evident. Specialists working with recent populations are continuously detecting changes in bat numbers, and in a good number of cases, unfortunately, towards a decrease in abundance. Only in Europe, more than a half of the species with a known trend in population numbers are decreasing. Actions to stop these trends will be more effective if they are taken under a better and deeper understanding of the dynamics of their populations including long-term fluctuations. The fossil record can provide key information about the response of extant species to the important environmental changes that took place in the Quaternary prior to human impact and thus foresee the success of the actions to be taken.

In the following sections we shall explain the usual procedure followed when studying fossil bats with the aim of obtaining past ecological information. We shall start explaining briefly the modifications that make the fossil assemblages (taphocoenosis) different from the original biocoenosis, followed by the current methods to identify bat species in fossil assemblages and how ecological information may be inferred from these remains, with examples chosen from published papers. Finally, a few examples of particular trends inferred from the Quaternary record of recent bat species in Europe are given.

FROM A LIVING COMMUNITY TO A FOSSIL ASSEMBLAGE

Bats are included within an informal group called "small mammals", which as a general reference, do not exceed 5kgweight (Merritt, 2010). Separation between "large mammals" and "small mammals" is very useful in palaeontology, since the fossilization of mammals usually differs depending on their size. Two main types of localities with bat fossils may be distinguished, those originated in lake or fluvial environments and those found in karsts (Figure1). Bat remains are less common in the first, whereas fissure fillings and cave deposits of karst environments are the most favourable sites to find bat fossils (Sigé et al., 1983). The reason for this difference is mainly that in fluvial and lake deposits, the small mammals found in the fossil assemblages are usually the result of predation, and are the fossilized bones and other non-digestible parts of ingested prey that are ejected by the predator as pellets or scats. Patterns in the degree of digestion and fracture inflicted on these remains depends on the type of predator, varying greatly according to different categories in which they are grouped according to these patterns, enabling frequently not only to confirm predation but also to identify the type of predator (Andrews, 1990). No predators are specialised on hunting bats, which are only occasional prey under opportunistic conditions. For this reason they are usually poorly represented in fluvial and lacustrine localities.

Guadix-Baza Basin-Orce   Sima del Elefante-Atapuerca

Figure 1. Examples of the two types of localities where bat fossils may be found. Left: ancient lake deposits of the Guadix-Baza basin (southern Spain), where small mammal fossils resulting from predation are found. (Photo: H.-A. Blain) Right: "Sima del Elefante" karstic deposits (Atapuerca, Spain), with abundant bat fossils both resulting from predation and from natural death within the karstic system. (Photo: J. Rubio)

On the other hand, bats are common in caves and rock fissures which are used as roosts; in these, many species roost during the day, hibernate and breed, sometimes grouping in high numbers. Natural death may overcome a bat while roosting, and after the carcass has undergone decay and is covered by sediment, the skeleton has a high probability of being preserved and ending up as a fossil. As a result, caves and fissures in karsts are the best places for finding bat fossils.

Preservation of Bat Fossils

Examples of complete or practically complete bat skeletons are well-known, usually found in "Lagerstatten" (fossil sites with exceptional preservation), such as Messel in Germany (Habersetzer et al., 1987) or the Green River Formation in Wyoming (Jepsen, 1966). The advantages of this exceptional preservation is that not only all anatomical elements of the skeleton are available, providing data to establish precise taxonomic relationships and evolutionary insight, but also ecological information otherwise difficult to obtain from non-exceptional fossils. Direct information about features such as prey, type of flight or echolocation development is obtained from these fossils (Habersetzer et al., 1994; Simmons et al., 1998, 2008).

Figure 2. Left: Bone breccia from Podlesice (Poland) showing a high density of bat remains. Right: Bat fossils collected in surface sediments of the Cave of Azokh (Ngorno Karabagh, Caucasus). The most common type of remains can be observed – mandibles, rostra, scapula and humeri. (Photos: P.Sevilla).

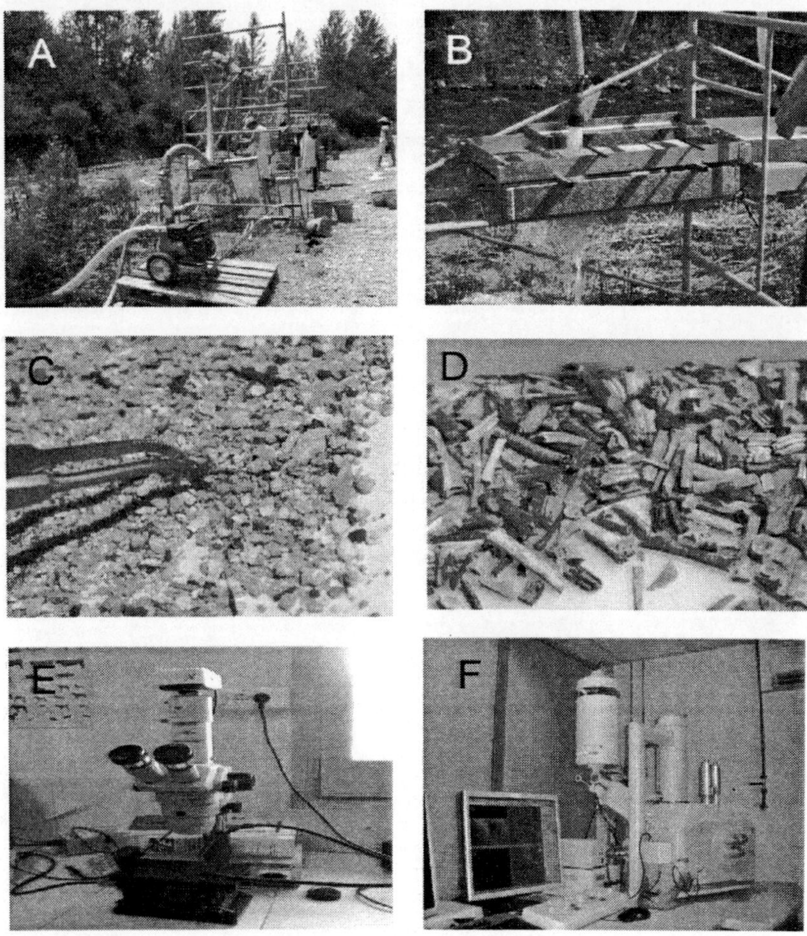

Figure 3. A: water screening in the field; B: a detail of the set of screen meshes that fraction sediment with fossils according to size; C: fossils being picked from the sediment after screen-washing and drying; D: a detail of the picked sediment with fossils of different small vertebrates; E: binocular microscope used for picking the smaller sized fraction and studying the material; F: Environmental Scanning Electron Microscope used to take high resolution pictures without damaging the fossils. (Photos: S. Bañuls).

However, as with other vertebrates, disarticulation of the skeleton usually takes place early during fossilization, isolated bones and teeth being the most common way in which bat fossils are found. The action of insects and scavengers, the trampling of other animals in the cave or the action of water circulating within the karstic system usually contribute to the dispersal of the bones after carcass decay. These processes also lead to the loss of the smaller and the most delicate bones. Consequently, mandibles, maxillae, teeth, and humeri are the most frequent remains by which bats are represented in a fossil assemblage (Figure2). Other parts of the skeleton such as scapulas, pelves, femurs, cochlea and fragments of phalanges are also variably represented in these assemblages. Adequate collecting methods (Figure3) are essential to prevent the loss of information; the use of 0.5 mm screens when sieving the sediments in the fossil localities guarantees that the bones of smaller species or deciduous teeth are not discarded; otherwise, a biased representation towards the more common or the larger species may occur, leading to inaccurate interpretations.

TAXONOMIC DETERMINATION OF BAT FOSSILS

Bat bones are easy to recognise from other small mammal bones in a fossil assemblage; adaptations to flight are evident in the light and slender long bones; the teeth present the typical pattern expected from their diet, insectivorous for the majority of bats. However, identification at generic and species level is not always possible, and depends on the particular elements of the skeleton available in the assemblage.

Among the most specialised parts of the skeleton of bats are the wings; its bones are typically elongate and light, easy to distinguish from the bones of other mammals. The long, slender phalanges, however, cannot be related to a particular species when found in an assemblage. On the other hand, the elbow joint has developed a specialised locking device which differs greatly between different groups of bats (Felten et al, 1973; Hand et al., 2009), enabling to identify the genera or species represented in an assemblage from their humeri .

Cranial and mandibular remains are also relatively resistant to the destructive processes during fossilization. If the tooth-row is preserved complete, the dental formula helps to determine the genus; the shape of particular teeth, the difference in size between the anterior premolars, degree of compression in the tooth row as well as general size enables species determination. Isolated teeth are the most common fossils of bats. In these cases species determination relies on the preservation of the most characteristic dental pieces (mainly the first two upper molars, and to a lesser extent, lower molars, fourth premolars and canines). Several papers dealing with genus and species determination of extant European species based on dental morphology have been published, enabling determination of this type of material in fossil localities (Menu and Sigé,1971; Sevilla, 1986; Menu and Popelard, 1987). Direct comparison with recent material is nevertheless always useful, since some species have undergone changes in size during the Pleistocene.

Types of Assemblages

As previously mentioned, bat fossils can be found in localities originated in different environments, but we shall focus our attention on the characteristics of assemblages developed within karstic systems, where their fossils are frequently abundant, leaving aside fluvial and lake deposits, where bat remains are less common. The bat fossils found in karstic localities usually, though not always, belong to animals that died in the cave (a "thanatocoenosis"). A "cave assemblage" is usually characterised by one or two dominant species and a variable number of accompanying species represented in lower numbers. Breeding colonies may be recognised when deciduous teeth are found in the sediment, from the presence of remains presenting a low degree of ossification or when unworn teeth are abundant. The guano accumulating under summer colonies with high numbers of individuals is very acidic and destroys bones (Kowalski, 1995). In these cases, the presence of this type of colonies may be inferred from the characteristics of the sediment – modified by the guano – in which material that might identify the species constituting the colony is rarely preserved. Remains of hibernating communities are usually recognised by lower share of individuals with intermediate degrees of wear in the teeth as compared to heavily worn (from old animals) or lightly worn teeth (from yearlings).

It is rather frequent to find assemblages in which bat remains are found together with those of other small vertebrates that do not live in caves. These assemblages are commonly interpreted as originated from the accumulation of the remains of the diet of a predator, evidenced by breakage patterns and digestion, and referred to as "prey assemblages" (Andrews, 1990) . Usually bats are rare components of pellets and scats of predators; evidence of digestion in the material would confirm their remains as belonging to the prey assemblage, as recognised in the material from Matuzka Cave (Rossina et al., 2006). Otherwise, the bat fossils might belong to cave bats that got mixed with the other vertebrate remains (a "mixed assemblage"). Bats coming from prey assemblages sometimes provide useful information about species uncommon in caves and contribute to our knowledge of the occurrence of these species in the past; the rare records of *Tadarida*, *Nyctalus* or *Eptesicus* species in Pleistocene and Holocene localities probably come from prey assemblages.

Inferring Bat Density and Diversity

Because of the processes involved in fossilization, the fossil record gives few opportunities to infer directly how many animals were present in a living community from those that died and became part of a fossil assemblage. In the case of bats found in a cave assemblage, where the possibility of bias towards a particular species is not probable, high numbers from a particular species can be taken as indicative of the presence of numerous individuals of that species. Since vertebrate skeletons usually are preserved disarticulate, a single individual may be represented in the assemblage by many different fossils. The use of indices such as the Minimum Number of Individuals (MNI) is generally used to counter the bias we would obtain if abundance were directly inferred from the number of identified remains (NISP) of a particular species (Lyman, 1994; Marshall and Pilgram, 1993). Though MNI does not give an absolute value of abundance, as we might obtain when working with living animals (Plug and Plug, 1990), it is a useful way for comparing species representation at different moments within the same locality, between different fossil localities, or to establish changes in abundance through time. Moreover, since fossil assemblages with bats are rarely constituted by a single species, MNI counts offer the possibility to establish reliable values of relative abundances of the different species that constituted a community.

CHANGES IN INFERRED ABUNDANCES ALONG A SEQUENCE

Sedimentation within caves and fissure filling can take thousands of years during which important climatic changes can have taken place. Temperature and humidity factors determine differences in the sediments of the cave, as well as in their fossil content. Through hibernation and flight, which provide bats with the means to face adverse conditions, these mammals are commonly considered as less sensitive to climate changes than other small vertebrates. However, many bat species are strongly dependent of particular foraging habitats, mainly determined by the type of vegetation in which their preferred prey can be found.

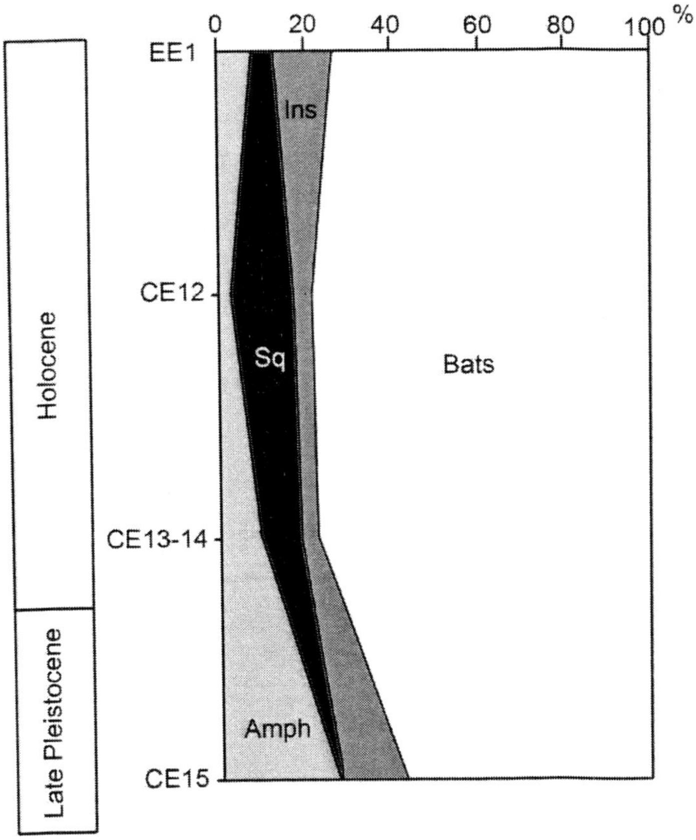

Figure 4. Variation in the relative abundances of bats in the sequence of Cova Colomera (Catalonia, Spain). The share of representation of bats in the assemblage increases significatively in layers CE13-14 and CE-12 from the early Holocene. The share of other small vertebrates in the assemblages are indicated by: Amph- amphibians; Sq- reptiles; Ins-Insectivores.

Climate changes in the past have caused modifications in local vegetation and secondarily have influenced the distribution and abundance of bats. Evidence of this is recorded in localities where a sedimentary sequence covering different climatic phases is preserved. For instance, during the cold climatic phases of the late Pleistocene a good part of Spain was covered cold steppe vegetation (Sánchez Goñi et al, 2002); with the climate improvement at the beginning of the Holocene, forest areas became more abundant and a higher variety of vegetation types extended through Iberia (Riera Mora, 2006). This change in the vegetation is encompassed with an increase in the abundance and diversity of bats recorded in the sedimentary sequences of caves such as El Mirador (Burgos, Spain) and Cova Colomera (Catalonia, Spain) which span from the Late Pleistocene to the early Holocene. Whereas in the Late Pleistocene layers remains of only a few widely tolerant species occur, such as *Myotis myotis*, *Miniopterus schreibersii* and *Pipistrellus pipistrellus*, the Holocene levels are enrichened with termophilous and forest species such as *Rhinolophus euryale*, *R. mehelyi*, and *Plecotus* species (López-García, 2008; López-García et al., 2008; Oms et al., 2009).

Similar examples are known in other European localities, where smaller sized *Myotis* species dominate the assemblages. For instance, the Komarowa Late Pleistocene-Holocene sequence in Poland (Ochman, 2003) is characterised by a change from the dominance of *M. daubentonii* in the Pleistocene layers to *M. bechsteinii* in the Holocene layers, accompanied by and increase in the abundance and diversity of bat remains. Similarly, Postawa (2004) concluded after studying the bat fossils of 11 caves with Holocene sediments in Southern Poland, that changes in species composition showed a high correlation with the climatic fluctuations and vegetation changes that took place during the Atlantic and Sub-boreal phases.

CHANGES IN GEOGRAPHIC DISTRIBUTION

Deposits in caves and fissure fillings do not always consist of a sequence representing different sedimentary episodes under different environmental conditions. A single filling event may be present, giving no information about previous or posterior conditions. The information provided by a single assemblage is still useful for the study of the dynamics of bat population in the past, since the fossils provide evidence of the occurrence of those species in the area at that time. Occasionally, records of species out of their recent distribution are reported, providing information of wider distributions in the past, or shifts caused by climate change. Among available examples of these changes in distribution are those of *Myotis dasycneme*, presently distributed in northern Europe, but occurring in localities of the middle and late Pleistocene in France (Mein, 1975, Clot et Duranthon, 1990), northern and central Italy (Tata and Kotsakis, 2005, Argenti et al., 2008) or the Caucasus (Sevilla, in press). Additional examples are given by the records of *Eptesicus nilssonii* in the middle and late Pleistocene in France (Mein, 1975, Noel et al., 2008) or *Nyctalus lasiopterus* in the Late Pleistocene of Catalonia (Spain) (López-García 2008, López-García et al., 2009) (Figure5)

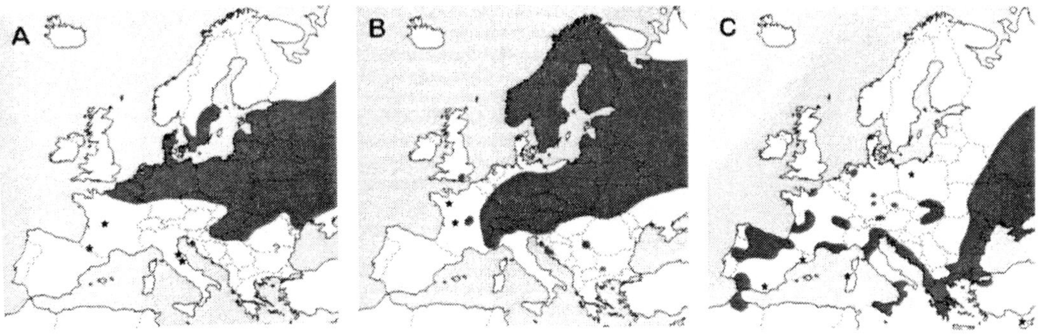

Figure 5. Examples of changes in the geographic distribution of bat species in the Pleistocene. Recent distributions are shaded in gray, records in fossil localities are marked by asterisks. A: *Myotis dasycneme*. B: *Eptesicus nilssonii*.. C: *Nyctalus lasiopterus*.

HUMAN INFLUENCE IN BAT ABUNDANCE AND DIVERSITY

Caves used as shelters by bats have been shared with humans in the past. In middle and late Pleistocene localities sedimentary sequences show a decline in bat abundances in the layers where there is evidence of intensive human occupation. Occasionally, seasonal use of the cave may be inferred; humans occupied the cave during the winter, leaving as evidence hearths, bones of large mammals with marks indicating butchering, and lithic implements, whereas bats sheltered in the cave during the summer, once free of human occupation. Such an example is described in the sequences of Denisova Cave in the Altai Mountains (Rossina, 2006), where Pleistocene levels previous to human occupation are characterised by an important representation of bat fossils in the cave with eight different species in the assemblages; coinciding with evidence of human presence in the cave during the late Pleistocene, bats numbers of all species decrease except for *Myotis blythii*, which probably only occupied the cave for breeding during summer, and established its winter colonies elsewhere. On thr other hand, the Matuzka Cave assemblages (Rossina et al., 2006) indicate regular and short visits of humans in the Upper Pleistocene, scarcely interfering with its bat communities, which varied influenced by climate and vegetation changes. Another example of Pleistocene human-bat interaction is described in the late Pleistocene layers of the Arene Candide Cave in Italy (Salari, 2010), in where abundance of bats is inversely correlated to large mammal remains (introduced in the cave by humans).

With the improvement of climate in the Holocene and abandonment of caves by humans, Holocene sequences in fossil localities show increased numbers of bat remains, with the few exceptions due to the use of rock shelters and small caves as sheepfolds. Along historic times many bat species have been in fact favoured by anthropogenic modification of landscape. Human constructions have provided new roosts, enabling species to colonise areas without natural shelters. The development of agriculture favoured the expansion of species foraging over low vegetation, but at the same time, the clearing of forest areas for agriculture has had a negative impact on the abundance of forest species. Evidence of these changes are recorded in the sediments of la Grotte Rochefort in North-Western France (Noel et al., 2008), with the lowermost excavated layers dating from the latest Pleistocene and the uppermost from the XIXth century.

Concluding Remarks

Since most of the bat species of the modern European fauna were already established in Europe approximately 2 Ma ago, the fossils preserved in Pleistocene and Holocene localities can provide a long-term insight of variations undergone by bat populations in abundance and distribution, providing a more complete basis to analyse and understand variations observed in recent populations. To this aim a good fossil record is necessary, covering different geographic areas in order to establish geographic variations and including well-dated sequences spanning periods under different environmental conditions to observe the consequences these variations had on bat populations. The relatively abundant data available from the European Quaternary has led to achievements such as the recognition of long-terms trends in certain species. Fossil evidence shows that some species common in Pleistocene and Holocene localities such as *Myotis myotis* or *Myotis nattereri*, common in the past, remain common at present, giving evidence of long-term stability in their populations. Other species, such as *Myotis bechsteinii* or *Miniopterus schreibersii*, also common in Pleistocene localities, are under a decreasing trend since the Holocene. Some uncommon species in Pleistocene localities, such as the *Nyctalus* species, *Tadarida teniotis*, or *Myotis capaccinii*, remain nowadays as rare species, indicating that, although uncommon, their populations are rather stable. Finally, examples of increasing numbers are found in species such as *Pipistrellus pipistrellus* or *Eptesicus serotinus*, rare in Pleistocene localities, but abundant nowadays, favoured by human activities.

Stable populations through the Pleistocene evidence ability to adapt to changing environments, whereas species that show sensibility to environmental changes, with varying representation in the fossil assemblages depending on climate or vegetation conditions are more vulnerable and must be the centre of special protection programs to prevent their extinction.

Acknowledgments

This chapter is part of projects CGL2009-07896/BTE and SGR2009-324. J.M.L.-G. has been supported by a postdoctoral grant from the Juan de la Cierva Subprogram (JCI-2009-04026), with the financial sponsorship of the Spanish Ministry of Science and Innovation. P.S. is a member of the GR58/08-910607 Research Group

References

Andrews, P. 1990. Owls, Caves and fossils. Predation, preservation and accumulation of small mammal bones in caves, with an analysis of the *Pleistocene Cave faunas from Westbury-sub-Mendip*, Somerset, UK. Natural History Museum. London.

Argenti, P., Kotsakis, T., Sabatini, F. 2008. *Myotis dasycneme* in a latest Pleistocene assemblage of Cittareale Cave (Rieti, Latium, Central Italy). *Geologica Romana*, 41, 117-123.

Clot, A., Duranthon, F. 1990. *Les Mammifères fossiles du Quaternaire dans les Pyrénées*. Museum d'Histoire naturelle de Tolouse. 159 pp.

Felten, H., Helfricht, A., Storch, G. 1973. Die Bestimmung der Europäischen Fledermausfaunen nach der distal epyphise des Humerus. *Senckenbergiana biologae* 54, 291-297.

Habersetzer, J., Storch, G. 1987. Klassifikation und funktionelle Flügelmorphologie paläogener Fledermäuse (Mammalia, Chiroptera).*Courier Forschungsinstitut Senckenberg,* 91, 117-150.

Habersetzer, J., Richter, G., Storch, G. 1994. Paleoecology of early middle Eocene bats from Messel, FRG. aspects of flight, feeding and echolocation. *Historical Biology*, 8(1-4), 235-260.

Hand, S.J., Weisbecker, V., Beck, R.M.D., Archer, M., Godthelp, H., Tennyson, A.J.D., Worthy, T.H. 2009. Bats that walk: a new evolutionary hypothesis for the terrestrial behaviour of New Zealand's endemic mystacinids. *BMC Evolutionary Biology,* 9,1471-2148.

IUCN 2010. IUCN Red List of Threatened Species. Version 2010. <*www.iucnredlist.org*>.

Jepsen, G.L.1966. Early Eocene Bat from Wyoming. *Science*, 154(3754), 1333-1339.

Kowalski, K., 1995. Taphonomy of bats (Chiroptera). *Geobios*, M.S. 18, 251-256.

López-García, J.M. 2008. *Evolución de la diversidad taxonómica de los micromamíferos en la Península Ibérica y cambios paleoambientales durante el Pleistoceno Superior*. PhD Thesis.

López-García J.M. Cuenca-Bescós G, Rosell J. 2008 Resultados del estudio de microvertebrados del Neolítico de la Cueva de El Mirador (Ibeas de Juarros, Sierra de Atapuerca, Burgos). In: Hernández MS, Solé JA, López JA (eds) Actas IV Congreso de Neolítico Peninsular. *Museo Arqueológico de Alicante, Alicante,* pp 338–344.

López-García, J.M., Sevilla, P. Cuenca -Bescós, G. 2009. New evidence for the greater noctule bat (*Nyctalus lasiopterus*) in the Late Pleistocene of western Europe. *Palevol,* 8,551-558

Lyman, R. L. 1987. On the analysis of Vertebrate Mortality Profiles: Sample Size, Mortality Type and Hunting Pressure. *American Antiquity*, 52 (1), 125-142

Marshall, F., Pilgram, T. 1993. NISP vs. MNI in Quantification of Body-Part Representation. *American Antiquity*, 58 (2), 261-269.

Mein, P. 1975. Les Chiroptères (Mammalia) du gisement Pléistocène moyen des Abimes de la Fage à Noailles (Corrèze). *Nouvelle Archives du Musée d'Histoire naturelle de Lyon,* 13, 57-67.

Menu, H., Popelard, J.B., 1987. Utilisation des charactères dentaires pour la détermination des vespertilioninés de l'ouest européen. *Le Rhinolophe* 4, 1-88.

Menu, H., Sigé, B., 1971. Nyctalodontie et myotodontie, importants caractères de grades évolutifs chez les Chiroptères entomophages. *Comptes Rendus de l'Academie des Sciences de Paris,* 272, 1735-1738.

Merritt, J.F. 2010. *The Biology of Small mammals.* John Hopkins University Press. Baltimore. 368 pp.

Noel, F., Chaut, J.-J. and Hinguant, S. 2008. 11000 ans d'histoire des chiroptères dans la Grotte Rochefort (Mayenne, France). In: Société française d'études et de protection des mammifères. *Symbioses* (n.s.) 21, 13-15.

Ochman, K. 2003. Late Pleistocene and Holocene bats (Chiroptera) from the Komarowa Cave (Cracow-Czestochowa Upland, Poland)- preliminary results. *Acta zoologica cracoviensia*, 46(1), 73-84.

Oms FX, Petit MA, Allué E, Bargalló A, Blain H-A, López-García JM, Martin P, Morales JI, Pedro M, Rodriguez A, Solé A (2009) Estudio trasndisciplinar de la fosa EE1 de la Cova Colomera (Prepirineo de Lleida): implicaciones domésticas y paleoambientales en el Bronce Antiguo del Noreste de la Península Ibérica. *Trabajos de Prehistoria* 66,123–144.

Plug, C., Plug, I. 1990. MNI Counts as Estimates of Species Abundance. *The South African Archaeological Bulletin,* 45 (151), 53-57.

Postawa, T., 2004. Changes in bat fauna during the Middle and Late Holocene as exemplified by Thanatocoenoses dated with 14C AMS from Kralów-Czestochowa Upland caves, Poland. *Acta Chiropterologica*, 6(2), 269-292.

Riera Mora, S. 2006. Cambios vegetales holocenos en la región mediterránea de la Península Ibérica: ensayo de síntesis. *Ecosistemas* 15(1):17-30.

Rossina, V.V., 2006. Bats as an Indicator of Human Activity in the Paleolithic, Using the Example of Denisova Cave, Northwestern Altai. *Paleontological Journal* 40 (4), 494-500

Rossina, V.V., Baryshnikov, G.F., Woloszyn, B.W., 2006. Dynamics of the Pleistocene bat fauna from the Matuzka Paleolithic site (Northern Caucasus, Russia) (Chiroptera). *Lynx*, 37, 229-240.

Salari, L. 2010. Late Glacial bats from the "M" layers of the Arene Candide Cave (Liguria, Italy). *Rivista italiana di Paleontologia e Stratigrafia*, 116(1), 119-138.

Sánchez Goñi, M.F., Cacho, I., Turon, J.-L., Guiot, J., Sierro, F.J., Peypouquet, J.-P., J.O. Grimalt, Shackelton, N.J. 2002. Synchroneity between marine and terrestrial responses to millennial scale climatic variability during the last glacial period in the Mediterranean region. *Climate Dynamics*, 19, 95-105.

Sevilla, P. 1986. Identificación de los principales quirópteros ibéricos a partir de sus dientes aislados. Valor sistemático de los caracteres morfológicos y métricos dentarios. *Doñana Acta Vertebrata*. 13, 111-130

Sigé, B., Legendre, S. 1983. L'Histoire des peuplements de chiroptères du bassin méditerranéen: l'apport comparé des remplissages karstiques et des dépôts fluvio-lacustres. *Mémoires de Biospéleologie*, 10, 209-225.

Simmons, N.B. Geisler, J.H. 1998. Phylogenetic Relationships of Icaronycteris, Archaeonycteris, Hassianycteris, and Palaeochiropteryx to Extant Bat Lineages, with Comments on the Evolution of Echolocation and Foraging Strategies in Microchiroptera. *Bulletin of the American Museum of Natural History*, 235, 1-182.

Simmons, N.B., Seymour, K.L., Habersetzaer, J., Gunnell, G.F. 2008. Primitive Early Eocene bat from Wyoming and the evolution of flight and echolocation. *Nature*, 451, 818-821.

Tata, C., Kotsakis, T. 2005. Italian fossil chiropteran assemblages: a preliminary report. *Geologia Alpina*, 2, 53-60.

In: Bats: Biology, Behavior and Conservation
Editors: J. L. Zupan and S. L. Mlakar, pp. 327-336
ISBN 978-1-61209-749-7

Chapter 11

MECHANISMS OF PROLONGED SPERM STORAGE IN FEMALE BATS

Amitabh Krishna[*] and Vikas Kumar Roy
Department of Zoology, Banaras Hindu University, Varanasi, India

Prolonged sperm storage is an adaptation to the reproductive cycle that is most highly developed in microchiroptera, where sperm may be stored up to ~198 days and still retain fertilizing capacity. The aim of this review is to describe the mechanism and molecules necessary for sperm to become efficiently stored in the female genital tract. Perpendicular orientation of sperm with their head towards the reproductive tract epithelium are characterized in all sperm-storing bat species so far investigated, suggesting that such relationships are an integral part of the mechanism of prolonged storage of sperm. Recent study suggests that the ultimate controls of the mechanism underlying sperm storage are hormonal. It has been demonstrated that sperm-storing female bats contain a high circulating level of androgen during the period of sperm storage. It is suggested that androgen creates a unique microenvironment that facilitates prolonged sperm storage. The identification of the specific molecules responsible for prolonged sperm storage may suggest a mechanism to keep sperm viable for a prolonged period for use in assisted reproductive techniques.

INTRODUCTION

The phenomenon of sperm storage by females is demonstrated in many species belonging to different animal groups, for example, insects (Baer et al., 2006; Adams and Wolfner, 2007), reptiles (Gist and Congdon, 1998; Yamanouye, 2004), birds (Das et al., 2006; Zhang, 2006), and mammals. Among mammals, prolonged sperm storage is an adaptation to the reproductive cycle that is most highly developed in hibernating microchiroptera from temperate latitude (Racey, 1975, 1979, 1982; Wang et al., 2008). Often, the female bat's ability to store sperm is an integral part of the species' reproductive strategy and can provide

[*] Corresponding Author: Dr. Amitabh Krishna, Professor, Reproductive Physiology Unit, Department of Zoology, Banaras Hindu University, Varanasi-221 005, India, E-mail: akrishna_ak@yahoo.co.in, Phone: 0542-6702527.

important advantages. Sperm storage typically occurs in females, whose gametogenic cycle is not synchronized with that of the male or in males that can retain spermatozoa beyond the cessation of spermatogenesis. Although, most examples of sperm storage are from vespertilionid and rhinolophid bats, a similar phenomenon has also been recorded in temperate zone molossid bats, *Mormopterus planiceps* (Crichton and Krutzsch, 1987). In temperate zone bats, mating activity commences in the autumn, spermatozoa are then stored in the female genital tract throughout the hibernating period. These sperm are used in the fertilization of ova released at ovulation after arousal in spring (Wimsatt, 1969; Racey, 1979). In bats, the period of sperm storage in the female genital tract ranges from 16 days, as in *Pipistrellus ceylonicus* (Gopalkrishna and Madhwan, 1971), to ~198 days, as in *Nyctalus noctula* (Racey, 1973). Sperm storage for relatively short periods has been reported in other mammal species (e.g dog, Doak et al., 1967; hare, Mortinet and Raynaud, 1974). Tropical bats generally store sperm for a much shorter period as compared with hibernating temperate zone bats. The interval between copulation and ovulation is up to five weeks in *Scotophilus heathi* (Krishna and Dominic, 1978) and 30-35 days in *P. mimus* (Krishna, 1984). The nature of specialization that has evolved in some bats to permit sperm survival for a prolonged period is still poorly understood. In the majority of female bats, sperm are stored in specialized organs or reservoirs, where sperm are retained for variable durations. This review describes different mechanisms proposed to explain the phenomenon of prolonged sperm-storage in female bats. The site of sperm storage and molecules necessary for sperm to become efficiently stored for prolonged period is also described briefly.

SITE OF SPERM STORAGE

Sperm storage in females is explained in part by the presence of specialized storage organs or regions in the female genital tract. The site of female sperm storage varies from periovarian spaces (Uchida and Mori, 1987) to the vagina, depending on the species (Racey, 1975). In many vespertilionid bats, the uterus may play an important role as a storage site for spermatozoa (Hiraiwa and Uchida, 1956; Racey and Potts, 1970). In some vespertilionid bats, viz. *Chalinolobus gouldii* (Kitchner, 1975) and *M. schreibersii fuliginosus*, the storage site is limited to the utero-tubal junction where intact spermatozoa are found in diverticulae formed by the mucosal folds (Mori and Uchida, 1980). The sperm are stored in utero-tubal junctions in *M. daubentoni* (Racey, 1975), *Myotis lucifugus* and *Myotis velifer* (Krutzsch et al., 1982). In *P. ceylonicus* and *S. heathi*, surviving spermatozoa are stored in both the uterus, and oviduct (Gopalakrishna and Madhwan, 1971; Krishna and Dominic, 1978). In *P. mimus*, comparatively, few spermatozoa are found in the uterus and most occur in the oviduct. In *Rhinolophus ferrum-equinum, R. hipposideros, R. ferrum equinum Nippon* and *Myotis nattereri*, the principal site of sperm storage is the oviduct (Racey, 1975; Mori et al., 1982). The storage organs/regions are thought to provide nourishment or protection for the sperm, allowing them to remain viable for longer periods (Smith and Yanagimachi, 1990; Krishna, 1997). In chickens, bats, boars, hamsters, and other mammals, a close association between sperm and the walls of the storage organs is observed (Smith and Yanagimachi, 1990; Suarez et al., 1991). Non-ciliated cells of storage organs (tubules) are rich in lipids, β-glycogen, mitochondria, and secretary granules (Tingari and Lake, 1973), suggesting that these cells produce secretions that bathe the sperm. Cells in the female storage organ may provide nutrients or metabolites to sperm to maintain viability or encourage quiescence during prolonged storage.

MECHANISM OF SPERM STORAGE

Studies have been made to understand the biology of sperm longevity in bats by morphological (Fawcett and Ito, 1965; Uchida and Mori, 1972; Racey, 1975; Krutzsch et al., 1982; Mori et al., 1982; Uchida et al., 1984; Krishna, 1997); metabolic (Crichton et al., 1981, 1982, 1993; Krutzsch et al., 1984); experimental (Lambert, 1981; Uchida and Mori, 1987), hormonal (Roy and Krishna, 2010) and nutritional approaches. Although, several theories have been offered to explain sperm storage, none have been unequivocally substantiated.

Low body temperature facilitates sperm-storage: Prolonged storage of viable spermatozoa within the female genital tract is well known in various poikilothermic animals (Hartman, 1939), but among mammals, the phenomenon observed in bats is unique and more extensively investigated. Pagenstecher (1859) first drew attention to the fact that spermatozoa are stored in the uterus of the Pipistrellus bat, *Pipistrellus pipistrellus* during hibernation. Subsequent reports of sperm storage in the reproductive tract of both male and female vespertilionid and rhinolophid bats were mainly from temperate regions. In the hibernating bats from Europe, North America and Japan, there is a long interval between copulation and ovulation and fertilization, and during this period, sperm remained stored, which coincides with the period of hibernation (Wimsatt, 1960; Uchida, 1957). Such studies suggest a direct functional relationship between sperm storage and torpor (Wimsatt, 1960, 1969). Torpor was also considered to be important for sperm storage, reducing loss or degradation of spermatozoa (Racey, 1973), and inhibiting ovulation (Wimsatt, 1969). Hartman (1933) suggested that low body temperature of the hibernating bat reduces sperm motility and thereby favors their survival. Discovery of sperm-storage phenomenon in the tropical vespertilionids, *Tylonycteris pachypus* and *T. robustula* (Medway, 1972), *Pipistrellus ceylonicus* (Gopalakrishna and Madhavan, 1971) and *Scotophilus heathi* (Krishna and Dominic, 1978), suggests that prolonged hibernation, which characterizes temperate zone species, is not a necessary pre-adaptation for sperm-storage.

Unique relationship between spermatozoa and storage organ: Perpendicular orientation of stored spermatozoa with their heads towards the reproductive tract epithelium cells have been demonstrated in most Chiropteran showing prolonged sperm storage Figure1A (Racey, 1975; Krishna, 1984). Special relationships exist between spermatozoa and female storage organs in all sperm-storing bat species so far investigated, suggesting that such relationships are an integral part of the mechanism of the prolonged storage of sperm (Racey, 1979; Uchida and Mori, 1987). This relationship is limited to the area around the entrance of each utero-tubal junction. (Rasweiler et al., 2010). Similar relationships between spermatozoa and storage organs have also been described in other vertebrates that retain sperm for prolonged periods (e.g. garter snake: Hoffman and Wimsatt, 1972; guppy: Jallabert and Billard, 1969; birds: Van Key et al., 1967; dog: Doak et al., 1967; shrew: Pearson, 1944). During storage, the sperm heads are either embedded in the uterine and oviductal epithelia or remain in close contact with microvilli of the host cells. Some spermatozoa actually indent the epithelial cells to the point that their heads are enveloped in epithelial cells cytoplasm (e.g. *Pipistrellus kuhli*: Andreuccetti et al., 1984; *Myotis lucifugus* and *M. velifer*: Krutzsch et al., 1982; Racey et al., 1975). Such a relationship was shown in *Pipistrllus kuhli*, to resist trypsin digestion but succumbed to treatment with EDTA and hyaluronidase, suggesting that attachment is mediated by proteoglycans or Ca++ rather than a protein (Andreuccetti et al., 1984). According to Racey (1975), the epithelial cells of the storage site actively provide the spermatozoa with nutrients for their prolonged survival. For example, glycogen in the uterine epithelium of *P. abramus* (Nakano, 1928), *P. pipistrellus* (Racey, 1975), *Myotis lucifugus* and *M. velifer* (Crichton et al., 1981) or lipids in cumulus cells in Rhinolophus ferrum-equinum (Oh et al., 1985)..

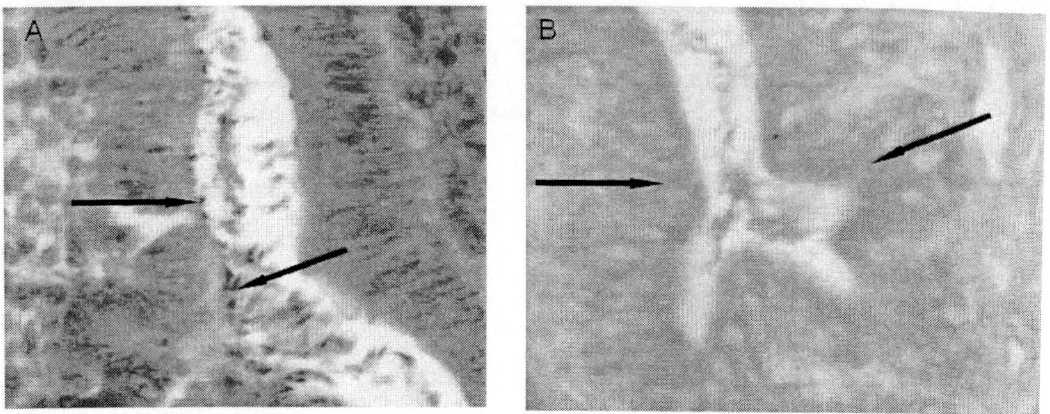

Histology of Utero-tubal Junction (UTJ)

Immunolocalization of Androgen Receptor (AR) in UTJ

Figure 1.

Furthermore, the association of spermatozoa with the epithelium of the storage organ probably protects them from various disintegrative factors (e.g. muscular movements of the uterus). Certain enzymes of glycolysis have been located in the sperm as well as in the uterus (Racey, 1975). Besides nutritional support, other factors required for long-term maintenance of spermatozoa might be a mechanism that acts to suppress sperm metabolism, thereby reducing demand for available substrates. In this regard, Hunter et al., (1971a,b) observed that the seminal fluid of *M. lucifugus* contains a protein that is inhibitory to uterine motility and phagocytic system. Zinc has been shown to temporarily inhibit the motility of both sperm and polymorphonuclear neutrophilic leucocytes. It has been suggested that such factors would also favor prolonged sperm storage (Crichton et al., 1982).

Krutzsch and Crichton (1991) suggested that bat spermatozoa resist capacitation and the acrosome reaction when cultured under conditions that are known to induce these events in other mammals. Furthermore, they have failed to reveal condensing bat sperm heads within zona-free hamster ooplasm using *in vitro* fertilization procedure, even after exposure to calcium ionophore A23187, and various proteases. Thus, it was suggested that the unusual longevity, stability and resistance of bat spermatozoa to *in vitro* fertilization could be due to unique modifications of their plasma membrane. A recent study in a hibernating bat (*Myotis velifer*), however, has failed to demonstrate any modification in plasma membrane lipid composition of the stored spermatozoa (Crichton et al., 1993). Further studies are needed to find out the unique features of prolonged sperm longevity of bats.

Hormonal regulation of sperm storage: It appears that the ultimate control of the mechanism underlying sperm storage may be hormonal. Racey (1972) ovariectomized *Pipistrellus pipistrellus* during hibernation, causing detrimental effects on the survival of stored sperm after two weeks as compared with control bats. Circulating levels of androgens that are required for the maintenance of sperm integrity and viability in the male are known for only a few bat species. Circulating androstenedione may be present in elevated levels in females storing sperm (Crichton, 2000). Androgen receptors (ARs, Figure1B) and androgen-binding proteins (ABPs) are present in the epithelial cell lining of the utero-tubal junction (storage organ) during the period of sperm storage in Scotophilus *heathi* (Roy and Krishna, 2010). The results of a recent study further confirm the role of androgen in sperm storage. The female bat treated with anti-androgen (Flutamide) during the period of sperm-storage caused the disappearance or loss of storing spermatozoa together with a significant decline in androgen level. This suggests that the decline in androgen activates changes in the epithelial cells that may promote degradation and disposal of spermatozoa (Jones, 2004). Recent immunocytochemical studies revealed the presence of increased expression of Bcl2 in the epithelial lining of utero-tubal junction of *S. heathi* during the period of sperm-storage. Treatment with testosterone *in vitro* caused a dose-dependent significant increase in Bcl2 expression, whereas treatment with flutamide together with testosterone caused a significant decrease in Bcl2 expression in utero-tubal junction of *S. heathi*. The utero-tubal junction also expresses caspase 3 together with Bcl2. The expression of caspase 3 decreased significantly during January, the period of sperm-storage, but increases during February prior to fertilization. This study provides evidence suggesting that androgen promotes sperm-storage by stimulating Bcl2 synthesis in the site of sperm storage. It may be hypothesized that a balance between Bcl2 and caspase 3 selects the healthier sperm for survival and subsequently utilized for fertilization.

Nutritional support and metabolic changes: Recent studies from our laboratory and the earlier study of Racey (1975) suggested the role of glucose and fatty acid as nutrient substances to the storing sperm and provided evidence of the transfer of these metabolites from uterine cells to stored spermatozoa. Racey (1975) provided a comprehensive light microscopical report on the histochemical analysis of the presence of lipids (phospholipids), proteins and carbohydrates in sperm storing tissues of both the male and female bat, *Pipistrellus pipistrellus*. Subsequent studies at the electron microscope level have confirmed the presence of glycogen in the sperm storage site (oviduct, utero-tubal junction), (Krutzsch et al., 1982; Andreuccetti et al., 1984; Krishna, 1997) giving rise to speculation that this may be an energy source for stored sperm. Enzymes of glycolysis have also been located in the sperm as well as in the uterus (Racey, 1975). Recent studies from our laboratory showed decreased levels of GLUT 4 and 8 in the utero-tubal junction suggesting decreased glucose availability during the period of sperm storage. Further studies showed a significant increased level of hormone-sensitive lipase and carnitine in the utero-tubal junction suggesting increased fatty acid availability during the period of sperm storage, which may be utilized by sperm during the storage period. Increased carnitine also maintains the integrity of sperm during storage. Besides nutritional support, suppressed sperm metabolism may be another important factor required for long-term maintenance of sperm, thereby reducing the demand for available substrates. Hunter et al., (1971 a,b) observed that the seminal fluid of *M. lucifugus* contains a protein that is inhibitory to uterine sperm motility. The metallic ion Zn is one such substance to inhibit the activity of sperm.

Significance of Sperm Storage

Storage of sperm by bats allows mating at the most propitious time and birth of young in the most favorable season (Fenton, 1984). Female bats in temperate-zones may have gained an advantage by storing sperm for longer periods, because this ensures that gestation could be initiated as soon as hibernation ended or whenever opportunity arises (Neubaum and Wolfner, 1999). Female sperm storage enables mating and ovulation to occur at widely separated intervals (Briskie and Montogomerie, 1993). When sperm are stored, the frequency of matings can be minimized. Mating requires time for courtship and copulation; it may disrupt feeding and increase exposure to predation, disease, and parasite transmission (Thornhill and Alcock, 1983; Hunter et al., 1993). Sperm storage in bats may also provide the basis for synchronized parturition and lactation in species where copulation is relatively asynchronous (Racey, 1982). Reproductive synchrony may provide advantages for the survival of young by diluting predation risks (Ims, 1990a,b). An important consequence of sperm storage is sperm competition (Parker, 1970). Sperm competition is the competition between sperm from different males to fertilize the eggs of a single female. Potentially, sperm competition can result in the healthiest sperm being used for fertilization.

Conclusion

Storage of sperm is an important strategy of the female reproductive cycle of bats. Normally, in mammals that don't store sperm, large numbers of white blood cells gather and phagocytize the sperm cells. Therefore, sperm storage is a remarkable process because "foreign" cells, which are normally destroyed by the immune system, are not only tolerated, but perhaps even nourished for several months in the female (Sharifi et al., 2004). It appears that sperm storage may be possible by activity that tends to prevent both fertilization and premature cell death. The demonstration of increased secretion of BCL2 (survival factor) by utero-tubal junction suggests its role in sperm viability. Decreased sperm motility due to insufficient availability of nutrients (e.g. glucose, lipids) may be a possible feature of the sperm storage mechanism and needs further investigation. We have not discussed the likelihood that sperm DNA-repair mechanisms might also be involved. The presence of heat shock protein on stored spermatozoa of boars (Boilard et al., 2004), thus its role in sperm viability, requires further investigation.

Acknowledgment

AK highly acknowledges the financial assistance from the Department of Science and Technology (SR/SO/AS-03/2010), New Delhi and Centre of Advance Study, University Grant Commission, New Delhi.

REFERENCES

Adams, E.M. and Wolfner, M.F. (2007). Seminal proteins but not sperm induce morphological changes in the *Drosophila melanogaster* female reproductive tract during sperm storage. *J.Insect Physiol.*, 53, 319-331.

Andreuccetti, P. Angelini, and F. Taddei, C. (1984).The interactions between spermatozoa and uterine epithelium in the hibernating bat, *Pipistrellus kuhli natt. Gamete Research*, 10, 67-76.

Baer, B. Armitage, S. and Boomsma, J. J. (2006). Sperm storage induces an immunity cost in ants. *Nature*, 41, 872–875.

Briskiej, V. and Montgomerie, R. (1993). Patterns of sperm storage in relation to sperm competition in passerine birds. *Condor*, 95, 442-454.

Crichton, E. G. Krutzsch, P. H. and Chvapil, M. (1982). Studies on prolonged spermatozoa survival in Chiroptera. II. The role of zinc in the spermatozoa storage phenomenon. *Comp. Biochem. Physiol.*, 71A, 71-77.

Crichton, E. G. and Krutsch, P.H. (1987). Reproductive biology of the female Little Mastiff Bat *Mormopterus planiceps* (Chiroptera : Molossidae) in southeast Australia. *Am. J. Anat.*, 178, 369-386.

Crichton, E. G. Krutzsch P. H. and Yanagimachi. R. (1993). Stability of the sperm plasma membrane of hibernating bats (*Myotis velifer*) compared with other mammals. *J. Reprod. Fertil.*, 97, 1-4.

Crichton, E. G. Krutzsch, P. H. and Wimsatt, W. A. (1981). Studies on prolonged spermatozoa survival in Chiroptera. I. The role of uterine free fructose in the spermatozoa storage phenomenon. *Comp. Biochem. Physiol.*, 70A, 387-395.

Crichton, E.G. (2000). Sperm storage and fertilization. In: E.G. Crichton, and P.H. Krutzsch (Eds.), *Reproductive biology of bats (*pp295-320). London, Academic Press.

Das, S. C. Isobe, N. and Yoshimura, Y. (2008). Mechanism of Prolonged Sperm Storage and Sperm Survivability in Hen Oviduct: A Review. *Am. J. Reprod. Immunol.*, 60, 477–481.

Doak, R. L. Hall, A. and Dale. H. E. (1967). Longevity of spermatozoa in the reproductive tract of the bitch. *J. Reprod. Fertil.*, 13, 51–58.

Fawcett, D.W. and Ito, S. (1965). The fine structure of bat spermatozoa. *Am. J. Anat.*, 116, 567-610.

Fenton, M. B. (1984). Sperm competition? The case of vespertilionid and rhinolophid bats. In R.L. Smith (Ed.), *Sperm Competition and the Evolution of Animal Mating Systems* (pp 573-587). New York, Academic Press.

Gist, D. H. and Congdon, J. D. (1998). Oviductal sperm storage as a reproductive tactic of turtles. *J. Exp. Zool.*, 282,526–534.

Gopalakrishna, A. and Madhavan, A. (1971). Survival of spermatozoa in the female genital tract of the Indian vespertilionid bat, Pipistrellus ceylonicus chrysothrix (Wroughton). *Proc. Indian Acad. Sci.*, 73, 43-49

Hartman, C. G. (1933). On the survival of spermatozoa in the female genital tract of the bat. *Q. Rev. Biol.*, 8, 185-193.

Hrtman, C. G. (1939). Ovulation, fertilization and the transport and viability of eggs and spermatozoa. In E. Allen, C. H. Danforth, and E. A. Doisy, (Eds.), *"Sex and Internal Secretions"*(2nd Ed., pp. 630-719)., Baltimore, Williams and Wilkins.

Hiraiwa, Y. and Uchida, T. (1956). Fertilization in the bat *Pipistrellus abramus abramus* (Temminck). III. Fertilizing capacity of spermatozoa stored in the uterus after copulation in the fall. *Sci. Bull. Fac. Agrie.* Kyushu Univ. 31, 565-574.

Hoffman, L. H. and Wimsatt, W. A. (1972). Histochemical and electron microscopic observations on the sperm receptacles in the garter snake oviduct. *Am. J. Anat.*, 134 71-96.

Hunter, A. G. Barker, L. D. S. Johnson, W. L. Fahning, M. L. and Schultz, R. H. (1971a). Antigenicity, toxicity and cross-reactions of male bat (*Myotis lucifugus*) reproductive organs. *J. Reprod. Fertil.*, 24, 171-177.

Hunter, A.G. Johnson, W.L. Barker, L.D.S. Fahning M.L. and Schultz, R.H. (1971b). Bat seminal vesicle protein: its characterization and physiological properties. *J. Rerod. Fertil.*, 24, 179-186.

Hunter, F. M. Petrie, M. Otronen, M. Birkhead, T. and Møller, A. P. (1993). Why do females copulate repeatedly with one male? *TREE*, 8, 21–26.

Ims, R. A. (1990a). The ecology and evolution of reproductive synchrony. - *Trends Ecol. Evol.*, 5, 135-140.

Ims, R. A. (1990b). On the adaptive value of reproductive synchrony as a predator-swamping strategy. - *Am. Nat.*, 136, 485-498.

Jalabert, B. and Billard, R. (1969. Etude ultrasrtuctural du site de conservation des spermatozoids dans l'ovaire de *Poecilia reticulata* (Poisson Teleosteen). *Annales de Biologie Animal, Biochimie and Biophysique.*, 9, 273-280.

Jones, R. (2004). Sperm survival versus degradation in the mammalian epididymis: A Hypothesis. *Biol. Reprod.*, 71, 1405-1411.

Kitchener, D. J. (1975). Reproduction in female Gould's Wattled Bat *Chalinolobus gouldii* (Gray) (Vespertilionidae) in Western Australia. *Aust. J. Zool.*, 23, 29-.42.

Krishna, A. and Dominic, C.J. (1978). Storage of spermatozoa in the female tract of the vespertilionid bat, *Scotophilus heathi. J. Reprod. Fertil.*, 54, 319-321.

Krishna, A. (1984). Storage of spermatozoa in the female genital tract of the Indian pigmy pipistrelle bat, *Pipistrellus mimus Wroughton. Archs. Biol.. Bruxelles.*, 95, 223-229.

Krishna, A. (1997). The relationship between spermatozoa and epithelium of the female genital tract during sperm storage in the greater yellow bats (*Scotophilus heathi*): the light and electron microscopic observations. *Proc. Natl. Sci. Counc. Repub. China B.*, 21, 31-36.

Krutzsch, P. H. Crichton, E. G. Lennon, D. L. F. Stratman, F. W. and Carter, A. (1984). Studies on prolonged spermatozoa survival in Chiroptera. III. Preliminary data on carnitine. *Andrologia*, 16, 34—37.

Krutzsch, P.H. and Crichton, E.G. (1991). Fertilization in Bats. In B.S. Dunbar, and M. O. 'Rand (Eds.), *A Comparative Overview of Mammalian Fertilization* (pp 137-149). New York, Plenum Press.

Krutzsch, P.H. Crichton, E.G. and Nagle, R.B. (1982). Studies on prolonged spermatozoa survival in Chiroptera: a morphological examination of storage and clearance of intrauterine and cauda epididymal spermatozoa in the bats *Myotis lucifugus* and *M. velifer*. Am. J. Anat., 165,421-434.

Lambert, H. (1981). Temperature dependence on capacitation in bat sperm monitored by zona-free bat ova. *Gamete Research*, 4, 525—533.

Martinet, L., and Raynaud, F. (1974). Survie prolongée des spermatozoïdes dans l'utérus de la Hase: explication de la superfoetation. In: E. S. E.Hafez, and C. G.Thibaut (Eds.), *Sperm Transport, Survival and Fertilizing Ability* (pp. 295–308). Paris, INSERM.

Medway, L. (1972). Reproductive cycles of the flat-headed bats *Tylonycteris pachypus* and *T. robustula* (Chiroptera: Vespertilionidae) in a humid equatorial environment. *Zool. J. Linnean Soc.*, 51, 33-61.

Mori, T. and Uchida, T.A. (1980). Sperm storage in the reproductive tract of the female Japanese longfingered bat, *Miniopterus schreibersii fuliginosus. J. Reprod. Fertil.*, 58, 429-433.

Mori, T. Oh, Y.K. and Uchida, T.A. (1982). Sperm storage in the oviduct of the Japanese greater horseshoe bat, *Rhinolophus ferrumequinum nippon. J. Fac. Agr.,* Kyushu Univ., 27, 47-53.

Nakano, O. (1928). Über die Verteilung des Glykogens bei den zyklischen Veränderungen in den Geschlechtsorganen der Fledermaus, und über die Nahrungsaufnahme der Spermien in den weiblichen Geschlechtswege. *Folia anat. jap.*, 6, 777-828.

Neubaum, D.M. and Wolfner, M.F. (1999). Wise, winsome, or weird? Mechanisms of sperm storage in female animals. *Curr. Top. Dev. Biol.*, 41, 67-97.

Pagenstecher, H.A. (1859). Über die Begattung von Vesperugo pipistrellus. Verh. naturh.-med. *Ver. Heidelb.*, 1, 194-195.

Parker, G. A. (1970). Sperm competition and its evolutionary consequences in the insects. *Biological Reviews*, 45, 525–567.

Pearson, O.P. (1944). Reproduction in the shrew (Blarina brevicauda Say). *Am. J. Anat.*, 75, 39-93.

Racey, P. A. and Potts, D.M. (1970). Relationship between stored spermatozoa and the uterine epithelium in the pipistrelle bat (*Pipistrellus pipistrellus*). *J. Reprod. Fertil.*, 22, 57–63.

Racey, P. A. (1979). The prolonged storage and survival of spermatozoa in Chiroptera. *J. Reprod. Fertil.*, 56, 391-402.

Racey, P. A. (1982). Ecology of bat reproduction. In: TH Kunz (Ed.), *Ecology of Bats* (pp-57-104). New York, Plenum Press.

Racey, P.A. (1973). The viability of spermatozoa after prolonged storage by male and female *European bats. Period. Biol.*, 75, 201-205.

Racey, P.A. (1975). The prolonged survival of spermatozoa in bats. In: J. G. Duckett, and P. A. Racey (Eds.), *The biology of the male gamete* (pp-385-416). London, Academic Press.

Rasweiler, J. J. IV Badwaik, N. K. and Mechineni, K. V. (2010). Selectivity in the transport of spermatozoa to oviductal reservoirs in the menstruating fruit bat, *Carollia perspicillata. Reproduction*, 140, 743-757.

Roy, V. K. and Krishna, A. (2010). Evidence of androgen-dependent sperm storage in female reproductive tract of *Scotophilus heathi. Gen. Comp. Endocrinol.*, 165,120–126.

Sharifi, M. Ghorbani, R. and Akmali, V. (2004). *Reproductive cycle in Pipistrellus kuhlii (Chiroptera. Vespertilionidae) in Western Iran.* Mammalia, 68, 323–327.

Smith, T.T. and Yanagimachi, R. (1990). The viability of hamster spermatozoa stored in the isthmus of the oviduct: the importance of sperm-epithelium contact for survival. *Biol. Reprod.*, 42, 450-457.

Suarez, S.S. Katz, D.F. Owen, D.H. Andrew, J.B. and Powell, R.L. (1991). Evidence for the function of hyperactivated motility in sperm. *Biol. Reprod.*, 44, 375-81.

Thornhill, R, and Alcock, J. (1983). *The evolution of insect mating systems*. Cambridge (MA): Harvard University Press. 547 p.

Tingari, M. D. and Lake, P. E. (1973). Ultrastructural studies on the uterovaginal sperm-host glands of the domestic hen, *Gallus domesticus*. *J. Reprod. Fertil.*, 34, 423–431.

Uchida, T.A. (1957). Fertilization and hibernation in bats. Heredity, Tokyo, 11, 14-17.

Uchida, T. A. and Mori, T. (1987). Prolonged storage of spermatozoa in hibernating bats. In M. B. Fenton, P. A. Racey, and J. M. V. Rayner (Eds.), *Recent Advances in the Study of Bats* (pp 351-356). Cambridge, Cambridge University Press

Uchida, T. A. Mori, T. and Oh, Y.K. (1984). Sperm invasion of the oviducal mucosa, fibroblastic phagocytosis and endometrial sloughing in the Japanese greater horseshoe bat, *Rhinolophus ferrumequinum*. *Cell Tiss. Res.*, 236, 327-331.

Uchida, T.A. and Mori, T. (1972). Electron microscope studies on the fine structure of germ cells in Chiroptera. 1. Spermiogenesis in some bats and notes on its phylogenetic significance. *Sci. Bull. Fac. Agrie Kyushu Univ.*, 26, 399-418.

Uchida, T.A., and Mori, T. (1987). Prolonged storage of spermatozoa in hibernating bats. In: M.B. Fenton, P.A. Racey, and J. M. V. Rayner (Eds.), *Recent Advances in the Study of Bats* (pp-351-365). Cambridge, Cambridge University Press.

Van krey, H. P. Ogasawara, F. X. and Pangborn, J. (1967). Light and electron microscopic studies of possible sperm gland emptying mechanisms. *Poultry. Sci.*, 46, 69–78.

Wang, Z. Liang, B. Racey, P. A. Wang, Y. and Zhang, S. (2008). Sperm Storage, Delayed Ovulation, and Menstruation of the Female Rickett,s Big-Footed Bat (*Myotis ricketti*). *Zool Studies*, 47, 215-221.

Wimsatt, W. A. (1960). Some problems of reproduction in relation to hibernation in bats. *Bull. Mus. comp. Zool. Harv.*, 124, 249-270.

Wimsatt, W.A. (1969) Some inter-relations of reproduction and hibernation in mammals. *Symp. Soc. exp. Biol.*, 23, 511-549.

Yamanouye, N. Silveira, P.F. Abdalla, F.M. Almeida-Santos, S.M. Breno, M.C. and Salomao, M.G. (2004). Reproductive cycle of the Neotropical *Crotalus durissus terrificus*: II. Establishment and maintenance of the uterine muscular twisting, a strategy for long-term sperm storage. *Gen. Comp. Endocr*, 139, 151-157.

Zhang, Y.Y. (2006). Semen characterization and sperm storage in *Cabots Tragopan. Poultry. Sci.* 85, 8 92-898.

INDEX

A

B

D

E

F

G

H

I

J

K

L

M

N

O

P

Q

R

S

T

U

V

W

X

Y

Z